# PROCEEDINGS

OF THE

# 1993 INTERNATIONAL CONFERENCE

ON

# PARALLEL PROCESSING

August 16 – 20, 1993

**Vol. II Software**

**Alok N. Choudhary and P. Bruce Berra, Editors**
**Syracuse University**

Sponsored by

**THE PENNSYLVANIA STATE UNIVERSITY**

**CRC Press**
**Boca Raton Ann Arbor Tokyo London**

The papers appearing in this book comprise the proceedings of the meeting mentioned on the cover and title page. They reflect the authors' opinions and are published as presented and without change in the interest of timely dissemination. Their inclusion in this publication does not necessarily constitute endorsement by the editors, CRC Press, or the Institute of Electrical and Electronics Engineers, Inc.



ISSN 0190-3918
ISBN 0-8493-8983-6 (set)
ISBN 0-8493-8984-4 (vol. I)
ISBN 0-8493-8985-2 (vol. II)
ISBN 0-8493-8986-0 (vol. III)
IEEE Computer Society Order Number 4500-22


Printed in the United States of America 1 2 3 4 5 6 7 8 9 0
Printed on acid-free paper

Additional copies may be obtained from:

CRC Press, Inc.
2000 Corporate Blvd., N.W.
Boca Raton, Florida 33431

## Preface

Interest in parallel processing continues to be high; this year we received 477 papers from 28 different countries, representing a multiplicity of disciplines. In order to accommodate as many presentations as possible and yet maintain the high quality of the conference, we have accepted papers as regular and concise. Regular papers offer well-conceived ideas, good results, and clear presentations. While concise papers also present well-conceived ideas and good results, they do so in fewer pages. The table below summarizes the number of submissions by area:

| Area | Submitted | Accepted | |
|---|---|---|---|
| | | Regular | Concise |
| Architecture | 166 | 16 (9.6%) | 42 (25.3%) |
| Software | 154 | 16 (10.4%) | 39 (25.3%) |
| Algorithms/Applications | 157 | 21 (13.4%) | 31 (19.7%) |
| Total | 477 | 53 (11.1%) | 112 (23.5%) |

This year the program activity was structured slightly differently than in past years since we had a program chair and three co-program chairs. With this organization we worked as a team and assigned papers to the three primary areas in somewhat of a balanced mode. Then each of the co-program chairs was largely responsible for their area. Each submitted paper was reviewed by at least three external reviewers. To avoid conflicts of interest, we did not process the papers from our colleagues at Syracuse University. These were handled by Professor Chita Das of Pennsylvania State University. We are very grateful for his professional and prompt coordination of the reviewing process.

We can truly state that the refereeing process ran much more smoothly than we had originally expected. We attribute this to two main reasons. The first is that we sent one copy of each paper outside of the United States for review; these reviews were thorough and timely. The second reason is that we used e-mail almost exclusively. We would like to express our sincere appreciation to the referees for their part in making the selection process a success. The quality of the conference can be maintained only through such strong support.

This year's keynote speaker is Dr. Ken Kennedy, an internationally known researcher in parallel processing. In addition, we have two super panels chaired by Dr. H.J. Siegel and Dr. Kai Hwang who are also internationally known in parallel processing. We are honored to have Dr.'s Kennedy, Siegel and Hwang sharing their visions on parallel processing with us.

Finally, the success of a conference rests with many people. We would like to thank Dr. Tse-yun Feng for his valuable guidance in the preparation of the program.In addition many graduate and undergraduate students aided us in the process and we gratefully acknowledge their help. We would like to thank Faraz Kohari for his help with the database management functions and Emily Yakawiak for putting the proceedings together. We gratefully acknowledge the New York State Center for Advanced Technology in Computer Applications and Software Engineering (CASE) and the Department of Electrical and Computer Engineering for their support.

P. Bruce Berra, Program Chair
C.Y. Roger Chen, Co-Program Chair-Architecture
Alok Choudhary, Co-Program Chair-Software
Salim Hariri, Co-Program Chair-Algorithms/Applications

The Case Center
Electrical and Computer Engineering
Syracuse University
Syracuse, NY 13244

## LIST OF REFEREES - FULL PROCEEDINGS

Abdennadher, N.
Aboelaze, M. A.
Abu-Ghazaleh, N.
AbuAyyash, S.
Agarwal, A.
Agarwal, M.
Agha, G.
Agrawal, D. P.
Agrawal, P.
Agrawal, V.
Alijani, G.
Amano, H.
Andre, F.
Annaratone, M.
Antonio, J. K.
Anupindi, K.
Applebe, B.
Arif, G.
Ariyawansa, K. A.
Arthur, R. M.
Arunachalam, M.
Atiquzzaman, M.
Audet, D.
Avalani, B.
Ayani, R.
Ayguade, E.
Babb, J.
Babbar, D.
Bagherzadeh, N.
Baglietto, P.
Baldwin, R.
Banerjee, C.
Banerjee, P.
Banerjee, S.
Barada, H.
Barriga, L.
Basak, D.
Baskiyar, S.
Bastani, F.
Bayoumi, M. A.
Beauvais, J.
Beckmann, C.
Bellur, U.
Benson, G. D.
Berkovich, S.
Berson, D.
Bhagavathi, D.
Bhuyan, L.
Bhuyan, L. N.
Bianchini, R.
Bic, L.
Bitzaros, S.
Blackston, D.
Boku, T.
Bordawekar, R.
Boura, Y.
Bove Jr., V. M.
Brent, R. P.
Breznay, P.
Brown, J. C.
Burago, A.
Burkhardt, W. H.
Butler, M.
Cam, H.
Capel, M.
Casavant, T. L.
Casulleras, J.
Cavallaro, J.
Celenk, M.
Chalasani, S.
Chamberlain, R.
Chandra, A.
Chandy, J.
Chang, M. F.
Chang, Y.
Chao, D.
Chao, J.
Chao, L.
Chase, C.
Chase, C. M.
Chen, C.
Chen, C. L.
Chen, D.
Chen, K.
Chen, S.
Chen, Y.
Cheng, J.
Cheong, H.
Chiang, C.
Chien, C.
Chiueh, T.
Choi, J. H.
Christopher, T.
Chu, D.
Chung, S. M.
Clary, J. S.
Clifton, C.
Cloonan, T. J.
Cook, G.
Copty, N.
Cortadella, J.
Costicoglou, S.
Crovella, M.
Cukic, B.
Cytron, R.
Dahlgren, F.
Dandamudi, S.
Das, A.
Das, C.
Das, C. R.
Das, S.
Das, S. K.
Datta, A. K.
Davis, M. H.
Debashis, B.
Defu, Z.
Delgado-Frias, J. G.
Demuynck, M.
Deshmukh, R. G.
Dimopoulos, N. J.
Dimpsey, R.
Douglass, B.
Dowd, P. W.
Drach, N.
Du, D. H. C.
Duesterwald, E.
Duke, D. W.
Durand, D.
Dutt, N.
Dutt, S.
Dwarkadas, S.
Edirisooriya, S.
Efe, K.
El-Amawy, A.
Elmohamed, S.
Enbody, R. J.
Erdogan, S.
Erdogan, S. S.
Eswar, K.
Etiemble, D.
Evans, J.
Exman, I.
Farkas, K.
Felderman, R. E.
Feng, C.
Feng, W.
Fernandes, R.
Field, A. J.
Fienup, M. A.
Fraser, D.
Fu, J.
Ganger, G.
Garg, S.
Genjiang, Z.
Gessesse, G.
Gewali, L. P.
Ghafoor, A.
Ghonaimy, M. A. R.
Ghosh, K.
Ghozati, S. A.
Gibson, G. A.
Gokhale, M.
Goldberg
Gong, C.
Gonzalez, M.
Gornish, E.
Granston, E.
Greenwood, G. W.
Grimshaw, A. S.
Gupta, A. K.
Gupta, S.
Gupta, S. K.
Gurla, H.
Gursoy, A.
Haddad, E.
Hamdi, M.
Hameurlain, A.
Han, Y.
Hanawa, T.
Hao, Y.
Harathi, K.
Harper, M.
Hauck, S.
Heiss, H.

Helary, J.
Helzerman, R. A.
Hemkumar, N. D.
Hensgen, D.
Ho, C.
Holm, J.
Hong, M.
Horng, S. J.
Hossain
Hsu, W.
Huang, C.
Huang, G.
Huang, S.
Hughey, R.
Hui-Li
Hulina, P. T.
Hurley, S.
Hwang, D. J.
Hwang, K.
Hwang, S.
Hyder, S.
Ibarra, O. H.
Iqbal, A.
Ito, M.
Iwasaki, K.
Jadav, D.
Jaffar, I.
Jain, R.
Jaja, J.
Jamil, S.
Jazayeri, M.
Jha, N. K.
Jiang, H.
Jin, G.
Jing, W.
Johnson, T.
Jorba, A.
Jou, T.
Jun, Y.
Jung, S.
Kaeli, D. R.
Kain, R. Y.
Kale, L.
Kanevsky, A.
Kannan, R.
Karl, W.

Katti, R.
Kaushik, S. D.
Kavianpour, A.
Kaxiras, S.
Kee, K.
Kee, K. K.
Keleher, P.
Kelly, P. A. J.
Keryell, R.
Kessler, R.
Khan, J.
Khokhar, A. A.
Kim, G.
Kim, H.
Kim, H. J.
Kim, K.
King, C.
Kinoshita, S.
Klaiber, A.
Klaiber, K.
Kobel, C.
Koelbel, C.
Koester, D.
Kong, X.
Koppelman, D. M.
Kornkven, E.
Kothari, S.
Kothari, S. C.
Kravets, D.
Kremer, U.
Kreuger, P.
Krieger, O.
Krishnamoorty, S.
Krishnamurthy, E. V.
Ku, H.
Kudoh, T.
Kuhl, J.
Kumar, B.
Kumar, V.
Kuo, S.
Kurian, L.
Kyo, A.
Kyo, S.
LaRowe, R.
Ladan, M.
Lai, H.

Laszewski, G. v.
Lauria, A. T.
Lavery
Lee, D.
Lee, E.
Lee, G.
Lee, H. J.
Lee, K.
Lee, M.
Lee, O.
Lee, P.
Lee, S.
Lee, S. Y.
Leung, A.
Levine, G.
Li, A.
Li, J.
Li, Q.
Li, Z.
Liang, D.
Lin, A.
Lin, C.
Lin, F.
Lin, R.
Lin, W.
Lin, W. M.
Lin, X.
Lin, Y.
Lisper, B.
Liszka, K.
Liu, G.
Liu, J.
Liu, Y.
Llosa, J.
Lombardi, F.
Lopez, M. A.
Louri, A.
Lu, B.
Lu, P.
Ludwig, T.
Lyuu, Y.
Mackenzie, K.
Maggs, B.
Mahgoub, I.
Mahjoub, Z.
Makki, K.

Mannava, P. K.
Mao, A.
Mao, S.
Marek, T. C.
Maresca, M.
Marinescu, D.
Martel, C.
Masuyama, H.
Matias, Y.
Matloff, N.
Matsumoto, Y.
Mawnava, P. K.
Mayer, H.
Mazuera, O. L.
McDowell, C.
McKinley, P.
Melhem, R.
Menasce, D.
Mendelson, B.
Meyer, D.
Michael, W.
Mitchell, C. D.
Mohamed, A. G.
Mohapatra, P.
Moreira, J. E.
Mouhamed, M. A.
Mounes-Toussi, F.
Moyer, S.
Mrsic-Flogel, J.
Mukai, H.
Mukherjee, B.
Mullin, L.
Mullin, S.
Mutka, M.
Mutka, M. K.
Myer, H.
Nair, V. S. S.
Nakamura, T.
Nang, C. M.
Nassimi, D.
Natarajan, C.
Natarajan, V.
Ni, L.
Ni, L. M.
Nico, P.
Noh, S.

Novack, S.
Nutt, G.
Obeng, M. S.
Oehring, S.
Oh, H.
Okabayashi, I.
Okawa, Y.
Olariu, S.
Olive, A.
Omer, J.
Omiecinski, E.
Oruc, A. Y.
Ouyang, P.
Paden, R.
Panda, D.
Panda, D. K.
Pao, D.
Parashar, M.
Park, B. S.
Park, C.
Park, H.
Park, J.
Park, K. H.
Park, S.
Park, Y.
Parsons, I.
Patnaik, L. M.
Pears, A. N.
Pei-Yung-Hsiao
Perkins, S.
Peterson, G.
Petkov, N.
Pfeiffer, P.
Phanindra, M.
Picano, S.
Pinto, A. D.
Pissinou, N.
Podlubny, I.
Poulsen, D.
Pourzandi, M.
Pradhan, D. K.
Pramanick, I.
Prasanna, V. K.
Pravin, D.
Puthukattukaran, J. J.
Qiao, C.
Raatikainen, P.
Radia, N.
Rafieymehr, A.
Raghavan, R.
Raghavendra, C. S.
Raghunath, M. T.
Rajagopalan, U.
Ramachandran, U.
Ramany, S.
Ravikumar, C.
Ray, S.
Reese, D.
Reichmeyer, F.
Rigoutsos, I.
Ripoli, A.
Robertazzi, T.
Rokusawa, K.
Rowland, M.
Rowley, R.
Roysam, B.
Ruighaver, A. B.
Ryan, C.
Saghi, G.
Saha, A.
Sajeev, A.
Salamon, A.
Salinas, J.
Sarikaya, B.
Sass, R.
Schaeffer, J.
Schall, M.
Schoinas, I.
Schwederski, T.
Schwiebert, L.
Seigel, H. J.
Seo, K.
Sha, E.
Shah, G.
Shang, W.
Shankar, R.
Sharp, D.
Sheffler, T. J.
Sheikh, S.
Shekhar
Shen, X.
Sheng, M. J.
Sheu, J.
Sheu, T.
Shi, H.
Shin, K. G.
Shing, H.
Shiokawa, S.
Shirazi, B.
Shoari, S.
Shu, W.
Sibai, F. N.
Siegel, H. J.
Simms, D.
Sinclair, J. B.
Singh, S.
Singh, U.
Sinha, A.
Sinha, A. B.
Sinha, B. P.
Slimani, Y.
Snelick, R.
So, K.
Soffa, M.
Son, S.
Song, J.
Song, Q. W.
Srimani, P. K.
Su, C.
Suguri, T.
Sun, X.
Sundareswaran, P.
Sunderam, V.
Sussman, A.
Sy, Y. K.
Szafron, D.
Szymanski, T.
Takizawa, M.
Tan, K.
Tandri, S.
Taylor, V. E.
Tayyab, A.
Temam, O.
Thakur, R.
Thapar, M.
Thayalan, K.
Thekkath, R.
Torrellas, J.
Tout, W.
Traff, J. L.
Trahan, J.
Tsai, W.
Tseng, W.
Tseng, Y.
Tzeng, N.
Ulusoy, O.
Unrau, R.
Vaidya, N.
Vaidyanathan, R.
Valero-Garcia, M.
Varma, G.
Varma, G. S. D.
Varman, P. J.
deVel, O.
Verma, R. M.
Vuppala, V.
Wagh, M.
Wahab, A.
Wakatani, A.
Wallace, D.
Wang, C.
Wang, D.
Wang, D. T.
Wang, H.
Wang, H. C.
Wang, M.
Watts, T.
Weissman, J. B.
Wen, Z.
Wills, S.
Wilson, A.
Wilson, D.
Wilton, S. F.
Wilton, S. J. E.
Wittie, L.
Wong, W.
Wu, C.
Wu, C. E.
Wu, H.
Wu, J.
Wu, M.
Wu, M. Y.
Xu, C.
Xu, H.

Xu, Z.
Yacoob, Y.
Yalamanchili, S.
Yamashita, H.
Yang, C.
Yang, C. S.
Yang, Y.
Yen, I.
Yeung, D.
Youn, H. Y.
Young, C.
Young, H.
Young, H. C.
Youngseun, K.
Yousif, M.
Yu, C.
Yu, C. S.
Yu, S.
Yuan, S.
Yuan, S. M.
Yum, T. K.
Zaafrani
ZeinElDine, O.
Zeng, N.
Zhang, J.
Zhang, X.
Zhen, S. Q.
Zheng, S. Q.
Zhou, B. B.
Zhu, H.
Ziavras, S. G.
Zievers, W. C.
Ziv, A.
Zubair, M.

## AUTHOR INDEX - FULL PROCEEDINGS

Volume I = Architecture
Volume II = Software
Volume III = Algorithms & Applications


Abraham, S. I-274
Abraham, S. G. II-95
Agarwal, A. I-2
Agrawal, D. P. I-303
Agrawal, D. P. III-284
Ahamad, M. I-324
Ahamad, M. I-332
Ahmad, I. I-219
Al-Hajery, M. Z. III-209
Al-Mouhamed, M. II-25
Aly, K. A. I-150
Ananthanarayanan, R. I-324
Appelbe, B. II-246
Armstrong, J. B. III-37
Ayani, R. I-171
Azevedo, M. M. I-91
Bagherzadeh, N. I-91
Baglietto, P. I-282
Baiardi, F. I-340
Banerjee, P. I-263
Banerjee, P. II-134
Banerjee, P. II-30
Banerjee, P. III-133
Barriga, L. I-171
Bataineh, S. I-290
Batcher, K. E. I-105
Batcher, K. E. III-209
Beauvais, J. II-130
Bechennec, J. I-72
Bhagavathi, D. III-192
Bhagavathi, D. III-307
Bhuyan, L. N. I-132
Bhuyan, L. N. II-193
Bhuyan, L. N. III-153
Bhuyan, L. N. III-23
Bic, L. II-25
Biswas, P. II-68
Blackston, D. T. III-201
Boura, Y. M. III-175
Brent, R. P. III-128
Breznay, P. T. I-307
Brooks III, E. D. II-103
Browne, J. C. II-58
Bryant, R. E. III-29
Cappello, F. I-72
Chamberlain, R. D. III-289
Chandy, J. A. I-263
Chang, Y. I-132
Chao, L. II-231
Chauhan, P. S. III-94
Chen, C. III-107
Chen, L. III-124
Chen, T. II-273
Chen, Y. I-184
Cheng, B. H. C. II-125
Cheng, J. II-265
Cheung, A. L. II-21
Chiueh, T. I-20
Cho, S. H. I-51
Chung, H. I-47
Chung, M. J. I-299
Chung, S. M. I-175
Cloonan, T. J. I-146
Cohen, D. I-39
Cohen, W. E. II-217
Copty, N. III-102
Dahlberg, T. A. III-284
Dahlgren, F. I-56
Das, C. R. I-110
Das, C. R. I-210
Das, C. R. I-254
Das, C. R. III-175
Das, S. K. I-311
Davis, E. W. I-202
Day, K. III-65
DeSchon, A. I-39
Delaplace, F. I-72
Deplanche, A. II-130
Dietz, H. G. II-217
Dietz, H. G. II-47
Dighe, O. M. I-158
Ding, J. II-193

Dowd, P. W. I-150
Drach, N. I-25
Dubois, M. I-56
Durand, M. D. I-258
Ebcioglu, K. II-283
Efe, K. III-311
Eisenbeis, C. III-299
Eswar, K. II-148
Eswar, K. III-18
Etiemble, D. I-72
Evans, J. D. III-271
Felderman, R. I-39
Feng, C. III-153
Feng, G. II-209
Fernandes, R. I-315
Fernandes, R. I-319
Field, A. J. I-189
Fijany, A. III-51
Finn, G. I-39
Fortes, J. A. B. III-248
Fox, G. III-102
Frank, M. I. I-232
Frankel, J. L. II-11
Fricker, C. I-180
Fu, J. W. C. II-87
Fuchs, W. K. I-138
Fuchs, W. K. I-64
Gerasch, T. E. II-260
Germain, C. I-72
Ghafoor, A. I-128
Ghafoor, A. I-219
Gherrity, M. III-37
Giavitto, J. I-72
Gill, D. H. II-260
Gokhale, M. II-188
Gong, C. II-39
Goutis, C. E. III-2
Grant, B. K. II-217
Grossglauser, M. I-154
Gupta, A. III-115
Gupta, R. II-39
Gupta, S. K. S. II-301
Gurla, H. III-192
Gurla, H. III-307
Halliday, H. II-292
Hameurlain, A. III-258
Han, S. Y. I-51
Han, Y. III-223
Hanna, J. II-201
Harikumar, S. III-280
Harrison, P. G. I-189
Hearne, J. II-292
Helman, D. III-90
Hensgen, D. A. II-269
Hoag, J. E. II-103
Hoel, E. G. III-47
Hokens, E. I-206
Horng, S. III-57
Hovland, P. D. II-251
Hsiung, T. I-290
Hsu, W. J. I-299
Hsu, Y. I-163
Huang, C. II-148
Huang, C. III-18
Huang, C. -H. II-301
Huang, S. S. III-73
Hurson, A. R. II-68
Hwang, D. J. I-51
Hwang, K. III-215
Hyder, S. I. II-58
Ibarra, O. H. III-77
Igarashi, Y. III-223
Iyer, B. R. III-276
JaJa, J. II-2
JaJa, J. III-90
Jalby, W. I-180
Jalby, W. I-258
Jang, J. III-236
Jayasimha, D. N. II-107
Jazayeri, M. I-340
Jha, N. K. I-246
Johnson, T. II-223
Jusak, D. S. II-292
Kacker, R. II-2
Kai, B. I-336
Kale, L. V. III-196
Kalns, E. T. II-175
Kanevsky, A. I-315
Kanevsky, A. I-319
Kao, T. III-57
Kaushik, S. D. II-301
Kavi, K. II-68

Kavianpour, A. II-297
Kervella, L. I-258
Keryell, R. II-184
Kessler, R. R. III-271
Kim, J. I-110
Kim, S. III-37
Kim, Y. D. I-51
Klaiber, A. C. II-11
Kohli, P. I-332
Koriem, S. M. I-214
Koufopavlou, O. G. III-2
Kranz, D. I-2
Krieger, O. II-201
Krishnamurthy, E. V. III-124
Krishnamurthy, G. II-47
Krishnan, S. III-196
Kuchlous, A. I-83
Kuhl, J. G. I-197
Kumar, A. III-23
Kumar, M. I-241
Kumar, V. III-115
Kyo, S. II-236
Kyriakis-Bitzaros, E. D. III-2
Lai, H. III-184
Lai, T. III-149
Lakshmanan, B. II-246
Latifi, S. I-91
LeBlanc, R. J. I-324
Lee, C. III-159
Lee, D. I-268
Lee, H. I-100
Lee, K. Y. III-167
Lee, P. II-161
Li, H. I-282
Li, H. II-140
Li, L. I-175
Li, Z. II-112
Lilja, D. J. II-112
Lin, R. III-307
Lin, W. III-107
Lin, W. III-227
Lin, X. I-294
Lin, Y. III-184
Ling, N. II-73
Liszka, K. J. I-105
Liu, G. III-167
Liu, H. III-73
Liu, J. I-100
Liu, Y. I-163
Lombardi, F. III-141
Lombardi, F. III-153
Lopez, M. A. I-307
Louri, A. I-206
Lu, H. I-345
Lu, Z. III-227
Lyon, G. II-2
Macleod, I. III-124
Manimaran, G. I-83
Marek, T. C. I-202
Maresca, M. I-282
Martens, J. D. II-107
McCreary, C. L. II-260
McKinley, P. K. I-294
McKinley, P. K. II-288
McMurdy, R. K. II-279
Mehra, P. III-263
Melhem, R. II-39
Meyer, D. G. II-103
Mizoguchi, M. II-236
Mohapatra, P. I-110
Mohapatra, P. I-210
Montaut, T. I-258
Moon, S. II-241
Moon, S. II-283
Morvan, F. III-258
Mourad, A. N. I-138
Mourad, A. N. I-64
Mufti, S. II-301
Mukherjee, B. II-205
Muthukumarasamy, J. I-237
Nandy, S. K. III-94
Narayan, R. III-94
Natarajan, V. I-2
Nation, W. G. III-37
Neiger, G. I-332
Neri, V. I-72
Nguyen, T. N. II-112
Ni, L. M. I-294
Ni, L. M. I-77
Ni, L. M. II-125
Ni, L. M. II-175
Ni, L. M. II-251

Nichols, M. A. I-274
Nichols, M. A. III-37
Nicolau, A. II-120
Novack, S. II-120
Nutt, G. J. II-77
Ohring, S. I-311
Okawa, Y. I-336
Okazaki, S. II-236
Olariu, S. III-192
Olariu, S. III-307
Oruc, A. Y. III-159
Palermo, D. J. II-30
Paris, N. II-184
Park, H. III-236
Patel, J. H. II-87
Patnaik, L. M. I-214
Peir, J. K. I-12
Peterson, G. D. III-289
Pfeiffer, P. II-188
Picano, S. II-103
Potlapalli, Y. R. I-303
Potter, T. M. I-47
Pradhan, D. K. I-118
Pramanik, S. II-213
Prasanna, V. K. III-236
Raghavendra, C. S. III-280
Ramachandran, U. I-237
Ramaswamy, S. II-134
Ranade, A. III-201
Ranka, S. I-241
Ranka, S. III-102
Ravikumar, C. P. I-83
Ravikumar, S. I-237
Reeves, A. P. II-21
Richards, G. W. I-146
Ritter, G. X. III-98
Robertazzi, T. G. I-290
Roy-Chowdhury, A. III-133
Roysam, B. II-279
Saab, D. G. I-138
Saab, D. G. I-64
Sadayappan, P. II-148
Sadayappan, P. II-301
Sadayappan, P. III-18
Sadayappan, P. III-94
Saghi, G. III-248
Salinas, J. III-141
Samet, H. III-47
Schwan, K. II-205
Schwing, J. III-192
Schwing, J. L. III-307
Sevcik, K. C. II-140
Seznec, A. I-25
Sha, E. H. II-231
Shah, G. I-237
Shang, W. I-30
Shankar, R. III-102
Sharma, D. D. I-118
Sharma, M. II-82
Sharma, S. II-301
Sharp, D. W. N. III-82
Sheffler, T. J. III-29
Shen, W. III-192
Sheu, J. II-273
Shi, H. III-98
Shih, T. II-73
Shin, K. G. I-227
Shing, H. I-77
Shirazi, B. II-68
Shu, W. II-167
Siegel, H. J. I-274
Siegel, H. J. I-349
Siegel, H. J. III-248
Siegel, H. J. III-37
Sims, D. L. II-269
Sinclair, J. B. III-276
Smith, T. J. II-260
Snelick, R. II-2
So, K. I-12
Soffa, M. L. II-156
Sridhar, M. A. III-280
Srimani, P. K. III-231
Stenstrom, P. I-56
Stumm, M. II-140
Stumm, M. II-201
Su, C. I-227
Su, E. II-30
Subbaraman, C. P. III-244
Sun, X. III-10
Swami, A. III-253
Tan, K. I-345
Tandri, S. II-140

Tang, J. III-276
Tang, J. H. I-12
Tayyab, A. B. I-197
Temam, O. I-180
Temam, O. III-299
Thazhuthaveetil, M. J. I-254
Tomko, K. A. II-95
Toteno, Y. I-336
Tout, W. R. II-213
Trahan, J. L. III-244
Trefftz, C. II-288
Tripathi, A. III-65
Tsai, H. III-57
Tsai, T. II-161
Tseng, Y. III-149
Turner, S. W. II-125
Tzeng, N. I-96
Tzeng, N. II-209
Unrau, R. II-201
Vaidyanathan, R. I-158
Vaidyanathan, R. III-244
Varman, P. J. III-276
Veidenbaum, A. V. I-184
Verma, R. M. III-73
Vernon, M. K. I-232
Visvanathan, V. III-18
Visvanathan, V. III-94
Wagh, M. D. II-82
Wah, B. W. I-30
Wah, B. W. III-263
Wang, H. C. III-215
Wang, J. I-241
Wang, M. I-274
Wang, M. III-37
Wen, Z. III-205
Werth, J. F. II-58
While, R. L. III-82
Whitfield, D. II-156
Wijshoff, H. III-299
Wills, D. S. I-154
Wilson, L. III-192
Wu, C. I-47
Wu, C. E. I-163
Wu, M. II-167
Xu, H. II-175
Yajnik, S. I-246
Yang, J. I-219
Young, H. C. III-253
Yousif, M. S. I-254
Yu, C. I-110
Zhang, J. III-192
Zhang, J. III-307
Zheng, Q. III-77
Zheng, S. Q. I-158
Zhou, B. B. III-128

TABLE OF CONTENTS
VOLUME II - SOFTWARE

Preface ..... iii
List of Referees ..... iv
Author Index - Full Proceedings ..... viii

SESSION 1B: MODELS/PARADIGMS ..... II-1

(R): Using Synthetic-Perturbation Techniques for Tuning Shared Memory Programs (Extended Abstract) ..... II-2
*Robert Snelick, Joseph JaJa, Raghu Kacker, and Gordon Lyon*
(R): Comparing Data-Parallel and Message-Passing Paradigms ..... II-11
*Alexander C. Klaiber and James L. Frankel*
(C): Function-Parallel Computation in a Data-Parallel Environment ..... II-21
*Alex L. Cheung and Anthony P. Reeves*
(C): Automatic Parallelization Techniques for the EM-4 ..... II-25
*Lubomir Bic and Mayez Al-Mouhamed*

SESSION 2B: COMPILER (I) ..... II-29

(R): Automating Parallelization of Regular Computations for Distributed-Memory Multicomputers in the PARADIGM Compiler ..... II-30
*Ernesto Su, Daniel J. Palermo, and Prithviraj Banerjee*
(R): Compilation Techniques for Optimizing Communication on Distributed-Memory Systems ..... II-39
*Chun Gong, Rajiv Gupta, and Rami Melhem*
(R): Meta-State Conversion ..... II-47
*H.G. Dietz and G. Krishnamurthy*

SESSION 3B: TOOLS ..... II-57

(R): A Unified Model for Concurrent Debugging ..... II-58
*S.I. Hyder, J.F. Werth, and J.C. Browne*
(C): PARSA: A Parallel Program Scheduling and Assessment Environment ..... II-68
*Behrooz Shirazi, Krishna Kavi, A.R. Hurson, and Prasenjit Biswas*
(C): VSTA: A Prolog-Based Formal Verifier for Systolic Array Designs ..... II-73
*Nam Ling and Timothy Shih*
(C): A Parallel Program Tuning Environment ..... II-77
*Gary J. Nutt*
(C): Decremental Scattering for Data Transport Between Host and Hypercube Nodes ..... II-82
*Mukesh Sharma and Meghanad D. Wagh*

SESSION 4B: CACHE/MEMORY MANAGEMENT ..... II-86

(R): Memory Reference Behavior of Compiler Optimized Programs on High Speed Architectures ..... II-87
*John W. C. Fu and Janak H. Patel*
(R): Iteration Partitioning for Resolving Stride Conflicts on Cache-Coherent Multiprocessors ..... II-95
*Karen A. Tomko and Santosh G. Abraham*

(C): Performance and Scalability Aspects of Directory-Based Cache Coherence in Shared-Memory Multiprocessors ........ II-103
*Silvio Picano, David G. Meyer, Eugene D. Brooks III, and Joseph E. Hoag*

(C): Compiling for Hierarchical Shared Memory Multiprocessors ........ II-107
*J.D. Martens and D.N. Jayasimha*

SESSION 5B: MAPPING/SCHEDULING ........ II-111

(R): Efficient Use of Dynamically Tagged Directories Through Compiler Analysis ........ II-112
*Trung N. Nguyen, Zhiyuan Li, and David J. Lilja*

(C): Trailblazing: A Hierarchical Approach to Percolation Scheduling ........ II-120
*Alexandru Nicolau and Steven Novack*

(C): Contention-Free 2D-Mesh Cluster Allocation in Hypercubes ........ II-125
*Stephen W. Turner, Lionel M. Ni, and Betty H.C. Cheng*

(C): A Task Allocation Algorithm in a Multiprocessor Real-Time System ........ II-130
*Jean-Pierre Beauvais and Anne-Marie Deplanche*

(C): Processor Allocation and Scheduling of Macro Dataflow Graphs on Distributed Memory Multicomputers by the PARADIGM Compiler ........ II-134
*Shankar Ramaswamy and Prithviraj Banerjee*

SESSION 6B: COMPILER (II) ........ II-139

(R): Locality and Loop Scheduling on NUMA Multiprocessors ........ II-140
*Hui Li, Sudarsan Tandri, Michael Stumm, and Kenneth C. Sevcik*

(R): Compile-Time Characterization of Recurrent Patterns in Irregular Computations ........ II-148
*Kalluri Eswar, P. Sadayappan, and Chua-Huang Huang*

(C): Investigating Properties of Code Transformations ........ II-156
*Deborah Whitfield and Mary Lou Soffa*

(C): Compiling Efficient Programs for Tightly-Coupled Distributed Memory Computers ........ II-161
*PeiZong Lee and Tzung-Bow Tsai*

SESSION 7B: SIMD/DATA PARALLEL ........ II-166

(R): Solving Dynamic and Irregular Problems on SIMD Architectures with Runtime Support ........ II-167
*Wei Shu and Min-You Wu*

(R): Evaluation of Data Distribution Patterns in Distributed-Memory Machines ........ II-175
*Edgar T. Kalns, Hong Xu, and Lionel M. Ni*

(C): Activity Counter: New Optimization for the Dynamic Scheduling of SIMD Control Flow ........ II-184
*Ronan Keryell and Nicolas Paris*

(C): SIMD Optimizations in a Data Parallel C ........ II-188
*Maya Gokhale and Phil Pfeiffer*

SESSION 8B: RESOURCE ALLOCATION/OS ........ II-192

(R): An Adaptive Submesh Allocation Strategy For Two-Dimensional Mesh Connected Systems ........ II-193
*Jianxun Ding and Laxmi N. Bhuyan*

(C): A Fair Fast Scalable Reader-Writer Lock ........ II-201
*Orran Krieger, Michael Stumm, Ron Unrau, and Jonathan Hanna*

(C): Experiments With Configurable Locks for Multiprocessors ........ II-205
*Bodhisattwa Mukherjee and Karsten Schwan*

(C): On Resource Allocation in Binary n-Cube Network Systems ........ II-209
*Nian-Feng Tzeng and Gui-Liang Feng*
(C): A Distributed Load Balancing Scheme for Data Parallel Applications ........ II-213
*Walid R. Tout and Sakti Pramanik*
(C): Would You Run It Here... Or There? (AHS: Automatic Heterogeneous Supercomputing) ........ II-217
*H.G. Dietz, W.E. Cohen, and B.K. Grant*

SESSION 9B: TASK GRAPH/DATA FLOW ........ II-222

(R): A Concurrent Dynamic Task Graph ........ II-223
*Theodore Johnson*
(C): Unified Static Scheduling on Various Models ........ II-231
*Liang-Fang Chao and Edwin Hsing-Mean Sha*
(C): Dataflow Graph Optimization for Dataflow Architectures - A Dataflow Optimizing Compiler ........ II-236
*Sholin Kyo, Shinichiro Okazaki, and Masanori Mizoguchi*
(C): Increasing Instruction-level Parallelism through Multi-way Branching ........ II-241
*Soo-Mook Moon*
(C): Optimizing Parallel Programs Using Affinity Regions ........ II-246
*Bill Appelbe and Balakrishnan Lakshmanan*

SESSION 10B: DATA DISTRIBUTION/PARTITIONING ........ II-250

(R): A Model for Automatic Data Partitioning ........ II-251
*Paul D. Hovland and Lionel M. Ni*
(C): Multi-Level Communication Structure for Hierarchical Grain Aggregation ........ II-260
*D.H. Gill, T.J. Smith, T.E. Gerasch, and C.L. McCreary*
(C): Dependence-Based Complexity Metrics for Distributed Programs ........ II-265
*Jingde Cheng*
(C): Automatically Mapping Sequential Objects to Concurrent Objects: The Mutual Exclusion Problem ........ II-269
*David L. Sims and Debra A. Hensgen*
(C): Communication-Free Data Allocation Techniques for Parallelizing Compilers on Multicomputers ........ II-273
*Tzung-Shi Chen and Jang-Ping Sheu*

SESSION 11B: MISCELLANEOUS ........ II-278

(C): Improving RAID-5 Performance by Un-striping Moderate-Sized Files ........ II-279
*Ronald K. McMurdy and Badrinath Roysam*
(C): On Performance and Efficiency of VLIW and Superscalar ........ II-283
*Soo-Mook Moon and Kemal Ebcioglu*
(C): Efficient Broadcast in All-Port Wormhole-Routed Hypercubes ........ II-288
*Philip K. McKinley and Christian Trefftz*
(C): Implementing Speculative Parallelism in Possible Computational Worlds ........ II-292
*Debra S. Jusak, James Hearne, and Hilda Halliday*
(C): System-Level Diagnosis Strategies for n - star Multiprocessor Systems ........ II-297
*A. Kavianpour*
(C): On Compiling Array Expressions for Efficient Execution on Distributed-Memory Machines ........ II-301
*S.K.S. Gupta, S.D. Kaushik, S. Mufti, S. Sharma, C.-H. Huang, and P. Sadayappan*

Table of Contents - Full Proceedings ........ A1

# SESSION 1B

## MODELS/PARADIGMS

# Using Synthetic-Perturbation Techniques for Tuning Shared Memory Programs*
# (Extended Abstract)

Robert Snelick
Joseph JáJá†
Raghu Kacker
Gordon Lyon

National Institute of Standards and Technology
Gaithersburg, Maryland 20899

## Abstract

*The Synthetic-Perturbation Tuning (SPT) methodology is based on an empirical approach that introduces artificial delays into the MIMD program and captures the effects of such delays by using the modern branch of statistics called design of experiments. SPT provides the basis of a powerful tool for tuning MIMD programs that is portable across machines and architectures. The purpose of this paper is to explain the general approach and to extend it to address specific features that are the main source of poor performance on the shared memory programming model. These include performance degradation due to load imbalance and insufficient parallelism, overhead introduced by synchronizations and by accessing shared data structures, and compute time bottlenecks. We illustrate the practicality of SPT by demonstrating its use on two very different case studies: a large image processing benchmark and a parallel quicksort.*

## Introduction

Today's multiprocessors provide unprecedented performance potential, yet all too often the actual performance obtained is far less impressive. The inherent complexity of parallel programs makes it far more difficult to capture *true* performance measurements on multiple-instruction stream, multiple-data stream (MIMD) architectures. In the absence of MIMD performance tools, obtaining reasonable parallel program performance is no small undertaking. Our objective here is to explain, extend, and apply a technique that gives the programmer useful performance information and is portable across machines as well as architectures. The technique works equally well in both shared memory and message passing environments. This work emphasizes the SPT techniques for shared memory programs.

Many existing tools [4, 6, 7, 8] focus on capturing performance metrics via monitoring. Performance metrics for parallel programs can provide an overwhelming amount of internal detail that is difficult to relate to performance bottlenecks. Our approach identifies sources of performance degradation via a sensitivity analysis which links program bottlenecks directly to the source code. Synthetic-Perturbation Tuning (SPT)[1] introduces the notion of inserting user-induced artificial delays into the source code and capturing the effect of such delays by employing design of experiments techniques.

Performance statistics have long been used to improve program execution efficiencies [5]. The most common statistics are frequency counts and timings for segments of code. Segments can be procedures or smaller entities, such as pieces of straight line code. Simple and intuitive to use, execution profiles reveal program bottlenecks that impede execution.

The advent of the MIMD parallel system raises two challenges to conventional profiling. The first problem is an exploding state space. The second is the coupling among profile states caused by parallel execution. Conventional profile statistics require a much deeper interpretation in MIMD. With separate threads of execution working on a joint com-

*Partially sponsored by the Advanced Research Projects Agency, Task No. 7066, Amendment 03.

†Also, Department of Electrical Engineering, Institute for Advanced Computer Studies, and Institute For Systems Research, University of Maryland, College Park, MD. 20742.

putation, it is natural that communication and constraints must exist among threads. Interdependencies are manifest as latencies–a wait for a message, a pause prior to writing some shared variable. Because latencies are generated by circumstances of the system and program, they are not easily estimated. Latencies can range from negligible to devastatingly large.

Existing techniques collect performance statistics in a number of ways. In the taxonomy shown in Table 1, the first row shows two common methods of defining events to be recorded. The first method is **periodic sampling** (I), which is tied to a clock and is based on collecting statistics.

The second method of **fixed triggering** (II) uses identifiable locations or patterns, which when reached or matched, define an event. For instance, a special procedure call upon entry to a segment of executable statements will record information about the program at that point.

| I. **Periodic Sampling** | II. **Fixed Triggering** |
|---|---|
| a. **Traces** | b. **Histograms** |

Table 1: Simple Taxonomy of Performance Techniques.

The bottom row of the table gives common types of recorded information. A **trace** (a) often comprises a record of a location in code or a configuration of a subsystem plus a time-stamp. A constant stream of traces is generated as system execution proceeds, and from this data much important behavior can be reconstructed. Unfortunately, the stream is often hard to manage because of its magnitude. Special collection hardware may be needed to handle the volume of data[9].

**Histograms** (b) are an accumulative approach that demands little extra bandwidth. The number of invocations of a procedure, the overall time spent in a loop– these are of type (b), histogram or profile statistics. Because histogram information accumulates, they demand far less storage or bandwidth than do traces. The cost is a loss of detail, since time is not generally recorded except as an accumulated amount. No detailed times are kept of individual events.

Tools *gprof* and *quartz* are of type I-b. The VLSI instrumentation chip *MultiKron*[9] supports either II-a or II-b. *MTOOL*, triggered by basic program blocks, builds histograms and is therefore of type II-b. A type I-a is uncommon, since periodic random sampling yields an erratic set of data.

## Description of Technique

Synthetic-Perturbation Tuning (SPT) is an empirical approach that treats an MIMD program as a *black box* with input parameters and outputs. The SPT approach introduces synthetic perturbations (i.e., *artificial delays*) into source code segments and relies on (for design and analysis) a modern branch of statistical theory called *design of experiments* (DEX) [2, 3]. DEX provides an efficient methodology for determining the relative sensitivity of the MIMD program to synthetic perturbations. SPT focuses the programmer's attention on the potential problem areas in the program. An important step in this methodology is to identify which segments of code are candidates for improvements. The identified code segments are termed *bottlenecks*. Each bottleneck is ranked quantitatively according to its sensitivity to synthetic perturbation. Such a list is called an **SPT Rank.** An SPT rank is a guide that can be used to improve (tune) the corresponding code segments.

The SPT premise is that if the program is highly sensitive to source code perturbations in a code segment (i.e., delay has a clearly detrimental effect on performance), then source code improvements to that segment will have an opposite (positive) effect. In the next section, a justification of this premise as it applies to shared memory programs is provided. In what follows, we briefly describe the generic SPT methodology for tuning an MIMD program.

**1. Determine objective and define test conditions.** To perform a set of SPT experiments, the user must define a set of test conditions. Test conditions include the source code implementation, data set, and machine. SPT's analysis applies to the defined test conditions. If these conditions change, a new set of SPT experiments and analysis may need to be performed. Based on our experience, given a source code and a machine, results for similar data sets are usually consistent.

**2. Choose candidate code segments.** A candidate code segment can be any section of code. Typically it is a function declaration, function call (usually for synchronization, e.g., a send protocol or a locking mechanism), critical section, or a loop construct.

**3. Insert Perturbations.** Each candidate code segment is instrumented with a delay option (*delay* or *no delay*). *No delay* leaves the code unperturbed. *Delay* takes the form of a function call that performs a specified number of instructions. The call does not alter the natural path of the program or the values of its variables. An example of how a delay might be implemented is given in the following pseudo C source code block:

```
while(v--){/* factor 12 *//* begin original code */
#if F12                   /* begin spt code      */
   spt_delay (delay_value);
#endif                    /* end spt code        */
          :
        Code
          :
}                         /* end original code   */
```

*spt_delay()* is a function that performs a specified number of synthetic instructions corresponding to *delay_value*. The looping block **while**(v--) { ...} is a designated code segment and referred to as *factor 12 (F12)*. The statistical term **factor** is used to represent a candidate code segment. The duration of the artificial delay is an important aspect. Ideally, the delay should be long enough so that it can easily be distinguished from noise and short enough so as not to produce unnecessarily long program execution times.

**4. Design experimental plan.** Once the candidate code segments are determined, an experimental plan can be developed. There is no theoretical limit on how many distinct factors (source code segments) can be investigated on a given SPT iteration. A variety of schemes can be used for designing an experimental plan[2]. A small $2^3$ factor complete factorial example is given to illustrate the ideas of experimental designs. A $2^n$ plan indicates that the experiment has n factors each at 2 levels. Here we have $n = 3$ factors (called, for example *F1, F2*, and *F3*) and 2 levels (*no delay* (−) and *delay* (+)) for each factor.

Suppose we have a MIMD program (call it *Xprog*) with three suspected bottleneck locations, *F1, F2*, and *F3* that correspond to certain code segments within *Xprog*. *F1* represents a *for loop* in the function *func_Y*, *F2* represents a *critical section* in the function *func_Z*, and *F3* represents a *while loop* in the function *func_Z*. With a three factor complete factorial plan, there are $2^3 = 8$ possible delay patterns each indicated by a row in Table 2. In DEX terminology each delay pattern is a **treatment**.

**5. Run experiments according to plan and record a response.** Each treatment or version of the program is run and the corresponding response is recorded. The response can be any useful measurement; typically the response is the total program execution time. The treatments are usually run in a random order.

| Treatment | Factors | | | Response |
|---|---|---|---|---|
| | *F1* | *F2* | *F3* | |
| 1 | − | − | − | 17.05 |
| 2 | + | − | − | 17.08 |
| 3 | − | + | − | 23.19 |
| 4 | + | + | − | 23.34 |
| 5 | − | − | + | 19.62 |
| 6 | + | − | + | 19.71 |
| 7 | − | + | + | 25.61 |
| 8 | + | + | + | 25.71 |

Table 2: $2^3$**Complete Factorial Design for** *Xprog*.

| Rank | Factor | M Effect † | Routine | Construct |
|---|---|---|---|---|
| 1 | F2 | 6.10 | func_Z() | *while* |
| 2 | F3 | 2.49 | func_Z() | *crit sect* |
| 3 | F1 | 0.09 | func_Y() | *for* |

† Standard Error of Main Effects: ±0.06

Table 3: **SPT Rank of Main Effects for** *Xprog*.

**6. Analyze Results.** The object of data analysis is to evaluate the main effects associated with each factor. The **main effect** of a factor is the difference between two average responses, one corresponding to the treatments which have the (+) level of the factor and the other corresponding to the treatments which have the (−) level of the factor. The computed values of the main effects are subsequently used to produce an SPT ranking of the factors (Table 3).

The first column of the SPT rank gives the standing of the corresponding code segment. A higher rank indicates a higher sensitivity to artificial delays (e.g., *F2* is most sensitive to the delay). Column 2 gives the factor number which provides a reference back to the source code location represented by the factor. The main effects column gives the sensitivity levels of the corresponding code segments as well as an estimate of the standard error[1]. The actual numbers are not as important as their relative magnitudes. Column 4 describes which function the section of code resides in. The last column indicates what type of construct the code segment is. By surveying Table 3 we can conclude that factor *F2* is the most significant. This code segment should be given first priority in the tuning effort.

**7. Improve bottlenecks and determine performance.** An SPT rank gives a list of potential bottlenecks. The bottlenecks so identified may or may not be improvable. Investigation begins with

[1] The standard error of the main effect is evaluated by treating high order interactions as errors from noise (see[2], page-327).

the higher ranked bottlenecks since they possess the greatest potential for improvement. These bottlenecks can be pursued further with SPT to gain more information about the bottlenecks or an attempt can be made to improve them.

## SPT Applied to Shared Memory Programs

The emergence of shared memory multiprocessors in the past decade has given rise to a substantial effort in designing and analyzing software for these machines. The performance of a shared memory program depends crucially on several interrelated factors such as the amount of parallelism used, the degree to which the work load is balanced among the processors, the contention over shared resources (interconnection network, bus, memory), and the overhead incurred by synchronization. In the rest of this section, we describe our approach for finding bottlenecks related to each of these aspects as they arise in a shared memory program. In the next section, we illustrate the use of these techniques on two case studies, the *Image Understanding Benchmark*, and a parallel version of the *quicksort* algorithm.

### Degree of Parallelism

A typical MIMD program contains a mix of scalar, serial, vector, and parallel operations. A section of code with insufficient parallelism is a bottleneck if its execution time is significant compared to the overall execution time of the program. Such a bottleneck can be detected only if the performance of the program is analyzed as a function of the number of processors involved. In fact, by Amdhal's law, for a given program, it is the execution time of the serial portions that will ultimately determine the speed of the program as the number of processors increases (and the input size is held constant). Our method is based on an extension of this observation.

We insert artificial delays into the sections of code under investigation. We then perform the design of experiments on successively scaled-up versions of the system. As the number of processors increases, the effects of the parallel code will become less important while the effects of the serial code will become more significant.

Consider for example a section of code that multiplies an $n \times n$ matrix $A$ by a vector $x$ to generate the vector $y = Ax$. Partition $A$ as follows

$$A = \begin{bmatrix} A_1 \\ A_2 \\ \vdots \\ A_p \end{bmatrix}$$

where each $A_i$ is of size $(n/p) \times n$, $n/p$ is assumed to be an integer, and $p$ is the number of processors available. The following section of the code corresponds to the computation $A_i x$ performed by the $i$th processor

```
for j = (i - 1)n/p + 1 to i n/p do
    y(j) := 0
    for k = 1 to n do
        SPT-delay
        y(j) := y(j) + A(j,k)x(k)
    end
end
```

The execution time of this section of code is proportional to $\frac{n^2}{p}(\Delta + 2t_{fp})$, where $\Delta$ is the SPT delay time, and $t_{fp}$ is the time it takes to execute a floating-point add or multiply (assumed to be equal for simplicity). Hence the effect of the SPT delay is a net increase of $\frac{n^2}{p}\Delta$ in the total execution time; thus, it represents a factor whose effect is a decreasing function of $p$. Therefore, the effect of the parallel code becomes less important as the system is scaled-up.

### Load Balancing

The speedup achieved by a parallel program is primarily due to the development of threads of execution that can be run concurrently. This can be done either by using functional or data decomposition present (explicitly or implicitly) in an existing algorithm, or by developing a new algorithm that has a higher degree of (functional or data) parallelism. With functional decomposition, each processor is responsible for executing a different function, and hence the distribution of the loads among the processors is completely dependent on the computational requirements of these functions. Similarly, data decomposition can result in some processors having to handle much larger amounts of data than the rest of the processors.

A load balancing problem can be viewed as insufficient parallelism that, in general, arises dynamically. The insertion of artificial delays followed by an SPT analysis allows us to determine each section of the code that generates a significant load imbalance. Notice that an SPT delay will cause the processor with the heaviest load to run even slower and hence its SPT effect will be significant.

In the Image Understanding benchmark that we study in the next section, we use this technique to predict the load imbalance that is caused by the procedure to determine the connected components of an image. In this case, the processor assigned to handle the background pixels has much more work to do than the remaining processors. Without analyzing the procedure, our SPT analysis was able to determine the load imbalance resulting from this procedure and to predict its importance as the number of processors increases.

### Critical Sections and Synchronization

Processors executing a shared memory program may waste a substantial amount of time trying to enter a critical section ("busy wait") or trying to synchronize their activities. SPT can be used to provide information concerning any significant overhead incurred in a critical section or at a synchronization point. We start by handling critical sections.

The insertion of an artificial delay into a critical section allows us to perform an SPT analysis similar to the previous two cases. We claim that, for a critical section that represents a significant bottleneck in the program, its SPT effect will become more important as we scale-up the system. In fact, the overall contribution of the delays tends to be cumulative with respect to the number of processors that are trying to access the critical section.

As for synchronization, we cannot use the technique in a straightforward way. However we can extend it as follows. For each synchronization barrier, we insert two types of perturbations, one immediately before the barrier and the other immediately after the barrier. The perturbation *FB1* inserted before the barrier consists of an artificial critical section, while the perturbation *FB2* inserted after the barrier consists of an artificial critical section followed by an artificial barrier. The justification of the perturbation *FB2* is as follows. The critical section delay in *FB2* is an obvious bottleneck since all the released threads try to execute it at once whereas the artificial barrier ensures that the new program is functionally identical to the original one. We then run our experiments and compare the effects of *FB1* and *FB2*. If their effects are about the same, we can conclude that the synchronization cost is marginal. The argument is that in this case *FB1* is also being pressed for execution by many threads, which is indicative of how threads arrive at the barrier– all together – a good parallel execution. As the difference in the two effects increases, the synchronization cost increases. Threads that arrive one-by-one at *FB1* will not find it much of a bottleneck and hence its effect will be lower than that of *FB2*. It follows that by comparing the effects of *FB1* and *FB2*, we will be able to diagnose a barrier being used efficiently. This method is applied in the next section to a quicksort program that contains several synchronization points and is shown to identify properly the costly synchronizations.

## Case Studies

### Image Processing Benchmark

In this section we present a practical shared memory tuning example based upon a large image processing benchmark. The test code is the *Image Understanding Benchmark* for parallel computers developed at the University of Massachusetts at Amherst[10]. The benchmark was described as a "complex benchmark that would be almost impossible to tune" [10]. Using SPT, we demonstrate how important bottlenecks were identified and subsequently analyzed and improved.

The benchmark was designed to test common vision tasks on parallel architectures. It consists of a model-based object recognition problem, given two sources of sensory input, intensity and range data, and a collection of candidate models. The intensity image is a $512 \times 512$ array of 8-bit pixels, while the depth image consists of a $512 \times 512$ array of 32-bit floating point numbers. The models contain rectangular surfaces, floating in space, viewed under orthographic projection. Added to the configuration is both noise and spurious nonmodel surfaces. The benchmark's task is to recognize an approximately specified 2 1/2-dimensional "mobile" sculpture in a cluttered environment. The sculpture is a collection of 2-dimensional rectangles of various sizes, brightnesses, orientations, and depths.

The experiments are performed on both a ten processor and twenty-six processor Sequent Symmetry. The Image Understanding Benchmark package comes with a number of data sets and their corresponding outputs. The example presented here uses test set number two. The benchmark consists of more than 50 procedures and has approximately 3500 lines of C code.

We selected 31 factors (loops, function declarations, and critical sections) as potential candidates for bottlenecks based on code inspection. An experimental plan is selected to handle the large number of code segments that need to be investigated. The

| Rank | Factor | M Effect † | Routine | Construct |
|---|---|---|---|---|
| 1 | 17 | 6.03 | Grad Magn | *for* |
| 2 | 26 | 5.46 | Med Filt | *while* |
| 3 | 2 | 5.26 | Conn Comp | *for* |
| 4 | 1 | 3.94 | Conn Comp | function |
| 5 | 4 | 3.84 | Conn Comp | *while* |
| 6 | 25 | 2.01 | Med Filt | *for* |
| 7 | 20 | 1.67 | Match | function |
| 8 | 29 | 1.36 | Probe | *for* |
| 9 | 6 | 0.64 | Ext Cues | *for* |

† Standard Error of Main Effects: ±0.04

Table 4: **SPT Rank for** *Image Benchmark*, **8 Processors.**

image benchmark is instrumented with an SPT delay for each factor. The treatments are run in a random order and the overall execution time of the program is recorded as the response.

Table 4 lists the main effects of the nine highest-ranked factors of the image processing benchmark running on 8 processors.

This initial set of experiments indicates that the three top ranked procedures (Gradient Magnitude, Median Filtering, and Connected Components) represent major bottlenecks. Hence tuning the corresponding code segments should be given first priority. Notice that none of the top ranked factors involves a critical section or a synchronization barrier. Therefore the emphasis of the tuning effort should concentrate on increasing the efficiency of the serial sections within the loops (corresponding to factors $F17$, $F26$ and $F2$), or better balancing the load among the processors, or increasing the degree of parallelism. Since factor $F17$ was ranked highest, we concentrated initially on the corresponding code segment.

The Gradient Magnitude procedure performs a standard $3 \times 3$ Sobel operation on the depth image. The section of code within the loop corresponding to factor $F17$ is quite inefficient. After removing multiplications by zeros, and reducing the total number of remaining multiplications, the execution time of the procedure improved 300%. At this point, the relative ranking of the procedure dropped to 8 with 8 processors(Table 8).

Our next task was to consider the Median Filtering procedure. While we were attempting to tune this procedure, we discovered that the procedure generated erroneous results. At this time we switched our efforts to tuning the third procedure, Connected Components. This procedure assigns a unique label to each contiguous collection of pixels having the same intensity level value. To gain a better understanding, we ran our experiments using 2, 8, and 24 processors. Tables 5, 6, and 7 show the resulting rankings of the major factors (on the original code) as a function of the number of processors.

| Rank | Factor | M Eff † | Routine | Construct |
|---|---|---|---|---|
| 1 | 17 | 24.04 | Grad Magn | *for* |
| 2 | 26 | 22.18 | Med Filt | *while* |
| 3 | 25 | 8.27 | Med Filt | *for* |
| 4 | 2 | 5.26 | **Conn Comp** | *for* |
| 5 | 1 | 4.50 | **Conn Comp** | function |
| 6 | 4 | 4.34 | **Conn Comp** | *while* |

† Standard Error of Main Effects: ±0.04

Table 5: **SPT Rank for** *Image Benchmark*, **2 Processors.**

| Rank | Factor | M Eff † | Routine | Construct |
|---|---|---|---|---|
| 1 | 17 | 6.03 | Grad Magn | *for* |
| 2 | 26 | 5.46 | Med Filt | *while* |
| 3 | 2 | 5.26 | **Conn Comp** | *for* |
| 4 | 1 | 3.94 | **Conn Comp** | function |
| 5 | 4 | 3.84 | **Conn Comp** | *while* |
| 6 | 25 | 2.01 | Med Filt | *for* |

† Standard Error of Main Effects: ±0.04

Table 6: **SPT Rank for** *Image Benchmark*, **8 Processors.**

It is immediately clear that there is a serious load balancing problem; the three factors (*F1, F2, F4*) corresponding to Connected Components have gradually moved to the very top of the table as the number of processors increased. A close examination of the procedure confirms our suspicion. One processor is assigned to handle the background pixels and hence ends up doing most of the work. A completely different scheduling policy or a completely new algorithm is required before a significant improvement can be made. Even by making slight modifications, we were able to improve the performance of this procedure.

| Rank | Factor | M Eff † | Routine | Construct |
|---|---|---|---|---|
| 1 | 2 | 5.43 | **Conn Comp** | *for* |
| 2 | 1 | 3.95 | **Conn Comp** | function |
| 3 | 4 | 3.93 | **Conn Comp** | *while* |
| 4 | 17 | 2.14 | Grad Magn | *for* |
| 5 | 26 | 2.03 | Med Filt | *while* |
| 6 | 20 | 1.55 | Match | function |

† Standard Error of Main Effects: ±0.12

Table 7: **SPT Rank for** *Image Benchmark*, **24 Processors.**

| Rank | Factor | M Effect † | Routine | Construct |
|---|---|---|---|---|
| 1 | 26 | 5.57 | Med Filt | *while* |
| 2 | 2 | 5.30 | Conn Comp | *for* |
| 3 | 1 | 4.06 | Conn Comp | function |
| 4 | 4 | 3.93 | Conn Comp | *while* |
| 5 | 25 | 2.04 | Med Filt | *for* |
| 6 | 20 | 1.64 | Match | function |
| 7 | 29 | 1.23 | Probe | *for* |
| 8 | 17 | 0.59 | Grad Magn | *for* |
| 9 | 6 | 0.59 | Ext Cues | *for* |

† Standard Error of Main Effects: ±0.04

Table 8: **SPT Rank for** *Improved Image Benchmark*, **8 Processors.**

We now show the results of the SPT analysis when performed on our improved version. We have modified the Gradient procedure as indicated earlier and have made some simple modifications to the Connected Components procedure. Table 8 shows a summary of the SPT analysis when performed on our improved version. On eight processors, our version runs 18.2% faster than the original version. Notice that the Gradient procedure is no longer a significant bottleneck (rank=8 on 8 processors, and rank=17 on 24 processors).

## Parallel Quicksort

The image processing benchmark provided insights on how SPT can be used to handle large applications. It successfully detected code inefficiencies and a load imbalance. However, synchronization and critical sections did not play a significant role. In this section, we discuss a parallel version of the *quicksort* algorithm and illustrate how SPT can be used to address bottlenecks due to synchronization and critical sections.

The test code is a parallel implementation of Hoare's *quicksort* algorithm[11]. *Quicksort* is a scheme that is based on partitioning a given list into two sublists relative to a selected member of the list, called the *pivot*. Elements of the list are rearranged such that all elements smaller than the pivot are to the *left* of the pivot and all elements greater than the pivot are to the *right* of the pivot. There are several ways of choosing the pivot to induce approximately equal partitions. We refer to a such partitioning step as a *pass*. Hence after a pass, the pivot value is positioned in its sorted order. This procedure is then applied recursively to each sublist. Once a sublist becomes small enough, it can be sorted by using a simple sorting routine, say selection sort or bubble sort.

A simple way to parallelize the quicksort procedure is to allocate newly-created sublists to available processors. A sublist assigned to a processor is then partitioned into two sublists by that processor. The allocation of sublists to processors is controlled by a shared stack. An idle processor asks for a sublist from the shared stack. To insure that no two processors take possession of the same sublist, the stack access is controlled by a critical section.

The following is a skeleton of the program code for a simple implementation of *quicksort*.

```
Initializations;
Put list on stack;
barrier(); /* barrier #1 */
while(stack is not empty) {
  barrier(); /* barrier #2 */
  lock(stack_lock);
   if(stack is not empty)
     pop();
  unlock(stack_lock);
  Select pivot; partition list into sublists L1 and L2;
     if(|L1| > |L2|) {
       lock(stack_lock);
         push(L2);
         push(L1);
       unlock(stack_lock); }
     else {
       lock(stack_lock);
         push(L1);
         push(L2);
       unlock(stack_lock); }
  barrier(); /* barrier #3 */
}
```

Our tuning effort of *quicksort* begins by investigating the cost of synchronization. There are three synchronization points, denoted as *barrier()*. Our SPT objective is to find out if processes are arriving at widely dispersed times, and hence causing many processors to idle for a significantly long period of time. Our investigation follows the treatment method presented earlier for synchronization points. For each synchronization barrier, two types of perturbations are inserted, one immediately before (*FB1*) and the other immediately after (*FB2*). The method is illustrated by the following code segment:

```
lock(spt_lock_1);            /*           */
   spt_delay(spt_delay);     /*    FB1    */
unlock(spt_lock_1);          /* treatment */

barrier();                   /* original barrier */

lock(spt_lock_1);            /*           */
   spt_delay(spt_delay);     /*    FB2    */
unlock(spt_lock_1);          /* treatment */
barrier();                   /*           */
```

| Paired Factor | Main Effect | Difference † |
|---|---|---|
| barrier pair 1 | *FB1:* 0.16 | 0.06 |
| | *FB2:* 0.22 | |
| barrier pair 2 | *FB1:* 14.22 | 1.00 |
| | *FB2:* 15.22 | |
| barrier pair 3 | *FB1:* 6.78 | 8.56 |
| | *FB2:* 15.34 | |

† Standard Error of the Difference: ±0.21

Table 9: **Paired Effects for *Quicksort's* Barriers.**

The three synchronization barriers are instrumented as shown above. This implementation demands six factors, two for each barrier tested. The experiments proceed as before; an experimental plan is created and tested. Table 9 shows the results. The leftmost column of Table 9 identifies the barrier. The second column gives the calculated main effect for each factor. The individual main effects are meaningless in isolation and must be paired up and compared to obtain the proper information. The last column, which contains the difference of each of the paired factors, gives an indication of the cost associated with each synchronization. Remember that the treatment *FB2* shows the effect of an ideal barrier application, and is very sensitive to delay. If the paired effects are about the same, we conclude that the synchronization cost is marginal. As the difference in the two effects increases, the synchronization cost increases. (Recall that *FB1* has less effect on straggling threads.)

In spite of its simplicity, this example illustrates the effectiveness and the generality of the SPT approach. The difference shown for the first synchronization barrier indicates that almost all processors arrive there at the same time. This is clearly the case since only one processor is responsible for the initialization phase and the rest crowd around the barrier. Used only once, the effects also show that this barrier is not very important to performance. The second synchronization barrier is not needed since the processors are already synchronized at the beginning of each pass. The test confirms what algorithm inspection tells us. The third row of the table indicates that the third synchronization barrier is costly compared to the other two synchronization barriers. This is because processors are working on different-length sublists (or no sublist at all) and hence arrive at the third synchronization point at widely different times. The barrier deserves some attention.

To alleviate the problem of synchronization at the end of the **while** loop, we rewrite the code following the skeleton shown next. The resulting improvement in performance is substantial (78% ).

| Rank | Factor | M Effect † | Routine | Construct |
|---|---|---|---|---|
| 1 | 1 | 13.82 | main() | *crit sect 1* |
| 2 | 3 | 2.86 | main() | *crit sect 3* |
| 3 | 2 | 2.62 | main() | *crit sect 2* |

† Standard Error of Main Effects: ±2.15

Table 10: **SPT Rank for *Quicksort's* Critical Sections.**

```
Initializations;
Put list on stack;
barrier();
for() {
  lock(stack_lock); /* CS1 */
    if(stack is not empty){
      pop();
    }
  unlock(stack_lock);
  if(!qsort_done) {
    Select pivot; partition list into sublists L1 and L2;
    if(|L1| > |L2|) {
      lock(stack_lock); /* CS2 */
        push(L2);
        push(L1);
      unlock(stack_lock);
    } else {
      lock(stack_lock); /* CS3 */
        push(L1);
        push(L2);
      unlock(stack_lock);
    }
  }
}
```

In the next experiment, SPT's objective is to obtain the relative importance (detrimental effect) of the three new critical sections (CS1, CS2, CS3). A delay is inserted in each critical section. An experimental plan is developed and run. Table 10 shows the SPT performance information for each critical section. Our SPT analysis shows that the first critical section dominates. At this point, we remove the other two factors from further consideration, and perform a complete SPT analysis that includes the factor (labelled $F1$) of the critical section $CS1$. Six additional code segments are selected to be tested along with this critical section. These are the procedures: partition_list(), bubble_sort(), swap(), push(), pop(), and select_pivot(). Since the delay for the critical section and regular code segments are equivalent, they can be compared. Table 11 shows the results obtained. Clearly factors *F3* and *F7* dominate the overall performance. Based upon this data, we examine the procedure partition-list() which calls the swap() procedure. Removing the calls to swap() and inserting its code into partition_list() resulted in an additional 23% improvement of the execution time of quicksort.

| Rank | Factor | M Effect † | Routine | Construct |
|---|---|---|---|---|
| 1 | F3 | 29.01 | part_list() | *while loop* |
| 2 | F7 | 8.94 | swap() | function |
| 3 | F4 | 1.26 | push() | function |
| 4 | F6 | 0.69 | bub_sort() | *while loop* |
| 5 | F1 | 0.14 | main() | *crit sect 1* |
| 6 | F2 | 0.11 | sel_pivot() | function |
| 7 | F5 | 0.09 | pop() | function |

† Standard Error of Main Effects: ±0.39

Table 11: **SPT Rank for** *Quicksort.*

## Conclusion

We have described the tuning methodology SPT, Synthetic-Perturbation Tuning, that is based on a branch of statistics called design of experiments. The main purpose of this methodology is to identify performance bottlenecks present in MIMD programs. SPT provides the basis of a powerful tuning tool that is scalable and portable across machines and architectures. We also considered in some detail the sources of poor performance on the shared memory model, and showed how these issues can be adequately captured using SPT. In particular, we developed simple techniques to detect the relative overhead of synchronization barriers and of accessing shared data structures. Two detailed case studies were then discussed and their bottlenecks analyzed using our methodology. Significant improvements were made based on the results of the SPT analysis.

We are currently refining and extending our methodology in several directions. In particular, we are analyzing approaches to measure the performance of memory hierarchy in a shared memory environment, and the communication overhead present in a message passing environment. Additional large case studies are currently being examined using SPT. Our future plans include the development of automated tools for performing the SPT analysis and reporting the appropriate information to the user.

A minor disadvantage of our methodology is the amount of experimentation necessary to perform the analysis. However, we believe that tuning MIMD programs is a highly nontrivial task requiring the capture of many parameters and their interactions. Simpler schemes are likely to fail in one aspect or another. The mathematical basis of our method provides a solid foundation upon which we can build general tuning techniques that are applicable across machines and architectures.

## References

[1] G. Lyon, R. Snelick, R. Kacker. "TPT: Time-Perturbation Tuning of MIMD Programs." Proceeding of the 6th International Conference on Modelling Techniques and Tools for Computer Performance Evaluation, Edinburgh, Scotland, September 1992, 211-224.

[2] G. Box, W. Hunter, J. Hunter, Statistics for Experimenters (1978), John Wiley and Sons Inc., New York.

[3] R. Jain. The Art of Computer Systems Performance Analysis. J. Wiley & Sons (New York, 1991), 720 pp.

[4] T. Anderson and E. Lazowska. "Quartz: A Tool for Tuning Parallel Program Performance." Processings, SIGMETRICS 1990 Conference, May 1990, 115-125.

[5] S. Graham, P. Kessler, and M. McKusick. "Gprof: A Call Graph Execution Profiler." "Proceeding, ACM SIGPLAN Symposium on Compiler Construction, June, 1982.

[6] A. Goldberg and J. Hennessy. "Performance Debugging Shared Memory Multiprocessor Programs with MTOOL." In Proceedings Supercomputing, pp. 481-490, Nov. 1991.

[7] M. Martonosi, A. Gupta, T. Anderson. "MemSpy: Analyzing Memory System Bottlenecks in Programs." Performance Evaluation Review, Vol. 20, No. 1, June 1992.

[8] H. Burkhart and R. Millen. "Performance-Measurement Tools in a Multiprocessor Environment." IEEE Transactions on Computers, Vol. 38, No. 5, May 1989.

[9] A. Mink and R. Carpenter. "Operating Principles of MULTIKRON Performance Instrumentation for MIMD Computers." Natl. Inst. of Standards and Technology, Gaithersburg, MD., NISTIR-4737; March 1992. 23 p.

[10] C. Weems, E. Riseman, A. Hanson, A. Rosenfeld. "The DARPA Image Understanding Benchmark for Parallel Computers." Journal of Parallel and Distributing Computing 11, 1-24, 1991.

[11] C. Hoare. "Quicksort." Computer Journal 5, 1(January 1962), 10-15.

# Comparing Data-Parallel and Message-Passing Paradigms

Alexander C. Klaiber
University of Washington
Seattle, WA 98105

James L. Frankel
Thinking Machines Corporation
Cambridge, MA 02142

Abstract — *The data-parallel programming model provides high-level abstraction, race-free and deterministic execution, and the illusion of lockstep execution — features that greatly simplify the task of writing and debugging applications. The message passing mode is typically faster, but very low level and hard to program. We implemented an event-driven circuit simulator using both models. The data-parallel version was developed faster, and was smaller. Although the circuit simulator is a poor fit to the data-parallel model, a few key optimizations and language extensions give the data-parallel version over 70% of the performance of the message-passing version. We conclude that with proper compiler support, data-parallel languages are a viable and attractive alternative even to handcrafted, low-level message-passing codes.*

## 1 Introduction

The use of massively parallel machines over recent years has outpaced the development and acceptance of corresponding high-level programming models and languages; most parallel applications are still written in existing sequential languages, using simple message passing libraries for communication. This approach is like programming in assembly language: it has the potential for highest performance since it operates close to the actual machine hardware. However, message passing programs are notoriously difficult to debug, with subtle race conditions, deadlock and nondeterministic execution. The programmer must take care of details that are irrelevant to the application, such as managing message buffers, avoiding race conditions or keeping track of messages in transit. Furthermore, the message-passing model does not present a clear performance model to the programmer.

Finally, routines in message-passing libraries have to be general enough to handle a wide range of communication patterns — hence for any given program, they are usually too general and thus slower than they would be if they could exploit compile-time knowledge about the program's communication pattern. High-level parallel languages may be able to provide the compiler with such information and thus actually achieve *better* communication performance than is easily feasible for message-passing codes [1].

Researchers have recognized the problems inherent in the message-passing model and many high-level parallel programming models and languages have been proposed, answering the fundamental design questions in many different ways. We believe many of these models to be viable approaches to successful high-level programming languages of the future. However, in this paper we concentrate specifically on *data-parallel languages* which exhibit the following characteristics:

**Imperative Style** — Programmers are familiar with this style, so it is likely to be used until another model (e.g. functional or logic programming) yields more efficient implementations.

**Explicit Parallelism** — Humans are still much better than compilers at making high-level decisions about data partitioning and parallelism. In addition, explicitly stating parallelism in the program provides a clearer performance model to the programmer.

**Explicit Communication** — The same considerations hold for communication operations: the programmer has ultimate control over data allocation, and potentially expensive communication operations show up explicitly in the source code, again clarifying the performance model.[1]

**Single Thread of Control** — The programmer is presented with the illusion that all virtual processors execute code synchronously. In other words, no matter how many processors are used, *conceptually* there is a single program counter. This makes understanding and debugging programs significantly easier. A compiler can achieve good performance by using any implementation that preserves the synchronous model's semantics; i.e. compiled code does not actually need to run in lockstep.

All parallelism in a data-parallel language is achieved by performing identical (or similar) operations on large numbers of data elements at the same time. This differs from control parallelism where parallelism ex-

[1] There are data-parallel languages that do not use explicit communication operations, for example FORTRAN-90.

ists in the form of independently executing threads of control. Data-parallel languages can be implemented efficiently on both shared-memory and distributed-memory architectures [8, 4, 3], providing the added advantage of portability across platforms. Data-parallel languages are free of race conditions and have a deterministic execution model, hence they are essentially as easy to debug as a sequential program. Thus, the data-parallel programming model provides an attractive alternative to the message-passing model.

It is often argued that data-parallel languages are suited to only a limited class of applications, and that their performance suffers dramatically for applications outside this area. In this paper, we investigate this claim by examining the relative performances of data-parallel and message-passing implementations of a parallel digital circuit simulator. This application poses a challenge to the data-parallel model, mainly because it exhibits low activity levels across virtual processors, and has deeply nested conditional statements and irregular communication patterns. These points will be discussed further throughout section 4.2. Indeed, a straightforward implementation of the simulator in the C* language [2, 12] yields performance significantly worse than a handcrafted message-implementation. However, we show that several language extensions and compiler techniques can enhance performance of the C* version, achieving over 70% of the performance of the message-passing version.

The rest of this paper is organized as follows. In section 2, we give an overview of the C* language in which the data-parallel version of the simulator was written, as well as a general outline of how to compile C* for a distributed-memory parallel machine. In section 3, we describe the circuit simulator and investigate some of its algorithmic properties and their impact on the performance of the data-parallel implementation. In section 4 we contrast the message-passing and the data-parallel implementations and then show a series of language enhancements and compiler techniques that improve performance of the C* version. In section 5, we report on our experiences developing the two implementations. Conclusions and directions for further work are given in sections 6 and 7, respectively.

# 2 The C* Data-Parallel Language

Several languages featuring data-parallel execution have been proposed over the last years [10, 6, 5, 9]. For this paper, we chose C* [12] as a representative. Significant work has been done on compilation of an older version of C* (defined in [9]) for both distributed-memory and shared-memory multiprocessors [8, 4, 3]. For a detailed description of the current language, the reader is referred to [2]; we give a brief overview here.

## 2.1 Language Overview

C* distinguishes between *scalar* and *parallel* data. Parallel data can be thought of as an array of data elements, distributed among the physical processors. A *shape* declaration describes the rank, layout and number of ***positions*** of parallel data, where each position contains one data element and is assigned one *virtual processor*. The programmer selects a *current shape* using the `with` statement. Primitive operations such as addition, when applied to parallel data, are executed in parallel for each position in the current shape. The following example declares two 100 × 100 matrices of doubles and performs elementwise addition.

```
shape [100][100]MatrixShape;
double:MatrixShape a, b;
with (MatrixShape)
    a = a + b;
```

Control flow is C* is sequential, i.e. from the programmer's point of view, conditional branches, procedure calls, etc. are followed by all processors. In fact, virtual processors behave as if they were executing code synchronously. Parallel operations can be *contextualized* inside a `where` statement by specifying a parallel boolean expression that determines which virtual processors are "active". Conceptually, a `where` statement first executes the `where` branch, then the `else` branch, with parallel operations restricted to the virtual processors on which the condition evaluates to `true` and `false` respectively.[2] The following example computes the square root of the absolute value of each element in the matrix `a`.

```
with (MatrixShape)
    where (a >= 0.0) b = sqrt(a);
        else b = sqrt(-a);
```

Communication in C* is performed by *send* or *get* operations, which are written as *left index* expressions, using a syntax reminiscent of array references. The code below transposes matrix $a$ by sending each element $a_{i,j}$ to position $a_{j,i}$. Matrix $b$ is transposed using an equivalent (but not necessarily equally efficient) *get* operation. The parallel expression `pcoord(i)` evaluates to a position's index along dimension $i$ of the current shape, i.e. for C*'s row-major layout, `pcoord(0)` yields an element's row number in the matrix and

[2] The semantics of C* specify that scalar code inside branches of a `where` statement is always executed, i.e. independent of the condition in the `where`. This is in keeping with C*'s global model of execution [2].

pcoord(1) yields the column number. Send and get operations are atomic in that they behave as though all data elements were sent simultaneously.

```
with (MatrixShape) {
    [pcoord(1)][pcoord(0)]a = a;
    b = [pcoord(1)][pcoord(0)]b;
}
```

In addition, C* provides powerful *reduction* operations by overloading the C language's "embedded assignment" operators. A simple example that computes the sum of all elements in matrix a is given below.[3]

```
double elementSum;
with (MatrixShape)
    elementSum = (+= a);
```

Functions in C* can take parallel arguments and return parallel results. For example, the parallel version of the sin function is declared as

```
double:current sin(double:current);
```

where current is a reserved word referring to the shape that is in effect at the time of the function call. The syntax for calling parallel functions is the same as that for scalar functions; e.g. to compute the sine of all non-negative elements in matrix a, one would write

```
with (MatrixShape)
    where (a >= 0.0) a = sin(a);
```

Given the synchronous model of execution, C* programs do not exhibit race conditions, and execution does not depend on nondeterministic events such as message arrival orders. The simple programming model of sequential control flow coupled with deterministic execution makes programming and debugging of C* programs almost as easy as for purely sequential languages. While requiring that all communication be explicit in the code does place additional burden on the programmer,[4] it provides the programmer with a clear performance model, exactly because any potentially expensive communication operations are clearly visible in the code. One drawback of C* (or similar languages) is that sometimes the synchronous semantics overly constrain a solution, as it is not possible to express arbitrary asynchronous operations. This may be exacerbated by naïve compilation, and later in this paper we will discuss compiler techniques that alleviate this problem.

[3] More complex reduction and scan operations are available using function call syntax, i.e. the language provides no special operators for them.

[4] This is in contrast to other programming models, such as dataflow or shared-memory, which try to hide details of data distribution and related communication from the programmer.

## 2.2 Code Generation Strategy

In this section, we outline how to compile C* for distributed-memory parallel machines; for a more detailed description, the reader is referred to [3]. Basic code generation for C* is fairly straightforward. The key insight is that virtual processors can execute independently of each other as long as they do not communicate. Only communication operations and assignments to scalar variables, which can be considered a form of communication, may require synchronization between the processors. The general approach to compilation then is to split up a procedure's code into *code blocks* wherever communication operations are encountered and insert a barrier synchronization between the blocks. Each of the code blocks is then surrounded by a virtual processor loop (*vp loop* for short) which executes the code block once for every virtual processor. Parallel variables are implemented as arrays (which contain one entry per virtual processor), distributed among the physical processors.

Communication operations that are nested inside where statements complicate this approach. C* requires that all virtual processors sending or receiving data participate in communication operations. This requires that these operations be "pulled out" to the top level of a procedure, so they are executed unconditionally. Figure 1 outlines the code generated for a conditional statement. Note that another parallel variable, temp has to be introduced to save the current context across the communication statement.

In a naïve approach, parallel C* functions are compiled as follows: the generated code for the function includes its own vp loops as required, and the context in effect at the call site is made available to the function either as a parameter or as a global variable. A call to such a function may of course not be put inside a vp loop. Hence, like communication operations, calls to parallel function must be hoisted out of vp loops, requiring the compiler to split code blocks. In addition, the context in effect at the call site must be stored in an array of boolean such that it can be passed to the parallel function. These transformations can introduce significant overhead, and we will discuss code generation for parallel functions in more detail in section 4.2.2.

Previous work by Hatcher and Quinn [3] has focused mainly on optimizations aimed at reducing the cost and frequency of synchronization and communication operations. While such optimizations are absolutely crucial, we found during our study that they are not always enough to achieve performance competitive with hand-coded message passing programs. In particular, we believe more work is required to re-

```
where (condition) {
    par_stmt_1;                     /* a parallel operation              */
    communication;                  /* some C* communication operation */
    par_stmt_2;                     /* a parallel operation              */
}
```

C* code with communication code nested inside conditional.

```
foreach vp {
    temp[vp] = condition[vp];       /* compute & save the condition    */
    if (temp[vp]) par_stmt_1[vp];   /* first parallel statement        */
}
communication;                      /* executed unconditionally        */
foreach vp {                        /* resume VP loop and where stmt   */
    if (temp[vp]) par_stmt_2[vp];   /* second parallel statement       */
}
```

Generated code has communication hoisted outside the conditional.

Figure 1: Hoisting communication operations.

duce the overheads due to vp emulation and other bookkeeping operations. High-level code transformations are needed to reduce unnecessary serialization and code block splitting.

## 3 Parallel Circuit Simulation

As a benchmark for our study we chose a distributed event-driven digital circuit simulation based on the Chandy-Misra algorithm. This applications poses several challenges to a compiler for a data-parallel language, as we will see throughout the rest of this paper.

The Chandy-Misra algorithm is most naturally specified as a set of asynchronous processes. Each process is assigned some subset of the gates in the circuit that is to be simulated; changing the partitioning affects absolute but not relative performance of the different implementations. Gates that are ready to be evaluated (for details, see [7]) are termed *active* gates. All processors continuously evaluate their active gates, i.e. some input events for an active gate are consumed and a new event may be produced, representing changes to the gate's output. Any new events are sent to the gate's *fanout* (the set of gates in the circuit to which the sending gate's output is connected) immediately as they are produced.

To decide whether a gate is active, the Chandy-Misra algorithm uses only information local to that gate. This local information is incomplete and overly conservative, and hence the simulation may occasionally get *deadlocked*, i.e. even though the simulation could make progress, no gates are considered active and therefore no new events can be computed according to the algorithm outlined above. At such points, the simulator needs to gather some global information in order to recover from the deadlock and continue the simulation. Our implementations use information about the minimum timestamp of all pending events in the system to recover from such deadlocks.

The message-passing version, taking advantage of asynchrony, is able to delay deadlock resolution until required, whereas the C* version performs resolution on every iteration, essentially implementing deadlock avoidance. For our benchmark circuit, the "lazy" method employed in the message-passing version performed best, though [11] indicates that the optimal choice of deadlock recovery strategy depends on the kind of circuit being simulated; details are beyond the scope of this paper.

The simulation algorithm usually proceeds independently per gate, sending only point-to-point messages between gates. When deadlock is detected, all processors synchronize in order to gather global information and resolve the deadlock, after which the processors run independently again.

Digital circuit simulation has several properties that make it difficult for a data-parallel implementation to achieve satisfactory performance.

- In a digital circuit, typically only very few gates

(on the order of 0.1% to 1%) are active at any given time [11]. This implies that in a C* implementation, most virtual processors are inactive, i.e. not doing any useful work. However, the vp loops still iterate over them and therefore for such *sparse contexts*, vp emulation can become a noticeable overhead.

- Our simulator performs little computation, but there are many nested conditional statements and function calls. These are difficult to compile efficiently, and indeed our benchmarks show that much time is lost due to operations maintaining the current context and due to overhead in parallel function calls.

- The communication pattern of the simulator is fairly irregular and cannot be easily implemented using the one-to-one communication primitives offered by C*. Our results demonstrate that more powerful communication operations are needed to achieve acceptable performance.

- Due to the low number of events generated per unit time, and the compact encoding of event messages, the simulator does not require high communication bandwidth. However, communication *latency* does have a significant impact on the execution time. It is therefore desirable to overlap computation and communication in order to hide these latencies. C* does not provide any primitives for asynchronous message passing, so overlap occurs only where inserted by the compiler.

We will examine these issues in more detail in section 4.2, after briefly describing the two competing implementations.

# 4 Performance Comparisons

We implemented the simulator in both data-parallel and message-passing style. We did not have access to Hatcher and Quinn's compiler on the CM-5 nor did the alpha-test version of Thinking Machines' CM-5 compiler lend itself to easy modification. Hence in order to evaluate the different optimizations and language features proposed in this paper, we hand-compiled the C* versions into C code, which was then was compiled using the same compiler and linked with the same communication libraries as used for the message-passing version. Note that the C* communication routines could be dramatically optimized by taking advantage of compile-time information, an option not directly available for the message-passing version. Future work will explore this avenue.

## 4.1 Message-Passing Implementation

The message-passing implementation is a fairly straightforward translation of the Chandy-Misra algorithm into C: the gates in the circuit are partitioned among the processors and each processor repeatedly iterates over its gates, evaluating those that are active and sending new events to other gates as soon as they are generated. For efficiency, we decided to use asynchronous rather than synchronous message passing routines, which allows us to overlap communication and computation. The drawback of this choice is that our code has to do extra work managing message buffers, and requires extra care to avoid race conditions and deadlock. To further improve performance, the list of active gates is updated incrementally.

## 4.2 Data-Parallel Implementations

### 4.2.1 First Data-Parallel Implementation

Perhaps surprisingly, the Chandy-Misra algorithm is fairly easy to formulate in C*, by dividing it up into three phases. In the first phase, all active gates are evaluated once, and new events are produced. Then, in a communication phase, all new events are sent to their destination processor. In the last phase, all received events are incorporated into the appropriate event queues.

We hand-generated C code following the compilation strategies outlined in section 2.2. Figure 4 shows that the performance of this first implementation (top solid line) is unacceptable when compared to the hand-coded message-passing version (bottom dotted line). On a single processor, this C* version is over 12 times as slow as the message-passing version. A significant fraction of the overhead is due to the interaction of vp emulation and the way calls to parallel procedures are handled; we attack this problem by introducing *elemental functions*.

### 4.2.2 Elemental Functions

In section 2.2, we mentioned that calls to parallel functions are moved to the outside of vp loops, requiring the compiler to split code blocks. While it may seem that the cost of splitting code blocks should be minimal — after all, it seems that only another vp loop needs to be introduced — we found that the performance penalty can actually be much higher than expected. Consider the C* code at the top of figure 2 which contains a call to the parallel `sin` function. The generated C code, shown at the bottom of the figure, shows that by moving the function call to the outside

```
double:S a, b, temp;
extern double:current sin(double:current);

where (a != 0) {
    temp = 1/a;
    temp = sin(temp);        /* parallel function call: split code block */
    b = 2*temp;
} /* assume temp is dead now */
```

```
double a[NUM_VPs], b[NUM_VPs], temp[NUM_VPs];
boolean _ctx[NUM_VPs];
extern void parallel_sin(boolean context[], double x[], double result[]);

for (vp=0; vp<NUM_VPs; ++vp) {
    _ctx[vp] = (a[vp] != 0);     /* compute context for function  */
    if (_ctx[vp])                /* start of "where" body         */
        temp[vp] = 1/a[vp];      /* temp must be stored in memory */
}
parallel_sin(_ctx, temp, temp); /* all VP's execute the call      */
for (vp=0; vp<NUM_VPs; ++vp) {
    if (_ctx[vp])                /* resume the "where" body       */
        b[vp] = 2*temp[vp];      /* use of temp                   */
}
```

Figure 2: Naive compilation of parallel function call.

of the vp loop, we have introduced significant overhead in several places:

- With two code blocks, we also need two vp loops, each of which has to check the current context. A third vp loop is executed inside the `parallel_sin` function, and it too will check the current context.
- Since the function call requires the current context to be passed along, it needs to be explicitly stored in an array of booleans, `_ctx[]` in figure 2.
- The variable `temp` (assumed dead at the end of the C* code fragment) now needs to be stored in memory, for use in the function call and the second vp loop.
- Similarly, opportunities for register allocation have been reduced, resulting in higher memory traffic and worse caching behavior.
- Reduced basic block sizes eliminate opportunities for basic-block optimizations.
- The overhead due to vp loops is proportional to the number of virtual processors, whereas the amount of "useful" work is proportional to the number of *active* virtual processors. Hence, in programs with sparse contexts, the vp loop overhead can overwhelm the time spent doing actual work. This is the case in our circuit simulator.

As a (partial) remedy to these inefficiencies, we introduce the notion of an *elemental function*. Elemental functions are simple scalar functions that, when applied to parallel data, are executed once for each position in the current context.[5] To preserve data-parallel semantics, communication is disallowed inside elemental functions, and scalar assignment may be performed only on local variables.[6] In other words, the effects of an elemental function must not be visible across virtual processor boundaries. Figure 3 shows the code that is generated for the previous example when an elemental instead of a parallel version of the `sin` function is used. There is now only one vp loop instead of the three that were needed before (two in the user's code, one in the parallel `sin` function), the context

[5] This is similar to the way functions were handled in the old version of C* [9].

[6] Assignment to global variables effectively constitutes a communication operation, and thus is not allowed. Local variables can be "promoted" to parallel variables and hence may be assigned to elementally.

```
for (vp=0; vp<NUM_VPs; ++vp) {
    if (a[vp] != 0) {
        double temp = 1/a[vp];          /* temp has been demoted to scalar */
        temp = elemental_sin(temp);     /* call elemental function */
        b[vp] = 2*temp;
    }
}
```

Figure 3: Better compilation using elemental functions.

does not need to be stored in memory, and the parallel variable `temp` has been *demoted* to a scalar and as such can be easily assigned to a register. The code does execute one function call for every active virtual processor (as opposed to one single call to the parallel function), but this cost is small compared to the savings.

The circuit simulator contains many calls to parallel functions, most of which could be executed elementally. Most of these calls occur nested inside several **where** statements, which introduces all the overheads of computing and saving the current context as outlined above. Further, recall that the overhead due to vp loops is proportional to the total number of virtual processors, whereas the amount of "useful" work is proportional to the number of *active* virtual processors. In our benchmark, the percentage of active processors is extremely low, on the order of 1%, hence the performance advantage of elemental functions is tremendous, yielding a speedup over the first C* version of 2× for one processor, less for more processors (see figure 4). While similarly large improvements will not be seen for programs with a higher percentage of active virtual processors, we find this optimization to be essential for good C* performance.

Note that it is very easy for a compiler to detect whether a given parallel function can be converted to an elemental one — it only needs to check for the presence of communication operations, assignments to scalar global variables or calls to non-elemental functions.

#### 4.2.3 More Powerful Communication

C* offers powerful point-to-point communication primitives and several predefined reduction operators. However, neither of these fit the communication pattern found in the circuit simulator. At the end of the evaluation phase, several gates will have computed new events. Each of these gates has to send the new event to some number of other gates, the *fanout*, as defined by the connectivity of the simulated circuit. Similarly, most gates have more than a single input, so a gate may receive more than one event during the communication phase (one per input). Implementing this many-to-many communication pattern with C*'s point-to-point *get* and *send* operations is awkward.

Let $d$ be the maximum fanout of any gate with a new event. Since C* can send to only one destination at a time, $d$ separate communication steps will be needed, one per fanout link. However, since gates can have more than one input, there can be more than one event destined for the same gate during a given communication step. In C*, each virtual processor can receive only one item at a time (collisions among data destined for the same virtual processor are resolved arbitrarily), so we need a tournament algorithm to resolve these collisions within each of the $d$ communication steps described above.

In the worst case, this algorithm requires $\max\{fanout\} \times \max\{fanin\}$ separate communication operations. In practice, this number is much smaller (a small constant times $\max\{fanout\}$) due in part to the very low activity levels in the simulation, but the performance is still unacceptable. This situation is particularly dissatisfying since *conceptually*, the simulation algorithm just requires one single many-to-many communication operation.

Clearly, primitives are needed that allow the programmer to express the communication operation without introducing unneeded orderings (and hence serialization of the compiled code). A first step is to provide a `sendToQueue` operation which functions very much like a point-to-point send, except that collisions due to multiple messages being sent to the same position are resolved by saving all of them in a queue at the receiver. C* already provides a similar mechanism, where simple operations (such as addition) may be performed on colliding messages. `sendToQueue` may be considered an extension of this mechanism, with concatenation as the combining operation. Using `sendToQueue`, the communication phase of the

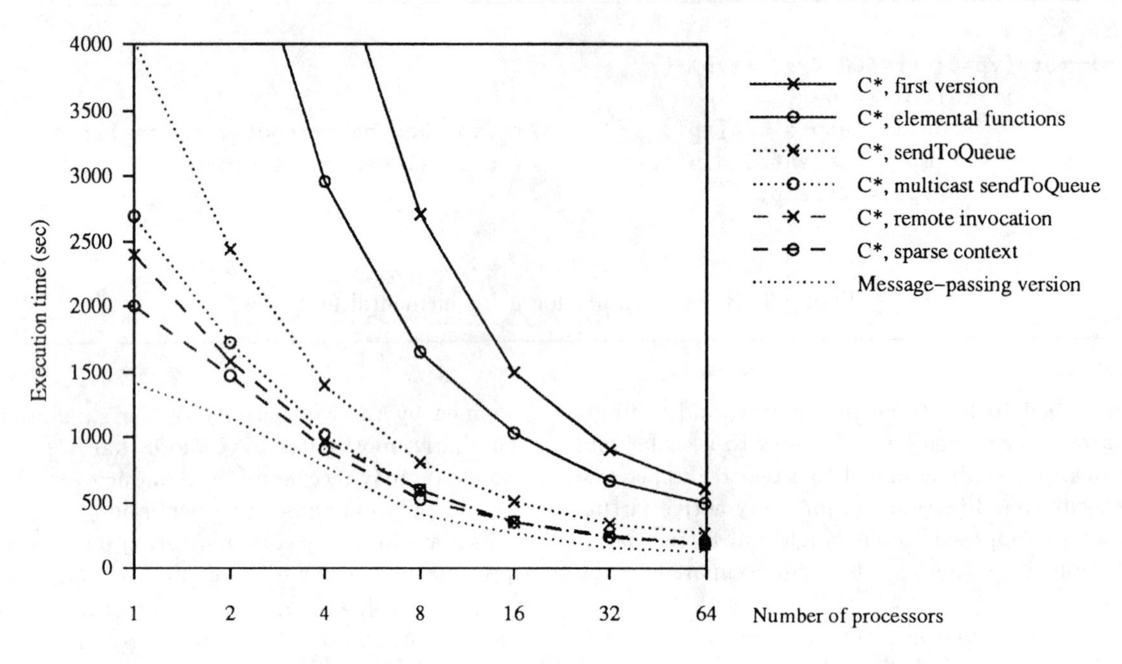


Figure 4: Performance of message-passing and various C* versions.

simulation becomes much more efficient: active gates can simply send their new event to the desired fanout gate. After this step is repeated *d* times (once for each fanout link), all gates can examine their queues and process any received events. Note that `sendToQueue` still appears to the programmer as a form of a data-parallel assignment operation; it is *not* an escape into the message-passing paradigm. Figure 4 shows that a 2× speedup is achieved by using `sendToQueue`.

While this is a much improved solution, performance is still not optimal: *d* separate send operations are performed in a loop (one for each fanout link), which according to the semantics of C* implies that they should be serialized — hence *d* synchronization operations would still be needed in the simulator's communication phase. In the current C* implementation on the CM-5, for each of the *d* send operations all processors have to wait until the network has been drained. This is necessary to ensure that the communication step has been completed. The disadvantage of this approach is that the pipelining capabilities of the network are not utilized across successive sends. This inefficiency can be eliminated by providing yet another communication primitive, a *multicast* version of `sendToQueue`: the sender specifies a data item to be sent and a *list* of destinations. Using this primitive, the *d* sends can be reduced to a single operation which sends messages at the maximum rate allowed by the network. Figure 4 shows a 1.4× speedup when the *multicast* version of `sendToQueue` is used.

One final inefficiency is due to the fact that all received messages are first saved in a queue and processed only after the communication step is complete. When the processing can be formulated as an elemental operation, the buffering step can be eliminated and messages processed as they arrive. We call this approach a *remote invocation*, in the spirit of RPC or active messages. Besides eliminating copying at the receiver, this also allows for some overlap of computation and communication, namely the sending of events and the processing of received events. Figure 4 shows that this results in another improvement. Again, note that remote invocation still appears to the programmer as a data-parallel operation, much like a "+=" reduction, except that a user-specified function is invoked in place of addition. It is not necessary to escape into the message-passing paradigm to provide this functionality.

One obvious question is exactly what communication primitives should be made available to the user. Since in general, one cannot foresee all users' possible needs in advance, there ought to be a mechanism that allows users to define their own communication primitives in a "safe" and structured way (safe meaning they should not violate the data-parallel model). How such a goal could be accomplished is an open question in language design.

### 4.2.4 Sparse Contexts

After applying all the optimizations described in the previous section, the generated code for the main loop of the simulator contains two barrier synchronizations and three vp loops that iterate over the active gates. As we have already pointed out, the number of active gates in the simulation at any given time is extremely low, on the order of 1% of the total number of gates. Hence, the cost of iterating over all gates becomes noticeable compared to the work done on the few active gates inside the vp loops, i.e. a great deal of time is wasted iterating over gates that are not active.

When users code their own equivalent of vp loops (as is the case when writing in a simple message-passing model), they can apply various techniques to reduce this overhead, such as incrementally updating a linked list of all active gates. In C*, however, the mechanisms implementing contexts and vp loops are fixed by the compiler and cannot be modified by the programmer. A possible remedy for this problem would be to allow the user to indicate that a given context is likely to be sparse. In that case, the compiler can choose alternative implementations for the vp loops and context checks.

As an experiment, we applied a simple optimization that only works when the same context is used in multiple vp loops. Note that in C*, this case is fairly common as code blocks get split by communication operations and the split blocks all use the same context. For example, the three vp loops in the core of the simulator all use the same context. We modified the code such that the first vp loop creates a linked list of all active gates. The next two vp loops then use this linked list to iterate over the active gates, rather than iterating over all gates and testing a boolean expression. As figure 4 shows, this last optimization produces another noticeable improvement.

## 5 Experience with the Models

In this section, we describe our experience designing and implementing the two different versions of the simulator. Overall, the C* version was developed in about half the time spent on the message-passing version, and it is shorter (2000 lines versus 3000). Most of the difference in development time is due to the large amount of time spent on debugging and tracking down race conditions in the message-passing version. The asynchronous communication routines require the programmer to explicitly manage message buffers, which further complicates the code and is very likely to cause race conditions or even deadlock. In fact, though the program has been been running successfully on different benchmarks, we still do not have high confidence that the last race condition has been eliminated.

The C* version was not plagued by any of these problems and debugging was very straight-forward, owing mainly to the synchronous model of execution. Several points are worth noting here. The C* version was written first, and much of its design was reused in the message-passing version. Also, while the author had previous experience with message-passing codes, it was his first non-trivial C* program. Had these conditions been reversed, surely the result would have been even more in favor of C*.

Thus, our experience clearly supports the claim that data-parallel languages are substantially easier to code and debug.

## 6 Conclusions

We have implemented a distributed event-driven simulator in both C* (a data-parallel language) and Standard C. For the latter, all communication operations are handcoded using a conventional message-passing library. This simulator presents a difficult test case for C* for three reasons:

- At any given time, only a few virtual processors (roughly 1% in our benchmarks) are active. As a consequence, the significance of vp loop overhead (relative to useful work) increases.
- The communication pattern is very irregular, and requires many-to-many communication, facilities that are not provided in the current version of C*.
- Communication performance is limited by communication latency, not bandwidth. The message-passing version is written in a way that allows each gate to send new events as soon as they are created, which helps to hide the network latency by overlapping communication and computation. The data-parallel version, however, may generate and send at most one event per gate, and furthermore communication and computation cannot be overlapped without help from the compiler.

The C* version was much easier to code and debug than the low-level message-passing version, which was plagued by difficult-to-find race conditions. Debugging of the C* version was significantly helped by its deterministic execution model. Overall, development time spent on the C* version was about half of that spent on the message-passing version. This supports the claim that the data-parallel model is easier to program than the message-passing model.

Not surprisingly, the performance of the first implementation of the simulator in C* was significantly worse than that of the carefully hand-coded message-passing version. However, we have shown that with the addition of a few language features and compiler optimizations, namely elemental execution, more powerful communication primitives, and sparse contexts, the C* version can achieve over 70% of the performance of the message-passing version.

Furthermore, high-level languages provide the compiler with extra information that can easily be used to optimize the communication operations [1]. To give a trivial example, no message-passing handshake protocol is needed in C* to insure that buffer space for incoming messages is available at the receiver (since that can be guaranteed at compile-time), thus reducing the latency of `send` operations compared to general message-passing operations. While we have not yet explored this possibility, we believe that such optimizations can significantly improve performance. Similar optimizations are near impossible to perform automatically on message-passing codes.

We believe that our experiments have shown that data-parallel languages are a viable alternative for message-passing codes, even for applications like circuit simulation that do not fit the data-parallel model of computation very well.

## 7 Future Work

We are currently studying more general optimization techniques in a more formal framework and also hope to evaluate them on a wider variety of benchmarks. For example, we believe that powerful procedure-level and interprocedural optimizations are an essential prerequisite for obtaining good performance from data-parallel languages. More aggressive optimizations are needed to detect and remove unnecessary serialization constraints, introduced by scalar or communication operations. Optimization of communication operations should be parametrized by a model of the target machine's interconnect and message-passing characteristics. In the same vein, our experiment should be repeated for machines other than a CM-5. We also need to examine what kinds of optimizations are possible with high-level information about a program's communication patterns.

Finally, there are several language issues worth exploring, such as support for elemental functions, parallel control flow, remote (cross-virtual processor) function invocation, and frameworks for letting users specify their own communication primitives, without violating the properties of the data-parallel model.

## References


[1] Edward W. Felten. The Case for Application-Specific Communication Protocols. Technical Report 93-03-11, University of Washington, Seattle, WA 98195, 1992.

[2] J. Frankel. C* Language Reference Manual. Technical Report, Thinking Machines Corp., 245 First St., Cambridge MA 02142, 1991.

[3] Philip J. Hatcher and Michael J. Quinn. *Data-Parallel Programming on MIMD Computers.* MIT Press, 1991.

[4] Philip J. Hatcher, Michael J. Quinn, and B. K. Seevers. Implementing a Data-Parallel Language on a Tightly Coupled Multiprocessor. In *Proc. 3rd Workshop Programming Languages Compilers Parallel Computers*, 1991.

[5] Harry Jordan. The Force. Technical Report ECE 87-1-1, Dept. of Electrical and Computer Engineering University of Colorado, January 1987.

[6] Charles Koelbel and Piyush Mehrotra. Compiling Global Name-Space Parallel Loops for Distributed Execution. *IEEE Transactions on Parallel and Distributed Systems*, 2(4):440–451, October 1991.

[7] Jadayev Misra. Distributed Discrete Event Simulation. *Computing Surveys*, pages 40–65, March 1986.

[8] M. J. Quinn, P. J. Hatcher, and K. C. Jourdenais. Compiling C* Programs for a Hypercube Multicomputer. In *Proc ACM/SIGPLAN PPEALS*, pages 57–65, 1988.

[9] J. R. Rose and G. L. Steele Jr. C*: An Extended C Language for Data Parallel Programming. In *Proceedings of the Second International Conference on Supercomputing*, volume ii, pages 2–16, 1987.

[10] M. Rosing, R. Schnabel, and R. Weaver. The Dino Parallel Programming Language. Technical Report CU-CS-457-90, Dept. of Computer Science, University of Colorado, April 1990.

[11] Larry Soule and Anoop Gupta. Characterization of Parallelism and Deadlocks in Distributed Digital Logic Simulation. In *26th ACM/IEEE Design Automation Conference*, pages 81–86, 1989.

[12] Thinking Machines Corp., 245 First St., Cambridge MA 02142. *C* Programming Guide, Version 6.0*, November 1990.

# Function-Parallel Computation in a Data-Parallel Environment

Alex L. Cheung
Sandia National Laboratories
Livermore, CA 94551

Anthony P. Reeves
School of Electrical Engineering
Cornell University
Ithaca, NY 14853

**Abstract**

*Asynchronous problems are those which may be decomposed into a set of independent sub-tasks which are suitable for concurrent execution. Th function-parallelism of these problems cannot normally be directly expressed using the data-parallel programming model. In this paper, data distribution strategies have been explored which allow an asynchronous problem to be implemented with function-parallelism in a data-parallel environment. When a problem can be implemented using both function-parallelism and data-parallelism, there are tradeoffs in using either approach. We have investigated the optimal balance between function-parallelism and data-parallelism for an asynchronous problem.*

## 1 Introduction

In our previous work [7, 6], data distribution strategies have been used to optimize performance for both regular and irregular problems. Although, data-parallel programming has proven to be suitable for many scientific problems, efficiencies often decline when the degree of data-parallelism is high. Some problems that exhibit function-parallelism characteristics may not be suitable for data-parallelism. These problems, commonly referred as *Asynchronous Problems* [9], usually consist of functionally independent tasks that can be executed concurrently without much overhead. These problems are sometimes performed on heterogeneous systems which comprise of specialized processors responsible for different computing tasks.

In this paper, an automatic data-distribution strategy for solving asynchronous problems with function-parallelism in a primarily data-parallel environment is considered. The strategy is implemented in four steps:

1. Model the target multiprocessor system based on primitive parallel operations.
2. Determine the possibilities for function-parallel implementation in the application.
3. Predict the performance of the application for different data distributions and select the best distribution.
4. Distribute the data to the processors and execute the program.

The first two steps are data-independent and can be performed at compile-time, however the third step may be data-dependent in which case it must be performed at run-time. The strategy optimizes the utilization of parallel resources by exploiting both the function and data-parallel aspects of the specified problem. A performance prediction model is used to aid the selection of the best data-distribution strategy. The strategy has been tested with an image processing "Generalized Hough Transform" application.

This paper is organized as follows. In section two, a theoretical performance analysis for function-parallel and data-parallel programming by data-distribution is considered. In section three, a performance prediction model for characterizing the target machine and predicting the cost of an algorithm is presented. Finally, in section four, results of running an example application using both programming models are presented.

## 2 Performance Analysis

### 2.1 Data-Parallel Execution

The performance or efficiency of the data-parallel execution depends on the overhead cost incurred by the data-parallel operations. The overhead cost is mainly composed of the interprocessor communication required by the data distribution. A detailed derivation can be found in [8]. Let $\alpha$ be the fraction of the execution time that is spent on computation, i.e., $T_{\phi\|_{\text{computation}}} = \alpha T_{\phi\|}$ and the efficiency also becomes, $\varepsilon_{\text{DP}} = \alpha$. Therefore, data-parallel performance can be characterized as both algorithmic and architectural dependent.

### 2.2 Function-Parallel Execution

Since the parallelism of a function-parallel execution is obtained by distributing independent tasks among processors, the performance depends on the task decomposition and the task assignment to the processors. The efficiency of the execution is optimized when the computation load across the processors is balanced.

A detailed derivation was given in [8]. The speedup achieved by a function-parallel execution is:

$$Sp_{\psi} = \frac{n}{(\beta+1) + \sigma n + (n - \beta - 1)\rho} \quad (1)$$

where $\rho$ is the fraction of a task that must be executed serially, $\sigma$ is the overhead factor and $\beta$ is the load imbalance factor.

The function-parallel execution is comparatively more algorithmic dependent. Although the overall performance depends on the balance of the sub-tasks on the processors, the communication cost of a system has little effect on the performance of an execution.

## 3 Performance Prediction

In order for a system to be able to automatically select an optimal strategy or amount of resources for a parallel execution, a performance prediction model is needed so that alternative strategies can be evaluated before execution. The performance of a parallel algorithm on a given parallel architecture can be predicted based on the performances of the primitive operations used in the language. One of the approaches is to use a machine performance characterization called *computation equivalence* [5].

### 3.1 Computation Equivalence

For a parallel operation $\gamma$ on a data structure with $m$ elements, the *computation equivalence* $c_{\equiv\gamma}(m)$ is defined to be the cost in terms of arithmetic operations of executing $\gamma$, i.e.,

$$c_{\equiv\gamma}(m) = \frac{t_\gamma(m)}{t_a(m)} \qquad (2)$$

where $t_\gamma(m)$ is the time to execute the operation $\gamma$ and $t_a(m)$ is the average time for a fundamental arithmetic operation applied to a data structure of the same size. More detailed description of the performance characterization concept and procedure is given in [5].

Let an algorithm consist of a sequence of $l$max sequential stages where each stage consists of $q_l$ parallel operations on data structures of size $m_l$. If the $i$th operation at stage $l$ of an algorithm is $\gamma_{l,i}$ and the computation equivalence for an operation $\gamma$ for a vector of size $m$ is given by $c_{\equiv\gamma}(m)$ then the time for the execution of the algorithm on a given parallel architecture is given by

$$T = \sum_{l=1}^{l\max} \sum_{i=1}^{q_l} c_{\equiv\gamma_{l,i}}(m_l) t_a(m_l) \qquad (3)$$

Performance prediction can be accomplished by pre-computing the computation equivalences of the primitive operations for a given system and storing them in a reference table. Equation 3 shows that the total cost of an algorithm is simply the sum of the costs of the primitive operations. Using the performance prediction model described above, one can predict the performance of a particular algorithm on a particular system and be able to select the appropriate amount of resources to obtain the optimal performance.

### 3.2 Combining Function-Parallelism and Data-Parallelism

The function-parallel execution is constrained by the possible task decomposition of the algorithm and the data-parallel execution is constrained by the performance of the architecture. Therefore, using solely one of the approaches may not give the optimal performance.

If the speedup of a function-parallel execution for $n$ processors is $Sp_{\mathrm{FP}}(n)$ and the speedup of a data-parallel execution for $n$ processor is $Sp_{\mathrm{DP}}(n)$, then the combined speedup of using both function-parallelism and data-parallelism is

$$Sp_{\mathrm{FP+DP}}(n) = Sp_{\mathrm{FP}}(n_f) Sp_{\mathrm{DP}}(n_d) \qquad (4)$$

where $n = n_f \times n_d$. Many combinations of $n_f, n_d$ exist which result in different performances for the algorithm. However, the speedup performances of the function-parallel and data-parallel executions can both be predicted using the computation equivalence model and therefore the optimal combined performance can also be predicted.

## 4 An Example Application

In this section, an image processing algorithm is used as an example application to demonstrate the performance prediction and data distribution strategies for both function and data-parallelism. Experiments are performed using the Paragon programming language [4] on a variety of testbeds including networks of workstations and a hypercube system. Both data and function-parallel models can be used with the same source code by employing different runtime data-distribution strategies.

### 4.1 The Generalized Hough Transform

An image processing algorithm the *Generalized Hough Transform* is used for a benchmark experiment. The Hough transform is a technique used in image processing for geometric object recognition [1, 3]. An object to be recognized can be described by a set of features. The Hough Transform for an object is specified by a reference table of feature relations $(r, \theta)$ pairs precomputed from an object prototype. More details of the algorithm can be found in [2].

The generalized Hough transform implemented in the Paragon language and was shown in [8]. The total cost of the Generalized Hough Transform can be predicted based on the pre-computed costs of the primitive operations and the number of features in the object. The computation equivalences of the primitive operations on a $256 \times 256$ array for a Sun Sparc IPC, a HP9000/720 and one node of an Intel iPSC/860 hypercube were also given in [8].

### 4.2 Experimental Results on serial processors

For the benchmark experiment, the Generalized Hough Transform is used for a character recognition task. Eight characters are to be identified in a page of text shown in Figure 1. The eight character recognitions can be performed as eight asynchronous, independent tasks that allow the possibilities of both function-parallel and data-parallel implementations. The features of the characters to be recognized are first extracted and stored in reference tables. The size of the image used is $256 \times 256$ [Performance for the single processor iPSC/860 case was extrapolated from an image size of $128 \times 128$ since a full $256 \times 256$ image exceeded the available memory]. On a serial processor, the character recognitions are performed as a series of Generalized Hough Transforms, as shown in Figure 2. The execution time for each transform is proportional to the number of features in the object and is shown in Figure 3.

The prediction on the Sparc processor over-estimates the execution time because multiple operations in a statement may be more efficiently handled by a compiler. However, on the HP9000 and the iPSC/860, the prediction is closer to the actual result. Overall, the prediction provides a reasonable approximation to the actual performance and can be used further in predicting the performance of parallel executions.

```
T H E R E   I S
W H I C H   S T
I F   E V E R
D I S C O V E R
W H A T   T H E
I S   F O R   A
I T   I S   H E
W I L L   I N S
```

Figure 1: A page of text used by Hough transform for character recognition

```
main(){
    /* Data declaration */
    ...
    /* read image into magnitude and direction */
    /* read reference tables into theta and dist */
    /* count contains number of features in a table */
    /* Perform eight transforms in series */

    for (i=0; i<8; i++)
        hough(magnitude,direction,dist[i],theta[i],count[i]);}
```

Figure 2: Serial Generalized Hough Transform in Paragon

## 4.3 Prediction on multiprocessors

The program shown in Figure 2 can be directly executed in data-parallel without any modification when multiprocessors are used. Furthermore, the program can be executed in function-parallel by replicating the image in the data declaration section. Since there is no data-dependency between the series of Hough transforms, the Hough transform function calls are automatically executed in function-parallel when multiprocessors are used. In effect, a function-parallel execution is performed in a data-parallel programming environment through the use of a non-conventional runtime data-distribution strategy.

### 4.3.1 Function-Parallel Execution

The performance using a function-parallel execution can be predicted based on the serial execution by distributing the tasks on multiprocessors. The task distribution which results in the lowest maximum cost on all processors is used as the basis for the optimal performance prediction. By assuming no serial code and no overhead cost, the

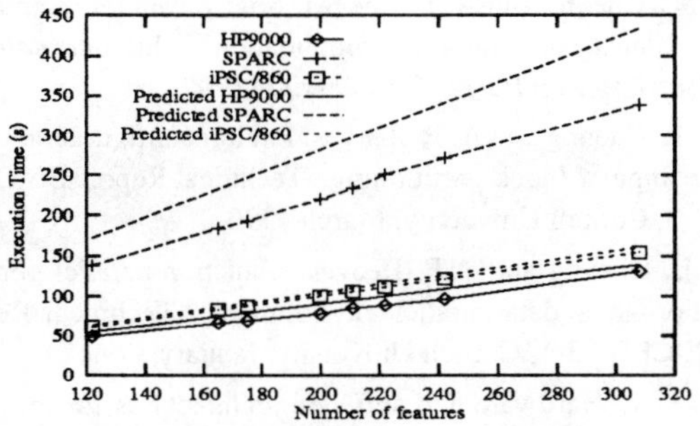


Figure 3: Execution Time of the Generalized Hough Transform on serial processors

optimal performance can be predicted solely based on the load-imbalance factor using Equation 1, with $\rho$ and $\sigma = 0$.

The load-imbalance factor, $\beta$, can be calculated for distributing tasks on the Sun, HP and iPSC processors using their respective serial costs. $\beta$ is simply defined as the relative deviation of the maximum cost on a processor to the average cost. The predicted optimal speedup based on Equation 1 and the experimental speedup are shown in Figure 4.

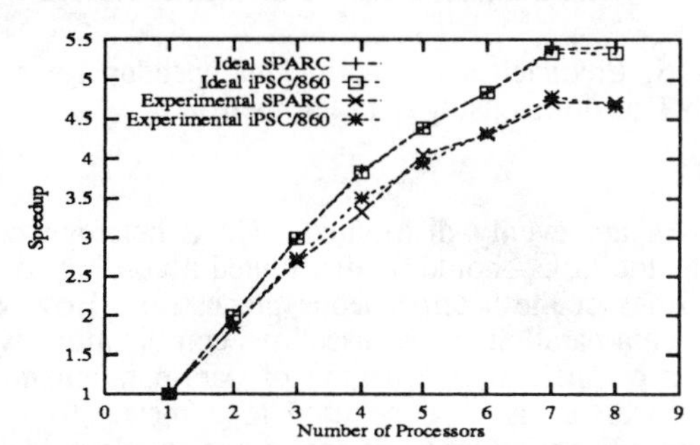


Figure 4: Predicted ideal and experimental speedup for eight Hough Transforms using Function-Parallelism

The predicted ideal speedups are only based on the difference in task sizes with the overhead ignored. As a result, the predicted speedup consistently over-estimates the performance. An almost linear speedup is predicted for less than 4 processors because processor load can be fairly easily balanced. The best speedup occurs at 7 processors rather than 8 because the largest task size exceeds the sum of the two smallest sizes. The experiment confirms that a higher degree of function-parallelism increases the load imbalance of a system and hence reducing the speedup improvement.

### 4.3.2 Data-Parallel Execution

The performance of the data-parallel execution can be predicted from the costs of the primitive operations on the multiprocessors again using the computation equivalence prediction model. The cost of an algorithm can be predicted from the sum of the costs of the operations as given in Equation 3.

The predicted and the experimental speedup for the eight character recognition tasks are shown in Figure 5. The predicted cost under-estimates the real performance mainly because the generic mapping function used by the prediction model is significantly less costly than the more complicated mapping function used by the Hough Transform. However, the cost of a mapping function is data-dependent and is difficult to estimate accurately. The experimental result also confirms that a high degree of data-parallelism is not always desirable because of a higher communication overhead.

### 4.3.3 Mixing Function-Parallel and Data-Parallel Executions

The performance of a function-parallel execution is mostly determined by the load balance in the system. On a homogeneous system, the load is balanced when

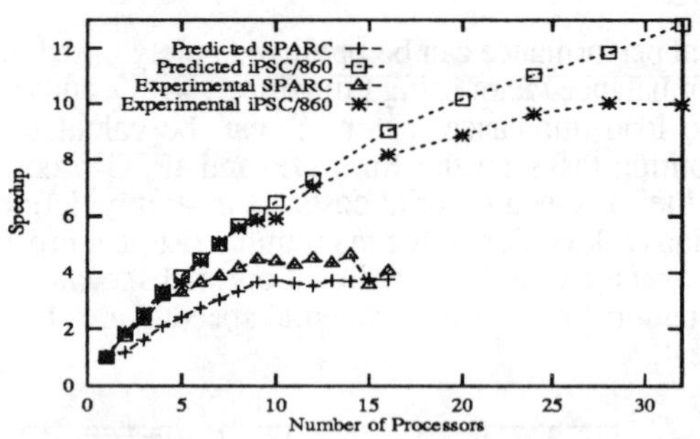


Figure 5: Predicted and experimental speedup for eight Hough Transforms using Data-Parallelism

the tasks are evenly distributed. On a heterogeneous system, the tasks should be distributed according to the capabilities of the heterogeneous processors. However, when data-parallelism is used in combination with function-parallelism, the degree of data-parallelism on each processor can also be used as a means for load balancing. For the purpose of analysis, a simple model of using the same degree of data-parallelism on all processors in a homogeneous system is used. By assuming the speedup for data-parallelism is consistent throughout the stages in an execution, the speedup for a system consisting of both function-parallel and data-parallel executions can be expressed simply in terms of the product of the two speedups as shown in Equation 4.

The optimal performance prediction of the Generalized Hough Transform on an iPSC system for $n = 8, 16, 32$ and 64 processors is shown in Figure 6. One clear observation from the results is that executing this algorithm using 4 function-parallel always gives the best speedup with any number of processors. When full function-parallelism is used, the optimal speedup achieved is 5.3 using 8 processors. When full data-parallelism is used, for example on 32 processors, the optimal speedup is 12.83. The best configuration, however, is 4 function-parallel and 8 data-parallel, which results in an optimal speedup of 21.75, a 70% improvement over full data-parallelism.

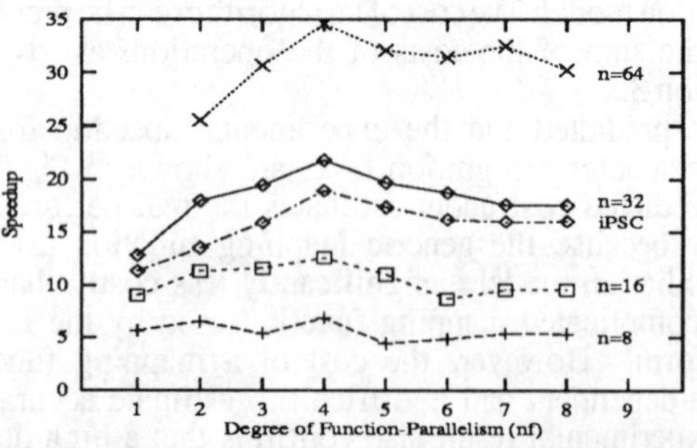


Figure 6: Predicted optimal and experimental speedup for eight Hough Transforms using both Function-Parallelism and Data-Parallelism on an iPSC/860

The experimental result for a 32-node iPSC/860 is also shown in Figure 6, indicated by plot labeled iPSC. A 65% improvement for a 4 function-parallel over full data-parallel is still obtained. The 4 function-parallel execution only benefits from a slight advantage (18%) over the 8 function-parallel execution because the maximum load imbalance for this particular example is only 50%, which is moderate for function-parallelism. An application with a highly diverse set of task sizes would result in a less efficient function-parallel execution and a better improvement of mixing both function-parallelism and data-parallelism.

## 5 Conclusion

In this paper asynchronous scientific problems were examined. These problems exhibit function-parallelism and generally cannot be solved efficiently using the data-parallel programming model. A novel data distribution scheme for implementing asynchronous problems with function-parallelism in a primarily data-parallel environment was developed. The technique involved distributing arrays to reside exclusively on a subset of processors to exploit function-parallelism. A performance prediction model was used to estimate the speedup performance achievable for a problem using either data-parallelism or function-parallelism. The data-parallelism, provides the most balanced load but with more costly communication overhead, while the function-parallelism incurs little communication overhead but at the cost of most unbalanced load. Runtime data distribution techniques and a performance prediction model for combining function-parallelism and data-parallelism in execution have also been developed. Experimental results indicate that for the given Hough transform application, the best performance is achieved with a mixture of both forms of parallelism.

## References


[1] D.H. Ballard. Generalizing the hough transform to detect arbitrary shapes. *Pattern Recognition*, 13(2):111--122, 1981.

[2] D.H. Ballard and C.M. Brown. *Computer Vision.* Prentice Hall, 1982.

[3] D. Casasent and R. Krishnapuram. Curved object location by hough transformations and inversions. *Pattern Recognition*, 20(2):181--188, 1987.

[4] A. L. Cheung and A. P. Reeves. The Paragon multicomputer environment: A first implementation. Technical Report EE-CEG-89-9, Cornell University, July 1989.

[5] A. L. Cheung and A. P. Reeves. High performance computing on a cluster of workstations. In *First International Symposium on High Performance Distributed Computing*, 1992.

[6] A. L. Cheung and A. P. Reeves. Sparse data representation for dense data-parallel computation. In *International Conference on Parallel Processing*, 1992.

[7] A. L. Cheung and A. P. Reeves. Fault reconfiguration using rectangular block partitioning. Technical Report EE-CEG-90-3, Cornell University, March 1990.

[8] A.L. Cheung and A.P. Reeves. Function-parallel computation in a data-parallel environment. Technical Report EE-CEG-93-1, Cornell University, January 1993.

[9] G. Fox. Hardware and software architectures for irregular problem architectures. *Unstructured scientific computation on scalable multiprocessors*, pages 125--160, 1992.

# Automatic Parallelization Techniques for the EM-4

Lubomir Bic and Mayez Al-Mouhamed
Department of Information and Computer Science
University of California, Irvine, CA 92717

## Abstract

This paper presents a *Data-Distributed Execution* (DDE) approach that exploits iteration-level parallelism in loops operating over arrays. It performs data-dependency analysis, based on which arrays are distributed over the different local memories. The code is then transformed to "follow" the data distribution by spawning each loop on all PEs concurrently but modifying its boundary conditions so that each operates mostly on the local subranges of the data, thus reducing remote accesses to a minimum. The approach has been tested on the EM-4 supercomputer by implementing several benchmark programs. The experiments show that high speedup is achievable by automatic parallelization of conventional Fortran-like programs.

## 1 Introduction

Distributed memory MIMD computers are among the most difficult to program, since independent processes or threads operating on their own memories and communication with other processes through message or remote memory access must be efficiently managed.

The objective of this paper is to demonstrate that the EM-4 multiprocessor [1], together with an automatic parallelization technique referred to as DDE (Data-Distributed Execution), which has originally been developed in the context of coarse-grain dataflow [2], offer an efficient computing environment in which large portions of scientific code can be parallelized using implicit parallelism.

This paper is organized as follows. The EM-4 architecture and its most important characteristics are described in Section 2. Section 3 presents the principle of DDE and the actual transformations applied to programs to extract parallelism. Section 4 presents the results of the benchmarks executed on the EM-4 and Section 5 concludes about this work.

## 2 The EM-4

The EM-4 distributed memory MIMD supercomputer [1, 3] has 80 PEs that are interconnected using a *direct connect topology* over an Omega network. Its impostant features are the *fast inter-processor communication* is and the support for *multithreading*.

To allow efficient multithreading, it is necessary to create threads and quickly switch among them by using the matching of operands required by dataflow [4]. For this, a 4-stage nested pipeline is used so that the outer 4-stage is is used for the dataflow mode and the inner 2-stage is used in sequential mode.

The first two stages perform the *direct matching* [3]. Each operand segment has an entry pointing to a dyadic instruction in the code segment. A pointer from the operand to the code segment is also created because distinct operand segments could be simultaneously pointing at the same code segment.

Stage 1 fetches the code segment pointed by each new packet. In stage 2, if the location addressed by the packet is empty, the packet is stored in that location and no further action is taken. If that location already contains an operand, it is marked empty and both operands are passed to the third stage, that performs the fetch and decoding of the instruction. Finally, the execution is performed by the fourth stage. The above cycle is repeated until an instruction indicates that sequential execution is to commence. At that time, no new packets are accepted by stage 1. Instead, stage 3 continues fetching subsequent instructions and passing them to stage 4 for execution in a normal von Neumann style. This mode continues until it is explicitly terminated by an instruction. Hence the EM-4 is capable of switching between data-driven and control-driven execution very efficiently.

The direct matching may be viewed as a mechanism for thread management [6] because it provides efficient means to: 1) execute sequences of control-driven instructions (threads) until termination or remote memory request, and 2) quickly switch to a new thread by using direct matching. Suspended threads are then resumed when the remote data becomes available. This approach is very useful to hide memory latency [5].

The EM-4 can be programmed using three distinct approaches: 1) a functional program is compiled into a dataflow graph of *strongly connected blocks*, 2) Using a library of threads, the user specifies their mapping and the distribution of data structures, and 3) using the proposed implicit parallelism with conventional languages. This last approach will be presented in the next section.

# 3 Data-Distributed Execution

The basic philosophy of DDE is to distribute the arrays over the PEs to minimize the amount of remote data transfer required during the execution of concurrent threads. We consider programs written in a conventional language, such as Fortran or Fortran-like c constructs.

At run time, each parallel loop is associated with two families of threads: 1) global threads (GT) are created by sub-dividing the range of the parallel iterator, and 2) partitioning each GT into local threads (LT). The GTs promote inherent parallelism and the LTs provide the PEs the opportunity to hide remote memory access (RMA) by performing context switching to ready LTs.

## 3.1 Analysis and Restructuring

Dependence analysis [9] is used to identify *loop-carried-dependencies* (LCD) that inhibit parallelization of the loop and lead to generation of a single scalar thread. Global threads will be created for loops having only *loop-independent-dependencies* (LID).

Reduction of the granule size of LCD loops is done by removing parallelizable code fragments using known techniques such that *loop distribution* and *partial parallelization*. These fragments are inserted closer to their data producer or consumer that belong to LID loops with the same loop headers. This causes immediate references to become subject to identical loop constraints. Next, the domain of each array that is indexed by the parallel iterator index is implicitly distributed across the PEs to yield the least number of remote memory accesses. Finally, a voting technique allows finding the most frequently used array distribution that becomes the global distribution.

Renaming [7] multiple write to the same variable is performed in order to make the code obey the *single assignment* principle, i.e. allows a value to be written only once. This eliminates possible race conditions and produces the correct result regardless of loop scheduling.

## 3.2 Transformations for Parallelism

We will use the generic program example in Figure 1 to illustrate the creation of global and local threads. The first step is to replace all array definitions (line 1) by a call to an *allocate() function*, which, at run time, performs a distributed allocation of the array

```
Sequential Code:

1    int A[][], B[][];
2    for (i = 0; i < n1; i++)
3      for (j = 0; j < n2; j++)
4        a[i][j]=some_comp(B[i][j],..);

Transformed Code:

5    A = allocate(ROW,...);
6    B = allocate(ROW,...);
7    for (p = 0; p < NO_PEs; p++)
8      fork(pe[p], i_loop, ...);
9      void i_loop(...)  {
10     lb = max(0, get_my_start_i(A));
11     ub = min(n1, get_my_end_i(A));
12     for (i = lb; i < ub; i++) {
13       for (j = 0; j < n2; j++) {
14         fork(self_pe, j_loop,...);
15   }
16   void j_loop(...)  {
17     value=some_comp(read_array(B,i,j),..);
18     write_array(A,i,j,value);
19   }
```

Figure 1: Program transformation

by sending requests to all PEs to allocate their own local subranges (lines 5-6). The type of distribution is determined, for each array, based on the preceding program analysis. Given an array, $A$, consider each access $A[i,j]$ within a loop. If this is a singly nested $i$ loop, or a nested loop with $i$ as inner index, then mark the access as a column access. Next, it counts the number of loops with row versus column accesses and choose the distribution having the highest frequency.

In Figure 1, the parameter *ROW* indicates that the arrays are to be distributed row-major, that is, each PE will be responsible for a certain subrange of the index $i$.

To implement DDE, each loop is started on all PEs concurrently so that each PE operates on a different subrange of the original loop. For nested loops it is first necessary to determine the loop nest that controls the array distribution. In most cases, all arrays accessed within a given loop will have been distributed along the same dimension. In this case, the index along which the arrays were distributed determines the loop level to be distributed. In the rare cases where not all arrays accessed within a given loop have been distributed along the same dimension, we count the number of array accesses along each dimension and select the most frequently used one to determine the loop level to be distributed.

In Figure 1, both arrays were distributed row-major and hence the *i_loop* (line 2) was chosen for distribution. It has been transformed into the function called *i_loop*() (line 9) and is spawned on all PEs using the loop shown on lines 7–8. This loop executes on $PE_0$ that is the master PE.

To make each PE operate on a different subrange, the code to compute the local lower and upper bounds (*lb*, *ub*) for the distributed loop is inserted (lines 10-11). This code, referred to as the *Range Filter*, accesses the header of the array the loop operates on and, from the recorded distribution information, computes the local subrange. The functions *get_my_start_i*() and *get_my_end_i*() represent the retrieval of the starting and ending *i_indices*, which are different for each PE. These are then combined using the *max* and *min* functions with the boundaries of the original loop, in this case, the values 0 and *n*1, respectively.

It is necessary to increase the level of parallelism within each PE by locally spawning the iterations of the next outer nest as separate LTs. This is analogous to the previous transformation except that the PE specified by the *fork*() primitive is the local PE. The *j_loop* becomes a separate function (lines 16-19) and is spawned for each *j* on the PE as a local thread (line 14).

The final transformation is to replace each reference to an array element by a call to the *read_array*() or *write_array*() function, which determines the location of the given element (local or remote) and performs the access. Given an array $A[n_1, n_2, \ldots]$, assume that the array is to be distributed along a given dimension $n_d$. We interpret $n_d$ as a binary number and use the leading $k$ bits as the PE number and the remaining bits as the local index. The number $k$ is determined by right-shifting $n_d$ until the result is smaller than the total number of PEs. The number $k$ is then stored in the array header and used by the access functions.

Programs are optimized by inlining the inserted functions, notably the *read_array* and *write_array*, and moving of invariant code outside of the loops. The schedule, resulting from the insertion of the various fork and barrier primitives, is also improved by moving loops that do not need to wait for a particular barrier in front of that barrier. Hence a form of a greedy schedule is implemented.

# 4 Results

This section presents the results of applying the proposed DDE approach to: 1) the conduction loop of the SIMPLE benchmark, and 2) the matrix multiply, and running the code on the EM-4.

## 4.1 SIMPLE

SIMPLE is a well-known benchmark program [8] that simulates the behavior of a fluid in a sphere, using the Lagrangian Formulation.

In this experiment, we have considered the *conduction function* which is the main and most difficult portion to parallelize. The code consists of a number of singly and multiply nested loops iterating over several 2-D arrays.

The resulting parallelism profile is shown in Figure 2. The measured speedup was 65 and the average idle time was 9.09%. The extracted parallelism was nearly 77 during most of the computation time. This parallelism profile is excellent because $PE_0$ and two other PEs – those holding the boundary rows – were idle during most of the computation time.

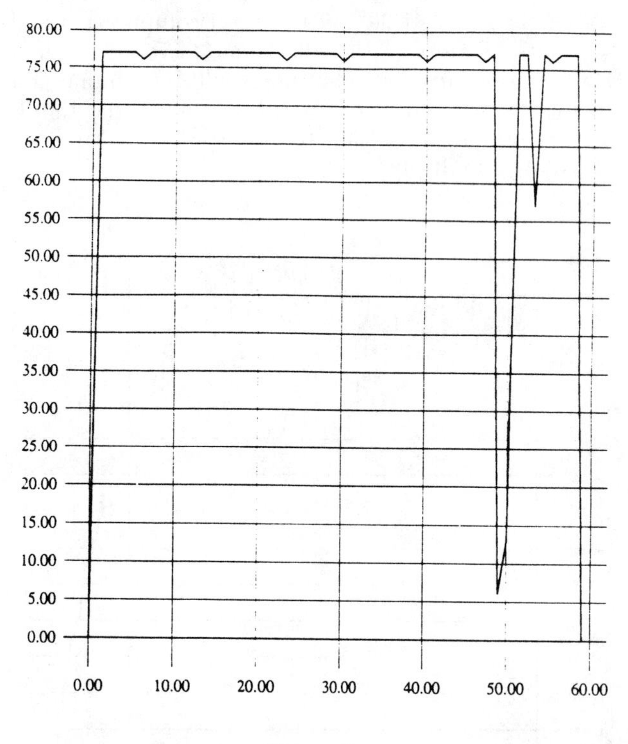


Figure 2

The drops in parallelism, resulting from barriers that could not be masked by other work, were steep, narrow, and few in number. The shape of the drop is a clear indication of the EM-4's superior communication network.

The small number of the drops and the fact that they do not extend all the way down to a single PE is an indication of the available parallelism in a typical scientific application. There were sufficient numbers of independent loops that could be run concurrently and thus mask the effect of much of the idle time resulting form barriers.

## 4.2 Matrix Multiply

Using the DDE apprach, we have parallelized the simple triply-nested loop of the matrix multiply. Three experiments were carried out for matrix sizes (multiples of 79) of $79^2$, $158^2$, and $316^2$, respectively. The parallelism profile for the second experiment (Figure 3) is representative of those of the other experiments. The obtained speedups were 7.5, 8.6, and 9.1 for each of the above matrix sizes, respectively.

The resulting speedup is quite modest, even when the matrix size is large. However, the speedup obtained by a manual parallelization was also low and hence the result indicates that the automatic parallelization approach performs very well.

The parallelism profile of the matrix multiply algorithm (Figure 3) shows a pronounced trailing edge that causes some load imbalance. This is surprising because of the regularity of the problem, the distribution, and the architecture. The trailing edge takes up on the order of 20% of the total computation and accounts of most of the idle time (11.7%) measured for this problem. This idle time could be eliminated if memory synchronization were available on the EM-4. Other computation could then partially overlap with the matrix multiply loop.

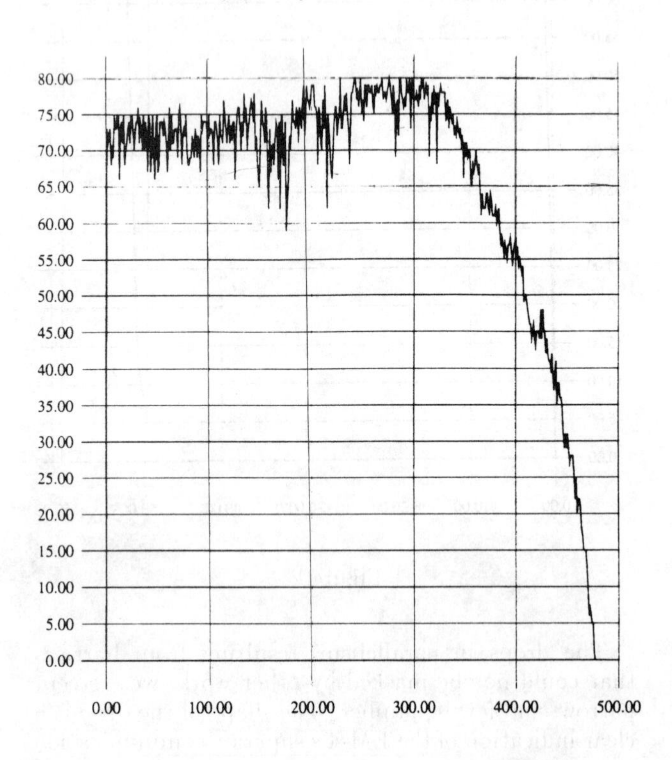


Figure 3

# 5 Conclusions

There are many real world applications that a hybrid machine like the EM-4 could exploit without requiring the labor-intensive and error-prone task of manual parallelization. Significant speedup can be achieved using Fortran-like programs that iterate over large data structures.

Automatic parallelization will, of course, not eliminate the need for the human involvement as was the case with the matrix multiply, where automatic parallelization of a given algorithm yielded only marginal speedup. Hence, the proposed approach is only one component of a parallel programming environment, which must take into consideration the user, the language, the compiler, the architecture, and the various development tools.

# References

[1] Sakai, S., Yamaguchi, Y., Hiraki, K., Kodama, Y., Yuba, T. *'An Architecture of a Dataflow Chip Processor', Proc. 16th Annual Int'l Symp. on Computer Arch., Jerusalem, Jun. 1989*

[2] Bic, L., Roy, J.M.A., Nagel, M. *'Exploiting Iteration-Level Parallelism in Dataflow Programs', 12th Int'l Conf. on Distributed Computing Systems, Yokohama, Japan, Jun. 1992*

[3] Sakai, S., Hiraki, K., Yamaguchi, Y., Kodama, Y., Yuba, T. *'Pipeline Optimization of a Dataflow Machine', Advanced Topics in Data-Flow Computing, Prentice-Hall, Ed. J-L. Gaudiot and L. Bic, 1991*

[4] Arvind, Bic, L., Ungere, T. *'Evolution of Data-Flow Computers', Advanced Topics in Data-Flow Computing, Prentice-Hall, Ed. J-L. Gaudiot and L. Bic, 1991*

[5] Arvind, Iannucci, R.A. *'Two Fundamental Issues in Multiprocessing', Proc. DFLVR Conf. on Parallel Processing in Science and Engineering, Bonn-Bad Godesberg, Germany, Jun. 1987*

[6] Sato, M., Kodama, Y., Sakai, S., Yamaguchi, Y., and Koumura, Y. *'Thread-Based Programming for the EM-4 Hybrid Dataflow Machine', Proc. 19th Annual Int'l Symp. on Computer Arch., Gold Coast, Australia, May 1992*

[7] Cytron, R., Ferrante, J. *'What's in a Name? -or- The Value of Renaming for Parallelism Detection and Storage Allocation', Proc. Int'l Conf. on Parallel Processing, Aug 1987, pp. 19-27*

[8] McMahon, F.H. *'The Livermore Fortran Kernels: A Computer Test of the Numerical Performance Range', UCRL-53745, Lawrence Livermore National Laboratory, Livermore, CA, Dec. 1986*

[9] Padua, Wolfe, M. *'Advanced Compiler Organization', Comm. ACM, Dec. 1989, pp. 1184-1201*

# SESSION 2B

## COMPILER (I)

# AUTOMATING PARALLELIZATION OF REGULAR COMPUTATIONS FOR DISTRIBUTED-MEMORY MULTICOMPUTERS IN THE PARADIGM COMPILER*

Ernesto Su, Daniel J. Palermo, and Prithviraj Banerjee
Center for Reliable and High-Performance Computing
University of Illinois at Urbana-Champaign
Urbana, IL 61801, U.S.A.
{ernesto, palermo, banerjee}@crhc.uiuc.edu

*Abstract – Distributed-memory multicomputers such as the Intel iPSC/860, the NCUBE/2, the Intel Paragon and the Connection Machine CM-5 offer significant advantages over shared-memory multiprocessors in terms of cost and scalability. Unfortunately, to extract all the computational power from these machines, users have to parallelize their existing serial programs, which can be an extremely laborious process. One major reason for this difficulty is the absence of a single global shared address space. As a result, the programmer has to distribute code and data on processors and manage communication among tasks explicitly. Clearly there is a need for efficient parallelizing compiler support on these machines. The PARADIGM project at the University of Illinois addresses these problems by developing a fully automated means to translate serial programs for efficient execution on distributed-memory multicomputers. In this paper we discuss parallelization of regular computations using symbolic sets as a representation for data accesses and communication.*

## 1. INTRODUCTION

One of the major advantages of distributed-memory multicomputers over shared-memory multiprocessors is that the former are much more easily scalable than the latter. Therefore, it is much easier to build massively parallel systems with distributed memory than with shared memory. However, lacking a global address space, distributed-memory machines are considerably more difficult to program than shared-memory machines. They rely on the programmer to perform important tasks whose quality directly affects the performance of the resulting program: to distribute data on different processors, and to implement interprocessor communication whenever a process needs data from another processor.

In recent years, significant research effort has been aimed at source-to-source parallel compilers for multicomputers that relieve the programmer from the task of communication generation, while the task of data partitioning remains a responsibility of the programmer [1, 2, 3, 4]. Based on the user-specified partitioning of data, these compilers can generate the corresponding parallel program for a multicomputer.

*This research was supported in part by the Office of Naval Research under Contract N00014-91J-1096, and in part by the National Aeronautics and Space Administration under Contract NASA NAG 1-613.

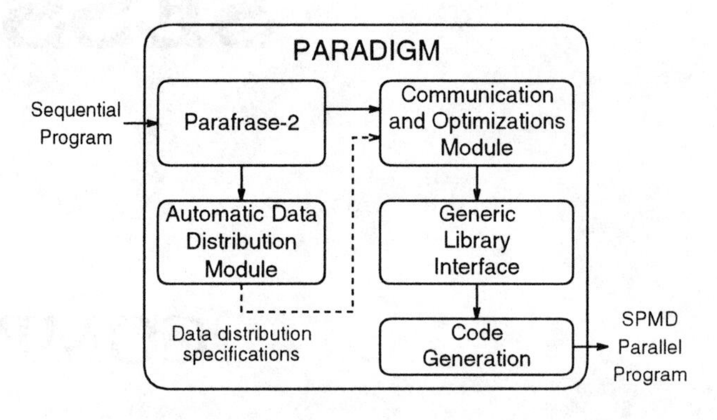


Figure 1: Compiler Overview

Our research aims at a compiler that can parallelize existing programs written in a common sequential language (currently FORTRAN 77) with little or no further input from the programmer. This means that the compiler must perform not only automatic communication generation, but also automatic data partitioning. Figure 1 shows a simplified view of the compiler, called PARADIGM (**PARA**llelizing compiler for **DI**stributed-memory **G**eneral-purpose **M**ulticomputers). Within the framework, *Parafrase-2* [5] is used as a preprocessing platform to parse the input program into an intermediate representation and to perform useful analysis and transformations. The Automatic Data Distribution Module determines how data is to be partitioned across processors and, based on this information, the Communication and Optimizations Module partitions the computations and generates communication accordingly. The compiler can also be easily retargeted to different machines through the use of a Generic Library Interface before code generation. The final output of PARADIGM is an SPMD (Single Program Multiple Data) parallel program with message passing.

Recently, the Automatic Data Distribution Module has been successfully developed and implemented by Gupta and Banerjee [6, 7]. It uses a constraint-based approach to choose data distributions that minimize both the computation and communication costs of a program for a particular target machine. For each target machine, there is a set of parameters describing relative computation and communication costs, so the decision algorithm is effectively isolated from the target architecture. This paper addresses the computation partitioning and communication generation aspects of PARADIGM.

Much research is being done to increase the functionality of PARADIGM. When fully implemented, it will be capable of performing all of the following tasks automatically:

- Generation of data partitioning specifications [6, 7].
- Partitioning of computations and generation of communication for data-parallel programs [8].
- Generation of high-level communication [9].
- Exploitation of functional parallelism [10, 11, 12].
- Support for multithreaded execution [13].

The rest of this paper is organized as follows. Section 2 describes the techniques used to analyze the access patterns of the input program. Section 3 outlines the approach taken in generation of communication using the information from the analysis phase. Section 4 presents a study of the performance of the compiler on FORTRAN 77 programs with regular computations to determine the applicability of our approach. Finally, conclusions are presented in Section 5.

# 2. ANALYSIS

Basically, our approach applies the *owner computes rule* which states that the processor owning a data item must perform all computations for that item. Any nonlocal data needed in the computation is obtained via interprocessor communication. A direct application of this rule without further optimizations leads to code with *runtime resolution* which explicitly computes the ownership and communication for each reference, resulting in extremely high computation and communication overheads. Using compile-time analysis and optimizations, it is possible to eliminate most of the overheads, generating much more efficient parallel programs [14].

## 2.1 Symbolic Sets

The computation partitioning, communication analysis, and optimization phases (all described later) require an efficient way of describing sets of iterations and regions of data. Our approach utilizes *symbolic sets*, which have many desirable characteristics.

The symbolic sets are represented as a lower and upper bound where each bound can be a single linear expression or a combination of two linear expressions (*min* or *max* operation, to represent masking operations on the sets). A multidimensional set has the following representation for each dimension:

$$\mathbf{set}_{dim} = [\lceil lower\ bound \rceil : \lfloor upper\ bound \rfloor\ [: step]\ [, cycle]]$$

$$\mathbf{bound} = \begin{cases} c_1 p + c_0 \\ min(c_1 p + c_0, d_1 p + d_0) \\ max(c_1 p + c_0, d_1 p + d_0) \end{cases}$$

where $c_i, d_i$, *step*, and *cycle* are all constants.

A single set representation is parameterized by the processor location ($p$) for each dimension of a virtual mesh, providing a uniform representation for every processor in the mesh, whether it is a boundary or an interior node. As a result, operations on the sets are very efficient because all processors in one or more dimensions can now be captured into a single symbolic set operation. In addition, symbolic sets and set operations are easily extended to an arbitrary number of dimensions, necessary for describing iteration sets in multiply nested loops as well as sections of multidimensional arrays.

Masks (*min*, *max*) in the set representation are used to express the upper and lower bounds of a given access. The optional *step* of a symbolic set corresponds to the addressing of elements within the range of the bounds. If not specified, then *step* is assumed to be one.[a] The *cycle* of a set is used to describe the replication of a block description through a cyclic distribution. Note that in global coordinates, the *cycle* can be viewed as a *mod* function on the iterations. In local coordinates, it can be viewed as an offset between replicated sub-block descriptors. In either case, the bound expressions for the transformations (to be defined later) are unaffected by the *cycle* parameter. For completeness, the *cycle* for a blocked distribution is the size of the entire space for the specific array dimension.

Both the data partitioning and the subscript reference affect the mapping of global iterations to the mesh of processors. Given a subscript reference $A(s(i))$ (with $s(i) = a_1 i + a_0$) as well as the block size $b$ of the (cyclic or blocked) distribution, a symbolic set representing the *global iterations* for which $A(s(i))$ is stored in a processor $p$ can be constructed as:

$$a_1 i + a_0 \in [bp + 1 : bp + b]$$

$$i \in [\lceil lbound(p) \rceil : \lceil ubound(p) \rceil]$$

$$\text{where} \quad lbound(p) = \left(\frac{b}{a_1}\right) p + \left(\frac{1 - a_0}{a_1}\right)$$

$$ubound(p) = \left(\frac{b}{a_1}\right) p + \left(\frac{b - a_0}{a_1}\right)$$

If $A(s(i))$ is a enclosed in the loop

```
DO i = l, u
    ··· A(s(i)) ···
END DO
```

then the symbolic set must also account for the bounds $l$ and $u$:

$$i \in [\lceil max(l, lbound(p)) \rceil : \lceil min(u, ubound(p)) \rceil]$$

Note that the loop bounds must be known at compile time (after constant propagation and induction variable substitution [5]). Figure 2a illustrates this set for the case of $l = 20$, $u = 480$, and an array $A(1 : 500)$ distributed by blocks (with $b = 63$) across eight processors. Figure 3a shows the same set for a blocked-cyclic distribution with a block size of $b = 10$.

### Symbolic Set Operations

To perform set operations on this representation, it is necessary to be able to perform comparisons between bound expressions. For linear expressions, comparisons can be made within a desired range to determine the relation between the two expressions. This linear comparison operation is extended to perform comparison of binary operations (*min*, *max*) of linear expressions. With symbolic

[a] Currently, only loop skips of 1 are allowed, but steps that are due to a non-unity coefficient of the loop index in the subscript expression are supported (see *Set Transformations*).

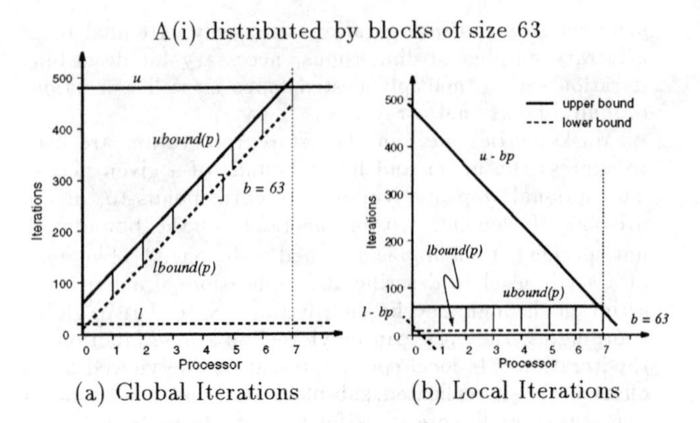


(a) Global Iterations (b) Local Iterations

Figure 2: Block Distribution

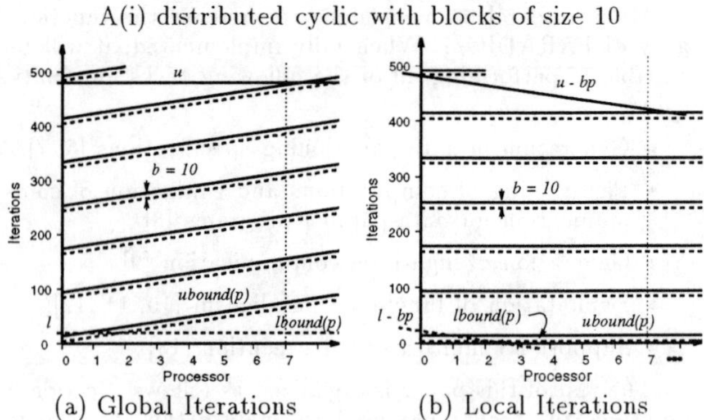


(a) Global Iterations (b) Local Iterations

Figure 3: Cyclic Distribution

comparisons defined between the available bound expressions, the following set operations are defined to operate on $n$-dimensional sets:

| | | |
|---|---|---|
| `is_subset` | true if $A$ is a subset of $B$ | $A \subseteq B$ |
| `are_disjoint` | true if $A$ and $B$ are disjoint | $A \cap B = \emptyset$ |
| `union` | $n$-D union of two sets | $A \cup B$ |
| `intersection` | $n$-D intersection of two sets | $A \cap B$ |
| `difference` | $n$-D difference of two sets | $A - B$ |
| `stretch` | *stretch* an $n$-D set $A$ over $B$ | |
| `location` | $n$-D *location* of $B$ wrt $A$ | |

Other than the basic set operations, two new ideas have also been introduced: *stretch* and *location*.

**Stretch** Stretching of a set is performed by combining the bounds of one set with another to generate a list of sets that surround the area common to both sets. This is used with $n$ 1-D difference operations to approximate an $n$-dimensional difference with the additional property of separating the resulting sets into the $3^n - 1$ neighbor regions that exist for an $n$ dimensional set (see Figure 4).

**Location** The *location* of a set is the $n$-tuple of $\{-, none, +\}$ direction vectors. Given two sets, this computes the relative relationship that exists between them.

The use of the location and stretch operations (as well as the application of the other operations) will become apparent during the actual analysis of reference and communication patterns.

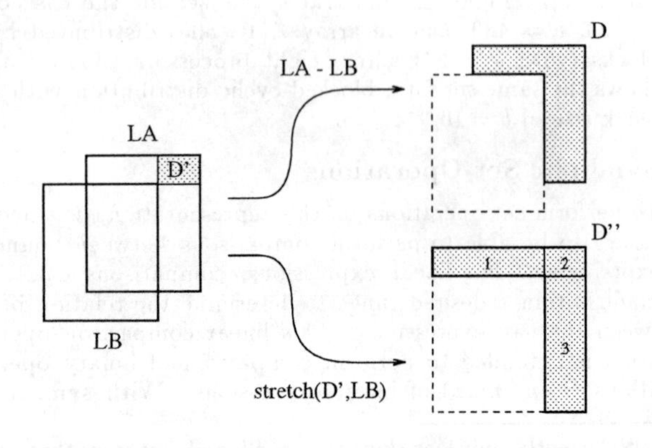


Figure 4: Comparison of Difference and Stretch

## Set Transformations

During analysis of the reference patterns in a program, symbolic sets are transformed among different domains. Initially, the array references are collected directly from the serial program in terms of the *global iteration space* (see Section 2.2). Altogether, four separate domains can be defined: global and local iterations, global and local indices. Transformations (and their inverses) between the different domains are defined as:

$$\begin{array}{lccc} & \text{Global} & & \text{Local} \\ & & \lambda_p(i) & \\ \text{Iterations} & \mathbf{GI} & \longrightarrow & \mathbf{LI} \\ & s(i)\downarrow & & \downarrow s(i) \\ \text{Indices} & \mathbf{GN} & \longrightarrow & \mathbf{LN} \\ & & \tau_p(i) & \end{array}$$

**Global Iterations to Global Indices** The subscript function $s(i)$ of an array reference transforms points in the global iteration space to the global index space (see Figure 5). Currently, $s(i)$ is allowed to be a linear function on a loop variable $i$:

$$s(i) = a_1 i + a_0 \tag{1}$$

To transform entire sets, the *step* must be scaled by $a_1$:

$$step' = a_1 \times step$$

Given a global iteration set $[u : v : w]$, the corresponding global index set is $[s(u) : s(v) : a_1 w]$.

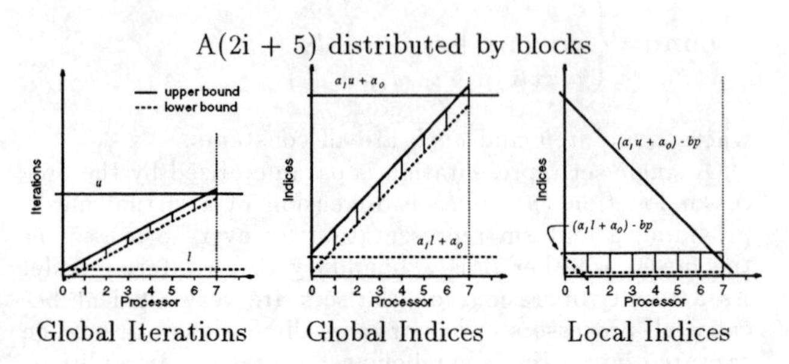


Global Iterations Global Indices Local Indices

Figure 5: Linear Reference Translations

**Global Indices to Local Indices** The function $\tau_p(x)$ transforms points in the global index space to indices local to a processor $p$ (see Figure 5):

$$\tau_p(x) = x - bp \tag{2}$$

where $b$ is the block size of distribution. Hence, a global index set $[u : v : w]$ can be transformed into a local index set $[\tau_p(u) : \tau_p(v) : w]$.

**Global Iterations to Local Iterations** The function $\lambda_p$ transforms points from the global iteration space to iterations local to a processor $p$ (Figures 2b and 3b):

$$\lambda_p(i) = i - \left\lfloor \frac{b}{a_1} p \right\rfloor \tag{3}$$

### 2.2 Access Iteration Sets

Let $R$ be an array reference enclosed in a loop, then the ACCESS set[(b)] of $R$ with respect to a processor $p$, denoted ACCESS$(R, p)$, is the set of global iterations for which $R$ accesses data owned by $p$. For example, for an array $A$, the ACCESS set of a reference $A(s(i))$ is:

$$\text{ACCESS}(A(s(i)), p) = [\lceil max(l, s^{-1}(1 + bp)) \rceil : \lfloor min(u, s^{-1}(b + bp)) \rfloor] \tag{4}$$

where $i$ is the loop variable, $l$ and $u$ are the lower and upper loop bounds (loop skip is 1), $b$ is the distributed block size of $A$, and $s(i)$ is the subscript function of the array reference.

The same concept can be extended for nested loops and multidimensional arrays by defining a subscript expression for each dimension as a function of the loop variables.

### 2.3 Computation Partitioning

An efficient implementation of the owner computes rule must avoid the runtime overhead of computing ownership. For computations inside a loop nest, this is done by reducing the loop bounds according to the Reduced Iteration Set [1] (RIS) of the loop, which is the union of the ACCESS sets of the $lhs$ references in the loop with respect to a processor $p$. The RIS represents the largest subset of the iteration space for which $p$ does some work in *every* iteration. Since the RIS is in the global iteration space, if used directly, a translation $\tau_p$ is required for the reference at every iteration, which is very costly:

```
DO i ∈ RIS
    A(τp(s(i))) = ···
END DO
```

A better approach is to apply $\lambda_p$ (Equation 3) to RIS to obtain the *Local* RIS (LRIS):

$$LRIS = \lambda_p(RIS)$$

A loop using the LRIS is much more efficient:

```
DO i ∈ LRIS
    A(s(i)) = ···
END DO
```

(b) ACCESS sets are called *Local Iteration Sets* in [1] which causes confusion since these sets describe global iterations.

If the ACCESS set of a particular $lhs$ in the loop is a *proper* subset of the RIS, then the corresponding statement must be *masked* [1] to make its execution conditional. Optimizations such as *mask merging* (consecutive assignment statements sharing the same mask) and *mask extraction* (multiple occurrences of a mask inside a loop are coalesced and extracted out of the loop) are performed automatically by the compiler [8].

### 2.4 Communication Analysis

To generate communication for regular computations, we can also apply symbolic set operations on the ACCESS sets of both the $lhs$ and $rhs$ references of assignment statements. This creates communication sets that encompass information about whether communication is needed, and if so under which iterations, between which pairs of processors, and what subsections of which arrays to send or receive.

#### Receive/Send Iteration Sets

The first communication sets to compute are the RECEIVE and SEND sets. These sets indicate which iterations require communication of data. For each statement $S$,

$$S : lhs = \mathcal{F}(rhs_1, rhs_2, rhs_3, \ldots)$$

RECEIVE$(rhs_i, p)$ is the set of iterations for which the execution of $S$ in processor $p$ requires nonlocal data due to the reference $rhs_i$. Similarly, SEND$(rhs_i, p)$ is the set of iterations for which reference $rhs_i$ is local to $p$, but $lhs$ is not, so $S$ is executed in a processor other than $p$ [1].

For each $rhs$ reference in each assignment statement, we must determine their RECEIVE and SEND sets. From the owner computes rule, the RECEIVE set is[(c)]

$$\text{RECEIVE}(rhs, p) = \text{ACCESS}(lhs, p) - \text{ACCESS}(rhs, p) \tag{5}$$

In other words, it is the set of iterations for which $lhs$ is local (and hence $p$ executes $S$) but $rhs$ is not available and must be *received* from another processor.

Similarly, the SEND set is given by

$$\text{SEND}(rhs, p) = \text{ACCESS}(rhs, p) - \text{ACCESS}(lhs, p) \tag{6}$$

As before, equations 5 and 6 can be extended to multidimensional arrays and meshes by maintaining one expression for each dimension.

#### In/Out Index Sets

To actually generate communication, the array subsections to be communicated must be known. The RECEIVE and SEND sets have to be translated from the *global iteration* space to the *local index* space of a processor $p$. Applying Equations 1 and 2, the IN and OUT local index sets are obtained from the RECEIVE and SEND iteration sets:

$$\text{IN}(A(s(i)), p) = \tau_p(s(\text{RECEIVE}(A(s(i)), p))) \tag{7}$$

$$\text{OUT}(A(s(i)), p) = \tau_p(s(\text{SEND}(A(s(i)), p))) \tag{8}$$

These sets can also be extended for multidimensional cases as before.

(c) Where "−" denotes either a set difference or a *stretch* operation as previously defined in Section 2.1.

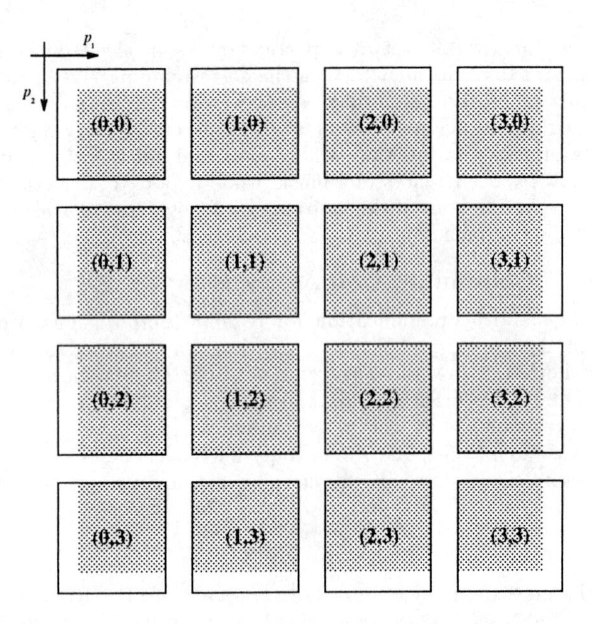


Figure 6: Iteration Space for Jacobi

### To/From Processor Sets

Both absolute and relative processor addressing is supported in the compiler. In general, computation of TO and FROM processor sets is addressed in the synthesis of high-level communication [9]. For nearest-neighbor communication, however, knowing the relative offset of either of the IN or OUT sets is sufficient to determine processor locations in the mesh. The *location* set operation is used to map index set offsets to the relative processor addresses.

### 2.5 Example: Jacobi's Iterative Method

We conclude this section with an illustration of some of the main ideas using Jacobi's Iterative method as an example (see Figure 7). The A and B arrays are distributed by blocks in both dimensions on a 2-dimensional processor mesh. The shaded region in Figure 6 represents the

```
program jacobi
parameter (np2 = 500, ncycles = 10)
real A(np2, np2), B(np2, np2)

np1 = np2 - 1
do k = 1, ncycles
    do j = 2, np1
        do i = 2, np1
            A(i,j) = (B(i - 1,j) + B(i + 1,j) + B(i,j - 1)
                              + B(i,j + 1)) / 4
        end do
    end do
    do j = 2, np1
        do i = 2, np1
            B(i,j) = A(i,j)
        end do
    end do
end do
end
```

Figure 7: Serial version of Jacobi's Iterative Method

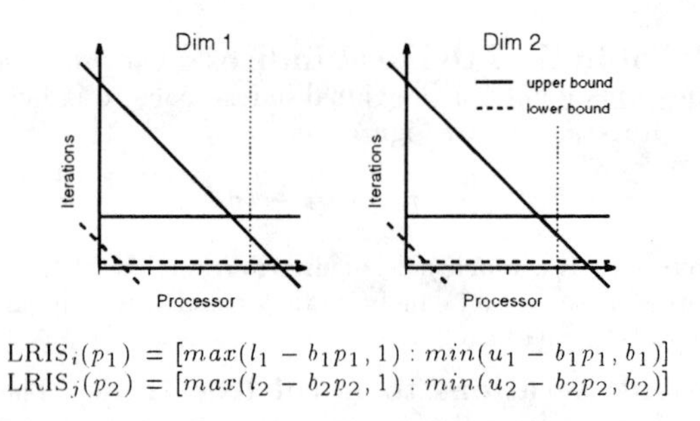


$$\mathrm{LRIS}_i(p_1) = [max(l_1 - b_1p_1, 1) : min(u_1 - b_1p_1, b_1)]$$
$$\mathrm{LRIS}_j(p_2) = [max(l_2 - b_2p_2, 1) : min(u_2 - b_2p_2, b_2)]$$

Figure 8: Local RIS for $i$ and $j$ Loops in Jacobi

global iteration space of the loop nest. In the figure, each processor is addressed by a pair $(p_1, p_2)$ which indicates its position in the processor mesh. The sets which describe the local RIS for each dimension are shown in Figure 8. Figure 9 shows sections of the array B that require communication determined by computing the IN and OUT sets (shown in Figure 10). Note that overlap regions are necessary for the communication described by these sets (this will be discussed in Section 3.2).

## 3. COMMUNICATION GENERATION

### 3.1 Communication Optimizations

Once the IN and OUT index sets have been computed, higher level decisions can be made to determine where to place the necessary communication. To reduce communication overhead, many optimization techniques can be applied during communication generation. Currently, these optimizations include message coalescing, vectorization, and aggregation [1]. Data dependence information is used to determine whether a given communication optimization is applicable, as well as exactly where the communication should be placed. Three lists of *communication descriptors* [8] (structures containing the IN/OUT and TO/FROM sets) are maintained for each loop header:

- BEFORE - before entering the loop
- TOP - at the start of each loop iteration
- BOTTOM - at the end of each loop iteration

These placements allow communication to be placed in the appropriate location with respect to any dependencies that may exist. If a reference is independent on a given loop variable, then its descriptor could be placed in the BEFORE list so that it would be performed only once over the entire loop. If a cross-iteration dependence exists then pipelining can be employed by splitting the send/receive operation between the TOP and BOTTOM lists.

**Message Coalescing** When statically analyzing the access patterns, redundant communications can be eliminated and the data can be reused from previous communication operations. By performing a union on overlapping index sets during placement of the communication descriptors, each element will be communicated at most once.

**Message Vectorization** Message vectorization combines element messages within a loop nest into larger *vectorized* groups of elements in a single message. By checking each loop nest from the innermost outward, the vectorization level is selected as the outermost nesting level that

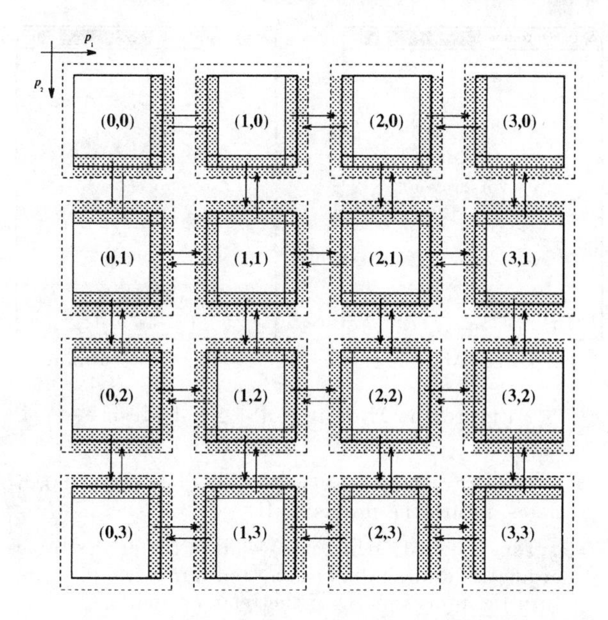


Figure 9: Communication with 4 Neighbors

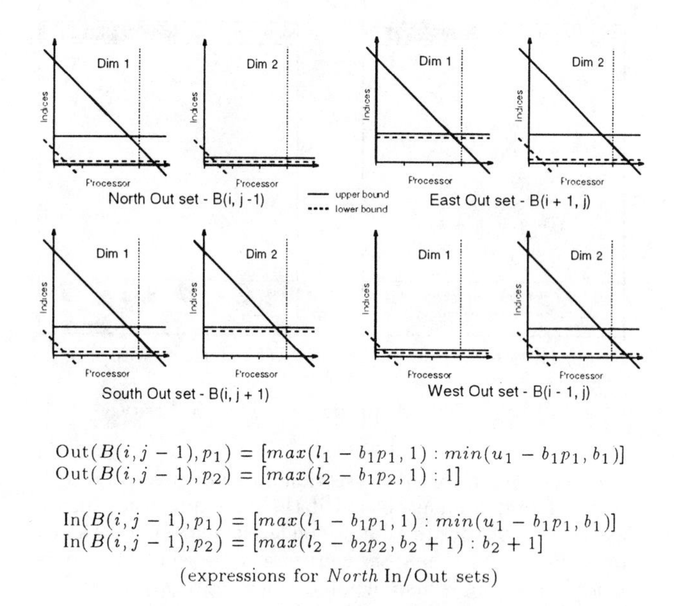


Figure 10: Communication Index Sets

does not contain any cross iteration dependence for the reference. The "itemwise" messages are *vectorized* as they are lifted out of the loop nests until they are placed on the communication descriptor at the vectorization level.

Coalescing vectorized messages now implies that any overlapping *regions* between messages are only communicated once. With symbolic sets, unions can be performed among entire vectorized messages to eliminate the redundant communication of the common regions.

**Message Aggregation** By grouping multiple messages with the same destination into a single message, the frequency of communication operations is further reduced. Communication operations which are placed at the same communication descriptor are first sorted by their destination. Messages being sent to the same destination can then be collected into a single communication operation.

### 3.2 Communication Summary Nodes

A *summary node* is constructed for each communication descriptor to facilitate the generation of communication. The index sets from the descriptors are analyzed to record the layout and length of the message for each dimension of the array. The mapping of array dimensions to the mesh is also recorded so that the array sections can be correlated with the mesh dimensions. Block distributed arrays can now be reduced, requiring analysis of the summary nodes for possible overlap regions.[d] Once overlap regions have been set, communication operations (along with possible packing operations) can now be generated from the summarized bound information. Finally, communication masks can be generated directly using the TO and FROM sets present in the communication descriptor.

[d] If cyclic distributed arrays were reduced, overlap regions could potentially exist between any of the sub-blocks, not simply at the boundaries, requiring further (complex) index translation.

#### Overlap Regions

Overlap regions need to be computed before any communication or buffering operations are generated. The extension of the bounds of an array directly affects the layout of the data and, therefore, whether or not packing needs to be performed with the communication.

To compute the overlap region for an array, the bounds of all corresponding IN index sets are examined. Since the array bounds in the symbol table may have already been reduced, direct comparison of these bounds with the target destination of the IN set will reveal the need for overlap areas. For each communication operation in the program, a destination region will be computed. If the symbol table entry for that reference is not large enough to enclose the region, the bounds are expanded for that entry. The need for overlap regions can be seen in Figure 9.

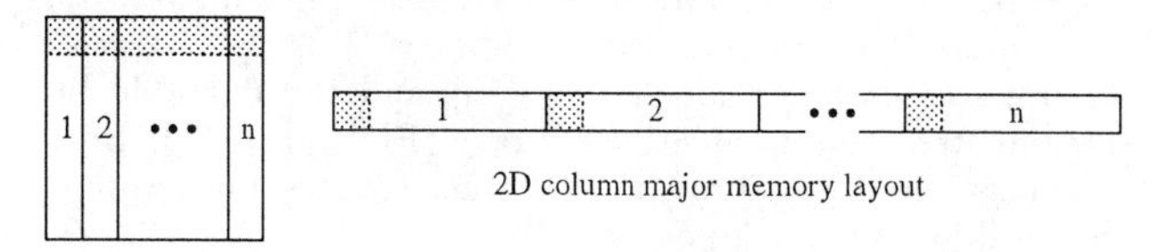


Figure 11: Buffering of a Non-Contiguous Region

#### Buffering Determination

It is also possible to detect the need for message packing by examining the communication summary nodes. Messages must be contiguous when they are sent and as they are received. In Figure 11, a two-dimensional array is shown with a column-major data layout (used in FORTRAN). It should be noted that the elements of a row in the array are located in memory such that each are separated by a distance equal to the column length. If a single row of the array were to be communicated, a contiguous intermediate buffer would be required.

The dimensions of the section to be communicated are analyzed in the order specified by its memory layout. For

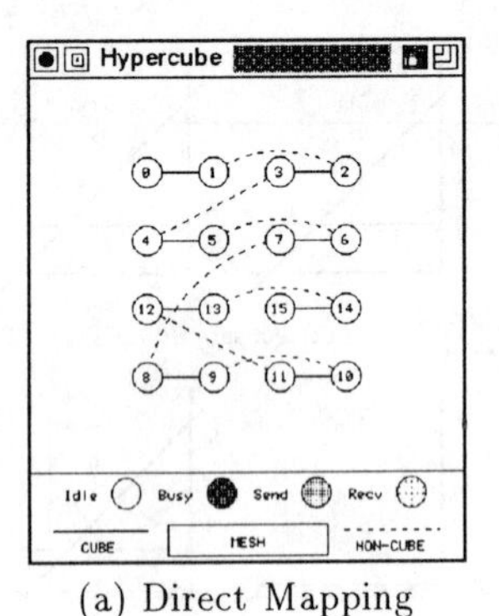


(a) Direct Mapping

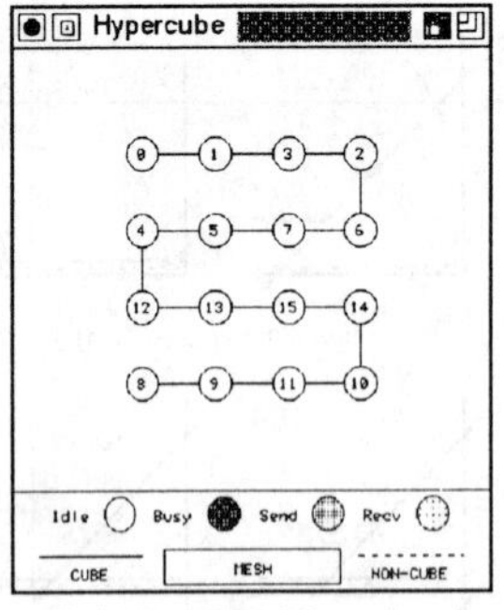


(b) Perfect Mapping

Figure 12: Mapping of a $1 \times 16$ Array

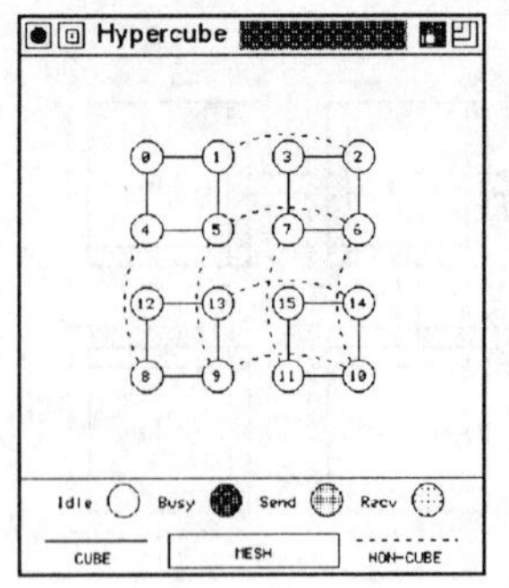


(a) Direct Mapping

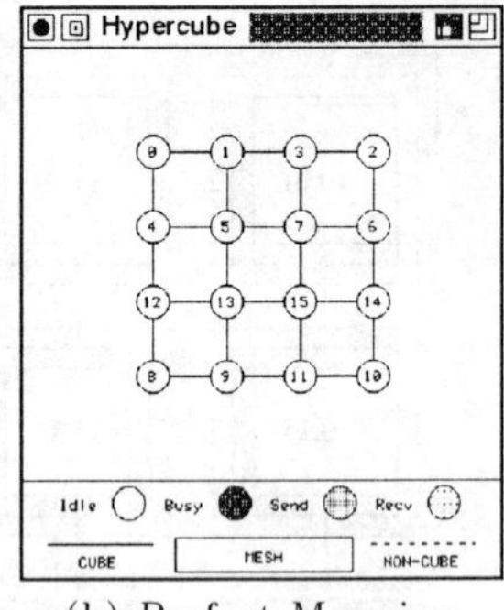


(b) Perfect Mapping

Figure 13: Mapping of a $4 \times 4$ Mesh

any given dimension, if the section does not cover the entire range of that dimension and if the next dimension in the layout accesses more than a single element (more than a unit-wide slice), then the entire section will have to be packed. Packing is also required when aggregating sections from different references into a single message regardless of their individual layouts.

### 3.3 Library Support

Library support is used to maintain machine independence (both for communication and mapping purposes) and to provide the pack and unpack operations. Also, by supporting certain communication libraries, performance evaluation through execution tracing is also made available.

#### Communication Library Support

Support for specific machines is provided through a generic library interface. For each supported communication library, abstract functions are mapped to the corresponding library-specific routines at compile-time. Since more information is carried internally than is required by any specific implementation, the mapping is obtained by selecting a subset of the available parameters. Both the communication operations, as well as a group of setup and cleanup statements are defined for each library.

Currently, library interfaces have been implemented for the Intel iPSC communication library [15], the Parasoft Express communication library [16], and the Portable Instrumented Communication Library (PICL) [17, 18]. Both Express and PICL provide execution tracing and support many different machines as well.

#### Processor Mapping

Internally, the compiler performs all analysis and communication generation for a virtual mesh topology. At run-time, this virtual mesh is mapped to the physical machine through the use of mapping functions. These functions are implemented for each target machine to efficiently map the mesh to the target architecture, maintaining the nearest neighbors of the virtual mesh as nearest neighbors on the machine whenever possible [19]. Processor mapping functions include:

- `gridinit(dim, num, nproc)` - Initialize mapping of a mesh of `dim` dimensions and `num[dim]` processors in each dimension (with `nproc` total available processors)
- `gridcoord(proc, coord)` - Given a processor ID, return the coordinates in the current mesh configuration
- `gridproc(coord)` - Given the processor mesh coordinates, return the processor ID
- `gridrel(coord, dim, rel)` - Given the processor coordinates, a mesh dimension, and a relative offset, return the processor ID of the relative node
- gridrel$n$(coord, rel$_1$, ...rel$_n$) - Given a processor number and a relative offset for **each** of $n$ mesh dimensions, return the processor ID

If the target machine has a hypercube topology, then it is possible to map a $D$-dimensional mesh onto it while maintaining **all** nearest-neighbor links as physical links. Even though communication latency is not much different for any two given nodes, mapping the mesh in nearest-neighbor fashion is important when communication rates are high. Contention will become a problem if communication is not distributed evenly (which is more likely if the mapping is **not** by nearest neighbor).

Figure 12 shows a direct mapping of a $16 \times 1$ mesh on a 16-node hypercube (which results in many non-neighbor links), and a *perfect* Gray code mapping (which preserves all physical neighbor links) Similarly, Figure 13 shows the direct and perfect mappings for a $4 \times 4$ mesh.

#### Communication Buffering

To facilitate the packing and unpacking of data for communication, two routines are also defined for each cardinality of arrays that may need packing:

- pack$_n$(a, byte, lb$_1$, ub$_1$, ...lb$_n$, ub$_n$, ls$_1$, us$_1$, sk$_1$, ...ls$_n$, us$_n$, sk$_n$, buf)
- unpack$_n$(a, byte, lb$_1$, ub$_1$, ...lb$_n$, ub$_n$, ls$_1$, us$_1$, sk$_1$, ...ls$_n$, us$_n$, sk$_n$, buf)

These functions are implemented separately for each of the possible degrees since the dimensions of arrays found in programs are very limited and a generalized pack/unpack would not be as efficient. The layout is specified by the size of the elements, the lower and upper bounds of the array (lb$_i$, ub$_i$), as well as the lower and upper bounds describing the section to pack (ls$_i$, us$_i$) for each dimension. A skip factor is also provided for each section description to support messages which require a stride between elements.

Library support can be seen in the resulting SPMD parallel program (see Figure 15 at the end of this paper) compiled from the serial version shown earlier in Figure 7.

Table 1: Performance Summary

| Test | N | Virtual Grid | Time (s) | Speedup | Efficiency |
|---|---|---|---|---|---|
| ADI | 1 | $1 \times 1 \times 1$ | 2.780 | 1.000 | 1.000 |
| | 2 | $1 \times 2 \times 1$ | 1.369 | *2.031* | *1.016* |
| | 4 | $1 \times 4 \times 1$ | 0.695 | *4.002* | *1.001* |
| | 8 | $1 \times 8 \times 1$ | 0.349 | 7.968 | 0.996 |
| | 16 | $1 \times 16 \times 1$ | 0.180 | 15.441 | 0.965 |
| Expl | 1 | $1 \times 1$ | 14.511 | 1.000 | 1.000 |
| | 2 | $2 \times 1$ | 7.580 | 1.914 | 0.957 |
| | 4 | $4 \times 1$ | 3.777 | 3.842 | 0.961 |
| | 8 | $8 \times 1$ | 1.902 | 7.629 | 0.954 |
| | 16 | $16 \times 1$ | 0.982 | 14.772 | 0.923 |
| Jacobi | 1 | $1 \times 1$ | 83.954 | 1.000 | 1.000 |
| | 2 | $2 \times 1$ | 37.566 | *2.235* | *1.118* |
| | 4 | $4 \times 1$ | 18.958 | *4.428* | *1.107* |
| | 8 | $4 \times 2$ | 10.250 | *8.191* | *1.024* |
| | 16 | $4 \times 4$ | 5.165 | *16.253* | *1.016* |

## 4. RESULTS

A group of small scientific program kernels [1, 20] are used to examine the performance of the output of the PARDIGM compiler on a real machine. The selected program fragments include an alternating direction implicit (ADI) integration (Livermore kernel 8), a 2-D explicit hydrodynamics routine (Livermore kernel 18), and Jacobi's iterative method. Array dimensions and loop bounds are determined at compile time with the constant propagation and induction variable passes in the compiler [5]. Both the virtual mesh configuration as well as the data partitioning of the arrays are automatically selected by the compiler [6] before generation of the SPMD program. The dimensions of the major arrays present in the programs are:

- ADI Integration - 1024 and $4 \times 1024 \times 2$ element arrays
- Explicit Hydrodynamics - $1024 \times 7$ element arrays
- Jacobi's Iterative Method - $500 \times 500$ element arrays

Using all optimizations, each of the test programs are compiled for an Intel iPSC/2 to examine the resulting performance over a range of processors. In Table 1, each program's partitioning configuration is shown along with the resulting execution time, speedup and efficiency. Superlinear speedup (due to cache effects associated with reduced array bounds and smaller local working sets) is italicized in the table. Both the execution time and speedup of the test programs are also shown in Figure 14.

Table 1 shows that both the ADI and Explicit Hydrodynamics programs have linear distributions for all machine sizes. This is due to the fact that the computation performed in both of these programs is mainly in a single dimension of the data arrays. The aligning capability of the distribution module of the compiler can correctly extract this information even though different array dimensions are used in different references to perform the computation.

It is interesting to notice in Table 1 that the Jacobi's iterative method has enough parallelism in two dimensions of the computation to allow for two-dimensional partitionings. Since communication becomes less expensive for 2-D partitionings for larger mesh sizes, the compiler has more freedom in selecting the mesh configuration for larger machines [14].

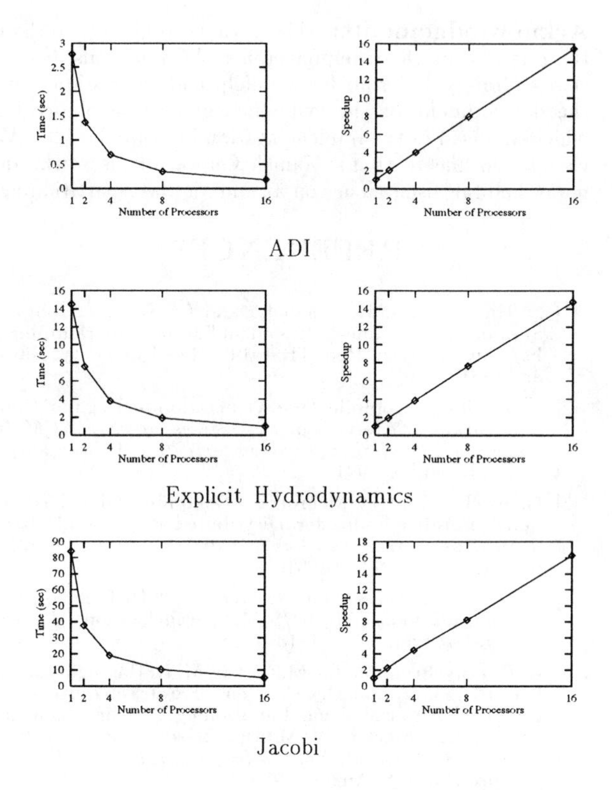


Figure 14: Performance of Test Programs

## 5. CONCLUSIONS

The purpose of this work is to automatically parallelize serial programs, efficiently mapping the abstract model to a given target machine. This is accomplished through the use of multidimensional partitioning schemes and processor mapping functions. Symbolic sets are heavily used to provide a uniform means of describing machine-wide access and communication patterns. Also, to provide machine independence, both the data distribution module as well as the library interface were developed such that the compiler could be easily retargeted.

Given the high efficiencies of the test programs, the applicability of this technique to regular computations is easily seen. Exactly how useful the techniques presented in this paper can be depends directly on the class of input programs that are to be parallelized. The current state of the compiler provides a useful testbed to develop further compiler optimizations for parallelization on distributed-memory multicomputers. For programs which show unbalanced computations, cyclic distributions will be added to the compiler to allow better load distributions. Research is also currently underway to support multithreaded execution [13] as well as to take advantage of functional parallelism [12] to address other classes of programs.

In conclusion, the two most complex tasks facing a user in parallelizing serial programs for a distributed-memory multicomputer are the distribution of the data and the implementation of interprocessor communication. In this paper, it has been shown that both of these tasks can be handled by the compiler for regular computations, thereby relieving the programmer of the burden of parallelization.

**Acknowledgements:** The authors would like to thank the referees for their helpful input. We would also like to thank Christy Palermo for her help and suggestions with the development of the symbolic comparison algorithms which are used in the implementation of symbolic sets. We would also like to thank Manish Gupta for his work, and many helpful discussions, on automatic data partitioning.

## REFERENCES


[1] S. Hiranandani, K. Kennedy, and C. Tseng, "Compiler Support for Machine-Independent Parallel Programming in Fortran D," Tech. Rep. TR90-149, Rice University, Februrary 1991.

[2] C. Koelbel, "Compile-Time Generation of Regular Communications Patterns," in *Proceedings of 1991 ACM International Conference on Supercomputing*, (Albuquerque, NM), November 1991.

[3] C. Koelbel and P. Mehrotra, "Compiling Global Name-Space Parallel Loops for Distributed Execution," *IEEE Transactions on Parallel and Distributed Systems*, vol. 2, pp. 440–451, October 1991.

[4] H. Zima, H. Bast, and M. Gerndt, "SUPERB: A tool for semi-automatic MIMD/SIMD parallelization," *Parallel Computing*, vol. 6, pp. 1–18, 1988.

[5] C. D. Polychronopoulos, M. Girkar, M. R. Haghighat, C. L. Lee, B. Leung, and D. Schouten, "Parafrase-2: An Environment for Parallelizing, Partitioning, Synchronizing and Scheduling Programs on Multiprocessors," in *Proceedings of the 1989 International Conference on Parallel Processing*, pp. II,39–48, August 1989.

[6] M. Gupta and P. Banerjee, "Demonstration of automatic data partitioning techniques for parallelizing compilers on multicomputers," *IEEE Transactions on Parallel and Distributed Systems*, vol. 3, pp. 179–193, March 1992.

[7] M. Gupta and P. Banerjee, "Compile-Time Estimation of Communication Costs on Multicomputers," in *Proceedings of 6th International Parallel Processing Symposium*, (Beverly Hills, CA), March 1992.

[8] E. Su, "Automating Parallelization of Regular Computations for Distributed-Memory Multicomputers," Master's thesis, Department of Electrical and Computer Engineering, University of Illinois at Urbana-Champaign, 1993.

[9] M. Gupta and P. Banerjee, "PARADIGM: A Compiler for Automated Data Partitioning on Multicomputers," in *Proceedings of 7th ACM International Conference on Supercomputing*, (Tokyo, Japan), July 1993.

[10] K. P. Belkhale and P. Banerjee, "Approximate Algorithms for the Partitionable Independent Task Scheduling Problem," in *Proceedings of the 1990 International Conference on Parallel Processing*, pp. 72–75, August 1990.

[11] K. P. Belkhale and P. Banerjee, "A Scheduling Algorithm for Parallelizable Dependent Tasks," in *International Parallel Processing Symposium*, pp. 500–506, 1991.

[12] S. Ramaswamy and P. Banerjee, "Processor Allocation and Scheduling of Macro Dataflow Graphs on Distributed Memory Multicomputers," *Proceedings of the 1993 International Conference on Parallel Processing*, August 1993.

[13] A. Lain, J. G. Holm, and P. Banerjee, "Compiler Transformations for Multithreading and Message-Driven Computation," submitted to the *1993 ACM International Conference on Supercomputing*, 1993.

[14] D. J. Palermo, E. Su, and P. Banerjee, "Evaluation of Communication Optimizations used in the PARADIGM Compiler," submitted to the *1993 ACM International Conference on Supercomputing*, 1993.

[15] Intel Corporation, *iPSC/2 and iPSC/860 User's Guide*, June 1990.

[16] Parasoft Corporation, Pasadena, CA, *Express Reference Guide for FORTRAN Programmers*, 1992.

[17] G. A. Geist, M. T. Heath, B. W. Peyton, and P. H. Worley, "PICL: A Portable Instrumented Communication Library, C reference manual," Tech. Rep. ORNL/TM-11130, Oak Ridge National Laboratory, Oak Ridge, TN, July 1990.

[18] M. T. Heath and J. A. Etheridge, "Visualizing the performance of parallel programs," *IEEE Software*, vol. 8, pp. 29–39, September 1991.

[19] G. Fox, M. Johnson, G. Lyzenga, S. Otto, J. Salmon, and D. Walker, *Solving Problems on Concurrent Processors*. Prentice Hall, 1988.

[20] F. McMahon, "The Livermore Fortran Kernels: A computer test of the numerical performance range," Tech. Rep. UCRL-53745, Lawrence Livermore National Laboratory, 1986.


```
program jacobi
character m$buf(500)
integer m$numdim, m$num(2), m$to(-1:1,-1:1), p(2)
real A(125, 125), B(0:126, 0:126)

m$numdim = 2   {number of mesh dimensions}
m$num(1) = 4   {number of processors on mesh dim. 1}
m$num(2) = 4   {number of processors on mesh dim. 2}
call m$gridinit(m$numdim,m$num,numnodes())
call m$gridcoord(mynode(),p)

m$to(-1,0) = m$gridrel2(p,-1,0)
m$to(0,-1) = m$gridrel2(p,0,-1)
m$to(0,1) = m$gridrel2(p,0,1)
m$to(1,0) = m$gridrel2(p,1,0)

do k = 1, 10
   if (p(2) .ge. 1) then
      call csend(0,b(1,1),500,m$to(0,-1),1)
   end if
   if (p(2) .le. 2) then
      call crecv(0,b(1,126),500)
   end if
   if (p(2) .le. 2) then
      call csend(1,b(1,125),500,m$to(0,1),1)
   end if
   if (p(2) .ge. 1) then
      call crecv(1,b(1,0),500)
   end if
   if (p(1) .ge. 1) then
      call f$pack2(b,4,0,126,0,126,1,1,1,1,125,1,m$buf)
      call csend(2,m$buf,500,m$to(-1,0),1)
   end if
   if (p(1) .le. 2) then
      call crecv(2,m$buf,500)
      call f$unpack2(b,4,0,126,0,126,126,126,1,1,125,1,m$buf)
   end if
   if (p(1) .le. 2) then
      call f$pack2(b,4,0,126,0,126,125,125,1,1,125,1,m$buf)
      call csend(3,m$buf,500,m$to(1,0),1)
   end if
   if (p(1) .ge. 1) then
      call crecv(3,m$buf,500)
      call f$unpack2(b,4,0,126,0,126,0,0,1,1,125,1,m$buf)
   end if
   do j = max(2 - 125 * p(2), 1), min(499 - 125 * p(2), 125)
      do i = max(2 - 125 * p(1), 1), min(499 - 125 * p(1), 125)
         A(i,j) = (B(i - 1,j) + B(i + 1,j) + B(i,j - 1)
                   + B(i,j + 1)) / 4
      end do
   end do
   do j = max(2 - 125 * p(2), 1), min(499 - 125 * p(2), 125)
      do i = max(2 - 125 * p(1), 1), min(499 - 125 * p(1), 125)
         B(i,j) = A(i,j)
      end do
   end do
end do
end
```

Figure 15: SPMD version of Jacobi's Iterative Method

# Compilation Techniques for Optimizing Communication on Distributed-Memory Systems

Chun Gong
gong@cs.pitt.edu

Rajiv Gupta
gupta@cs.pitt.edu

Rami Melhem
melhem@cs.pitt.edu

Department of Computer Science
University of Pittsburgh
Pittsburgh, PA 15260

## Abstract

*Communication overhead can significantly impact the performance during parallel execution of programs on distributed-memory systems. In this paper we describe optimizations that reduce communication overhead and avoid execution time delays caused by interprocessor communication. The optimizations include: (a) avoiding sequentialization caused by communication; (b) avoiding redundant communication; (c) overlapping the communication and the computation; and (d) combining several small messages sent to the same destination into one large message. We develop a data flow framework for collecting information necessary to apply the above optimizations to programs being compiled for* single-program, multiple-data *(SPMD) execution on a distributed-memory system. We are able to deal with the tradeoff between conflicting optimizations through this unifying data flow framework.*

## 1 Introduction

While a distributed-memory system can exploit massive parallelism due to its scalability, it is difficult to program such an architecture. The programmer must explicitly distribute data and must program the transfer of data among the processors. One approach considered by researchers provides a combination of language extensions and compilation techniques for programming such systems [1] [9] [10] [15]. In this approach, the user specifies the distribution of data by using language extensions and the compiler translates the program for *single-program multiple-data* (SPMD) execution. One approach to translation assigns a unique processor as the owner of each data element and then yields a program in which a processor executes a statement only if a data value computed by the statement is owned by the processor. If a processor needs a data from a non-local memory, a *receive* instruction must be executed by the processor and a *send* instruction must be executed by the processor that owns the data. These *receive* and *send* instructions are automatically inserted by the compiler.

The communication overhead latency is of primary concern during the SPMD execution of programs. First, the long communication latency may cause a processor to idle while waiting for the data it needs. Second, some communication schemes may sequentialize the execution of an otherwise parallel computation. Therefore for the SPMD approach to work effectively, optimizations must be performed to reduce the communication overhead. In this paper we develop compile-time techniques for optimizing this overhead. The goals of the optimizations are:

1. Avoiding sequentialization caused by communication;

2. Avoiding redundant communication;

3. Overlapping of communication and computation; and

4. Combining small messages sent to the same destination into larger messages.

Significant work has been done in optimizing communication [1] [6] [4] [11]. In this paper, we propose to perform all of the above optimizations in a unifying framework. This framework allows us to deal with the tradeoff between conflicting optimizations. We develop a data flow framework for collecting the information needed to perform the above optimizations. We focus on optimizing the communication resulting from references to distributed arrays. More specifically, a data flow analysis algorithm, which propagates portions of arrays, is used to determine points in the program at which communication can be performed. Using this information, optimization algorithms select appropriate communication points to achieve the above optimizations. Since arrays are used very often in scientific applications, significant performance improvement can be expected by optimizing array references. In order to develop efficient data flow analysis algorithms we treat an array reference as an integrated unit. All array elements corresponding to a reference are represented by a single pattern rather than being represented individually.

```
Do i = 2, 100                          DoAll i = 2, 100
  S(B[i-1], P_{i-1} → P_i)               S(B[i-1], P_{i-1} → P_i)
  R(B[i-1], P_i ← P_{i-1})               R(B[i-1], P_i ← P_{i-1})
  A[i] = B[i-1] + 1                    DoAll i = 2, 100
                                          A[i] = B[i-1] + 1
        (a)                                     (b)
```

Figure 1: (a) A forced sequential loop; (b) After optimization.

The remainder of the paper is organized as follows. In Section 2, we identify the communication optimizations. In Section 3, we present the data flow framework for collecting the information necessary to perform the optimizations. In Section 4, we provide an integrated communication optimization algorithm. In Section 5, we point out some extensions of the basic data flow analysis algorithm. We discuss related work in section 6 and give a summary in Section 7.

## 2 Communication Optimizations

We assume that a serial program is translated for SPMD execution, using the owner computes rule. In this approach each processor examines the statements in the program sequentially and for each statement it takes one of the following actions: (a) performs the operation indicated by the statement if it owns the data element whose value is being computed. Nonlocal operands are received from other processors; (b) sends local data to the processor that needs the data to execute the statement; or (c) skips the statement. In this approach communication instructions are introduced by the compiler. The communication primitives are non-blocking *send* and blocking *receive*. It is assumed that the instruction *send* transmits both the name and the value of the data to the destination. We will use the following notations for *send* and *receive* instructions:

S(name, $P_s \rightarrow P_d$)≡ $P_s$ Sends {name=val(name)} to $P_d$,
R(name, $P_d \leftarrow P_s$)≡ $P_d$ Receives {name=val(name)} from $P_s$.

If the same variable is sent twice with different values, they are delivered to the destination in the same order as they are sent. We consider the following optimization problems:

(A) Avoiding Sequentialization Caused by Communication: A key problem with the basic translation techniques for SPMD execution is that the data-owner processor sends the data to the data-consumer processor at the same point in the program at which the data is used [1]. As illustrated in [1] [6], this approach can sequentialize the execution of otherwise parallel code. For example, consider the loop in Figure 1 (a). In order to simplify the presentation, we assume that data array is distributed such that $A[i]$ and $B[i]$ are stored in the local memory of $P_i$. The loop in Figure 1 (a) will be executed sequentially in SPMD mode. Processor $P_{i-1}$ must finish executing the $(i-1)$th iteration before sending the value of $B[i-1]$ to $P_i$. Thus processor $P_i$, which is to execute the $i$th iteration, is kept

```
DoAll i=2, N                       DoAll i=2, N
  S(B[i-1], P_{i-1} → P_i)           S(B[i-1], P_{i-1} → P_i)
  R(B[i-1], P_i ← P_{i-1})           R(B[i-1], P_i ← P_{i-1})
  A[i] = B[i-1]+1                  ...
...                                DoAll i=2, N
DoAll i=2, N                         A[i] = B[i-1] + 1
  S(B[i-1], P_{i-1} → P_i)         ...
  R(B[i-1], P_i ← P_{i-1})         DoAll i=2, N
  C[i] = B[i-1]*d                    C[i]=B[i-1] * d
        (a)                                (b)
```

Figure 2: (a) The program with redundant communication; (b) After optimization.

```
Do i = 1, N                        Do i = 1, N
  B[i]=B[i]*C[i]                     B[i]=B[i]*C[i]
DoAll i = 2, N                     DoAll i = 2, N
  S(B[i],P_i → P_{i-1})              S(B[i],P_i → P_{i-1})
  R(B[i],P_{i-1} ← P_i)            ...
...                                DoAll i = 1, N-1
DoAll i = 1, N-1                     R(B[i+1],P_i ← P_{i+1})
  A[i] = A[i]+B[i+1]                 A[i]=A[i]+B[i+1]
        (a)                                (b)
```

Figure 3: (a) Avoiding sequentialization; (b) Separating send and receive.

waiting. If $B[i-1]$ is sent to $P_i$ before the loop starts execution, as shown in Figure 1 (b), then all iterations can be executed in parallel. In general, we would like to send data as early as possible, since the earlier the data is sent the less the chance that the consumer processor will be delayed and thus the better the chance that all processors can execute the program in parallel.

(B) Avoiding Redundant Communication:[6] Sometimes a value is used more than once by a processor. For example, in Figure 2 (a) $B[i-1]$ is used twice by processor $P_i$. Assuming that there is no definition for $B[i-1]$ between the two references, the value $B[i-1]$ may only be sent once. We want to identify this kind of redundant communications and eliminate all but the first communication, as shown Figure 2 (b).

(C) Overlapping Communication with Computation: In SPMD mode, *send* and *receive* are inserted at the same point in the program [1]. Since the *send* is non-blocking and the *receive* is blocking, the processor that executes a *receive* instruction will be blocked until the data arrives. Thus, inserting *send* and *receive* at the same point is likely to cause delays during execution since the latency of data communication in a distributed-memory system is large. This delay can be reduced by maximizing the overlapping of communication and computation. Such overlap may be achieved by inserting the *send* at an earlier point than the *receive*. Consider the example of Figure 3 (a), in which we insert both *send* and *receive* instruction at the same point (after $B[i]$ is defined). When processor

```
Do i = 2, N                     Do i=2, N
  S(B[i-1], P_{i-1} → P_i)        S(B[i-1],C[i-1],P_{i-1} → P_i)
  R(B[i-1],P_i ← P_{i-1})         ...
  A[i] = B[i-1] + d             Do i=2, N
...                               R(B[i-1],P_i ← P_{i-1})
Do i=2, N                         A[i]= B[i-1] + d
  S(C[i-1],P_{i-1} → P_i)       ...
  R(C[i-1],P_i ← P_{i-1})       Do i=2, N
  A[i] = A[i]+C[i-1]              R(C[i-1],P_i ← P_{i-1})
                                  A[i] = A[i]+C[i-1]
        (a)                             (b)
```

Figure 4: (a) A program with two communications; (b) After optimization.

$P_{i-1}$ executes the receive instruction, it may be idle, waiting for $B[i]$ to arrive. In Figure 3 (b), the *receive* is separated from the *send* to increase the overlapping of communication and execution. A *send* and its corresponding *receive* have to be separated in a manner which ensures that the execution of each *receive* matches the execution of a corresponding *send*.

(D) Combining Messages: In a distributed-memory system, there is a considerable setup overhead associated with the communication of each message. One way to reduce this overhead is to combine several small messages that need to be sent to the same destination into one large message so that the overhead of constructing and transmitting messages is amortized [15]. In Figure 4 (a), two values $B[i-1]$, and $C[i-1]$ are sent to $P_i$ in the form of two messages, while in Figure 4 (b) they are combined into one message.

## 3 Data Flow Framework

In order to implement the communication optimizations, we propose a data flow analysis algorithm that collects the necessary information needed to perform the optimizations. For each array reference requiring communication, we determine all points in the program at which the communication can be performed. We compute this information by identifying the communication associated with each statement and propagating it in the backward direction along all execution paths in the program. An array reference nested within a loop may reference different array elements during each loop iteration. We represent these group of elements as a single entity. Groups of array elements that must be communicated are propagated. If definitions of array elements are encountered, the propagation of those elements is discontinued. We may encounter a definition which only defines some of the array elements. Thus a reference to a group of elements may have to be split. In addition, we may encounter repeated communication of the same array elements. In such a situation we must combine repeated references. In the subsequent sections we describe the representation of array references, and their propagation including *splitting* and *combining* operations.

### 3.1 Representation of array references

We need a concise notation to represent an array reference when performing data flow analysis for arrays. In [7], the authors compared several representations of array references by their precision and complexity. We use an extension of one of these notations and the extension allows us to keep track of the communication pattern, that is, which source processor should send what data to which consumer processor. We define an array reference as a tuple:

$$u = < Name, D, Source(i_1, \ldots, i_n), Dest(i_1, \ldots, i_n) >,$$

where $Name$ identifies the array that is referenced, $D$ is the set of values that the index variables $i_1, \ldots, i_n$ take, $Source(i_1, \ldots, i_n)$ is a function indicating the array element that is used for the index point $(i_1, \ldots, i_n)$ in $D$, and $Dest(i_1, \ldots, i_n)$ is a function representing the processor that uses the array element. For a $m$ dimensional array, $Source$ will produce a vector of $m$ values. In general, both $Source$ and $Dest$ are functions of type: $I^n \rightarrow I^m$, where $I$ is the set of integers. We call the *range* of the source function the *range* of the array reference. For an array reference $u$, $Name_u$, $D_u$, $R_u$, $Source_u$ and $Dest_u$ denote the $Name$, the domain, the range, the source function and destination function of $u$, respectively.

Before we present the data flow equations, we define the operators *difference* $(-_{<>})$, *union* $(\cup_{<>})$ and *intersection* $(\cap_{<>})$ on sets of array references. These operators, which will be used in data flow equations, are different from simple set operators in that they require *splitting* and *combining* of array references. To explain the notation of *splitting* and *combining*, we consider a simple form of array reference, a one dimensional array in a singly nested loop. In this case, the array reference can be represented as

$$u = < B, (i = l, h, m), Source(i), Dest(i) >,$$

where $l$ is the initial value of the loop variable $i$, $h$ is its final value, and $m$ is its increment. The algorithm will do backward data flow analysis, trying to locate the definition for $B$. During the processing, the following cases will arise:

1. Elements $B[Source(l')]$,.., $B[Source(h')]$ are defined and these definitions reach the reference $u$. If $l < l' < h' < h$, then we have to split $u$ into two references with domain $(l, l'-1, m)$ and $(h'+1, h, m)$, respectively, and propagate them above the definitions. See Figure 5 (a).
2. Another reference $v = < B, (i = l', h', m), Source(i), Dest(i) >$, is encountered. If $l < l' < h < h'$, then we can combine $u$ and $v$ into one array reference $< B, (i = l, h', m), Source(i), Dest(i) >$. See Figure 5 (b).
3. The reference $u$ is in one branch of an *If* structure, $v = < B, (i = l', h', m >, Source(i), Dest(i) >$ is in the other branch and $l < l' < h <$

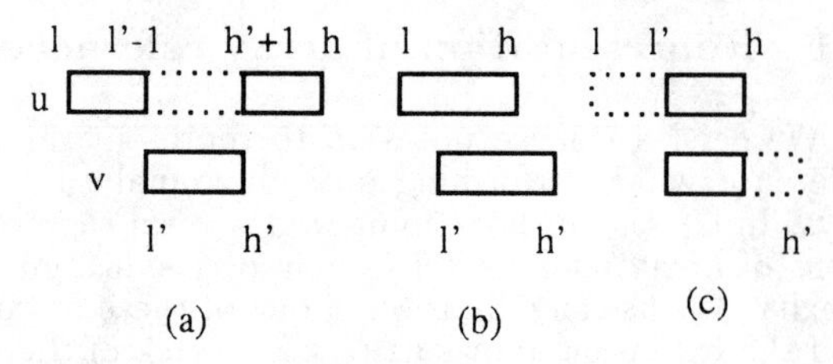


Figure 5: (a) Split operation; (b) Widen operation; (c) Narrow operation .

$h'$. Then we propagate only the common portion of the two references, $< B,(i = l',h,m), Source(i), Dest(i) >$, above the *If* structure. See Figure 5 (c).

In Section 3.3, we will study these cases in more detail. In the rest of this section, we introduce the following definitions for the general form of array references:

**Def. 1. Split(u,v):** Given two array references $u$ and $v$ such that $Name_u = Name_v$, this function returns the portions of $u$ that are not part of $v$.

**Def. 2. Widen(u,v):** Given two array references $u$ and $v$ such that $Name_u = Name_v$, $Source_u = Source_v$ and $Dest_u = Dest_v$, this function returns $w$ with $D_w = D_u \cup D_v$.

**Def. 3. Narrow(u,v):** Given two array references $u$ and $v$ such that $Name_u = Name_v$, $Source_u = Source_v$ and $Dest_u = Dest_v$, this function returns $w$ with $D_w = D_u \cap D_v$.

**Def. 4.** $U -_{<>} V$ is a set $W$ such that
$W = \{u | u \in U \& (Name_v \neq Name_u \; \forall v \in V)\} \cup$
$\{Split(u,v) | u \in U \& v \in V \& Name_u = Name_v\}$

**Def. 5.** $U \cup_{<>} V$ is a set $W$ such that
$W = \{u | u \in U \& (Name_v \neq Name_u \; \forall v \in V)\} \cup$
$\{v | v \in V \& (Name_u \neq Name_v \; \forall u \in U)\} \cup$
{Widen(u, v)— u∈ U& v∈ V& Name$_u$=Name$_v$&
$Source_u = Source_v \& Dest_u = Dest_v\}$

**Def. 6.** $U \cap_{<>} V$ is a set $W$ such that
$W = U \cap V \cup \{Narrow(u,v) | u \in U \&$
$v \in V \& Name_u = Name_v \&$
$Source_u = Source_v \& Dest_u = Dest_v\}$

In the next section, these operators are used to describe the data flow equations.

### 3.2 Data flow equations

The data flow analysis algorithm presented in this section is developed for the following language:

$$S \longrightarrow S;\ S \mid If\ E\ then\ S\ else\ S \mid L$$
$$L \longrightarrow Do\ i_1 = l_1, h_1, m_1; \ldots; i_n = l_n, h_n, m_n$$
$$ID_1[f(i_1,..,i_n)] = ..ID_2[g(i_1,..,i_n)]..$$

This language is motivated by those data parallel programming languages in which array references are crucial for achieving parallelism [16] [8] [13]. For example, Fortran 90 array expression can be expressed using the above language. Therefore although the above language is simple, it expresses an important class of programs involving data level parallelism.

The data flow algorithm computes four synthesized attributes $ref(S)$, $def(S)$, $in(S)$, $out(S)$ for each construct $S$:

- $in(S)$ ={ $u|u$ is an array reference that is encountered before the array elements are redefined by a statement along each path starting at the point before $S$};
- $out(S)$ = {$u|u$ is an array reference that is encountered before the array elements are redefined by a statement along each path starting at the point after $S$};
- $ref(S)$ = {$u|u$ represents array elements that are used by $S$ before redefinition}; and
- $def(S)$ = {$u|u$ represents array elements that are defined by $S$};

The data flow equations used to compute the above attributes are given as follows:

- $S \longrightarrow S_1;\ S_2$

$$\begin{aligned} ref(S) &= ref(S_1) \cup_{<>} \\ &\quad (ref(S_2) -_{<>} def(S_1)) && (1) \\ def(S) &= def(S_1) \cup_{<>} def(S_2) && (2) \\ in(S) &= in(S_1) && (3) \\ out(S_1) &= in(S_2) && (4) \\ out(S_2) &= out(S) && (5) \end{aligned}$$

Any array reference that is used by $S_1$ or used by $S_2$ but not defined by $S_1$ is considered as used by $S$. The array elements defined by $S$ are defined by at least one of $S_1$ and $S_2$.

- $S \longrightarrow If\ E\ then\ S_1\ else\ S_2$

$$\begin{aligned} ref(S) &= ref(S_1) \cap_{<>} ref(S_2) && (6) \\ def(S) &= def(S_1) \cup_{<>} def(S_2) && (7) \\ out(S_1) &= out(S) && (8) \\ out(S_2) &= out(S) && (9) \\ in(S) &= in(S_1) \cap_{<>} in(S_2) && (10) \end{aligned}$$

The set $ref(S)$ contains an array reference if it is used along both paths. An array reference in $def(S)$ is used along at least one of two paths.

- $S \longrightarrow Do\ i_1 = l_1, h_1, m_1; \ldots; i_n = l_n, h_n, m_n$
$ID_1[f(i_1,..,i_n)] = \ldots ID_2[g(i_1,..i_n)] \ldots$

$$\begin{aligned} D &= \{ranges\ of\ i_j | (1 \leq j \leq n)\} \\ ref(S) &= \{< ID_2, D, g(i_1,\ldots,i_n), \\ &\quad f(i_1,\ldots,i_n) >\} && (11) \end{aligned}$$

$$def(S) = \{< ID_1, D, f(i_1, \ldots, i_n), f(i_1, \ldots, i_n) >\} \quad (12)$$

$$in(S) = (out(S) -_{<>} def(S)) \cup_{<>} ref(S) \quad (13)$$

By the definition of $in(S)$, any array reference that is defined in $S$ cannot be in $in(S)$ and any array reference that is used in $S$ is in $in(S)$.

Using these equations, we can collect all *in* and *out* information by traversing the syntax tree.

### 3.3 Solving data flow equations

In this section, we describe algorithms for solving data flow equations. The programs that we can handle include an important class of programs: *singly nested loops* and *unidimensional arrays*. Furthermore, we make the following assumptions:

1. The loop step $m$ is always 1. This is not a crucial assumption since a loop can be converted into an equivalent form with step $m = 1$. We also assume that the loop bounds are compile time constants.

2. Both the source function and the destination function are shift functions, $Source(i) = i + c$ and $Dest(i) = i + c'$ where $c$, $c'$ are integer constants. We convert the destination function into an identity function by changing the domain to $(i = l + c', h + c')$ and the *Source* function to $Source(i) = i + c - c'$. Hence we assume that the destination function is the identity function.

We can also expand our framework to cover cases like *block* and *cyclic* data distribution. For block distribution, we can assign a new index value to the collection of elements that are in the same block; for cyclic distribution, we simply divide the whole array into several subarrays and give each subarray a new name. Under the above assumptions, both the domain and the range of an array reference will be sets of consecutive integers and therefore can be represented by two integers $D_u = (i = l, h)$ and $R_u = (l + c, h + c)$. In the remainder of this section, we will omit the destination function from the reference tuples since it is always assumed to be the identity function. The *Split, Widen* and *Narrow* operations for such array references are given in Figure 6, Figure 7 and Figure 8.

We apply the data flow analysis algorithm to the program in Figure 9 (a). The computation of the attributes for each statement are given as follows:

1. First, we initialize $out(S_7) = \{\}$;

2. By (11): $ref(S_7)$={<B,(i=1, 100), i-1>, <C, (i=1, 100), i+1>, <D, (i=1, 100), i+1>};
   By (12): $def(S_7)$={<A, (i=1,100), i>};
   By (13): $in(S_7)$={<B, (i=1, 100), i-1>, <C, (i=1, 100), i+1>,<D, (i=1, 100), i+1>};

3. By (4), (8) and (9): $out(S_5) = out(S_6) = in(S_7)$;

**Algorithm 1:** Splitting an array reference.
Input: $u =< B, (i = l, h), i + c >$ and
$v =< B', (i = l', h'), i + c' >$;
Output: A set of array references;

Split(u, v)
  If ($B' \neq B$) Then return $\{u\}$;
  Else
    case 1: $l + c < l' + c' \& h' + c' < h + c$
      return $\{< B, (i = l, l' + c' - c - 1), i + c >, < B, (i = h' + c' - c + 1, h), i + c >\}$;
    case 2: $l' + c' \leq l + c \& h + c \leq h' + c'$
      return $\{\}$;
    case 3: $l + c < l' + c' < h + c < h' + c'$ (or $l' + c' < l + c < h' + c' < h + c$)
      return $\{< B, (i = l, l' + c' - c - 1), i + c >\}$;
      (or $\{< B, (i = h' + c' - c + 1, h), i + c >\}$)
    case 4: $h + c < l' + c$ or $h' + c < l + c$
      return $\{u\}$;

Figure 6: Operation *Split*.

**Algorithm 2:** Combining two array references.
Input: $u =< B, (i = l, h), i + c >$ and
$v =< B, (i = l', h'), i + c >$
Output: A set of array references;

Widen(u, v)
  case 1: $l < l' \& h' < h$
    return $\{< B, (i = l, h), i + c >\}$;
  case 2: $l' \leq l \& h \leq h'$
    return $\{< B, (i = l', h'), i + c >\}$;
  case 3: $l < l' < h < h'$ (or $l' < l < h' < h$)
    return $\{< B, (i = l, h'), i + c >\}$;
    (or $\{< B, (i = l', h), i + c >\}$)
  case 4: $h < l'$ or $h' < l$
    return $\{u, v\}$;

Figure 7: Operation *Widen*.

**Algorithm 3:** Intersecting two array references.

Input: $u =< B, (i = l, h), i + c >$, $v =< B, (i = l', h'), i + c >$
Output: A set of array references;
Narrow(u, v)
  case 1: $l < l' \& h' < h$
    return $\{< B, (i = l', h'), i + c >\}$;
  case 2: $l' \leq l \& h \leq h'$
    return $\{< B, (i = l, h), i + c >\}$;
  case 3: $l < l' < h < h'$ (or $l' < l < h' < h$)
    return $\{< B, (i = l', h), i + c >\}$;
    (or $\{< B, (i = l, h'), i + c >\}$)
  case 4: $h < l'$ or $h' < l$
    return $\{\}$;

Figure 8: Operation *Narrow*.

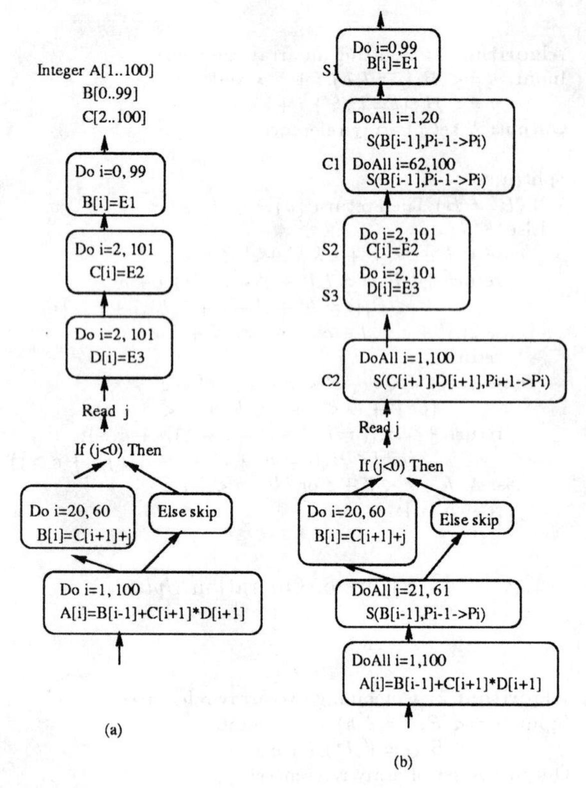


Figure 9: (a) The program; (b) After optimization.

4. By (11) and (12):$ref(S_5)=\{<C,(i=20,60),i+1>\}$; $def(S_5)=\{<B,(i=20,60),i>\}$; and by (13):
   $in(S_5)=\{<D,(i=1,100),i+1>,$
   $Split(<B,(i=1,100),i-1>,<B,(i=20,60),i>),$
   $Widen(<C,(i=1,100),i+1>,<C,(i=20,60),i+1>)\}$

   $=\{<B, (i=1, 20), i-1>, <B, (i=60, 100), i-1>,$
   $<C,(i=1,100),i+1>, <D,(i=1,100),i+1>\}$

5. $in(S_6) = out(S_6)$;

6. By (10): $in(S_4) = in(S_5)$;

7. The rest of the computation is straightforward:
   $in(S_3)=\{<C, (i=1, 100), i+1>,$
   $<B, (i=1, 20), i-1>,$
   $<B,(i=60+2,100),i-1>\}$
   $in(S_2)=\{<B, (i=1, 20), i-1>,$
   $<B, (i=62, 100), i-1>\}$
   $in(S_1) = \{\}$.

## 4 An Optimization Algorithm

In this section we develop an optimization algorithm which takes advantage of the information computed in the preceding section. In order to facilitate the implementation of the optimizations, we modify the data flow sets as follows. If an array reference is split during the computation of $in(S)$ then we modify $out(S)$ by rewriting the reference as the split portions, which are in $in(S)$, and the remaining portion. All sets of statements after $S$ are also modified similarly. These modifications can be carried out in a single top-down pass of the syntax tree. For example, in the program of Figure 9 (a), since $B$ is split at $S_5$ we rewrite the reference $B$ in $out(S_5) = out(S_6) = in(S_7)$ as: $\{< B,(i = 1,20),i-1 >,< B,(i = 21,61),i-1 >,< B,(i = 62,100),i-1 >\}$, which results from splitting $< B,(i = 1,100),i-1 >$. Next we insert the communication primitives into the program at the earliest point at which the communication can be performed. The earliest point for a communication can be found as follows:

For each array reference $u$, identify the statement $S$ such that $u \in out(S)$ but $u \notin in(S)$. This $S$ is the earliest point for the communication u.

This algorithm achieves the optimizations of *avoiding sequentialization, overlapping* and *avoiding redundancy*. In order to achieve the *combining of smaller messages*, we need to modify the above method. One observation is that we do not need to insert a data *send* at the earliest possible point. Sometimes it is better to insert it at a later point so it can be combined with another data *send*. For example, consider the program of Figure 9 (a) again. We can combine the sending of $C$ and $D$ if we do not insert the sending for $C$ at the earliest point. However, not inserting the *send* at the earliest point may affect other optimizations, such as *avoiding sequentialization* and *overlapping*. Therefore there is some interaction among the different optimizations and the degree to which various optimizations are applied must be balanced. One way of balancing these optimizations is to find, for each array reference $u$, a point above which the communication of $u$ results in complete overlapping of communication with computation. For this purpose, we introduce the following definition:

**Def. 7.** Given a program $P$, for any array reference $u$ in $P$, $S^u$ is the earliest possible point where the communication required by $u$ can be performed and $S_u$ is the point at which or before which we can perform the communication yet allowing enough overlapping.

We already know how to compute $S^u$. In order to compute $S_u$, suppose $u$ is referenced at point $S_1$ and $S_2$ is earlier than $S_1$. If between points $S_2$ and $S_1$ the amount of computation performed by the consumer processor is larger than some given threshold, $T$, then we can choose $S_2$ as $S_u$. The value of $T$ depends on the cost of communication in the system. It should be large enough to hide the communication latency by providing sufficient overlap of communication with computation. An outline of the optimization algorithm is given in Figure 10. Applying this algorithm on the program of Figure 9 (a), we get the program of Figure 9 (b) in which we only show where the *send* instructions are inserted. The *receive* instructions should be inserted at the points where the data are actually needed.

**Algorithm 4:** Optimization.
Input: A program $P$ with array reference information $U$ at each point;
Output: The optimized program.

Optimize(P, U)
For each statement $S$ of P
check the two array references sets before and after S: $in(S)$, $out(S)$
If there is $u \in out(S)$ but $u \notin in(S)$ then
let the point immediately after $S$ be $S^u$;
For each statement $S$
For each array reference $u$ referenced in $S$
find the point $S_u$ by backward searching;
For each array reference $u$
If $S^u$ is earlier than $S_u$ then
choose between $S^u$ and $S_u$ the best point $C_u$ at which $u$ can be combined into the largest message;
else choose $S^u$ as $C_u$;
For each array reference $u$ Do
insert a parallel loop at $C_u$ send elements of $u$.

Figure 10: Optimization Algorithm

## 5 Extensions of the Basic Data Flow Analysis Algorithm

Suppose the source function is a linear function of the form: $Source(i) = ai + c$ with $a \neq 1$. In this situation, the range of an array reference is no longer a set of consecutive integers, thus complicating the implementation of the *splitting* and *combining* operations. The implementations of these operations are even harder for arbitrary source and destination functions. In any case, we must adopt a conservative approach. Being conservative has the following implications. First, for *splitting* operation, if we do not know exactly which portion of $u$ is redefined by $v$, we should guarantee that no element of $u$ that is actually redefined by $v$ will be propagated. In this situation, existing data flow dependence tests [7] [14] [12] [5] can be used to determine whether two array references are independent or not. Based on the result, either the entire array reference is propagated or nothing is propagated. Second, for *combining* operation, if we are not sure that two array references $u$ and $v$ have the same communication pattern, we simply do not combine the array references. This impacts the performance of the data flow algorithm not its precision.

So far, we have considered only unidimensional arrays. For multi-dimensional arrays in nested loops, the situation becomes more complicated. In some situations, however, the loop can be collapsed and the array can be linearized. In these cases, the techniques of Section 4 can be applied. Obviously, this approach is not applicable to all loops. Thus, extensions are required to handle multidimensional arrays and multiply nested loops. For two dimensional arrays within doubly nested loops, the *Source* function produces a vector of two values. If the first value of $Source(i,j)$ is a shift function of $i$ and the second value of $Source(i,j)$ is a shift function of $j$, then we can perform *spliting* and *combining* operations on the two dimensions independently. Consider the following loops:

```
Do i=1, 60
  Do j=1,100
  S1: C[i,j] =..B[i,j]..
Do i=50, 100
  Do j=1, 100
  S2: A[i, j] = ...B[i, j]*C[i+1, j+2]
```

In this example, we can propagate the following two array references at point $S_2$ above the second loop:

$$u =< B, (i = 50, 100; j = 1, 100), (i; j) >, \text{ and}$$
$$v =< C, (i = 50, 100; j = 1, 100), (i + 1; j + 2) > .$$

The reference $u$ will be combined with the array reference at point $S_1$ resulting in $u' =< B, (i = 1, 100; j = 1, 100), (i; j) >$ and $v$ will be split resulting in $v' =< C, (i = 60, 100; j = 1, 100), (i; j) >$. We can do the same for $n$ dimensional arrays within $n$ nested loops.

Now let's consider a different situation involving arrays defined and referenced at different level of a nested loop.

```
S1: Do i=1, 100
S2:   Do j=1, 100
S3:     B[i,j]= Exp.
S4:     Do k=1, 100
S5:       A[j,i] = ... B[i,k]...
```

In this nested loop, we have an array reference $u$ and an array definition $v$. The problem here is that at point $S_3$ only one element of $u$ is redefined by $v$ and at point $S_1$ all elements of $u$ are redefined. For this case, the best communication scheme would be to send the whole $B$ before the point $S_1$ and then send $B[i,j]$ again after point $S_3$. Since $B[i,j]$ is sent twice, we need to receive it twice to get the right value and the first *receive* can be performed immediately after the second *send*.

```
Do i=1, 100
  Do j=1, 100
    Do k=1, 100
    S(B[i,k], P_{i,k} → P_{j,i})
Do i=1, 100
  Do j=1, 100
    B[i,j]= Exp.
    S(B[i,j], P_{i,j} → P_{j,i})
    R(B[i,j],P_{j,i} ← P_{i,j})
    Do k=1, 100
      R(B[i,k],P_{j,i} ← P_{i,k})
      A[j,i] = ... B[i,k]...
```

We need to develop some test conditions under which we can introduce limited amount of extra communication.

## 6 Related Work

Many approaches have been proposed to optimize the communication in SPMD execution. These include: *code reordering, loop interchange, loop reversal, loop elimination,* and *communication combining* [1] [6] [4] [10]. The purpose of these approaches is to extract as much parallelism as possible. Instead of inserting the communication at the earliest possible point, these approaches usually insert communication immediately before the computation. Moreover, all of these approaches assume that the *send* and *receive* instructions are inserted at the same point in the program. This does not allow effective overlapping of communication and computation. Gallivan and Jalby [3] were one of the first to formally treat the problem of optimizing data transfers in distributed-memory systems by **avoiding redundant communication** and **avoiding sequentialization**. In [11], the authors defined some array reference patterns and use pre-implemented routines to match those patterns and achieve communication optimization. However, only restricted classes of communication patterns can be optimized in this way. In [9], the authors proposed optimizations similar to the ones presented in this paper. Our unified framework, however, allows us to perform all the optimizations together and enables us to achieve the tradeoff between competing optimizations.

Data flow analysis has been successfully applied to allow code optimizations involving scalar variables. Only recently has the technique been applied to array variables. In [2], the author proposed a data flow analysis algorithm to collect information about array and scalar references and suggested that this information could be used for program verification and parallel program construction. We use data flow analysis for communication optimizations for which we need to collect the communication pattern information.

## 7 Summary

The purpose of the optimizations presented in this paper is to reduce communication and to increase parallelism. The contribution of this paper is twofold. First, we exploit the idea of separating the *send* and *receive* instruction to overlap communication with execution. Second, we present the data flow analysis technique for solving data communication optimization problems. We are currently investigating how to expand the framework to include intreprocedural analysis.

## References


[1] D. Callahan and K. Kennedy, "Compiling Programs for Distributed-Memory Multiprocessors," *The Journal of Supercomputing*, Vol. 2, 1988.

[2] P. Feautrier, "Dataflow Analysis of Array and Scalar References," *International Journal of Parallel Programming*, Vol. 20, January, 1991.

[3] K. Gallivan and W. Jalby, "On the Problem of Optimizing Data Transfers for Complex Memory System," *Proc. ACM International Conference on Supercomputing*, 1988.

[4] H. M. Gerndt and H. P. Zima, "Optimizing Communication in SUPERB," *CONPAR 90*, 1990.

[5] G. Goff, K. Kennedy, and C. Tseng, "Pratical Dependence Testing," *ACM SIGPLAN '91 Conference on Programming Language Design and Implementation*, 1991.

[6] R. Gupta, "Compiler Optimizations for Distributed Memory Programs," *Proc. of the Scalable High Performance Computing Conference*, 1992.

[7] P. Havlak and K. Kennedy, "An Implementation of Interprocedural Bounded Regular Section Analysis," *IEEE Trans. Parallel and Distributed Sys.*, Vol. 2, July, 1991.

[8] S. Hiranandani, K. Kennedy, C. Koelbel, U. Kremer, and C. W. Tseng, "An Overview of the Fortran D Programming System," *Proc. of the Fourth Workshop on Languages and Compilers for Parallel Computing*, 1991.

[9] S. Hiranandani, K. Kennedy, and C. W. Tseng, "Compiling Fortran D for MIMD Distributed-Memory Machines," *Communication of ACM*, Vol. 35, August, 1992.

[10] C. Koelbel, P. Mehrotra, and J. V. Rosendale, "Supporting Shared Data Structure on Distributed Memory Architectures," *Proceedings of the Second ACM SIGPLAN Symposium on Principles & Practice of Parallel Programming*, 1990.

[11] J. K. Li and M. Chen, "Compiling Communication-Efficient Programs for Massively Parallel Machines," *IEEE trans. on Parallel and Distributed Sys.*, Vol. 2, July, 1991.

[12] D. E. Maydan, J. L. Hennessy, and M. S. Lam, "Efficient and Exact Data Dependence Analysis," *ACM SIGPLAN '91 Conference on Programming Language Design and Implementation*, 1991.

[13] Thinking Machines Corporation, "CM Fortran Reference Manual," 1991.

[14] W. Pugh and D. Wonnacott, "Eliminating False Data Dependences Using the Omega Test," *Proceedings of the ACM SIGPLAN conference on Programming Language Design and Implementation*, 1992.

[15] M. J. Quinn and P. J. Hatcher, "Compiling SIMD Programs for MIMD Architecthres," *Proceedings of International Conference on Computer Languages*, 1990.

[16] "Fortran 90 Standard," 1991.

# META-STATE CONVERSION†

H. G. Dietz and G. Krishnamurthy
Parallel Processing Laboratory
School of Electrical Engineering
Purdue University
West Lafayette, IN 47907-1285
hankd@ecn.purdue.edu

Abstract — *In MIMD (Multiple Instruction stream, Multiple Data stream) execution, each processor has its own state. Although these states are generally considered to be independent entities, it is also possible to view the set of processor states at a particular time as single, aggregate, "Meta State." Once a program has been converted into a single finite automaton based on Meta States, only a single program counter is needed. Hence, it is possible to duplicate the MIMD execution using SIMD (Single Instruction stream, Multiple Data stream) hardware without the overhead of interpretation or even of having each processing element keep a copy of the MIMD code. In this paper, we present an algorithm for Meta-State Conversion (MSC) and explore some properties of the technique.*

## 1. Introduction

The differences between data parallelism (SIMD execution) and control parallelism (MIMD execution) are at least superficially quite large. In a data parallel program, parallelism is specified in terms of performing the same operation simultaneously on all elements of a data structure; this naturally fits the SIMD execution model. It is also easy to see that, because the abilities of a MIMD are a superset of the abilities of a SIMD, the data parallel model can be extended to MIMD targets [11] [7]. However, the control parallel model suggests that each processor can take its own path independent of all others, and this characteristic seems to require the multiple instruction streams possible only in MIMD execution. Control parallelism is impossible on a SIMD with only one instruction stream... or is it?

There are two basic approaches that might allow SIMD hardware to efficiently support a control parallel programming model: "MIMD emulation" and "meta-state conversion."

### 1.1. MIMD Emulation

Perhaps the most obvious way to make SIMD hardware mimic MIMD execution is to write a SIMD program that will interpretively execute a MIMD instruction set. In the simplest terms, such an interpreter has a data structure, replicated in each SIMD PE, that corresponds to the internal registers of each MIMD processor. Likewise, each PE's memory holds a copy of the MIMD code to be executed. Hence, the interpreter structure can be as simple as:

**Basic MIMD Interpreter Algorithm**

1. Each PE fetches an "instruction" into its "instruction register" (IR) and updates its "program counter" (PC).
2. Each PE decodes the "instruction" from its IR.
3. Repeat steps 3a-3c for each "instruction" type:

3.a Disable all PEs where the IR holds an "instruction" of a different type.

3.b Simulate execution of the "instruction" on the enabled PEs.

3.c Enable all PEs.

4. Go to step 1.

The only difficulty in implementing an interpreter with the above structure is that the simulated machine will be very inefficient.

A number of researchers have used a wide range of "tricks" to produce more efficient MIMD interpreters [9], [12], and [3]. However, some overhead cannot be removed:

1. Instructions must be fetched and decoded.
2. Instructions must be accessible to all PEs, hence, each PE typically will have a copy of the entire MIMD program's instructions. In a massively-parallel machine, this wastes a huge amount of memory.
3. There will be some overhead associated with the interpreter itself, e.g., the cost of jumping back to the start of the interpreter loop.

Although problems 1 and 3 merely slow the execution, the second severely restricts the size of MIMD programs. For example, the Purdue University School of Electrical Engineering has a 16K processing element MasPar MP-1 [1] with only 16K bytes of local memory for each PE. Even with very careful encoding, 16K bytes cannot hold a very large MIMD program.

Although meta-state conversion is more difficult to implement and more restrictive in its abilities, it can eliminate even these three overhead problems.

### 1.2. Meta-State Conversion

In MIMD execution, each processor has its own state. Although these states are generally considered to be independent entities, it is also possible to view the set of processor states at a particular time as single, aggregate, "Meta State." Using static analysis based on the timing described in [6], a compiler can convert the MIMD program into an automaton based on meta states.

Once a program has been converted into the form of a meta-state automaton, it is no longer necessary for each PE to

† This work was supported in part by the Office of Naval Research (ONR) under grant number N00014-91-J-4013 and by the National Science Foundation (NSF) under award number 9015696-CDA.

fetch and decode instructions, nor is it necessary that each PE have a copy of the program in local memory. Only the SIMD control unit needs to have a copy of the meta-state automaton; PEs merely hold data. Further, because there is no interpreter, there is no interpretation overhead. Literally, the meta-state automaton is a SIMD program that preserves the relative timing properties of MIMD execution.

However, just as interpretation has drawbacks, so too does meta-state conversion:

1. If there are $N$ processors each of which can be in any of $S$ states, then it is possible that there may be as many as $S!/(S-N)!$ states in the meta-state automaton. Without some means to ensure that the state space is kept manageable, the technique is not practical.
2. In execution, meta-state transitions are based on examining the aggregate of the MIMD state transitions for all processors.
3. Meta-state transitions are N-way branches keyed by the aggregate of the MIMD state transitions.
4. Dynamic creation of new processes is difficult to accommodate, since construction of the meta-state automaton requires that all possible MIMD states can be predicted at compile time.

Fortunately, we have developed a number of techniques that can control the state space explosion suggested above. Making meta-state transitions based on aggregate information is conceptually simple, but requires some hardware support, e.g., the "global or" of the MasPar MP-1 [1]. The efficient implementation of N-way branches is a difficult problem, but can be accomplished using customized hash functions indexing jump tables [5]. Unfortunately, the fully dynamic creation of processes seems to be impractical — but that is exactly the case in which the interpretation scheme works best. Consequently, this paper focuses on techniques to control the state explosion, and restricts the input MIMD code to be formulated as an SPMD program.

The second section of this paper presents the meta-state conversion algorithm, using an example to clarify the process. Section 3 discusses issues involving how the resulting meta-state automaton can be efficiently encoded for SIMD execution. In section 4, we discuss how the prototype implementation was constructed, and give a simple example of the output generated. Finally, section five summarizes the contributions of this work and directions for future study.

## 2. Meta-State Conversion

The meta-state conversion algorithm is surprisingly straightforward; perhaps it would be more accurate to say that it is familiar. The process of converting a set of MIMD states that exist at a particular point in time into a single meta state is strikingly similar to the process of converting an NFA into a DFA, as used in constructing lexical analyzers.

To begin, the code for the MIMD processes is converted into a set of control flow graphs in which each node (MIMD state) represents a basic block [2]. Each of these MIMD states has zero, one, or two, exit arcs. A MIMD state with no exit arcs marks the end of that process. A single exit arc represents unconditional sequencing (e.g., an unconditional branch), whereas two exit arcs respectively represent the "TRUE" and "FALSE" successors of that MIMD state (e.g., targets of a conditional branch). In addition, it is assumed that we know in which particular MIMD state each process begins execution; these states are called MIMD start states.

The set of MIMD start states forms the start state of the meta-state automaton. Since each MIMD start state may have up to two successors, each process may pick either of its two possible successors. If we further assume that there may be multiple processes in each MIMD state, it is further possible that *both* successors might be chosen. Hence, for a meta state that consists of one MIMD start state, there may be as many as three meta-state successors. In general, from $n$ MIMD start states, there could be as many as $3^n$ meta-state successors.

To clarify the operation of the algorithm, we will trace the algorithm's actions on a simple example. The framework for the example is the following SPMD code:

```
if (A) {
    do { B } while (C);
} else {
    do { D } while (E);
}
F
```

**Listing 1:** Example MIMD (SPMD) Code

It is assumed that all processors begin executing this code simultaneously and that processors computing different values for the parallel expressions $A$, $C$, and $E$ are the only sources of asynchrony (i.e., there are no external interrupts).

### 2.1. Construction of the MIMD Control-Flow Graph

Before meta-state conversion can be applied, the program must be converted into a form that facilitates the analysis. The most convenient form is that of a traditional control-flow graph in which each node represents a maximal basic block. Constructing the control-flow graph in the usual way, code straightening [2] and removal of empty nodes are applied to obtain the simplest possible graph. The result of this is figure 1. State 0 corresponds to block $A$, state 2 corresponds to $B$ followed by $C$, state 6 corresponds to $D$ followed by $E$, and state 9 corresponds to $F$.

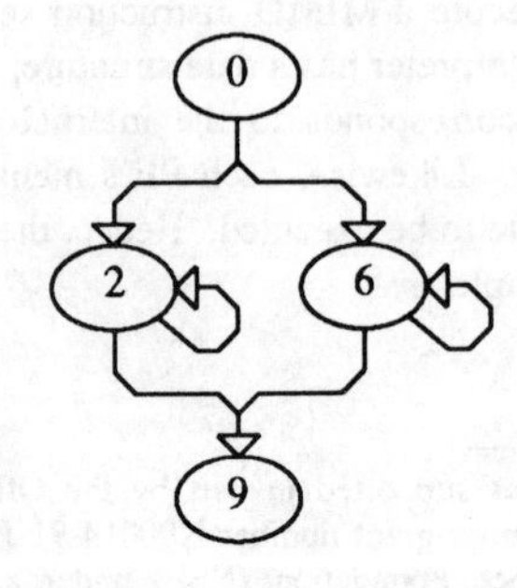


**Figure 1:** MIMD State Graph for Listing 1

### 2.2. Handling Of Function Calls

Although our example case does not contain any function calls, it is important that meta-state conversion be applicable to codes that contain arbitrary function calls — perhaps including recursive function invocations. Thus, we need some way to represent function call/return directly using control flow arcs in the MIMD state graph.

In the case of non-recursive function calls, it is sufficient to use the traditional solution of in-line expansion of the function code (i.e., of the MIMD state graph for the function body). Surprisingly, recursive function calls also can be treated using in-line expansion — and an additional "trick" that converts `return` statements into ordinary multiway branches.

Consider the following C-like code fragment in which the `main` program invokes the recursive function `g`:

```
main()
        ...
a:  g();
b:  ...
c:  g();
d:  ...
}

g()
{
        ...
    g();
e:  ...
}
```

**Listing 2:** Example Recursive Function Call

The only difficulty in in-line expanding `g` is that the target of any `return` statements in `g` is not known until runtime. However, at compile time we can compute the set of all possible `return` targets given that `g` was initially invoked from a particular position.

When in-line expanding the call to `g` from position `a`, we know that any `return` statements within `g` must return to either position `b` or `e`, and can replace the `return` statements with the appropriate multiway branch. Likewise, when in-line expanding `g` called from position `c`, `return` statements are translated into multiway branches targeting `d` or `e`. The result is a call-free control flow graph for the entire program; thus, the meta-state conversion algorithm can ignore the direct handling of function calls without loss of generality.

### 2.3. Base Conversion Algorithm

The following C-based pseudo code gives the base algorithm for meta-state conversion.

```
meta_state_convert(x)
set x;
{
 /* Given the start meta state x,
    generate the rest of the automaton
 */


 do {
  /* Mark this meta state as done */
  mark_meta_state_done(x);

  /* Add arcs to meta states y| x->y */
  reach(x, x, Ø);

  /* Get another meta state */
  x = get_unmarked_meta_state();

  /* Repeat for that meta state */
 } while (x != Ø);
}

int
reach(start, s, t)
set start, s, t;
{
 /* Make entries for all meta states
    t| start->t
 */
 if (s == Ø) {
  /* All MIMD state transitions from
     within start have been considered,
     hence, t must be a meta state
  */
  make_meta_state_transition(start, t);
 } else {
  /* Select a MIMD state and process
     its transition(s), recursing to
     complete the meta state
  */
  element e, next, fnext;

  e = [e| e ∈ s];
  s = s - {e};
  next = next_MIMD_state(e);
  fnext = next_MIMD_state_if_false(e);

  /* Take each path and both paths */
  if (next) {
   reach(start, s, t ∪ next);
   if (fnext) {
    reach(start, s, t ∪ fnext);
    reach(start, s, t ∪ next ∪ fnext);
   }
  } else {
   reach(start, s, t);
  }
 }
}
```

Applying the above algorithm to our simple example, the resulting meta-state graph is given in figure 2.

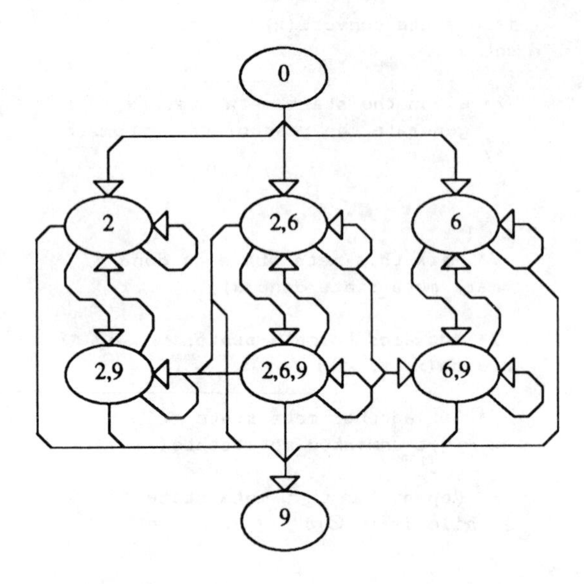


**Figure 2:** Meta-State Graph for Listing 1

### 2.4. MIMD State Time Splitting Algorithm

In the base conversion algorithm, we made the assumption that each MIMD state took exactly the same amount of time to execute. However, such an assumption is unrealistic:

- If each instruction is treated as a separate MIMD state, then reasonable size programs will generate unreasonably large automata. This makes the analysis for meta-state conversion much slower and also can result in an impractically large meta-state automaton. In addition, some computers have instruction sets in which even the execution time of different types of instruction varies widely.
- If instead we simply treat each maximal basic block as a MIMD state and ignore the differences in execution time between these blocks, this can result in very poor processor utilization. For example, if a block that takes 5 clock cycles to execute is placed in the same meta-state as one that takes 100 cycles, then the parallel machine may spend up to 95% of its processor cycles simply waiting for the transition to the next meta state.

In other words, the meta-state automaton embodies an *execution time schedule* for the code, and it is necessary that the execution time of each block be taken into account if a good schedule is to be produced.

There are many possible ways in which timing information could be incorporated, but our overriding concern must be keeping the state space manageable, and this greatly restricts the choice. Clearly, the smallest MIMD state automaton results from treating each maximal basic block as a MIMD state; hence, this will be our initial assumption. As the conversion is being performed, we may be fortunate enough to have all the MIMD states merged into each meta state happen to have the same cost. If the costs differ, but do not differ by a significant enough amount, we can ignore the difference.

This leaves only the case of a meta state that contains MIMD states of widely varying cost, for example, the 5 and 100 cycle MIMD states mentioned above. The solution we propose is a simple heuristic that will break the 100 cycle MIMD state into an approximately 5 cycle MIMD state which is unconditionally followed by the remaining portion of the original 100 cycle state. Since this change might also affect the construction of other meta states that had incorporated the original 100 cycle MIMD state, the construction of the meta-state automaton is restarted to ensure that the final meta-state automaton is consistent.

The following pseudocode gives the algorithm for performing MIMD state splitting based on the variation in timing within a meta state. It would be invoked on each meta state as it is created.

```
flag
time_split_state(s)
set s;
{
 /* Determine if time imbalance between
    MIMD states within the meta state s
    is sufficient to time split the more
    expensive MIMD states to get better
    balance; this assumes that each MIMD
    state already has an execution time
    associated with it
 */
 flag didsplit;

 /* Ignore zero time components because
    you can't do anything about them
 */
 s = s - {e| e ∈ s, time(e) == 0};

 /* Get min and max MIMD state times */
 min = min_MIMD_state_time(s);
 max = max_MIMD_state_time(s);

 /* Is enough time wasted to be worth
    splitting?  Not if the difference
    between times is already at noise
    level (split_delta) or if the
    utilization is already sure to be
    greater than an acceptable
    percentage (split_percentage)
 */
 if ((min + split_delta) > max) {
  return(FALSE);
 }
 if (min > ((split_percent*max)/100)) {
  return(FALSE);
 }

 /* Splitting seems useful...  do it */
 didsplit = FALSE;
 while (s != Ø) {
  element e;

  e = [e| e ∈ s];
  s = s - {e};
  if (time(e) > min) {
   /* If possible, split this node into
      two nodes, the first with time
```

```
        ≈ min, the second with the
        remaining time...
     */
     ...
     didsplit = TRUE;
   }
  }

  return(didsplit);
}
```

The splitting of a state is illustrated in the next two figures. The relevant portion of the initial MIMD state graph is:

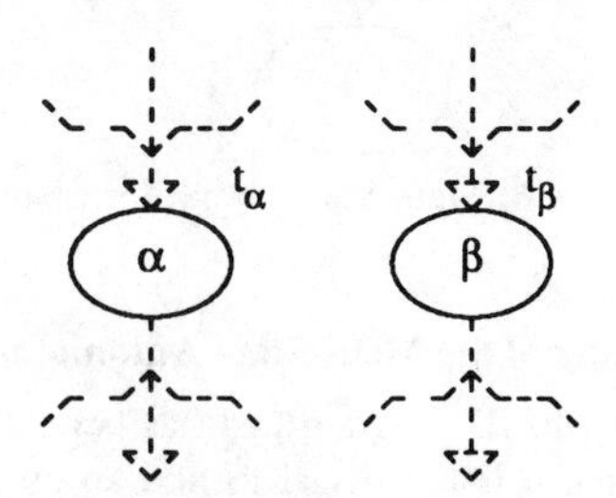


**Figure 3:** MIMD States Before Time Splitting

Suppose that meta-state conversion would combine states α and β and that β takes much longer to execute than α, i.e., $t_\alpha<t_\beta$. The state splitting algorithm would attempt to convert this portion of the state graph into:

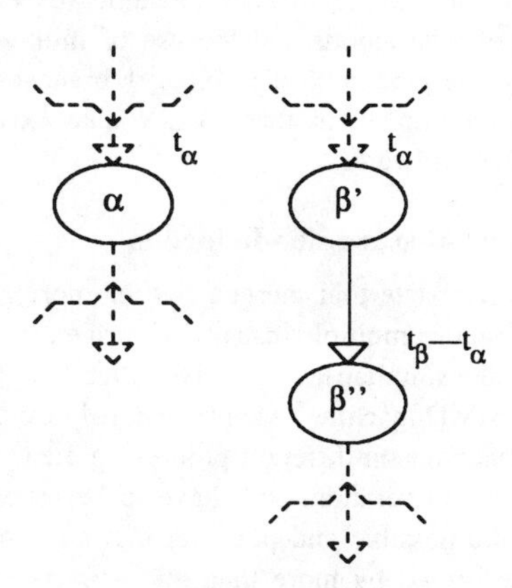


**Figure 4:** MIMD States After Time Splitting

Thus, states α and β' would be merged — without any idle time being introduced for either thread of execution.

### 2.5. Meta State Compression Algorithm

Despite the reduction in state space possible using maximal basic blocks and time splitting, the automata created can be very large. Hence, it is useful to find a way to reduce the upper bound on the number of meta states created.

Because MIMD nodes with zero or one exit arc can only increase the state space linearly, the explosion in meta state space is related to the occurrence of MIMD states that have two exit arcs. Each such MIMD state could contribute three meta states: the `TRUE` successor, `FALSE` successor, and both successors. However, if there are many processes in any given MIMD state, it is easy to see that the most probable case is that of both successors. Further, the case of both successors can always emulate either successor, since it has the code for both. Thus, a very dramatic reduction in meta state space can be obtained by simply assuming that both successors are always taken.

```
int
reach(start, s, t)
set start, s, t;
{
 /* Make entries for all meta states
    t| start->t
 */

 if (s == Ø) {
  /* All MIMD state transitions from
     within start have been considered,
     hence, t must be a meta state
  */
  make_meta_state_transition(start, t);
 } else {
  /* Select a MIMD state and process
     its transition(s), recursing to
     complete the meta state
  */
  element e, next, fnext;

  e = [e| e ∈ s];
  s = s - {e};
  next = next_MIMD_state(e);
  fnext = next_MIMD_state_if_false(e);

  /* Always take all possible paths... */
  if (next) {
   if (fnext) {
    reach(start, s, t ∪ next ∪ fnext);
   } else {
    reach(start, s, t ∪ next);
   }
  } else {
   reach(start, s, t);
  }
 }
}
```

Returning to our example code, the meta-state compression algorithm results in a graph with only two meta-states, compared to eight for the uncompressed graph:

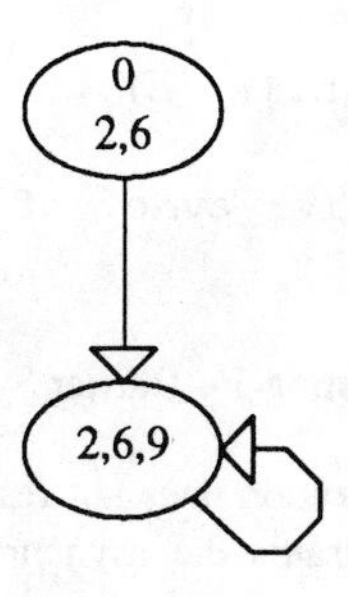


**Figure 5:** Compressed Meta-State Graph for Listing 1

Notice that meta-state transitions into compressed portions of the graph are unconditional; i.e., there is no need to use a `globalor` to determine what states are present. The disadvantage is that the average meta-state is wider, which implies that the SIMD implementation will be less efficient.

### 2.6. Barrier Synchronization Algorithm

While the above compression scheme produces very small automata, it does increase overhead somewhat in that each meta state becomes much more complex. Hence, it is useful to seek yet another method to reduce the state space — without adding to the complexity of each meta state. Careful use of barrier synchronization provides such a mechanism.

```
set
barrier_sync(s)
set s;
{
  /* If s is a meta state that contains a
     MIMD state which is a barrier
     synchronization point, then the
     barrier should prevent any
     transitions past that MIMD state.
     Hence, unless all processors have
     reached the barrier (i.e., every MIMD
     state within s is a barrier state),
     simply remove barrier states from s
  */
  set waits;

  /* Construct the set of MIMD barrier
     wait states within s
  */
  waits = {e| e ∈ s, is_barrier_wait(e) == TRUE};

  /* Has everyone reached the barrier? */
  if (waits == s) {
    /* Yes; go into all barrier state */
    return(waits);
  } else {
    /* No; remove barriers from s */
    return(s - waits);
  }
}
```

For example, consider modifying the code framework of listing 1 to contain a barrier sync at the end of the `if`:

```
if (A) {
    do { B } while (C);
} else {
    do { D } while (E);
}
wait;   /* barrier sync. of all threads */
F
```

**Listing 3:** Listing 1 + Barrier Synchronization

The barrier synchronization does not result in a runtime operation, but rather constrains the asynchrony as defined by the above algorithm. The result is a meta-state graph of the form:

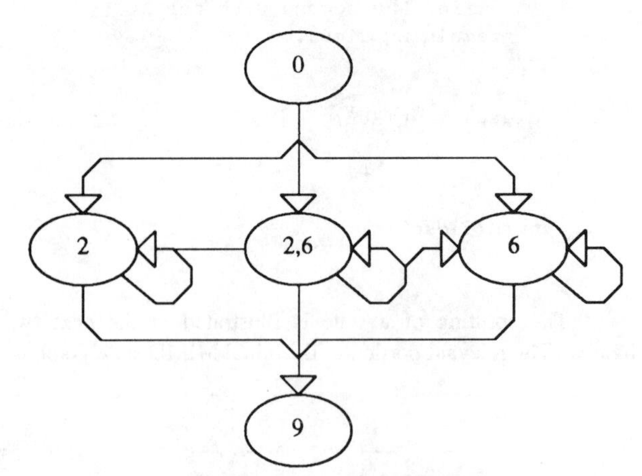


**Figure 6:** Meta-State Graph for Listing 3

## 3. SIMD Coding of the Meta-State Automaton

Given a MIMD program that has been converted into a meta-state graph, it is not trivial to find an efficient coding of the meta-state automaton for a SIMD architecture. The meta-state graph does reduce control flow to a single instruction stream, but that instruction stream would appear to execute different types of instructions in parallel — the meta-state graph employs a variation on VLIW semantics.

There are two aspects of the graph that mirror VLIW constructions[1]: the apparently simultaneous execution of different types of instructions and the use of multiway branches generated by merging multiple (binary) branches. Thus, we must efficiently implement these VLIW-like execution structures on SIMD hardware.

### 3.1. Common Subexpression Induction

Any meta state that merged two or more MIMD states effectively contains multiple instruction sequences that are supposed to execute simultaneously. Given that it is impossible for a traditional SIMD machine to simultaneously execute different types of instructions on different processing elements, it would appear that these operations will have to be serialized. However, it is quite possible and practical that any operations that would be performed by more than one sequence can be executed in parallel by all processors. Common subexpression induction (CSI) [4] is an optimization technique that identifies these operations and "factors" them out.

The CSI algorithm analyzes a segment of code containing operations executed by any of multiple threads (enabled sets of SIMD PEs). From this analysis, it determines where threads can share the same code and what cost is associated with induc-

[1] The meta-state graph is not suitable for execution on a traditional VLIW because which processing elements execute which instructions is determined statically for VLIW, but dynamically in the graph. I.e., the graph would be appropriate for a VLIW in which each processing element could select at runtime which instruction field it would execute, rather than having each processing element statically associated with a particular instruction field.

ing that sharing. Finally, it generates a code schedule that uses this sharing, where appropriate, to achieve the minimum execution time. Unfortunately, this implies that the CSI algorithm is not simple.

The algorithm can be summarized as follows. First, a guarded DAG is constructed for the input, then this DAG is improved using inter-thread CSE. The improved DAG is then used to compute information for pruning the search: earliest and latest, operation classes, and theoretical lower bound on execution time. Next, this information is used to create a linear schedule (SIMD execution sequence), which is improved using a cheap approximate search and then used as the initial schedule for the permutation-in-range search that is the core of the CSI optimization.

### 3.2. Multiway Branch Encoding

At the end of each meta-state's execution, a particular type of multiway branch must be executed to move the SIMD machine into the correct next meta state. Before discussing the encoding of these multiway branches, it is useful to specify the precise semantics of meta-state transitions, so that an optimal coding can be achieved. The following defines the possible types of meta-state transitions.

#### 3.2.1. No Exit Arc

A meta state without an exit arc is a terminal node, i.e., it represents the end of the program's execution. Thus, it is implicitly followed by a return to the operating system. There is no difficulty in generating code to implement this.

#### 3.2.2. Single Exit Arc

If there is a single exit arc from a meta state, the code for that meta state is is followed by a `goto` (aka, `jump`) to the code for the target meta state. Again, it is simple to generate an efficient coding.

Notice that all entries to compressed meta states fall into this category.

#### 3.2.3. Multiple Exit Arcs

If there are multiple exit arcs from a meta state, then the aggregate of the "pc" values for each of the processing elements must be used to determine the next state. For example, when, at the end of executing a meta state, some processing elements have "pc" value 2 and others have "pc" value 6, meta state {2,6} is the next state. In order to efficiently collect this aggregate, each possible "pc" value is assigned a bit; thus, a `globalor` of the "pc" values from all processors determines the aggregate.

#### 3.2.4. Multiple Exit Arcs Involving Barriers

The treatment of multiple exit arcs must be slightly adjusted if some, but not all, of the processing elements have reached a barrier at the time a meta state's execution completes. For example, in figure 6 the transitions from meta states 2, {2,6}, and 6 into 2, {2,6}, and 6 would not be sufficient if even one processing element had reached the barrier (i.e., meta state 9). Consequently, the processing elements are allowed to set their "pc" value to 9, but they are not permitted to enter meta state 9 unless all "pc"'s are 9.

This is accomplished by a simple check to see if (`globalor` pc) is contained within the set of all barrier states. If it is, then the state transition proceeds normally. Otherwise, the next meta state is determined by subtracting the set of all barrier states from the result of the `globalor`.

#### 3.2.5. Restricted Dynamic Process Creation

Although the completely static nature of meta-state conversion makes it impossible to efficiently support forking of new processes to execute different programs, a minor encoding trick can be used to implement a restricted form of dynamic process creation. This restricted type of `spawn` instruction looks just like a conditional jump, except the semantics are that both paths must be taken (i.e., the compressed meta state transition rule). One exit is taken by the original processes, the other by the newly created processes.

Initially, processing elements that are not in use would be given a "pc" value indicating that they are not in any meta state. When a `spawn(x)` instruction is reached by N processing elements, the original N processing elements do not change their pc values, but N currently-disabled processing elements are selected and their pc values are set to $x$. No other changes are needed, provided that the number of processes requested does not exceed the number of processors available.

Note further that processors that complete their processes early can be returned to the pool of free processors by simply executing a `halt` instruction to set their pc value to indicate that they are not in any meta state.

### 3.3. Allocation of Bits for "pc" Values

Although it is easy to implement each "pc" value by assigning a different bit to each MIMD state, this would result in impractically long bit strings for large meta state automata. Thus, although conceptually a different bit is used to represent each MIMD state, bits actually can be reused without changing the basic conversion algorithm. This bit allocation problem is similar to that of allocating registers to values where the number of values can be larger than the number of registers. However, unlike register allocation, there is no concept of "spilling" a bit position; if an allocation is not found using *maxbit* bits, we must increase the number of bits. Fortunately, it is unlikely that *maxbit* will need to be large; in our preliminary experiments, it never was necessary to use more than 8 bits.

Intuitively, there are just two rules that govern the reuse of a bit to represent multiple MIMD states:

1. No two MIMD states contained within the same meta state can be allocated the same bit. If this were violated, it would be impossible to tell which code within that meta state should be executed by each processing element.
2. No two meta states which are successors of the same meta state (i.e., which are sibling meta states) can be allocated the same bit pattern. If two siblings had the same bit pattern, the meta state automaton would not be able to decide which of these sibling meta states to execute.

The algorithm which we have implemented is given below. It applies rule 1 directly, but uses a safe approximation to rule 2. The approximation is simply that no two distinct MIMD states that appear in sibling meta states are allocated the same bit.

```
allocate_bits(maxbit)
int maxbit;
{
 /* Allocate bits to "pc" values, using at
    most maxbit bits.  Returns with ERROR
    if need more than maxbit bits.
 */
 set M, C, S;
 int pos, pat, n, c_n, bit[];

 M = {m| m ∈ meta states};

 /* Assign start state bit 0 */
 m0 = (start meta state ∈ M);
 bit[m0] = 2^0;

 for (m| m ∈ (M - m0)) {
  /* Set of all next meta states of m */
  C = {c| c ∈ M, m→c };
  S = {s| s ∈ C, bits of s have been allocated };
  C = C - S;

  /* OR patterns of all members of S */
  pat = OR(p| p is a pattern of some s ∈ S);
  n = # of distinct MIMD states in S;
  if (# of 1 bits in net_pattern ≠ n) {
   /* Some MIMD states have same bit pattern */
   return(ERROR);
  }

  if (C != Ø)) {
   /* There are some MIMD states whose
      bit patterns need to be allocated
   */
   for (c| c ∈ C) {
    if (c has more than 1 MIMD state) {
     bit[c] = OR(p| p is pattern of MIMD state∈c);
     c_n = # of MIMD states in c;
     if (# of 1 bits in bit[c] ≠ c_n) {
      return(ERROR);
     }
     pat = OR(pat, bit[c]);
    } else {
     /* If only 1 MIMD state, assign pattern */
     if (# of 0 bits in pat == 0) {
      /* No free bit to allocate */
      return(ERROR);
     }
     pos = position of first 0 bit in pat;
     bit[c] = 2^pos;
    }
   }
  }
 }

 return(NO_ERROR);
}
```

## 4. Implementation

The current prototype meta-state converter does not directly generate executable SIMD code from a MIMD-oriented language. Instead, it simply outputs a set of meta-state definitions. Each of these meta states must then be common subexpression inducted and the meta-state transitions (multiway branches) must be encoded using hash functions. However, these last two steps are implemented by two software tools developed earlier:

- A common subexpression inductor, described in [4].
- A hash function generator, described in [5].

Thus, in this paper we will confine the discussion to the implementation of the prototype meta-state converter. The meta-state converter was written in C using PCCTS [10] and is actually a modified version of the `mimdc` compiler described in [3].

### 4.1. The Input Language

The language accepted by the meta-state converter is a parallel dialect of C called MIMDC. It supports most of the basic C constructs. Data values can be either `int` or `float`, and variables can be declared as `mono` (shared) or `poly` (private) [11].

There are two kinds of shared memory reference supported. The `mono` variables are replicated in each processor's local memory so that loads execute quickly, but stores involve a broadcast to update all copies. It is also possible to directly access `poly` values from other processors using "parallel subscripting":

```
x[||i] = y[||j] + z;
```

would use the values of `i`, `j`, and `z` on this processor to fetch the value of `y` from processor `j`, add `z`, and store the result into the `x` on processor `i`. In addition to allowing use of shared memory for synchronization, MIMDC supports barrier synchronization [6] using a `wait` statement.

### 4.2. The Conversion Process

A brief outline of the prototype implementation is:

1. As the PCCTS-generated parser reads the source code, a traditional control-flow graph whose nodes are expression trees is built. This control-flow graph is constructed in a "normalized" form that ensures, for example, that loops are all of the type that execute the body one or more times, rather than zero or more (e.g., by replicating some code and inserting an additional `if` statement).
2. The control-flow graph is straightened and empty nodes are removed. This maximizes the size of the nodes.
3. The meta-state conversion algorithm is applied. Except for the handling of function calls, the prototype implements the full algorithm.
4. The resulting meta-state graph is straightened and output.

The current prototype implementation does not perform the final encoding of the meta-state automaton. Hence, a CSI tool [4] and a tool for finding hash functions [5] are applied by hand to produce the final SIMD code in MPL.

### 4.3. An Example

To illustrate how the prototype meta-state converter works, consider the MIMDC program presented in listing 4. This example has the same control structure given in listing 1, but is a complete program, so that the actual code generated can be given.

```
main()
{
    poly int x;

    if (x) {
        do { x = 1; } while (x);
    } else {
        do { x = 2; } while (x);
    }

    return(x);
}
```

**Listing 4:** Example MIMDC Program

Without compression or time cracking, the resulting meta-state SIMD automaton, written in MPL [8] for the MasPar MP-1 [1], is given in listing 5 (note that the algorithm in section 3.3 was not applied). The code within each meta state is simple SIMD stack code using MPL macros for each operation. The only surprising stack operation is `JumpF(x, y)`, which simply sets each processing element's `pc` equal to $2^x$ if the top-of-stack value is "FALSE" or to $2^y$ if it is "TRUE." The `apc` is simply the aggregate obtained by oring the values of all the individual `pc`s; the switch at the end of each meta state simply employs a customized hash function to ensure that the multiway branch is implemented efficiently. For example, at the end of meta state 0 (i.e., `ms_0`), instead of a `switch` on `apc` with cases for `BIT(2)|BIT(6)`, `BIT(6)`, and `BIT(2)`, a hash function is applied to make the case values contiguous so that the MPL compiler will use a jump table to implement the `switch`.

### 5. Conclusions

Although meta-state conversion is a complex and slow process, it does provide a mechanical way to transform control-parallel (MIMD) programs into pure SIMD code. Further, the execution of the meta-state program can be very efficient. In particular, fine-grain MIMD code is generally inefficient on most MIMD machines due to the cost of runtime synchronization, but synchronization is implicit in the meta-state converted SIMD code, and hence has no runtime cost.

While the prototype implementation demonstrates the feasibility and correctness of the meta-state conversion algorithm, it does not yet automate the process of generating the final SIMD code. Future work will integrate the code generation process and will benchmark performance on "real" programs.

### References


[1] T. Blank, "The MasPar MP-1 Architecture," 35th IEEE Computer Society International Conference (COMPCON), February 1990, pp. 20-24.

[2] J. Cocke and J.T. Schwartz, *Programming Languages and Their Compilers,* Courant Institute of Mathematical Sciences, New York University, April 1970.

[3] H.G. Dietz and W.E. Cohen, "A Control-Parallel Programming Model Implemented On SIMD Hardware," in Proceedings of the *Fifth Workshop on Programming Languages and Compilers for Parallel Computing*, August 1992.

[4] H.G. Dietz, "Common Subexpression Induction," Proceedings of the *1992 International Conference on Parallel Processing,* Saint Charles, Illinois, August 1992, vol. II, pp. 174-182.

[5] H.G. Dietz, "Coding Multiway Branches Using Customized Hash Functions," Technical Report TR-EE 92-31, School of Electrical Engineering, Purdue University, July 1992.

[6] H.G. Dietz, M.T. O'Keefe, and A. Zaafrani, "An Introduction to Static Scheduling for MIMD Architectures," Advances in Languages and Compilers for Parallel Processing, edited by A. Nicolau, D. Gelernter, T. Gross, and D. Padua, The MIT Press, Cambridge, Massachusetts, 1991, pp. 425-444.

[7] M. S. Littman and C. D. Metcalf, *An Exploration of Asynchronous Data-Parallelism,* Technical Report, Yale University, July 1990.

[8] MasPar Computer Corporation, *MasPar Programming Language (ANSI C compatible MPL) Reference Manual, Software Version 2.2*, Document Number 9302-0001, Sunnyvale, California, November 1991.

[9] M. Nilsson and H. Tanaka, "MIMD Execution by SIMD Computers," Journal of Information Processing, Information Processing Society of Japan, vol. 13, no. 1, 1990, pp. 58-61.

[10] T.J. Parr, H.G. Dietz, and W.E. Cohen, "PCCTS Reference Manual (version 1.00)," *ACM SIGPLAN Notices,* Feb. 1992, pp. 88-165.

[11] M.J. Phillip, "Unification of Synchronous and Asynchronous Models for Parallel Programming Languages" Master's Thesis, School of Electrical Engineering, Purdue University, West Lafayette, Indiana, June 1989.

[12] P.A. Wilsey, D.A. Hensgen, C.E. Slusher, N.B. Abu-Ghazaleh, and D.Y. Hollinden, "Exploiting SIMD Computers for Mutant Program Execution," Technical Report No. TR 133-11-91, Department of Electrical and Computer Engineering, University of Cincinnati, Cincinnati, Ohio, November 1991.

```
ms_0:
  if (pc & BIT(0)) {
    Push(0) LdL JumpF(6,2)
  }
  apc = globalor(pc);
  switch (((-apc) >> 5) & 3) {
  case 1: goto ms_2_6;
  case 2: goto ms_6;
  case 3: goto ms_2;
  }

ms_2:
  if (pc & BIT(2)) {
    Push(1) Push(0) LdL Push(12) StL
      Pop(2) Push(4) LdL JumpF(9,2)
  }
  apc = globalor(pc);
  switch (((-apc) >> 8) & 3) {
  case 1: goto ms_2_9;
  case 2: goto ms_9;
  case 3: goto ms_2;
  }

ms_9:
  if (pc & BIT(9)) {
    Push(4) LdL Ret(3)
  }
  /* no next meta state */
  exit(0);

ms_2_9:
  if (pc & BIT(2)) {
    Push(1) Push(0) LdL
    Push(12) StL Pop(2)
  }
  if (pc & (BIT(2) | BIT(9))) {
    Push(4) LdL
  }
  if (pc & BIT(2)) JumpF(9,2)
  if (pc & BIT(9)) Ret(3)
  apc = globalor(pc);
  switch (((-apc) >> 8) & 3) {
  case 1: goto ms_2_9;
  case 2: goto ms_9;
  case 3: goto ms_2;
  }

ms_6:
  if (pc & BIT(6)) {
    Push(2) Push(0) LdL Push(12) StL
    Pop(2) Push(4) LdL JumpF(9,6)
  }
  apc = globalor(pc);
  switch (((-apc) >> 8) & 3) {
  case 1: goto ms_6_9;
  case 2: goto ms_9;
  case 3: goto ms_6;
  }

ms_6_9:
  if (pc & BIT(6)) {
    Push(2) Push(0) LdL
    Push(12) StL Pop(2)
  }
  if (pc & (BIT(6) | BIT(9))) {
    Push(4) LdL
  }
  if (pc & BIT(6)) JumpF(9,6)
  if (pc & BIT(9)) Ret(3)
  apc = globalor(pc);
  switch (((-apc) >> 8) & 3) {
  case 1: goto ms_6_9;
  case 2: goto ms_9;
  case 3: goto ms_6;
  }

ms_2_6:
  if (pc & BIT(2)) Push(1)
  if (pc & BIT(6)) Push(2)
  if (pc & (BIT(2) | BIT(6))) {
    Push(0) LdL Push(12) StL
    Pop(2) Push(4) LdL
  }
  if (pc & BIT(2)) JumpF(9,2)
  if (pc & BIT(6)) JumpF(9,6)
  apc = globalor(pc);
  switch (((apc >> 6) ^ apc) & 15) {
  case 5: goto ms_2_6;
  case 8: goto ms_9;
  case 9: goto ms_6_9;
  case 12: goto ms_2_9;
  case 13: goto ms_2_6_9;
  }

ms_2_6_9:
  if (pc & BIT(2)) Push(1)
  if (pc & BIT(6)) Push(2)
  if (pc & (BIT(2) | BIT(6))) {
    Push(0) LdL Push(12) StL Pop(2)
  }
  if (pc & (BIT(2) | BIT(6) | BIT(9))) {
    Push(4) LdL
  }
  if (pc & BIT(2)) JumpF(9,2)
  if (pc & BIT(6)) JumpF(9,6)
  if (pc & BIT(9)) Ret(3)
  apc = globalor(pc);
  switch (((apc >> 6) ^ apc) & 15) {
  case 5: goto ms_2_6;
  case 8: goto ms_9;
  case 9: goto ms_6_9;
  case 12: goto ms_2_9;
  case 13: goto ms_2_6_9;
  }
```

**Listing 5: Meta-State Converted Example**

# SESSION 3B

## TOOLS

# A UNIFIED MODEL FOR CONCURRENT DEBUGGING †

S. I. Hyder, J. F. Werth and J. C. Browne‡
The University of Texas at Austin
Austin, Texas 78712

**Abstract:** *Events are occurrence instances of actions. The thesis of this paper is that the use of "actions", instead of events, greatly simplifies the problem of concurrent debugging. Occurrence instances of actions provide a debugger with a unique identifier for each event. These identifiers help the debugger in recording the event orderings. The recorded orderings indicate much more than a mere temporal order. They indicate the dependences that "cause" the actions to execute. A debugger can, then, collect the dependence information from the orderings of different instances of the same action, and deduce the conditions that govern the execution of the action. This provides a framework for representing and checking the expected behavior. Unlike existing approaches, we cover all parts of the debugging cycle. Our unified model, therefore, allows a single debugger to support different debugging facilities like execution replay, race detection, assertion/model checking, execution history displays, and animation.*

**Keywords**: Pomset, concurrent debugging, execution replay, race detection, animation, model checkers.

## 1.0 INTRODUCTION

Parallel programs typically express concurrency by adding synchronization constructs to the usual sequential text. This produces a complex entanglement of the concurrent considerations with the sequential considerations. A programmer, then, has to debug the synchronizations and communications in the presence of the already complex problem of debugging the flow of data and the flow of control of the sequential text. This makes the debugging of concurrent programs an extremely complex problem.

During debugging, a programmer forms some expectations about the execution behavior of the program. This expected behavior is, then, represented by an assertion/model. Checking of the actual execution against this assertion/model, reveals any unexpected behavior. Mapping of this behavior to the program, brings the programmer closer to the bug. A sequence of these interactions between the specified behavior of the program P, model of the expected behavior M, and the observed execution behavior E constitutes a debugging cycle. This cycle will be illustrated as P→ M→E → P.

† This work is supported in part by the research initiation grant program of IBM, Corp., to support interdesciplinary computer science research.

‡ E-mail: {hyder | jwerth | browne}@cs.utexas.edu.

Concurrent debugging becomes so complex because of the ambiguities that obscure the interactions between P, M and E in the debugging cycle. These ambiguities arise from the entanglement of the concurrent and sequential considerations in the representations used for P, M and E. As these ambiguities obscure the interactions, each part of the debugging cycle becomes a separate problem area. Existing debugging facilities that target separate parts of the debugging cycle, then, appear incompatible and even orthogonal. Incompatibility of these facilities forces the programmer to either use different facilities for different parts of the cycle or debug without them. On the other hand, it compels the debuggers to either constrain the range of behaviors that can be checked [9], [7]; or to tolerate the ambiguities in the observed behavior [8], [16]; or to demand extra programming effort [4], [12], [6], [1].

**Def. 1:** A *computation action* is a piece of program text that starts and/or ends with a synchronization statement.

Although concurrent debuggers define execution events to be the occurrence instances of program actions, they often leave the actions implicit. The thesis of this paper is that the complexity of concurrent debugging greatly simplifies when the set of actions, instead of events, is used to define the various behaviors used in debugging. Our unified model of concurrent debugging debugs the concurrent behavior using the relations on the set of computation actions; specified in the program, modeled in the expected behavior and observed in the execution behavior.

We use the abstraction of computation actions and the "causality" of their dependence relations to disentangle the concurrent considerations from the sequential. This decomposes the concurrent debugging problem into two almost disjoint problems that can be debugged at different levels. A programmer debugs the concurrent state (§ 3.3) at the upper level, where the only important concerns are the relations on the set of computation actions. Internal states of a computation action are not important at this level. They only become important when the programmer moves to the lower level, inside the action.

The set of computation actions serves as a basis for instrumenting the program. The debugger identifies the events as occurrence instances of actions. As an action can occur "multiple" number of times, the identity of the action and its instance number act together as a unique identifier (logical clock) for each event. Using these identifiers, our debugger records a partial order that we call the causal orderings (§ 3.1). In this ordering, the immediate predecessors and successors of a given event map to the dependences of the action corresponding to the event (§ 3.2).

A programmer's expectations consist of a set of events and an orderings on these events. An execution is erroneous if the expected events do not occur, or occur in some un-expected order. The programmer, therefore, describes the expected behavior with actions and the dependences that order the occurrences of those actions as events (§ 4.1). If the observed ordering relations do not match the conditions represented in the expected behavior, then a debugger has detected some unexpected behavior and raises the exception (§ 4.2).

Our approach differs from the existing ones in that it uses actions, instead of events. It is program oriented, instead of execution oriented. It records the causal orderings, instead of approximating them. It unifies all the parts of the debugging cycle. This overcomes the incompatibility of existing facilities, and allows one debugger to support different debugging facilities like execution replay, race detection, assertion/model checking, execution history displays, and animation.

### 1.1 Problems in Various Parts of the Cycle

Complexity of concurrent debugging has received much attention in recent years. A 1989 bibliography cited over 370 references [21]. Existing approaches complicate the problem by only covering a subpart of the debugging cycle; P → M → E → P. This makes them incompatible, and compels the user to debug the remaining parts of the cycle with separate debugging facilities. This often involves extra programming effort.

Execution environments typically provide processes or threads as executable units to run the program text. There is often a complex multiplexing of the program text among process/thread structure due to resource limitations, scheduler policies, and other constraints. Debugging support must resolve the timing ambiguities arising from such multiplexings. An *execution history display* [12], [4], [20], [14] helps in resolving some of these ambiguities. It, however, only provides a time-process graph representation of E. An *animation facility* [20], [6] is, then, needed to provide an instantaneous view of the E → P mapping. A textual representation of P is inadequate for supporting this mapping. Hence, extra user effort is needed to develop a graphical structure that can support animation.

Moreover, another debugger is needed to resolve the ambiguities arising from the inability to record the causal orderings in E → P part of the cycle [8], [18].

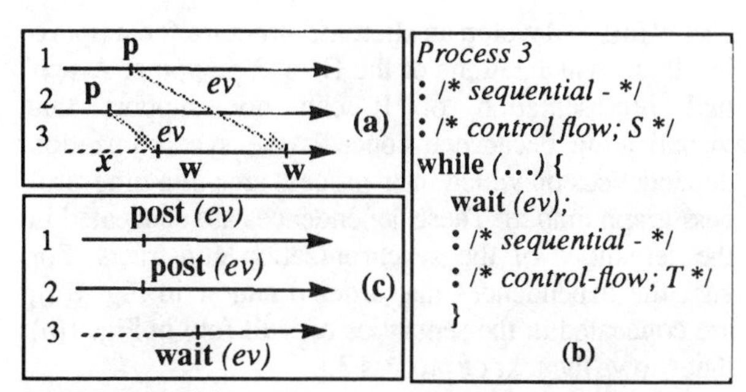


**FIGURE 1. (a) A time-process graph (b) Program text segment for process 3 (c) Event traces.**

Such a debugger only helps in modeling the race behaviors [18], and in automating *race detection*. An *assertion/model checker* [1] is, then, needed to help in modeling the expected behavior, and in automating its checking. However, a textual representation of P makes it difficult for the user to represent the expected behavior in P → M part of the cycle. The text does not allow the user to represent conditions about the concurrent state that involve event orderings. Hence, execution and *problem* oriented approaches [7] are used. They, however, demand extra user effort [1]. Furthermore, their use of events, instead of actions, creates additional problems in recognizing the expected behavior in M → E part of the cycle.

Finally, an execution replay facility is needed to make *cyclical* debugging possible [10].

**Mapping Ambiguities:** In E → P part of the cycle, a debugger has to map the events defined in the context of processes/threads of the execution environment on to the program text. However, ambiguities arise in mapping intra-process arcs of a time-process graph to their corresponding sequential text in the program. For instance, in Fig. 1(a) there is an ambiguity about the intra-process arc ***x*** of Process 3. ***x*** can either map to the sequential text S or to the sequential text T of Fig. 1(b). Event ***w*** that immediately follows ***x***, and maps to the synchronization statement **wait** *(ev)* is not of much help. **wait** *(ev)* is neither associated with S nor with T. The synchronization event simply sits at the boundary where a piece of text ends and another one starts.

Instead of letting a synchronization statement sit ambiguously on the border of two sequential text segments, we propose an abstraction that permanently associates the synchronization statements of a program with its sequential text segments. The abstractions resulting from this association is that of a *computation action* (Appendix). It disentangles the sequential control-flow considerations from the synchronization considerations.

*Animation facilities* [14], [20] often demand extra

user effort to develop an alternate structure for supporting their visualizations of the E → P mapping. A textual representation of P can not support this visualization because it conceals the synchronization dependences on which inter-process arcs of a time-process graph map to. These dependences are concealed in the semantics of the synchronization constructs. For e.g., the dependences that order **p** and **w** in Fig. 1(a), are concealed in the semantics of **wait** *(ev)* in Fig. 1(b) that shows the text of process 3.

We, therefore, use a graphical representation of P [15] whose nodes are the computation actions and whose arcs are their dependences (§ 2.0). Occurrence instances of computation actions are partially ordered. They provide a "pomset" representation of E that allows us to automatically generate the animation structure (§ 3.0).

**Ordering Ambiguities**: In P → E part of the cycle, a debugger should record the events and their orderings. Distributed systems often record the event orderings by exploiting the data dependences introduced by the send/receive of messages with the help of unique time-stamps (or identifiers) [13]. Shared memory debuggers that detect races [16], [8], however, ignore the data dependences introduced by the accesses to the shared synchronization variables. They, also, ignore the importance of unique identifiers or time-stamps for each event. Their recorded event traces, therefore, contain ambiguities about the inter-process orderings as shown in Fig. 1(c). There is ambiguity as to whether the **wait** *(ev)* of process 3 was fulfilled by the **post** *(ev)* of process 1 or 2. This necessitates the use of approximations [8], and leads to intractability [17]. Inability to record the order of accesses to shared objects further complicates the detection of *races* (simultaneous access to shared objects with at least one write), and affects the accuracy of detected races [16], [18].

Note, however, that if an event that write-accesses a shared object, appends its unique identifier to the object, then a later event access to that object can identify its "causal" predecessor (§ 3.1). This observation allows us to support execution replay, and to support race detection without any extra overhead (§ 5.0).

**Modeling Problems:** A *dbx* debugger is user-friendly. It allows the user to associate expected conditions with a program action. For instance, in a *dbx* command like "**when at** *stmt* **if** *condition*", the user associates a condition about the sequential state with a statement of interest. Later, the debugger allows the user to interactively follow the conditional progress of the execution that has been restricted to the interesting actions. Such a *program oriented* approach [7] is not possible with a "textual" representation of the concurrent program. Unlike *dbx* conditions, conditions about the concurrent state involve event orderings whose corresponding orderings are not visible in the textual representation. For instance, event orderings in an expected behavior like "$(w \wedge p)$ precede $w$" for Fig. 1(a), correspond to the dependences that are not visible in the textual representation of Fig. 1(b). Hence, assertion/model checkers [7], [14] adopt execution oriented approaches that use models like temporal logic, interleaving, partial order or automatas [9]. In P → M part of the cycle, therefore, a user has to exert extra effort to learn a new language, program the expected behavior and, then, debug it for *user errors* [1].

**Filtering Ambiguities:** In M → E part of the cycle, *assignment* and *resolution* problems [1] are typical of the ambiguities that arise during filtering and recognition of the expected behavior. As existing checkers are execution oriented, they use events in their representation of the expected behavior, and leave the actions implicit. This conceals the information that (i) events are actually multiple occurrences of actions, and (ii) the observed event orderings are the unrolling of the communication/synchronization structure of the program actions. For instance, a behavior like "**p** precedes w" gives no information to the debugger about the actions that correspond to events **p** and **w**. Ambiguities can, then, arise whenever more than one observed behavior fits the expected behavior. In Fig. 1(a), "**p** precedes w" can fit several behaviors; **p** of process 2 precedes the first **w** of process 1, **p** of process 1 precedes the second **w** of process 2, or **p** of process 2 precedes the second **w** of process 3. Such ambiguities restrict the range of checkable behaviors. This, in turn, restricts the range of behaviors that can be represented in M [9], [1].

The information about the actions like their statement line number or process id, could have resolved these ambiguities. We, therefore, use such information about the actions to simplify the filtering and recognition of the expected behavior (§ 4.2).

## 1.2 Example: Events as Occurrences of Actions

Fig. 2(b) graphically shows the actions $w$, $p$, and $q$, and their dependences that we denote by the ordered pairs $(w, w)$, $(p, w)$, and $(q, w)$. This is formalized later. $i$-th occurrence instances of the actions give us event identifiers $w_i$, $p_i$, and $q_i$. Our debugger records the event orderings by appending the identifiers to the shared state of the dependences (§ 3.1). The observed orderings, therefore, indicate those dependences of actions that "caused" the events to be ordered. So, event orderings of Fig. 2(a) indicate that they were caused by the dependences of Fig. 2(b): Ordering $p_1 < w_3$ was caused by the dependence $(p, w)$, ordering $q_1 < w_2$ was caused by the dependence $(q, w)$, and $w_1 < w_2$ and $w_2 < w_3$ were

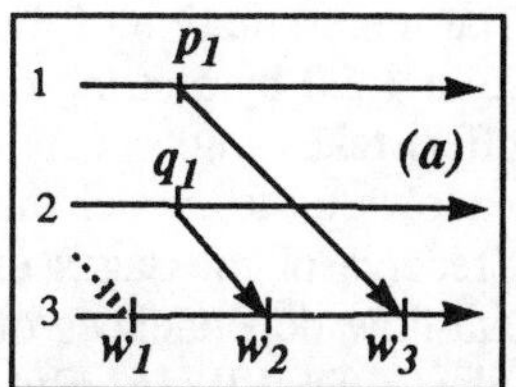


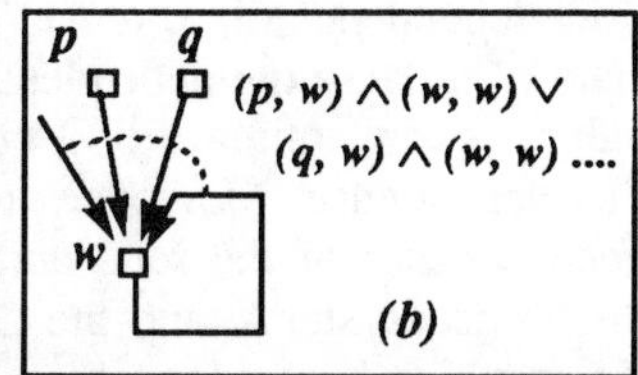


**FIGURE 2. Occurrences of actions; *p, q* & w. *p* & *q* are post *(ev)* of process 1 & 2. w is wait *(ev)* of process 3.**

caused by the dependence *(w, w)*.

Furthermore, the dependence information obtained from various instances of an action allows our debugger to infer the conditions that govern the action's execution. For example, the dependence information from $w_2$ and $w_3$ of Fig. 2(a) allows the debugger to infer that the condition "$((p, w) \wedge (w, w)) \vee ((q, w) \wedge (w, w))$" governs the execution of *w*. The immediate predecessors of $w_3$ indicate that the condition "$(p, w) \wedge (w, w)$" must have initiated the third instance of *w*. Similarly, immediate predecessors of $w_2$ indicate that the condition "$(q, w) \wedge (w, w)$" must have initiated the second instance of *w*. The "$\vee$" operator appears because either of the two conditions can initiate the execution of *w*. Now suppose that *p, q* and *w* represented the synchronization primitives **post** *(ev)* and **wait** *(ev)*. Then, the condition brings out the information implicit in the semantics of the primitives.

Therefore, a user can represent M as conditions on the dependences of actions. Then, representing (**w** $\wedge$ **p**) precede **w** simply requires that we select the actions corresponding to **p** and **w**, connect them with the appropriate dependences, and specify an $\wedge$ condition. This is shown below by the $\wedge$ condition on the dependences of *post(ev)* and *wait(ev)*.

$\wedge$ **post** *(ev)*

**wait** *(ev)*

## 2.0 THE SPECIFIED BEHAVIOR

Our model decomposes the problem of debugging a concurrent program into two levels. A programmer debugs the concurrent state at the upper level, where the only important concerns are the relations on the set of computation actions; specified in the program, modeled in the expected behavior and observed in the execution behavior. As execution occurrences of computation actions are atomic, internal states of a computation action are not important at this level. These states only become important when the programmer moves to the lower level, inside the computation action, to debug its internal sequential text.

We represent the data dependences that force the computation actions to execute in a particular order by ordered pairs. If $\Sigma_P$ denotes the set of computation actions, then:

**Def. 2:** *Data-flow dependences* are $F_P \subseteq \Sigma_P \times \Sigma_P$.

For instance, the dependence of a receipt of a message on its send, the dependence of a P of a semaphore on its V, or the dependence of a wait of a synchronization event on its post, represent such data-flow dependences. The write-read dependence on a shared synchronization variable, or on a message, forces the actions to execute in a particular order. Our motivation for representing the synchronization dependences as data-flow dependences comes from the language independence and machine independence goals of the CODE graphical programming environment [2], [15]. The data flow characterization of the synchronization and control-flow dependences in CODE allow the environment to support shared memory, as well as, distributed systems.

The set of ordered pairs $F_P$ gives a graphical representation $(\Sigma_P, F_P)$ whose nodes are the set of computation actions and whose arcs are the data-flow dependences. See Fig. 4(a). Intuitively, a computation action acts like a procedure whose input parameters are the input dependences and output parameters are the output dependences. It begins execution by obtaining a set of values from its input dependences. Then, it performs a sequential computation on this data. It ends its execution by putting a set of values on its output dependences.

Input dependences of an action *a* are given by the incoming arcs; $in(a) \equiv \{(x, a) \mid (x, a) \in F_P\}$. And, output dependences are given by the out-going arcs; $out(a) \equiv \{(a, x) \mid (a, x) \in F_P\}$. Conditions specified on the input dependences determine when to initiate the execution of a computation action, and conditions specified on the output dependences determine what follows its execution. The pre-condition that initiates the execution of a computation action is called an *input firing-rule*, and the post condition that follows its execution is called the *output firing-rule* [15].

**Def. 3:** An *input firing rule* $I_P(a)$ is a set of subsets of input dependences, i.e. $I_P(a) \subseteq 2^{in(a)}$. An *output firing rules* $O_P(a)$ is a set of subsets of output dependences; i.e. $O_P(a) \subseteq 2^{out(a)}$.

An input firing rule $I_P(a)$ is a condition in the disjunctive normal form (sum of products). Each element of $I_P(a)$ represents a disjunct, and is given by a subset of *in(a)*, input dependences of *a*. The state of a data-flow dependence (x, a) can be represented by a string of values denoted by [x, a]. A computation action is ready for execution if the state of all the dependences of an element $\iota \in I(a)$ are non-empty strings i.e. $\forall (x, a) \in \iota :: [x,a] \neq \varepsilon$. Then, input to a computation *a* is a set of suffix values detached from the state of dependences in $\iota$.

On completing its computation, ***a*** will catenate a set of output values as prefixes to the state of all the dependences given in some element $o \in O(a)$.

## 3.0 THE OBSERVED BEHAVIOR

We trace the execution occurrences of computation actions and their orderings in P → E part of the debugging cycle. Fig. 4(b) shows this information.

**Def. 4:** *A computation event* is an execution occurrence of some *computation action.*

The set of computation events is denoted by V. Not all the computation actions of a program may occur in an execution. A subset of computation actions that do occur are, then, collected in Σ. Thus, $\Sigma \subseteq \Sigma_P$.

An action can occur multiple number of times. Subscripts in Fig. 4(b) denote the multiple occurrences of actions. Set *V* of events is, thus, a set of multiple occurrences of actions, or a "multiset" of occurrences of actions. The function $\mu: V \rightarrow \Sigma$ maps each event of V to that action of Σ, of which it is an occurrence.

We support this mapping by associating an instance counter *u.i* with each *action* $u \in \Sigma_P$. The id-instance pair; *u* and *u.i*, provide a unique identifier for each event. We, therefore, denote the *i*-th occurrence of actions *u, v, w* ... as events $u_i$, $v_i$, $w_i$, ....

### 3.1 Causality of Data Flow Dependences

In order to record the orderings enforced by $F_P$, the debugger appends the unique identifier with the data shared through the data-flow dependences. Whenever an action *u* puts some data on its output dependence $(u, v) \in F_P$ in its *i*-th instance, it appends the identifier $u_i$ to the data. Similarly, whenever an action *u* begins its *i*-th execution by removing some data from its input dependence $(v, u) \in F_P$, it detaches the identifier appended to the data, and puts the detached identifier in a predecessor list denoted by *u.i.P.* The list contains such pairs for all the predecessor events that have "caused" the *i*-th instance of *u*. The traces contain records of the execution instances of each event. The trace record of an event $u_i$ contains the action id *u*, instance number *u.i*, and the predecessor list *u.i.P.* Also, see Table 1.

**Def. 5:** The orderings enforced by $F_P$ are $<^F \equiv \{(u_i, v_j) \mid u_i, v_j \in V \wedge u_i \in v_j.P\}$.

The transitive closure of $<^F$ results in an irreflexive partial ordering that constitutes the *causal orderings* $<^c$. The orderings $<^F$ simply reflect the "causality" of data-flow dependences.

**Lemma 1:** $u_i <^F v_j \Rightarrow (u, v) \in F_P$.

Consequently, if $u_i <^F v_j$, then there is some data shared between $u_i$ and $v_j$, namely, the state of the data-flow dependence $(u, v) \in F_P$. The representation of the state of a data-flow dependence in § 2.0 by a string of values (or an infinite FIFO buffer) takes into account this dependence. Moreover, it allows us to model the general cases of the send and receive of messages in distributed systems, and the data-flow dependences of the graphical/visual languages like CODE [2],[15]. But, the representation may create problems in modeling the synchronization primitives of shared memory systems. However, as seen in § 3.3, concurrent state does not depend upon the internal representation of the states of the data-flow dependences. It only depends upon the event orderings. Thus, we can represent the state of a synchronization dependence, with a string (or a buffer) of length one. It will denote the data dependence due to the shared synchronization variable (whose only permitted values are set or reset). The debugger records the causal orderings by appending and detaching the identifier to the shared synchronization variable.

### 3.2 Execution History Pomsets

As seen above, actions can occur multiple number of times as events, i.e. the set V of events is related to the set Σ of actions through the function $\mu: V \rightarrow \Sigma$. This effectively turns the poset $(V, <^c)$ into a pomset $(\Sigma, V, <^c, \mu)$ [19]. A POMSET is a Partially Ordered MultiSET of occurrences of **actions**, in much the same way as a **string** is a TOMSET; a Totally Ordered MultiSET of occurrences of **alphabets**. The pomset $(\Sigma, V, <^c, \mu)$ is called a causal pomset because $<^c$ are the causal orderings. It is instrumental in unifying our model because its expression of the concurrency properties is independent of the way time or events are modeled in a system [5].

A pictorial representation of the causal pomset is an *execution history display.* We can now explain why execution history displays are so helpful in debugging. They display the causal orderings of events. These orderings allow a programmer to determine the conditions that initiated and followed each execution instance of an action. From Lemma 1, an immediate predecessor of $u_i$ must map to an input dependence of ***u***; and from Def. 3, an element of the input firing rule of ***u*** is a subset of input dependences. Hence, immediate predecessors of an event $u_i$ inform the programmer about that element of the input firing-rule that initiated the ***i***-th instance of ***u***. Similarly, immediate successors of an event $u_i$ inform the programmer about that element of the output firing-rule that determined the condition following the ***i***-th execution instance of ***u***. If ${}^\bullet u_i \equiv \{(v, u) \mid v_j <^F u_i\}$ and $u_i^\bullet \equiv \{(u, v) \mid u_i <^F v_j\}$, then:

**Def. 6:** A causal pomset $(\Sigma, V, <^c, \mu)$ is *compatible* with the firing rules iff $\forall u_i \in V :: {}^\bullet u_i \in I_P(u) \wedge u_i^\bullet \in O(u)$.

This shows the compatibility of the immediate orderings of a given event with the firing rules specified on the immediate dependences of its corresponding action. In § 4.0, we extend this compatibility of immediate orderings with the immediate dependences to the compatibility of transitive orderings with the transitive dependences. This provides a framework for representing and checking the expected behavior.

### 3.3 Concurrent Execution State

There is non-determinism associated with the choices of the elements given in the input and output firing rules. An action can non-deterministically select different elements of a firing rule. In Fig. 2, the second instance of action $w$ can non-deterministically select any element from its *input firing rule*. During execution replay, a record of the causal orderings informs our debugger to select the right element of the firing rules for each execution instance of an action. Thus, it reconstructs the states of the previous execution. Using $<^c$ and following [13], we find a notion of concurrent state:

**Def. 7:** A *concurrent state* of an execution $(\Sigma, V, <^c, \mu)$ is a consistent cut-set of the poset $(V, <^c)$. A set $C \subseteq V$ is a consistent cut-set iff $e \in C \wedge e' <^c e \Rightarrow e' \in C$.

This definition is independent of the local state of an action. It is, also, independent of the contents of the messages exchanged by the actions. It only depends upon the order of events. Hence, distributed systems often reduce their roll-back and recovery overhead by only recording the event orderings, and not the content of messages or the checkpoints of the local states. Therefore, our execution replay facility exploits this definition to reduce the recording overhead (§ 5.1).

### 3.4 Animation

Animation provides an instantaneous view of the execution history. It is simply the process of displaying Def. 6 on an animation structure while traversing the execution pomset. An *animation structure* shows the assignment of actions to the executable units (processes/threads) of the execution environment. It can be considered as an elaborated form of the program structure. Fig. 3(c) shows such an animation structure where action $w$ of Fig. 3(a) replicates (or elaborates) into three actions; $w^1$, $w^2$, and $w^3$ running on different processes. We represent the run-time assignment of actions $\Sigma_P$ to the executable units by the set $\Sigma \subseteq \Sigma_P \times N$; where, $w^i$, $w^j \in \Sigma$ implies that $w$ is assigned to different executable units whose logical ids are $i$ and $j$. The actual identity of an executable unit is not important. The superscripts $i, j$ are only for distinguishing between the multiple copies.

During animation, the debugger traverses the execu-

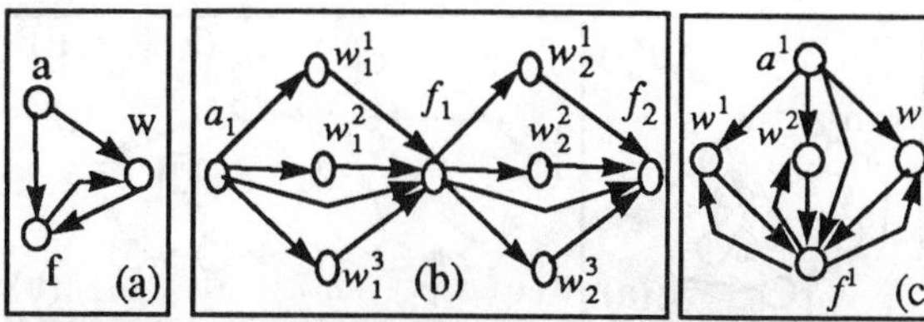


**FIGURE 3. (a) *Program structure*, (b) *pomset execution*, and (c) elaborated *animation structure.***

tion pomset. On encountering the $i$-th event instance of an action $u^k$, it highlights in the animation structure those input dependences of the action that correspond to the immediate predecessors of its $i$-th event instance. It, then, highlights the action $u^k$. Then, it highlights those output dependences of the action that correspond to the immediate successors of its $i$-th event instance.

We can automatically generate an animation structure from the execution pomset. Note that the structure shown in Fig. 3(c) is obtained by folding all the subsequent instances of actions in Fig. 3(b) to their first occurrences. This fulfills the requirement of a strong coupling between animation and execution history [14].

## 4.0 THE EXPECTED BEHAVIOR

In the P ⟶ M part of the cycle, a user represents the expected behavior by selecting some "interesting" actions $\Sigma_M$ from $\Sigma_P$ and, then, representing the expectations as conditions $I_M$, $O_M$ on their dependences $F_M$.

### 4.1 Representing Expected Behavior

A *dbx* debugger is closely coupled with the program because it compels the programmer to use only those objects that already exist in the program; e.g. it would not allow a user to specify a non-existent print variable. Taking cue from *dbx*, we closely couple our checker with the program and only allow the user to work with those objects that already exist in the program. This is in contrast to the existing checkers that can not verify if a user has supplied a non-existent order of events in the expected behavior.

Note that if a user expects that events $u_i$ and $v_j$ will be ordered in the execution, then their actions $u$ and $v$, must exhibit a (transitive) data-flow dependence in the program. That is, $u_i <^c v_j \Rightarrow (u, v) \in F_P{}^*$, where $F_P{}^*$ is a transitive closure of $F_P$. This also follows from Lemma 1. For instance, $m_1 <^c c_1$ in Fig. 4(b) corresponds to the transitive *data-flow dependence* between $m$ and $c$ of Fig. 4(a); both are shown by dotted lines. Thus, any pattern that is expected in an execution, must be the unrolling of a pattern already present in the program structure.

Thus, a user starts specifying expected behavior by selecting a subset $\Sigma_M$ of interesting actions from the program; $\Sigma_M \subseteq \Sigma_P$. Fig. 4(c) shows a selection of such actions. The user can then specify a dependence

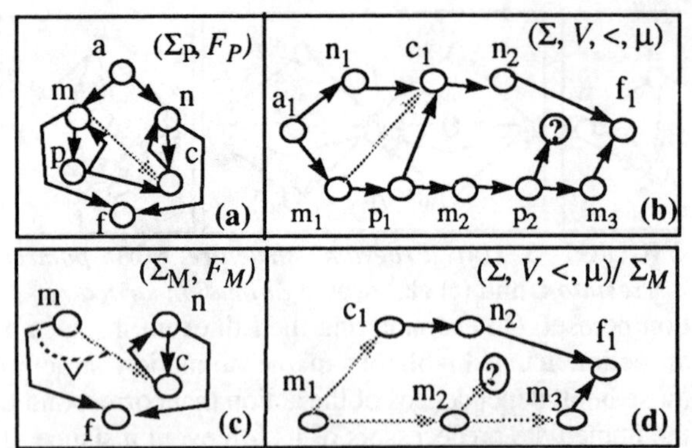


**FIGURE 4. (a) Program Structure. (b) Execution. (c) Expected Behavior. (d) Restricted execution.**

between the selected actions only if it corresponds to some dependence of $F_P$*. Some of such selected dependences are shown as $F_M$ in Fig. 4(c).

**Def. 8:** $F_M$ are the selected dependences from the transitive *data-flow dependences* $F_P$* restricted to interesting actions i.e. $F_M \subseteq F_P* / \Sigma_M$, where $\Sigma_M \subseteq \Sigma_P$.

An observed ordering like $(m_2, ?)$ in Fig. 4(d), that can not be mapped to a data-flow dependence from $F_M$ is, therefore, symptomatic of a bug!

In Fig. 4(c), the structure built on $\Sigma_M$ is the specification of the expected behavior. Akin to the conditions in a *dbx* command like "**when at** stmt **if** condition," a user can, then, provide firing rules $I_M$ and $O_M$ to further restrict the instances of "interesting" actions. Note that firing rules are conditions about the concurrent state, whereas *dbx* conditions are about the sequential state. Suppose the user specifies an "∨" *output firing rule* for action ***m*** in Fig. 4(c). Then, the checker can filter out instances $m_1$ and $m_3$ because they subscribe to the "∨" rule. But, will raise an exception for $m_2$ as it does not subscribe to the "∨" rule.

### 4.2 Recognizing the Expected Behavior

In *dbx*, a directive like "**when at** stmt . . . ." informs the debugger to make necessary preparations for the specified statement. It also informs it to ignore the rest of statements. Similarly, the selection of $\Sigma_M$ from $\Sigma_P$ informs the debugger to specially prepare for "interesting" actions $\Sigma_M$, and to safely ignore the "uninteresting" actions $\Sigma_P$ - $\Sigma_M$. Then, the debugger can filter out the uninteresting events and can restrict the execution to instances of $\Sigma_M$.

The restricted pomset $(\Sigma, V, <^c, \mu)/ \Sigma_M$ of Fig. 4(d) only contains the interesting events and their mutual orderings. The debugger establishes the orderings among interesting events by exploiting the fact that instances of interesting actions can only be ordered if there exists a mutual (transitive) dependence. In Fig. 4(d), events $m_1$ and $c_1$ are only ordered because of the transitive dependence that exists between actions ***m*** and ***c***. Note in Fig. 4(a) that the dependence goes through an intervening uninteresting action ***p***. Our debugger, therefore, establishes the orderings between instances of ***m*** and ***c***, by asking the uninteresting action ***p*** to relay the causality information that arrived from its predecessors, forward to its successors.

Unlike interesting actions, uninteresting actions do not trace their execution instances. Instead of sending the identifier of their current instance to their successors, they simply relay forward their predecessor lists. See Table 1. These lists keep getting relayed forward by the intervening uninteresting events until they land in the predecessor lists of the interesting events. Only then they are traced. Predecessor lists of interesting events, therefore, only contain the identifiers of their causally preceding interesting events. Execution is thus filtered to $(\Sigma, V, <^c, \mu) / \Sigma_M$.

**Table 1: Monitoring and tracing of actions.**

| Monitored Occurrences | Interesting Actions $u \in \Sigma_M$ | Uninteresting Actions $u \in \Sigma_P - \Sigma_M$ |
|---|---|---|
| ***u* sends to *v*** | **append**(*msg: u, u.i*); | **append** (*msg: u.i.P*); |
| ***u* receives *msg*** | *u.i.P* ∪**detach**(*msg*); | *u.i.P* ∪ **detach**(*msg*); |
| ***u* executed** | **trace** (*u, u.i, u.i.P*); *u.i := u.i + 1;* | *u.i := u.i + 1;* |

The structural information of M and the fact that the causal pomset is restricted to interesting actions, greatly simplifies recognition of the expected behavior. The debugger traverses the partial order, and tries to check if Lemma 1 and Def. 6 also hold for M. It checks whether each immediate successor $v_j$ of a given event $u_i$ corresponds to some successor $v$ of the action $u$ in the dependences $F_M$. Additionally, the debugger checks whether the immediate predecessors and successors of an event $u_i$ satisfy the input and output firing rules $I_M$ and $O_M$ for $u$. If an ordering $u_i <^F v_j$ fails to correspond to some dependence $(u, v) \in F_M$, or the immediate predecessors $^\bullet u_i$ or successors $u_i^\bullet$ fail to meet the expected conditions $I_M(u)$ or $O_M(u)$, then an error has been recognized. This happens for the unexpected orderings of $m_2$ in Fig. 4(d).

**Def. 9:** Event $u_i$ is in *error* if $u_i^\bullet \notin O_M(u) \vee {}^\bullet u_i \notin I_M(u)$.

*Thus, concurrent debugging is the process of following the unexpected orderings given by the erroneous events, in the direction of causality.*

## 5.0 SHARED DATA DEPENDENCES

Unlike dependences $F_P$ that force an ordering on the execution of actions, shared data dependences do not impose any particular orderings. We, therefore, model a shared data object by the set of actions that share it. The set of those objects is denoted by S.

**Def. 10:** The set of *shared data dependences* is $S \subseteq 2^{\Sigma_P}$. A *shared data dependence* is a set $D$ of *computation actions*, $D \subseteq \Sigma_P$ (or $D \in S$).

The actions that participate in a shared data dependence $D$ are classified into disjoint sets of *readers* and *writers*. *Readers* of $D$ are $\rho.D$ and *writers* are $\omega.D$; where $\rho.D \cup \omega.D \equiv D$, $\rho.D \cap \omega.D \equiv \phi$. In Fig. 5(a), shared data dependence $D = \{a, r, w\}$. For example, we may have the set of *readers*, $\rho.D = \{r\}$ and set of writers, $\omega.D = \{a, w\}$.

## 5.1 Execution Replay

The goal of an execution replay facility is to record enough information about the non-deterministic choices made by the events of an execution. During replay, the events can, then, be forced to make the previous choices. A record of causal orderings $<^F$ is sufficient to overcome the non-determinism associated with the choices of elements in a firing rule. But, there is another source of non-determinism. This is associated with shared data dependences that do not force the actions to execute in any particular order. As accesses to shared objects can take place in any order, the debugger must record the non-deterministic order of accesses to shared objects so that during replay the accesses can be forced to occur in the previously recorded order.

We record the order of accesses to shared objects with the same mechanism that we use in recording the $<^F$orderings (§ 3.1). But before we can do that, we need a protocol to ensure a valid serialization on the accesses to shared objects like the CREW (concurrent read exclusive write) protocol [10]. The protocol disallows simultaneous write-access to a shared object with other accesses. The debugger ensures a deterministic replay by implementing the protocol. Without the implementation, simultaneous accesses to a shared object can hinder a deterministic replay by corrupting the object and giving unpredictable results. Note that members $u$ and $v$ of a shared data dependence have a valid serialization if for every instance $u_i$ of a write-access and every instance $v_j$ of another access, either $u_i$ occurs before $v_j$, or $v_j$ occurs before $u_i$

Unlike [10] that uses versions of shared objects to record the orderings, we use our simpler mechanism for recording the order $<^D$ of accesses to a shared object $D \in S$. Each shared object has the identifier of the last instance of its write-access appended to it. Whenever an action accesses a shared object, it reads the identifier appended to the object, and places the identifier in the predecessor slot reserved for that object in its trace

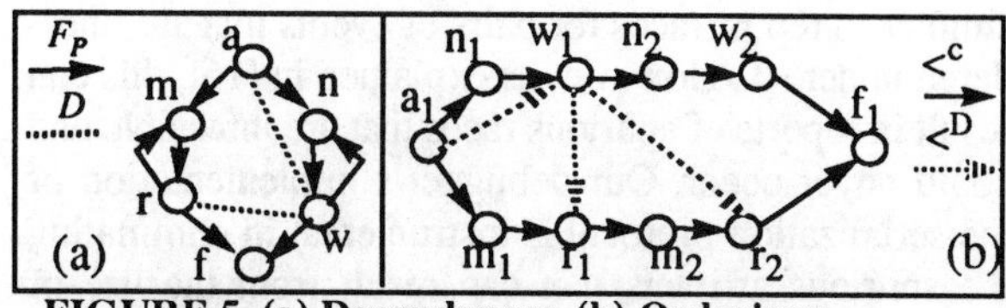


**FIGURE 5. (a) Dependences. (b) Orderings.**

event record. A writer, in addition to the above, replaces the identifier appended to the object with the identifier of its present instance.

In addition to the predecessor list $u_i.P$ for the dataflow dependences (§ 3.1), the trace record for each event now requires another predecessor list $u_i.S$ for shared data dependences. The list has a slot $u_i.S[D]$ for each shared data dependence $D$ in which a given action participates. Then, the identifier in the slot for $D$ in the predecessor list of an event record, determines the causal orderings $<^D$. There is, thus, an ordering relation $<^D$ for each $D \in S$ generated like $<^F$. The debugger, then, ensures the replay by forcing the events to occur in the pre-recorded order.

To simplify the following discussion, we assume that there is "one" shared data dependence $D \in S$. We now consider the strict partial orders due to $<^F$and $<^D$.

## 5.2 Race Detection

Simultaneous accesses (with at least one write) to a shared object can race with each other and can corrupt the shared data with unpredictable results. We detect races by identifying those pairs of events whose accesses to a shared object included at least one write and whose orderings were only due to the debugger's enforcement of the serialization protocol. For e.g., dotted arcs of Fig. 5(b) show the $<^D$ orderings for events $w_1$ and $r_1$ (and $w_1$ and $r_2$) that were forced by the debugger's serialization protocol. The events are otherwise unordered under $<^F$. Without the debugger's protocol, these $<^D$ orderings may not exist. Then, $w_1$ and $r_1$ (or $w_1$, $r_2$) can execute simultaneously with unpredictable results. Thus, all $<^D$ orderings observed under the serialization protocol, should be supported by $<^F$ as, for instance, the ordering between $a_1$ and $w_1$.

Although this technique detects the possibility of race for events $w_1$ and $r_1$ (and for events $w_1$ and $r_2$) of Fig. 5(b), it does not detect the possibility of race between events $w_2$ and $r_2$. To detect such races, we note that these events are unordered by both $<^D$ and $<^F$. Thus, our debugger will also signal the races for all those pairs of events that are unordered by $<^D$ and $<^F$, and that access a shared object with at least one write.

Implementing a serialization protocol, and recording the $<^D$ orderings of accesses to a shared object, may seem unnecessary for detecting races. It may appear simpler to report races for pairs of events that are unordered under $<^F$. However, as explained in [16], this can result in reports of spurious races that are infeasible and could never occur. Our debugger's implementation of the serialization protocol is instrumental in eliminating the spurious artifacts that can result from the use of shared objects that were corrupted by an earlier race. Furthermore, the record of $<^D$ helps in improving the accuracy of detected races by identifying other spurious races. Note that race detection in our model does not require any extra overhead. The record of $<^D$ and $<^F$ already exists for supporting execution replay.

## 6.0 CONCLUSIONS

The unified model of concurrent debugging presented in this paper covers all the parts of the debugging cycle. This overcomes the incompatibility of existing facilities, and allows our debugger to support different debugging facilities like *execution replay*, *race detection*, *assertion/model checking*, *execution history displays*, and *animation*. We, thus, show that the benefits of modeling the whole cycle are greater than a simple sum of its parts.

Our model uses the set of computation actions and the "causality" of their dependences to simplify the complexity of concurrent debugging. We show that it is easier for a debugger to record the orderings when events are considered to be the occurrence instances of actions. We, also, show that the debugger can obtain much more information by recognizing the underlying dependences of the observed orderings.

Our use of a program oriented approach for checking the model of expected behavior saves the user extra programming effort. It, also, simplifies for the debugger, filtering and recognition of the expected behavior.

Our model proposes a solution for the following needs highlighted in the panel discussions of [11]. It provides a theoretical framework for defining an ***error*** and explaining the process of concurrent debugging (J. Wilden)[1]. The framework explains precisely when and why a particular facility is needed (P. Bates). It eases implementation by separating the monitoring and presentation concerns. It is not specific to a particular language, or a particular system. It proposes the abstraction of *computation actions* to fulfill the need for a lower limit for the granularity of data collection (A. Tilberg). Such an abstraction is usually available in a graphical environment like CODE [2], [15]. However, as explained in the Appendix, some static analysis may be necessary to obtain the *computation actions* from a textual representation of a concurrent program.

1. Parenthesized names indicate the person who highlighted the need in the panel discussion of [11]

# References

[1] P. Bates, "Debugging heterogeneous distributed systems using event-based models of behavior," *ACM SIGPLAN Notices*, 24(1), (Jan '89), pp. 11-22.

[2] J. C. Browne, M. Azam, and S. Sobek, "A Unified Approach to Parallel Programming", *IEEE Software*, (Jul '89).

[3] D. Callahan and J. Subhlok, "Static Analysis of Low-Level Synchronizations," *ACM SIGPLAN Notices*, 24(1), (Jan '89)

[4] R. J. Fowler, T. J. LeBlanc and J. M. Mellor-Crummy, "An Integrated Approach to Parallel Program Debugging and Performance Analysis," *ACM SIGPLAN Notices*, 24(1), (Jan '89), pp. 163-73.

[5] H. Gaifman, "Modeling Concurrency by Partial Orders and Nonlinear Transition Systems," *Lecture Notes on Comp. Science #354*, (May '88), pp. 467–88.

[6] A. A. Hough and J. E. Cuny, "Perspective views: A Technique for Enhancing Parallel Program Visualization," *ICPP*, (Aug '90), pp. II.124–II.132.

[7] W. Hseush and G. E. Kaiser, "Modeling Concurrency in Parallel Debugging," *ACM Symp. on Princples & Practice of Parallel Prog.*, (March '90), pp. 11–20.

[8] D. P. Helmbold, C. E. McDowell, and J. Wang, "Analyzing traces with anonymous synchronization," *ICPP*, (Aug. '90), pp. II.70–II.77.

[9] A. A. Hough, "Debugging Parallel programs Using Abstract Visualizations," TR 91:53, CS Department, University of Massachusetts at Amherst, (Sep '91).

[10] T. J. LeBlanc and J. M. Mellor-Crummey, "Debugging Parallel Programs with Instant Replay," *IEEE Trans. on Computers*, C36 # 4, (Apr '87), pp. 471-81.

[11] T. J. Leblanc and B. P. Miller, Ed.s, "Workshop Summary; What we have learned and where we go from here?" *ACM SIGPLAN Notices, 24(1)*, (Jan, 89), pp. ix–xxii.

[12] T. J. LeBlanc, J. Mellor-Crummey & R. J. Fowler, "Analyzing Parallel Executions with Multiple Views," *J. of Paral. & Dist. Comp.* #9, (Jun '90), pp. 203-17.

[13] F. Mattern, "Virtual Time and Global States of Distributed Systems," *Parallel and Distributed Algorithms*, (1989), pp. 215–26.

[14] C. E. McDowell, D. P. Helmbold, "Debugging of Concurrent Programs," *ACM Computing Surveys*, 21(4), (Dec '89), pp. 593–622.

[15] P. Newton and J. C. Browne, "The Code 2.0 Graphical Programming Environment," *Supercomputing* '92 (Jul '92).

[16] R. H. Netzer and B. P. Miller, "Improving Accuracy of Data Race detection," *ACM SIGPLAN*

*Notices 26(7)*, (Jul '91), pp. 133-44.

[17] R. H. B. Netzer and B. P. Miller, "On the Complexity of Event Ordering for Shared Memory Programs," *ICPP*, (Aug '90), pp. II.93–II.97.

[18] R. H. Netzer and B. P. Miller, "What are Race Conditions? Some Issues and Formalism," TR91-1014, CS Dept. Univ. of Wisconsin (Mar '91).

[19] V. Pratt, "Modeling Concurrency with Partial Orders," *Internaltional Journal of Parallel Programming*, 15(1), (1986), pp. 33–71.

[20] C. M. Pancake and S. Utter, "Models for Visualization in Parallel Debuggers," *Supercomputing '89*, (Nov '89) pp. 627–36.

[21] S. Utter and C. M. Pancake, "A Bibliography of Parallel Debugging Tools," *ACM SIGPLAN Notices*, 24(10), (1989), pp. 24-42.

## Appendix: Computation Actions in a Textual Representation of a Program

A textual program contains three types of statements; blocking synchronization, signal synchronization and non-synchronization statements. From such a text, *static analysis* routinely extracts a synchronization-control-flow graph that contains three types of nodes and two types of arcs [14], [3]. Nodes represent blocking synchronization, signal synchronization, and control decision statements. While arcs represent inter-process synchronization dependences and intra-process control-flow dependences.

Fig. 6(a), (b), and Fig. 7(a), (b) show the synchronization control flow graphs obtained from a textual program of a PPL like extension of C. Intra-process control arcs labeled by ***a***, ***w*** and ***f*** correspond to the sequential text containing non-synchronization statements. The abstraction of computation actions, shown by the dotted ovals, permanently associates the synchronization statements with the sequential texts. It permanently associates a blocking synchronization with the sequential text that follows it, and associates a signal synchronization with the sequential text that precedes it. Thus, blocking synchronizations **wait** *(ev_i)* is associated with the text ***w*** that follows it, and blocking synchronization **c_wait** *(ct)* is associated with the text ***f*** that follows it. Also, *signal* synchronizations **c_set**(*ct*) is associated with the text ***w*** that precedes it, and signal synchronization **post** $(ev_i : i = 1..n)$ is associated with texts ***a*** and ***f*** that precede it.

Def. 11: A *computation action* is a block of a flow graph that:

1. may contain internal control flow provided the internal control structures (loop, if-then-else, etc.) do not contain any synchronization statement;
2. it may begin with a *blocking* synchronization, that must be the first statement of the block; and
3. it may end with a *signal* synchronization, that must be the last statement of the block.
4. it may contain more than one *blocking (signal)* synchronization provided all are together at the start (end) of the block with no other intervening statement.

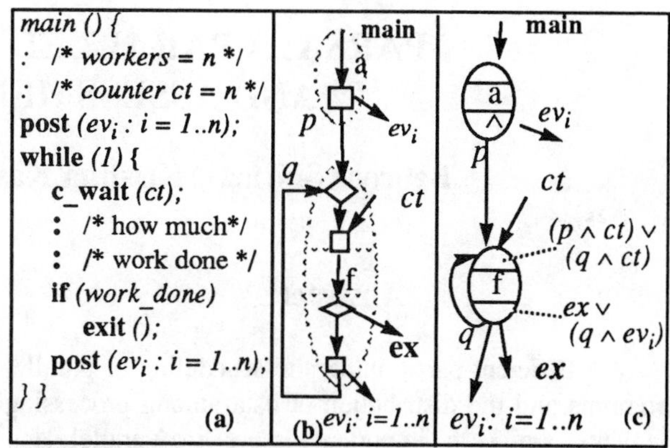


**FIGURE 6. (a) *main* starts *n* workers, and waits on a counter *ct*. On completion, each *worker* reports by incrementing *ct*. When count reaches *n*, *main* wakes up. If there is need for more work then, it signals the *workers* again, else it exits. (b) *Synchronization-control-flow* graph. (c) Actions *a* and *f*.**

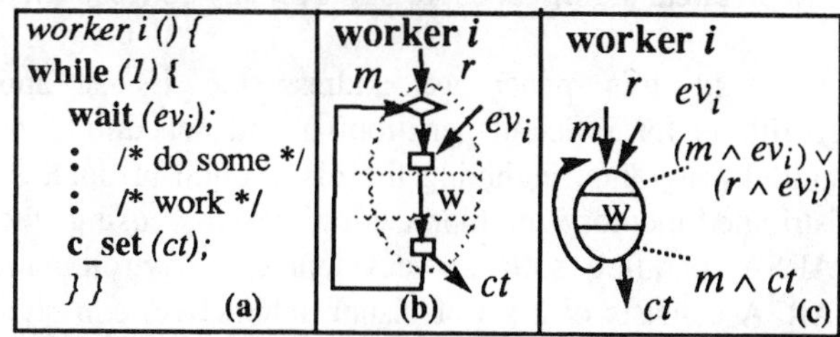


**FIGURE 7. (a) Text. (b) *Graph*; *m*, *r* are control dependences; *ct* and $ev_i$ are synchronization dependences. (c) Action *w* ; $(m \wedge ev_i) \vee (r \wedge ev_i)$ is *input-rule*, $m \wedge ct$ is the *output-rule*, and *w* is the sequential text.**

After the synchronization statements have been associated with their bordering sequential texts, we are left with the control flow decision nodes. The abstraction of firing rules subsumes these decision nodes as shown in Fig. 6(c) and Fig. 7(c). The set of *computation actions* $\Sigma_P$, then, characterizes the specified behavior. Hence, $\Sigma_P = \{a, w, f\}$; $F_P = \{(a, w), (a, f), (w, f), (f, w)\}$; $O_P(a) = \{\{(a, w), (a, f)\}\}$; $O_P(f) = \{\{exit\}, \{(f, w)\}\}$; $I_P(w) = \{\{(a, w)\}, \{(f, w)\}\}$; $I_P(f) = \{\{(f, f), (w, f)\}\}$.

a main $ev_i : i = 1..n$ f ct + worker i $ev_i$ ct w → a w f $(\Sigma_P, F_P)$

# PARSA: A PARALLEL PROGRAM SCHEDULING AND ASSESSMENT ENVIRONMENT[1]

**Behrooz Shirazi[2], Krishna Kavi[2], A.R. Hurson[3], Prasenjit Biswas[4]**

## Abstract

Efficient partitioning and scheduling of parallel programs and the distribution of data among processing elements are very important issues in parallel and distributed systems. Existing tools fall short in addressing the issues satisfactorily. On one hand, it is believed to be unreasonable to leave the burden of these complex tasks to the programmers. On the other hand, fully automated schedulers have been shown to be of little practical significance, or suitable only for restricted cases.

In this paper we address the issues and algorithms for efficient partitioning and scheduling of parallel programs, including the distribution of data, in distributed-memory multiprocessor systems, using the PARSA parallel software development environment. PARSA consists of a set of visual, interactive, compile-time tools that will provide automated program partitions and schedules whenever possible, while permitting the user to exert control over these operations for a better performance. The supported program assessment tool provides the users the opportunity to fine-tune the program and achieve their performance objectives.

## 1. Introduction and Background

With the wide-spread use of parallel and distributed systems in database, real-time, defense, and commercial applications, efficient partitioning and scheduling of parallel programs, to fully utilize available processing power, have become very important issues. Existing tools fall short in addressing the issues satisfactorily. Some strongly believe that it is unreasonable to expect the programmer to deal with the complex tasks of partitioning and scheduling the programs. However, fully automated schedulers have been shown to be of little practical significance, or suitable only for restricted cases. Many existing tools do not consider efficient data distribution, and thus, do not exploit the full potential of parallel execution. Likewise, many tools do not facilitate the collection of performance data as an integral part of the parallelizing environment.

The objective of this work is to address the issues and algorithms for efficient partitioning and scheduling of parallel programs, as well as distribution of data, in distributed-memory multiprocessor systems. To achieve this goal, we have developed a visual, interactive, compile-time environment, called PARSA (PARallel program Scheduling and Assessment), to assist the users in scheduling of their parallel programs on a target architecture. The interactive nature of PARSA allows choice of many scheduling methods as well as modifications to the generated schedules, in order to incorporate the user domain-specific knowledge in the scheduling process. In addition, PARSA provides an environment which allows the users to play "what if" type scenarios in order to evaluate, or fine-tune, their parallel programs and choose a suitable architecture for execution of their application. PARSA consists of a collection of tools to aid in:

- static partitioning of programs,
- scheduling of the tasks on available processors,
- distribution of data to minimize network and memory access delays, and
- assessing the program performance on the underlying parallel architecture.

There are several efforts which have demonstrated the usefulness of parallel program scheduling tools [App89, Bai90, Ber87, Ber88, Don87, Elr90, Gua89, Lo91, Pei86, Wu89]. In addition, there is a large body of literature on static partitioning and scheduling of parallel programs [Shi92a, Shi93]. The features that distinguish our work from similar tools include: (1) more accurate estimations of execution times of the tasks and communication delays at compile time; (2) choice of many partitioning and scheduling tools; and, (3) built-in visual assessment tools for performance evaluation and tuning of parallel programs before execution.

[1]This work is in part supported by a grant from the State of Texas Advanced Technology Program (grant no. 003656-080) and a grant from NSF (grant no. CDA-9300252).

[2]The University of Texas at Arlington, Department of Computer Science & Engineering, Arlington, TX 76019-0015.

[3]Pennsylvania State University, Department of Electrical and Computer Engineering, University Park, PA 16802.

[4]Cyrix Corporation, 2703 N. Central Expressway, Richardson, TX 75080.

## 2. PARSA Environment

Although great progress has been made in static partitioning, scheduling, and programming aid tools for parallel programming, there are still a number of issues and problems which need to be addressed in this area. We first briefly identify these issues and then discuss how they are alleviated in PARSA:

*Incorporation of domain-specific knowledge*: Due to being NP-complete problems, static partitioning and scheduling methods are often based on general heuristics and rules of thumb [Shi92a, Shi93]. Provisions to allow the programmers to i) incorporate their domain- specific knowledge and ii) visually observe the approximate performance of the program, can result in a better schedule.

*Accurate estimations of execution times and communication delays*: Estimating the execution time of the tasks at compile time can be difficult due to conditional, iterative, and recursive constructs. In addition, accurate pre-runtime estimations of the communication delays are often impractical due to the dynamic load on the network links at execution time.

*Data distribution*: In distributed memory multiprocessor systems, the task (function) schedulers should be coupled with efficient data distribution methods. Otherwise, the overhead caused by network and memory access delays (due to data access) can become prohibitive.

*Performance assessment*: A tool to evaluate the performance of a parallel program at compile time lets the programmer tune the environment by considering different scheduling and allocation schemes or different parallel architectures before the actual execution.

PARSA consists of a set of inter-related tools, as depicted in Figure 1. We will discuss the details of the development of each tool and indicate how their synergistic cooperation will achieve our goals. It should be noted that PARSA is developed on a Sun Sparc-station, running Unix, using X-Window and Motif graphical tools.

2.1.1. Application specification tool:

This tool converts an application program into its equivalent Directed Acyclic Graph (DAG) representation. The DAG nodes represent fine-grain program instructions. Currently, this tool accepts programs written in SISAL language [LLN85]. The SISAL programs are then compiled into IF1 - Intermediate Form 1, which is an acyclic graphical language (in textual form) developed by the Lawrence Livermore National Lab as an intermediate form for SISAL language [LLN85]. The major advantage of using SISAL and IF1 is that their functional, hierarchical structure makes them suitable for dataflow analysis, partitioning, and merging of the program tasks for the purpose of scheduling. In IF1, primitive instructions, such as add or multiply, are represented by simple nodes. Compound nodes represent complex constructs, such as loops and conditionals, in a hierarchical fashion. The application specification tool generates the DAG equivalent of the IF1 code and displays it visually. The tool has been implemented under Motif and can graphically display IF1 codes [Shi92b]. This graphical environment has a hierarchical structure; i.e., by double clicking on a compound node, the tool displays the internal structure of that node. We are currently extending the application specification tool to cover C programs as well. The process consist of converting C programs to single-assignment C and from there to IF1.

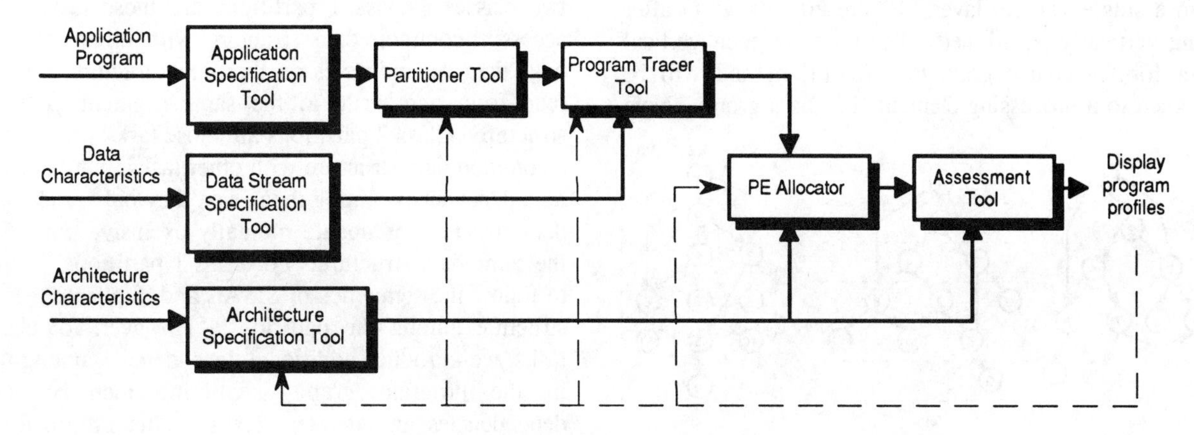


Figure 1: Block diagram of PARSA.

2.1.2. Partitioner Tool:

This tool merges together the fine-grain DAG nodes, from the application specification tool, into coarser-grain partitions. Partitioning into coarse-grain tasks is necessary in order to avoid excessive message passing and data access delays among the dependent fine-grain tasks that should be executed sequentially. The partitioner tool is based on the Vertically Layered Partitioning (VLP) method [Hur90] developed by the authors. We briefly discuss the VLP method through an example.

Figure 2(a) depicts an example DAG (G) in which nodes labeled 1-11 identify different primitive operations. Here G=<*N*,*A*>, where *N* is the set of nodes and *A* is the set of arcs. First, a modified topological sort is performed to transform G into disjoint horizontal layers such that the nodes in each layer can be performed in parallel and the layers are linearly ordered with respect to the precedence constraints among the nodes. Figure 2(b) shows the transformed graph.

To determine the appropriate vertical layers of the transformed graph, we first identify a critical path of this graph (defined as one of the longest paths from the root node to an exit node). The execution time of simple nodes can be easily estimated, given the instruction execution times of the underlying architecture. However, we encounter a problem in computing the length of a critical path due to conditional and loop constructs. An approximate critical path is determined by assigning a probability $P_{ij}$ (of taking that arc) to each arc $a_{ij}$ for all $a_{ij} \in A$. The expected execution time of the successors of a conditional node is then computed as a function of $P_{ij}$ and their execution times [Hur90].

Identification of an approximate critical path allows the nodes in each horizontal layer (Figure 2(b)) to be rearranged so that nodes which lie on a critical path form a single vertical layer. Figure 2(c) shows G after being vertically partitioned. The nodes in each vertical layer form a coarse grain task (partition) and will be assigned to a processing element (PE) as a group. Note that the execution time of each partition has been estimated at this point as well.

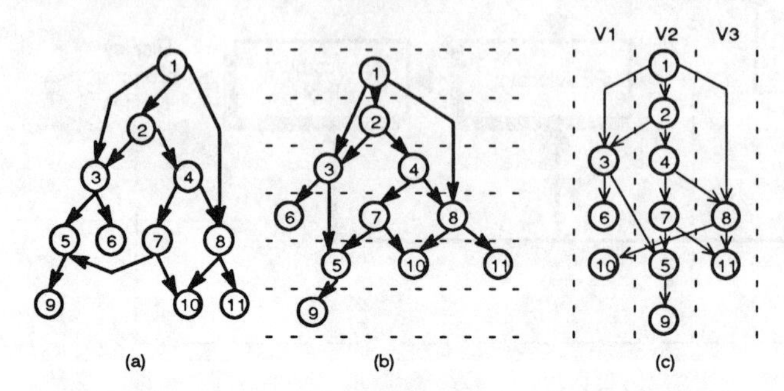


Figure 2: DAG partitioning.

2.1.3. Data Stream Specifier:

The data stream specifier is a tool which allows the user to define the logical layout of the input data. It assumes a sequential virtual address space for the data. The tool begins by prompting the user to specify the type of input data. It then maps the data sequentially into the virtual address space. For example, *integer A, integer Array[1:100,1:50]*, and *real B*, will be mapped into virtual addresses 0, 1 through 5000, and 5001, respectively. This address information will later be used for the distribution of data among the physical memory of the PEs.

2.1.4. Program Tracer Tool:

The purpose of this tool is to distribute the data among the PEs with the goal of minimizing memory access delays for data fetch operations. It should be noted that the single-assignment, functional semantics of SISAL and IF1 require duplication of data structures among the processes that consume them. However, implementation of such semantics can become quite memory inefficient and expensive. For example, in many applications, such as matrix multiplication, multiple processes simply read data from the same data structure and it is not necessary to duplicate the same data structure among all these processes. Thus, to the extent possible, the goal of the program tracer tool is to share data structures among processes and to allocate the data structures in physical memories such that memory access delays due to structure accesses are minimized. This is achieved by tracing the data structure access patterns for the generated partitions. Once the access patterns are established, the partitions are classified into two classes. Class 1 partitions are those tasks which access a common data structure with other tasks, but have data dependencies on that data structure; i.e., they read from and write to the same segment of a data structure. Class 2 partitions are those tasks which access a common data structure with other tasks, but there is no data dependency among them; i.e., they only read from a data structure or update mutually exclusive portions of the same data structure. For class 1 partitions, we have to follow the semantics of SISAL and duplicate the data structure among the partitions. However, for class 2 tasks, we introduce additional dependencies among them in the program graph, signifying memory access dependencies among these tasks. This information is later used in the program allocator tool to either allocate these tasks to the same processor or neighboring PEs with the purpose of reducing memory access delays.

2.1.5. Architecture Specification Tool:

This is a graphical interactive tool for building a database of detailed hardware specifications for the underlying parallel architecture. Different system parameters, such as instruction timings, memory access timings, communication delays, and interconnection topology, will be stored in a database. The users will have the option of using the database built-in architectures (such as a hypercube, mesh, ring, etc.), modifying the built-in systems, or defining their own architectures. Naturally, the users have the option of specifying many of the system parameters such as the number of processors, the interconnection topology, etc.

2.1.6. PE Allocator Tool:

The purpose of this tool is to allocate (schedule) the program trace (program and data) among the PEs, using the information supplied by the program tracer and architecture specification tools (Figure 1). The goal of the allocator is to achieve a minimal execution time with as small of a communication or memory access delay overhead as possible.

The scheduling method implemented in PARSA is a novel list scheduling method which features accurate estimation of the communication delays, taking into account the load on the network links [Shi91, Shi92b]. In the implemented method, we use a simple heuristic called Heavy Node First (HNF) [Shi90]. In HNF, the graph nodes are prioritized level by level, beginning with the starting nodes and going towards the exit nodes, and at each level, the node with the highest execution time (heaviest) is given a higher priority. Figure 3(a) depicts a sample DAG in which the number inside each node represents the execution delay of that node (in units of time) and the number on each link represent the communication delay for message passing between the two nodes on one communication link. Note that this communication delay does not include the delays due to link contention or for going through multiple links. According to HNF, the nodes of the DAG of Figure 3(a) are prioritized as: B, A, C, D, F, E, G, J, I, H, K, L, and M.

During the scheduling process, a task $T$ should be assigned to a PE which can execute it at the earliest possible time. This time depends on the current load assigned to the PEs and the amount of time needed for this task to receive its input parameters from its predecessors. Let $DSM_{i,T}$, Desirable Starting Moment for executing task $T$ on $PE_i$, be the time that $T$ receives all its input messages when assigned to $PE_i$. Then, the Actual Starting Moment (ASM) of a task $T$ assigned to $PE_i$ is defined as:

$$ASM_{i,T} = MAX(LOAD(PE_i), DSM_{i,T}),$$

where $LOAD(PE_i)$ is the load already assigned to $PE_i$. For $n$ PEs, the most suitable PE for Task $T$ will be the one with minimum ASM:

$$MIN(ASM_{i,T}), \text{ for } i=1,2, \ldots, n.$$

If $n$ is too large, one may consider a subset of the processors (e.g., $m$ PEs) in computation of the $MIN(ASM_{i,T})$, where the $m$ processors are the $m$ nearest neighbors of $PE_i$.

For example, consider scheduling DAG of Figure 3(a) on the multiprocessor of Figure 3(b). Assume that nodes A, B, C, D, and F are already assigned to the PEs, as shown in Figure 3(c). According to HNF, we select E for assignment next and must determine the suitable PE for its assignment. We need to compute all possible ASM's for E and select the PE which results in the least ASM. We have: $ASM_{0,E}$=15, $ASM_{1,E}$=25, $ASM_{2,E}$=25, and $ASM_{3,E}$=20. Thus, $PE_0$ is the most suitable PE for task E since it can begin its execution at the earliest possible time according to the current schedule. However, note that if we assign E to $PE_0$, then there will be a message communication between tasks A and E from time 5 to time 15 on Link 0. Thus, this message is scheduled in the Gantt chart of Figure 3(d). If sometimes between the times 5 and 15, there is another message which requires use of Link 0, it should be delayed due this link contention. The link schedules; e.g., Figure 3(d), will become the basis of the routing tables which will be used at run-time for message routing.

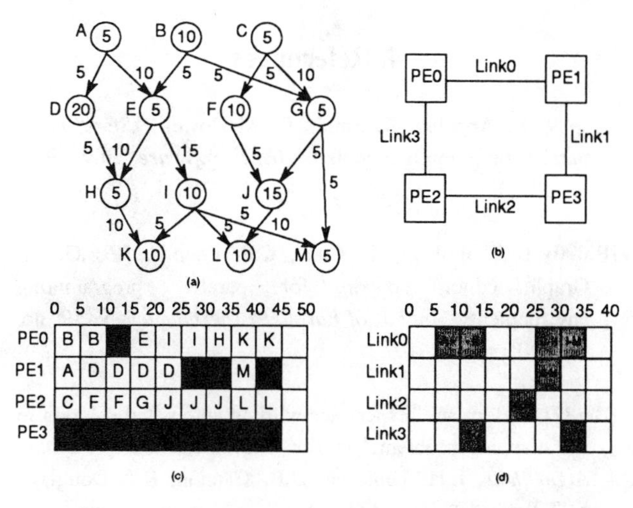


Figure 3: An example of program allocation.

2.1.7. Assessment Tool:

The purpose of this tool is to display the simulated run time behavior of the scheduled program at compile time. This is accomplished by the simulation of the underlying parallel architecture (output of the architecture specification tool) using the scheduled program trace as the input. The output of this tool consists of: (i) animated system resources' (e.g., processors and network subsystem) utilization as a function of the simulated time; (ii) memory and network latencies; (iii) starting and ending times of the tasks on each processor; (iv) length of the system queues; and, (v) overall statistics such as total simulated execution time, processors idle time, possible speed-ups, etc. This information is presented in a graphical (under Motif) hierarchical manner, giving the programmer the option of requesting more detailed information from level to level. At this point the programmer can study the expected run time behavior and performance of the program. If the performance is not satisfactory, the programmer can modify the program partitions, modify the schedule, or experiment with a different architecture (dashed lines in Figure 1).

## 3. Conclusion

This paper introduced PARSA as a software tool for parallel program partitioning and scheduling in distributed memory multiprocessor systems. The distinguishing features of PARSA include more accurate compile-time estimations of execution times and communication delays; choice of many partitioning and scheduling tools; and, a built-in graphical assessment tool for evaluation and tuning of parallel programs before execution.

## 4. References

[App89] B. Appelbe, K. Smith, C. McDowell, "Start/Pat: A parallel-programming toolkit," *IEEE Software*, July 1989, pp. 29-38.

[Bai90] D.A. Bailey, J.E. Cuny, C.P. Loomis, "ParaGraph: Graph editor support for parallel programming environments," *Int'l J. of Parallel Programming*, v. 19, no. 2, 1990, pp. 75-110.

[Ber87] F. Berman, "Experience with an automated solution to the mapping problem," in *The Characteristics of Parallel Algorithms*, L.H. Jamieson, D.B. Gannon, R.J. Douglass, MIT Press, 1987, pp. 307-334.

[Ber88] T. Bemmerl, "An integrated and portable tool environment for parallel computers," *Int'l Conf on Parallel Processing*, 1988, pp. Vol. II-50-53.

[Don87] J.J. Dongarra and D.C. Sorensen, "SCHEDULE: Tools for developing and analyzing parallel FORTARN programs," in *The Characteristics of Parallel Algorithms*, L.H. Jamieson, D.B. Gannon, R.J. Douglass, MIT Press, 1987, pp. 363-394.

[Elr90] H. El-Rewini and T.G. Lewis, "Scheduling parallel program tasks onto arbitrary target machines," *J of Parallel and Distributed Computing*, v. 9, 1990, pp. 138-153.

[Gua89] V.A. Guarna, D. Gannon, D. Jablonowski, A. Malony, Y. Gaur, "Faust: An integrated environment for parallel programming," *IEEE Software*, July 1989, pp. 20-27.

[Hur90]. A. Hurson, B. Lee, B. Shirazi, M. Wang, "A program allocation scheme for dataflow computers," *Int'l Conf. on Parallel Processing,* 1990, pp. I-415-423.

[LLN85] "An Intermediate Form Language IF1," Lawrence Livermore National Laboratory reference manual, 1985.

[Lo91] V.M. Lo, et al, "OREGAMI: Tools for Mapping Parallel Computations to Parallel Architectures," *Int'l J. of Parallel Programming*, v. 20, no. 3, 1991, pp. 237-270.

[Pei86] J.K. Peir, D. Gajski, "CAMP: A programming aide for multiprocessors," *ICPP*, 1986, pp. 475-482.

[Shi90] B. Shirazi, M. Wang, G. Pathak, "Analysis and evaluation of heuristic methods for static task scheduling," *J of Parallel and Distributed Processing*, Vol. 10, 1990, pp. 222-232.

[Shi91] B. Shirazi, M.F. Wang, B. Lee, and A.R. Hurson, "Accurate communication cost estimation in static task scheduling," *HICSS-24*, 1991, pp. 10-16.

[Shi92a] B. Shirazi, A.R. Hurson, "Scheduling and load balancing," *Journal of Parallel and Distributed Computing*, Special issue on scheduling and load balancing, Dec. 1992.

[Shi92b] B. Shirazi, K. Kavi, A.R. Hurson, P. Biswas, "PARSA: a parallel program scheduling and assessment environment," Tech Report, Dept of CSE, UTA, Dec. 1992.

[Shi93] B. Shirazi, A.R. Hurson, "Scheduling and Load Balancing," *Hawaii Int'l Conf on Systems Sciences*, Software Track on Scheduling and Load Balancing, 1993.

# VSTA: A PROLOG-BASED FORMAL VERIFIER FOR SYSTOLIC ARRAY DESIGNS

*Nam Ling* and *Timothy Shih*
Computer Engineering Department
Santa Clara University, Santa Clara, CA 95053, U.S.A
email: nling@SCU.bitnet phone: (408)554-4794

***Abstract***[1] - *Special purpose formal design verifier for a specific class of architecture has the advantage of being able to exploit the attributes of that architecture class to produce efficiency in the design verification process. Such development is important due to the fact that architecture design verification using general purpose theorem prover is usually extremely time consuming. This paper briefly presents a Prolog-based verifier VSTA that we developed for formal design verification of systolic array architectures. VSTA is based on a formalism we developed earlier (called STA) and its associated design specification and verification procedures. The verifier exploits systolic attributes for fast design verification.*

## 1. INTRODUCTION

As systolic arrays become more complex in terms of cell structure, interconnection topology, and data flow, the need for using formal design verification has become obvious in guaranteeing the correctness of complex designs with respect to the upper functional level specifications. Although several mapping techniques have already been developed to produce correct-by-construction arrays from given algorithms, those techniques can only be applied to a limited class of algorithms. In fact, many systolic arrays reported are designed by ad hoc or systematic, but not necessarily formal, techniques. In view of this, there has been a growing amount of research work in formal design verification for systolic arrays. In spite of contributions like Melhem [1], Hennessey [2], Purushothaman [3], and a few others (a detailed discussion is provided in [4]), the amount of work in this area is still rare. A few techniques are implemented on semi-automated tools (e.g. the use of general purpose Boyer-Moore theorem prover by Purushothaman [3]).

In our earlier work, we have developed a special purpose formalism, called Systolic Temporal Arithmetic (STA), for specifying and verifying systolic array architectural designs at the array level [4]. STA exploits the nature of systolic architecture to provide elegant notations and efficient verification procedures [4]. In view of the fact that architecture design verification using general purpose theorem prover is usually extremely time consuming, we have decided to develop a special purpose formal design verifier to implement STA so as to exploit the attributes of systolic array architecture to produce efficiency in the verification process.

Due to the limitation in paper length, STA formalism is not reviewed in this paper. We concentrate on presenting a suitable Prolog-based verifier VSTA that we developed to automate our techniques. Our verifier is designed with the following goals in mind: 1. Ease of encoding, debugging, and manipulating the steps of execution; 2. Quality and correctness of the verification process; and 3. Fast execution time.

## 2. A BRIEF REVIEW OF SYSTOLIC DESIGN VERIFICATION TECHNIQUES

Interested readers can refer to [4] for a detailed discussion of STA and several design specification and verification techniques. This section only provides a brief review so that later discussion can be appreciated.

A formal verification of a systolic array at the array architectural level is a mathematical process for checking whether the array level architecture realizes the mathematical expression of the function that it is supposed to realize. The verification process comprises of the proving of the correctness of the following *correctness formula*:

$$S_S(n), S_F, S_I(n) \vdash S_O(n)$$

for a systolic array of size $n$. Here $S_S$, $S_F$, $S_I$ and $S_O$ represent the structural design specification (description) of the array (types of cells used and how they are connected), the functional specification of each array cell (PE), the array input behavior specification, and the intended array output behavior specification, respectively. That is, the correctness of the array means that if the array is built according to the structural specification, if the array inputs satisfy the input specification (includes initial conditions), and each array component and connection operates in faithfulness to its functional specification, then the array outputs are guaranteed to satisfy the intended output specification. Since this output specification constitutes the mathematical expression of the upper level function with appropriate array operation cycle delays, proving of the correctness of this implication will mean that the array realizes the function. Our specification (description) is written in declarative STA sentences. The preciseness of the description is followed from the precise semantics of STA [4] used.

We apply inductive techniques to verify the correctness formula. Due to the repeatability, regularity, and locality nature of systolic arrays, we found that inductive techniques are very efficient and the number of steps does not depend on the array size.

Let $P(n)$ be the correctness formula for an array of size $n$, proving the correctness of $P(n)$ involves showing $P(n_0)$, $n_0 < n$ (or $P(N)$, for very large $N$), and showing $P(k+1)$ (or $P(k-1)$), assumed $P(k)$, $k \geq n_0$ (or $k \leq N$). Besides regular

1. The work is supported in part by NSF under Grant No. MIP-9010385.

mathematical induction, structured induction (stronger), double induction (for 2-D or 3-D arrays with several indices), and reverse induction (proof proceeds backward) are also developed (see [4] for details). Although applying inductions to proving 1-D arrays are not difficult, applying them to 2-D or 3-D arrays and automating the proofs with logic programming techniques are not as trivial.

## 3. A PROLOG-BASED SYSTEM

In this paper, we concentrate on discussing VSTA, the Prolog-based verifier we developed for automating the proof techniques. Prolog, being the most popular logic programming language, has served as a powerful tool for proving the correctness of hardware (e.g. [5]). Prolog is adopted due to the following factors: 1. Its usefulness and similarity to STA in representing bodies of facts in predicate forms (hence efficiency in encoding, understanding, and debugging). 2. Its power in symbolic manipulation. Prolog's pattern matching automatic backtracking mechanisms are very useful for implementing logical inferences. 3. Its ability to complete a proof within a reasonable amount of time and its implementation minimizes the unnecessary lower abstraction layers. 4. Its popularity and its wide acceptance for lower level module and circuit specifications and verifications. This allows the forming of a multilevel reasoning system.

In our verifier, temporal variables are encoded using Prolog list structure. Slight dissimilarity between Prolog and STA is bridged by a few operator definitions in Prolog. The representation of STA axioms, rules, and theorems are expressions in Prolog. They can be decomposed and built up by Prolog predicates. These are represented as abstract objects in VSTA, and can be manipulated easily by our verifier.

## 4. THE BASIC REASONING STEPS OF VSTA

Formal specification of a systolic design is input to the verifier in three different forms (clauses). If both the antecedents and consequent exist in a logic implication, the resulting clause is an "inference rule". Cell functional specification falls into this category (in an implication, cell input description and cell type form the antecedents and the functional description of the cell output is the consequent). If a clause has only the consequent part, the clause is a "fact". Structural specification and input behavior specification fall into this category. On the other side, if a clause has only the antecedent part, the clause is a "question". The output specification to be proved is provided by the user as a question to the verifier. The input to our verifier thus consists of (1) declaring constrained facts (quantified predicate-type specifications), (2) defining inference rules (implication-type specifications), and (3) asking questions (a yes/no question on output specification). Our Prolog-based verifier is a man-machine interactive tool using induction, backward chaining, and rewriting to perform a proof of the goal. Normalization techniques are applied to normalize temporal variables. The output specification is formatted as a yes/no question input to the verifier and is treated as the goal to be proved. Our verifier takes care of different kinds of goals. Backward proof takes place by matching a goal with the consequent of a rule; the antecedents become the subgoals. This unfolding process (matching each goal/subgoal with a consequent and replacing the consequent part by the corresponding antecedents and these antecedents become new subgoals) repeats until sufficient number of facts are matched, in which case the proof of that goal is completed. Since the STA specification involves some bound quantifier, a fact (or the antecedents of a rule) may have constraints. If the constraints are satisfied, the fact is valid and thus it can be used in a deduction step.

## 5. AUTOMATION OF INDUCTIVE PROOF BY VSTA

VSTA uses backward chaining to implement induction. The goal of proving $P(n)$ is set as a yes/no question and is divided into proving two subgoals: proving $P(n_0)$ (or $P(N)$) and $P(k+1)$ (or $P(k-1)$), assumed $P(k)$. Each subgoal is then treated as goal with the corresponding antecedents treated as subgoals to be proved. This process is done recursively, matching goals against consequents, setting up antecedents as subgoals, and backtracking in case of failure. If all the subgoals can be satisfied or validated, the goal is proved and the answer to the yes/no question will be yes. For the purpose of illustration, a simple example of this is depicted as a proof tree shown in Figure 1(b) for verifying a 1-D systolic array for matrix-vector multiplication (Figure 1(a)). At certain points in the tree, the subgoals are simply the constrained facts (i.e. structural and input specifications themselves), and are therefore satisfied, or auxiliary predicates, which are validated. These are shown as leaf nodes in the tree. Cell functional specifications are used as rules of inference to help set up subgoals. Our verifier allows the user to select appropriate rules in a deduction step to improve efficiency. The fact that Prolog representation is close to our STA notation in representing bodies of facts in predicate forms makes user control

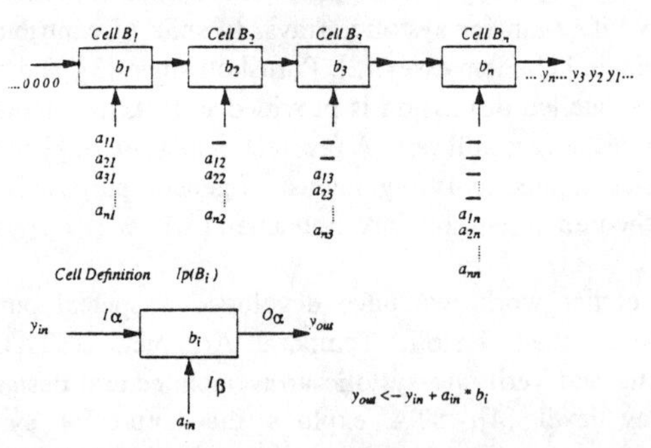


Figure 1(a). 1-D systolic array for matrix-vector multiplication

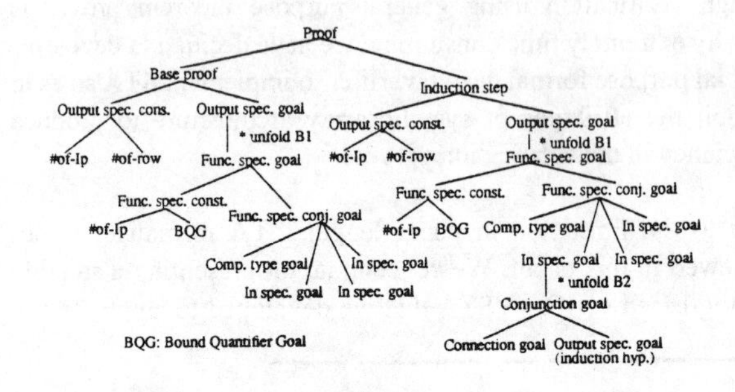


Figure 1(b). A proof tree for the array of Figure 1(a)

and debugging easy.

Referring to the proof tree of Figure 1(b), the root of the proof tree is our goal (i.e. output specification). Proof of this goal consists of proving the base and the induction step (the two branches from the root). The base proof is a constrained goal consisting of constraints (left branch) and a goal body. The constraints (the size of the base case) is a conjunction of subgoals which checks the number of the inner product (Ip) cells and the number of input cycles. These subgoals are satisfied due to the structural specification and the input specification applied to the base case. The goal body of the base proof (its right branch) is the output specification part of $P(n_0)$ in the predicate form. The verifier matches this predicate with a consequent in the implication of the cell (component) function specification ((B1) in the Figure). This implication is treated as an inference rule and the verifier then unfolds this functional specification, replacing the consequent with the corresponding antecedents (with proper substitution) as subgoals to be proved. This is called the functional specification goal, which is again, a constrained goal. The constraint subgoal is similarly satisfied by the quantification aspect of the component type specification. The goal body of the constrained goal is a conjunction goal of four elements: one component type goal and three input specification goals. These four elements are constrained facts given in the specifications, and are therefore satisfied. Hence the base proof is completed. The induction step proof (proof of $P(k+1)$) is done in the same manner except that in one of the input specification goal we have another unfold. This unfold is due to the fact that the input to the level of cells forming the array of size $k+1$ depends on the output of the array of size $k$. This is done by unfolding the function specification of an interconnection, which is simply to transmit signals from one end to another without alteration. This is given as (B2) in the Figure. This function specification is applied to the interconnections which connect the array of size $k$ to cells to form the array of size $k+1$. The antecedents of this implication requires the proof of the existence of appropriate interconnections (connection goal on its left branch) and correct output value from the array of size $k$ (output spec. goal for size $k$ on the right branch). The connection goal is satisfied by structural connectivity specifications while the output specification goal is validated by the inductive hypothesis (correctness of $P(k)$).

In summary, the proof tree shows the decomposition of the proof procedure down to the leaf level. Each formula at the leaf level is either a constrained fact or a goal that can be validated by auxiliary predicates. Since all the leaf goals are satisfied (constrained facts) or valid (auxiliary predicates), the proof is completed. Our Prolog verifier uses depth-first search in the proof process.

## 6. VSTA USER INTERFACE PROTOTYPE

The observable part of the interface splits into six main windows, as shown in Figures 2 and 3. In Figure 2, the upper left window displays the structural design of the array entered by the user. The lower left window displays the function of each cell (PE) used and the function of an interconnection. The upper right window displays the input behavior of the array entered by the user. The lower right window displays the intended output behavior of the array entered by the user. The upper window in Figure 3 displays the VSTA verifier output. Important intermediate steps of the proof process are displayed. The lower window is reserved for the user to enter intermediate decisions to guide the verifier and to select appropriate rules to improve efficiency.

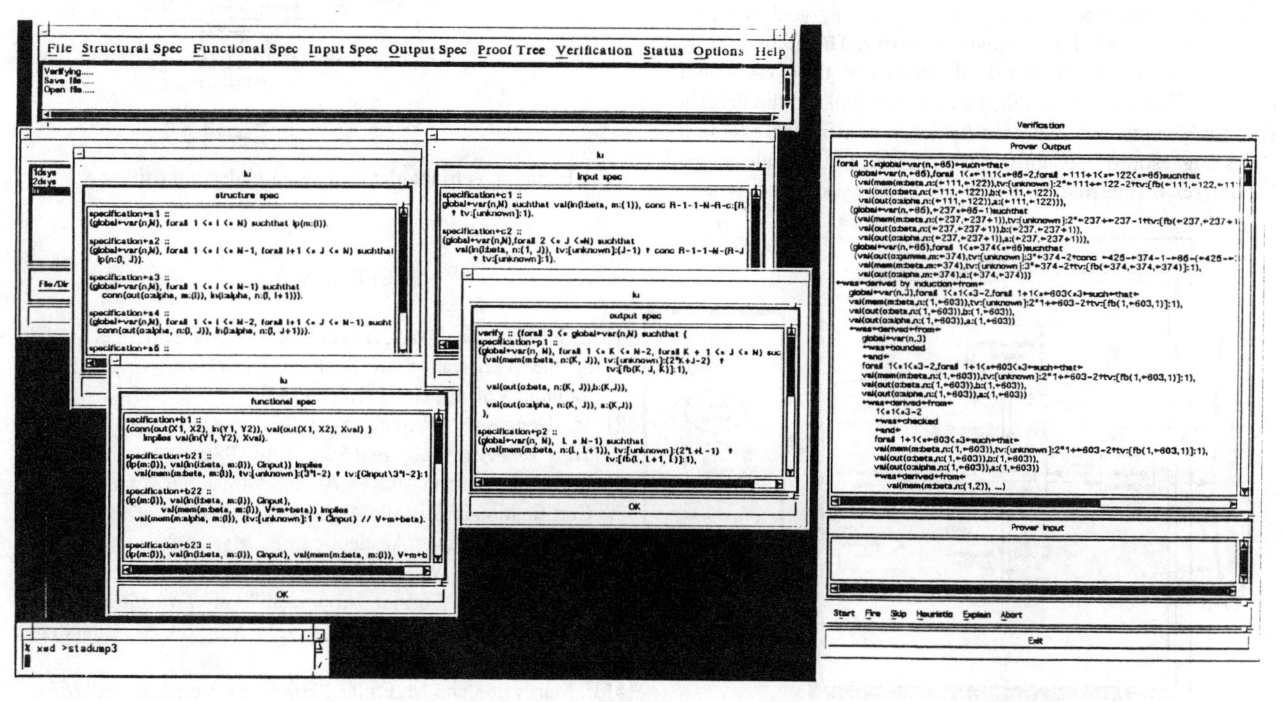


Figure 2. VSTA user interface windows for design specification

Figure 3. VSTA windows for design verification

These windows allow user to change his/her design (structural description and cell functions), as well as input behavior pattern until a correct design (one that matches the intended output behavior) is derived. Through the windows, the user may drive a proof process as well as view the consequences of the steps he/she initiated.

## 7. THE STRUCTURE OF VSTA

The structure of our verifier VSTA is given as in Figure 4. Figure 4 also shows some additional features that are added to form a simple expert system. The user provides three kinds of information to VSTA: the specifications which describe the architecture, the strategy decision which is interactively constructed and selects an appropriate rule to apply in an unfolding step, and the architecture specific heuristics which suggest temporal variable values used in the unfolding process.

VSTA consists of several subprocessors. Each of the subprocessors handles a specific task. For instance, an induction engine generates the base goal and the induction step goal and passes them to the main processor which has a central control routine and communicates with three databases holding STA specifications for the architecture, STA rules, and STA axioms. The central control routine picks a goal (the first goal in the goal list to be solved), passes it to one of the subprocessors, which either solves the goal (in the cases of the three testers), or transforms the goal to a new goal (in the case of the temporal value normalizer and the unfolding rule solver). The three testers receive informations from the STA specification database and the STA axiom database and send messages back to the main processor to indicate whether a goal is solved. The temporal value normalizer normalizes a temporal value if it is not normalized; this process transforms the existing goal to a new goal, which is passed to the main processor. The normalization process invokes the STA rule database and the STA axiom database. Finally, the unfolding rule solver looks at the domain specific heuristic rule database (which is constructed based on user's suggestion) and unfolds the conclusion of a rule to the hypotheses of the rule. The new goal (the hypotheses) is then passed back to the main processor of VSTA. These processes are executed semi-automatically. At some steps the user is asked to give decisions of using a specification or a rule and to provide domain specific heuristics in the process.

Figure 4. The structure of VSTA

## 8. CONCLUSION

The displays shown in Figures 3 and 4 indicate the process of verifying the design correctness of a triangular systolic array, as shown in Figure 5, for LU decomposition. The array decomposes a matrix $C$ into a lower triangular matrix $A$ and an upper triangular matrix $B$. Proving the correctness of the array is not trivial due to its complicated data flow and topology. STA strategy for this proof is given in [4]. The proof is based on regular induction, reverse induction, and rewriting.

The execution of the proofs are performed in SICStus Prolog on our Sun workstations. For a typical array design, it takes a total elapsed time of less than 5 minutes, and an execution time of less than 30 seconds, if excluding time of interaction. Such verifier can also be extended to perform a wider class of pipelined and parallel architecture in later stages.

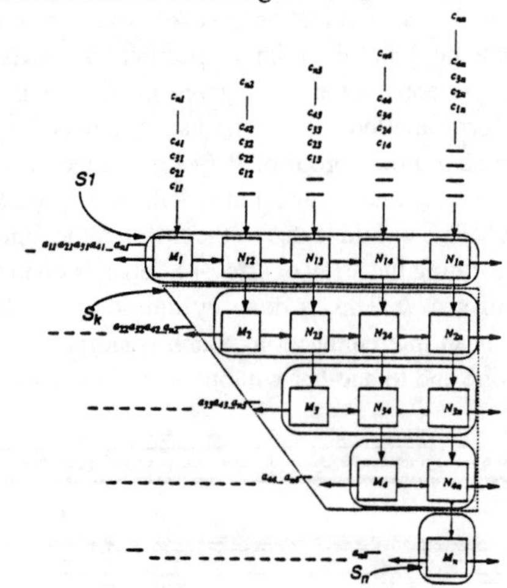

Figure 5. A triangular array for LU decomposition

## REFERENCES


[1] R. G. Melhem and W. C. Rheinboldt, "A Mathematical Model for the Verification of Systolic Networks," *SIAM J. of Comput.*, Vol. 13, No. 3, Aug. 1984.

[2] M. Hennessy, "Proving Systolic System Correct," *ACM Trans. on Prog. Lang. and Syst.*, July 1986.

[3] S. Purushothaman and P. A. Subrahmanyan, "Mechanical Certification of Systolic Algorithms," *Journal of Automated Reasoning*, Kluwer Academic Pub., Mar. 1989.

[4] N. Ling and M. A. Bayoumi, "From Architecture to Algorithm: A Formal Approach," in *Transformational Approaches to Systolic Design*, G. M. Megson, ed., Chapman & Hall, London, 1992.

[5] F. Maruyama and M. Fujita, "Hardware Verification," *IEEE Computer Mag.*, pp. 22-32, Feb. 1985.

# A PARALLEL PROGRAM TUNING ENVIRONMENT

Gary J. Nutt
University of Colorado
Department of Computer Science
Boulder, Colorado
nutt@cs.colorado.edu

Abstract — *A parallel program tuning environment is described to assist shared memory application programmers. The tools rely on a spectrum of underlying support and analysis tools, and many use interactive visual presentations. The value of this work is perceived to be the incorporation of a broad spectrum of tuning tools within a consistent environment that provides fundamental measurement services.*

## INTRODUCTION

Designing and implementing *effective* parallel programs is proving to be a difficult task. The difficulties result not only from the development of effective algorithms that take advantage of data placement and computing cycles, but also from the number of variables in the underlying computing platform. An efficient program must have an effective algorithm that matches the platform configuration. While contemporary abstract machines (e.g., the hardware, operating system, and runtime system) provide access to a relatively uniform set of facilities, the performance characteristics of the implementations of the facilities must generally also be known by the application programmer. For example, the programmer may be able to construct a more efficient matrix multiplication implementation if he knows the size of the cache memory in a shared memory system than if he is oblivious to its size.

Thus the parallel application programmer must be able to observe almost arbitrary performance characteristics of the execution of his program from a number of different perspectives as determined by his specific needs at any time. This suggests that performance tools to support parallel application development must either be defined explicitly for each application domain (or worse, for each programmer), or that the tools must be comprehensive, flexible, and configurable to support specific domains and programmers. The distinction is essentially one of expert systems versus toolkits.

Our experience with individual tools also strongly supports the utility of visualization when addressing parallelism. Irregular concurrency is difficult to comprehend because it inherently has a "thread" for each part of the operation; the observer must be able to synthesize the collective actions of the "threads", e.g., by finding an intellectual lever for abstracting away the details so that one can see the more general pattern of behavior. We have found that this kind of knowledge sometimes requires that the designer be able to obtain a *qualitative* understanding of the behavior of the program before focusing on the *quantitative* details. For example, animation introduces a new dimension to the perception of the performance of a program — the resulting observations tend to be qualitative rather than quantitative. While the animation is useful for identifying general trouble spots in the computation, it is less useful for determining the specific cause. We rely on visualization to gain a qualitative understanding of performance and on traditional measures for observing the performance of the program quantitatively.

A parallel program tuning environment (PPTE) is most useful if it provides a comprehensive set of performance tools which provide a broad (but complementary and consistent) set of views of the execution. Like all other tools, the tuning tools must be easy for the application programmer to use. This is the goal of the tuning environment described in this paper.

The PPTE framework supports either expert systems or toolkits, and tools for both qualitative (visual) and quantitative observations of the performance for shared memory machines. We have been building components for several years, and we are now integrating them into a consistent environment that could be used to design expert system performance tools or tailored environments from generic toolkits. In this paper we provide a high level description of the PPTE, including a brief description some of the tools.

## BACKGROUND

There is a widely held belief that the paucity of software tools is a substantial barrier to the use of parallel machines to solve general problems in science, e.g., see [16]. Researchers that address this area tend to base their approach around *traces* of the programs execution that can be analyzed after the subject program has executed [1,3-5,7,14]. The critical components of the tool are the instrumentation component and the analysis tool itself — some researchers concentrate on the instrumentation and others on the tool. A set of subtle, new problems are introduced when one attempts to use traces to represent the load provided by a parallel program, e.g, the order of event occurrences can be changed by resource utilization and/or nondeterminacy, the tracing tool execution dilation may change the observed behavior, etc. The PPTE has an underlying testbed that has been carefully designed to address most of these concerns [9].

There are radically different perspectives from which to view performance, ranging from performance metric displays to animation. Some of the interface to our PPTE have been influenced by the success of Paragraph and others as tools for tuning a parallel program, e.g., see [12]. Paragraph has illustrated that even modest development efforts at building such a tool can have tremendous impact on the domain programmer, provided that the tool has been designed by/for the programmer rather than in a vacuum. Other aspects of the interface are influenced by our own work with control flow models of parallel computations [6,15]. Our goal for performance visualization is to provide the appropriate view — performance metric or control flow — from the domain programmer's perspective (as opposed to the system perspective).

## THE PPTE FRAMEWORK

The goal of our PPTE is to allow application programmers to tune an application for a shared memory multiprocessor on various configurations of the machine using only a uniprocessor system for the study. Further, the PPTE should enable the application programmer to view the performance from a very broad range of perspectives, including quantitative and qualitative views, visual and numeric reports, metric-based and flow graph views, etc.

The PPTE presumes the existence of a parallel program written using a sequential programming language with a parallel programming package (e.g., C programs that use a threads package), a specification for the target execution architecture (i.e., combination of hardware, operating system, and runtime system), and a uniprocessor with a similar operating system and a processor that executes the target instruction set.† An instrumented version of the program is traced on the uniprocessor, producing a trace with abstract events; the resulting trace is bound to a specific execution architecture, then is used to drive various tuning tools in the PPTE.

A *causal event trace* lists the sequence of certain events in the order in which they occurred on the instrumented specimen; the precise nature of the events is determined by the nature of the instrumentation. A *timestamped event trace* incorporates the causal order, and also adds the virtual time at which each event occurred during the execution. Causal traces can be used to define the load due to the program for some model of the execution architecture; timestamped traces are most useful for direct analysis, since they already incorporate the resultant performance behavior. Both causal and timestamped traces are used in the PPTE.

† We have implemented our PPTE for the C/C threads tools but have also used it on Fortran/Parmacs programs by using a Fortran-to-C conversion package, and by mapping Parmacs calls to C threads calls.

Figure 1 is a block diagram of the components in the PPTE. The environment includes tools to instrument the target program, to execute it to produce an abstract causal trace, to bind the trace to a causal or timestamped trace, and to provide a wide set of views of the performance of the program on a specified execution architecture.

### PEET and SPAE

The Symbolic Program Abstract Execution SPAE trace generator is based on Larus's AE tracing facility [13]; AE is used to collect trace data from sequential C programs while SPAE uses the same mechanism to collect abstract trace data from parallel programs written in C and using the C threads library. SPAE instruments the target program so that it will issue normal events such as memory references, and *abstract events* relating to concurrent operation, e.g., that the program is spawning a new thread, that it is attempting to obtain a lock, etc. The instrumented program can be executed serially on a uniprocessor, causing normal events and abstract events to be saved as an *abstract causal trace*. The AE technique separates information that can be determined about the trace at compile time from information that can only be obtained at runtime; the figure illustrates this by indicating that part of the abstract causal trace is derived directly from the SPAE compiler and part of it from the execution of the program — see [11] for more details.

Both SPAE and AE record arbitrary events in the abstract trace; SPAE also recognizes events related to parallel activity, particularly critical events relating to thread management in C thread calls. Second, SPAE keeps track of execution contexts based on calls to the C threads library; whenever a new thread is created, the instrumented library code emits an abstract event to identify that occurrence. Similar abstract events are emitted whenever a thread call might cause a context switch or a thread to be destroyed. The result is that whenever an uninstrumented program would create or change a context, it does so by calling the thread library; in the instrumented version the thread library has been changed to generate an abstract event to that effect in the trace.

### Binding the Trace

The abstract causal trace is a sequential causal trace from the uniprocessor execution, with abstract events inserted whenever logically concurrent operations were encountered. The abstract causal trace could be bound to a *specific causal trace* by simulating the behavior of a particular parallel execution architecture on each abstract event, translating its effect into a set of parallel event traces. Notice that this simulation essentially performs two tasks: it converts abstract events into specific events, then establishes an order on these events with the specific events that exist in the abstract causal trace. After such a simulation, there will be as many specific causal traces as there were processors in the target execution architecture, and no specific causal trace will contain any abstract events.

Translation from abstract to specific causal traces implies that the *trace binder* in the figure simulates the behavior of the runtime thread systems on the target execution architecture, i.e., the trace binder simulates the thread scheduler, lock contention, barriers, and condition variables. In the figure we indicate this by showing how the PEET performance testbed uses SPAE traces to bind abstract traces to specific traces, i.e., PEET is itself an environment for using the trace data produced by SPAE. The trace binder reads the abstract trace, identifies abstract events, then invokes the scheduler/lock manager simulation to have it model the activity. The scheduler may simulate the architecture itself, or it may invoke a more detailed architectural simulation, e.g., one that models cache behavior. When the behavior for an event has been defined, it is fed back to the trace binder. (PEET also employs feedback between the trace binder and the C threads runtime system to literally replace the C threads scheduler and lock management implementations by the simulated scheduler and lock manager in the trace binder. This allows (requires) PEET to gather traces on the fly by directing the execution of the program. This requires that the program be executed each time it is analyzed, but eliminates the need for storing massive trace files.

Specific causal traces are used to drive three classes of performance visualization tools: timestamped trace analyzers, trace-driven flow graph animator/interpreters, and arbitrary high level trace-driven simulations.

A specific timestamped trace can be derived from the specific causal trace by a trace-driven simulation that introduces resource utilization metrics. The simulator uses the causal trace to define the load on the target system, then simulates resource utilization of the resulting execution architecture; e.g., a cache simulator would introduce delays related to cache misses and the implied data movement in the memory hierarchy. The resulting timestamped trace is a serialized audit trail of the performance of the program that can be analyzed to present the performance data using numeric reports or performance visualization tools. Notice that the simulator not only introduces time to the trace, it also filters event occurrences so that only the appropriate events are passed to the presentation tools.

It is sometimes difficult to infer characteristics of the flow of control from conventional performance visualization displays. Instead, a flow graph or precedence graph model may be a more useful view of the parallel program's execution. Petri nets (and Petri net variants) are often used to represent control flow aspects of a parallel program. The tuning environment uses the Olympus systems [15] to provide a visualization tool based on trace-driven Petri nets. Causal traces are used to provide additional constraints on the flow of tokens in a timed Petri net; this results in a view that illustrates the parallelism inherent in the parallel program correlated with the execution of parallel segments constrained by available resources.

Performance visualization tools and Petri net models may not provide precisely the view that the analyst needs to tune the program; the final class of visualization tools uses specific simulators to produce performance reports for specific aspects of the program and platform. In this case, the causal trace provides a load (as in the case of the generic timing simulator used for the performance visualization toolset), but the simulator and its performance reports are arbitrary.

The underlying assumption in the design of the PPTE is that the tool designer cannot always predict the appropriate view of performance data that will be of the most use to the application programmer. The PPTE provides a framework for gathering essential performance data on a uniprocessor, then processing that data with the tools that are most useful to the programmer. It is essential that these tools be easily invoked, so that there are few barriers to their use; that is part of the challenge in building a useful tuning environment. Our approach has focused on providing a useful set of built-in tools rather than in providing a mechanism for customizing tools. However, we believe that the environment is a prerequisite to such visualization toolkits; we will continue our investigation in parallel program tuning with such toolkits once the environment is sufficiently easy to use.

## CONCLUSIONS

The PPTE represents the convergence on a particular set of base facilities (PEET) for obtaining trace data, and on using various parts of PEET to support different trace analyses. PEET is a particularly valuable part of the PPTE because of its ability to produce specific causal traces for multiprocessors while executing the PPTE in a uniprocessor environment The measurement and modeling tools that we have been developing for many years have been adapted to use PEET specific causal traces, illustrating how one can define a uniform mechanism for composing diverse performance tools into a single PPTE.

Various workers have added critical components to the PPTE. For example Farber and Grunwald built a trace binder and architecture simulator in the course of a study of cache behaviors [8]. Grunwald also adapted his simulation package so that other architecture simulators could use the PEET tracing facilities [10].

We believe that the utility of performance tuning tools depends heavily on the breadth of views that can be offered, and the ease with which the tools can be used.

The PPTE does not provide any of the tools per se, but it provides the infrastructure for which those tools can be developed with far less effort than if they were to developed without it. In some cases we have built tools that represent this philosophy, (Olympus tools, the ProVis tool [2]) while in other cases our position is speculative, but based on other experiences (high level trace-driven simulators).

Our intent is to continue to use the existing PPTE facilities by refining our work in trace-driven control flow (Petri net) simulation, performance visualization tools and toolkits, and in selective high level simulation tools. At this point we believe that performance tuning can only take a major step forward by using a PPTE like the one we have built, and then by applying visualization toolkit technology to the environment.

**ACKNOWLEDGEMENTS**

The various projects that are parts of the parallel program tuning environment have been built by the author and many other people. First, PEET is due to work with Tony Sloane, Dirk Grunwald, Dave Wagner, and Ben Zorn, otherwise known as the Parallel Program Measurement Group. Grunwald and Phillip Farber built the architectural simulators. Olympus and related tools were built with many other people over several years, notably Bruce Sanders, John Hauser, Steve Elliott, Adam Beguelin, Isabelle Demeure, Jeff McWhirter, and Mohammad Amin. ProVis was built by Casey Boyd, Mike Jones, and Mike Thielen. Various parts of this work have been supported by NSF, U S West Advanced Technologies, Bull System Automatique, and others.

**REFERENCES**

1. A. Agarwal, J. Hennessy and M. Horowitz, ''Cache Performance of Operating System and Multiprocessing Workloads'', *ACM Transactions on Computer Systems 6*, 4 (November 1988), 393-431.
2. C. Boyd, M. Jones and J. Thielen, ''Visualizing the Performance of Parallel Programs: Interface Design using Task-Centered Walkthroughs'', University of Colorado Department of Computer Science unpublished note, June 1992.
3. E. A. Brewer, C. N. Dellarocas, A. Colbrook and W. E. Weihl, ''PROTEUS: A High-Performance Parallel-Architecture Simulator'', Massachusetts Institute of Technology, MIT/LCS/Tech. Rep.-516, September 1991.
4. R. G. Covington, S. Madala, V. Mehta, J. R. Jump and J. B. Sinclair, ''The Rice Parallel Processing Testbed'', in *Proceedings of the ACM Sigmetrics Conference*, May 1988.
5. H. Davis, S. R. Goldschmidt and J. Hennessy, ''Multiprocessor Simulation and Tracing Using Tango'', in *Proceedings of the ICPP*, May 1991.
6. I. M. Demeure and G. J. Nutt, ''Collected Papers on VISA and ParaDiGM'', University of Colorado, Department of Computer Science Technical Report CU-CS-488-90, August 1990.
7. S. Eggers, D. Keppel, E. Koldinger and H. Levy, ''Techniques for Efficient Inline Tracing on a Shared-Memory Multiprocessor'', in *Proceedings of the ACM Sigmetrics Conference*, May 1990.
8. P. G. Farber, ''Analysis of a Shared Bus Multiprocessor Memory System Using Trace Driven Simulation'', University of Colorado Department of Computer Science, M. S. Thesis, 1991..
9. D. Grunwald, G. Nutt, A. Sloane, D. Wagner and B. Zorn, ''A Parallel Execution Evaluation Testbed'', University of Colorado Department of Computer Science Technical Report No. CU-CS-560-91, November 1991.
10. D. Grunwald, ''Awesime: An Object Oriented Parallel Programming and Simulation Systems'', University of Colorado Department of Computer Science Technical Report No. CU-CS-552-91, 1991.
11. D. Grunwald, G. Nutt, A. Sloane, D. Wagner and B. Zorn, ''A Testbed for Studying Parallel Programs and Parallel Execution Architecture'', in *Proceedings of MASCOTS 93*, January 1993, 95-106.
12. A. H. Hayes, M. L. Simmons and D. A. Reed, *Workshop Summary Parallel Computer Systems: Software Performance Tools*, NSF and Department of Energy, Santa Fe, NM, October 1991.
13. J. R. Larus, ''Abstract Execution: A Technique for Efficiently Tracing Programs'', *Software -- Practice and Experience 20*, 12 (December 1990), 1241-1258.
14. C. L. Mitchell and M. J. Flynn, ''The Effects of Processor Architecture on Instruction Memory Traffic'', *ACM Transactions on Computer Systems 8*, 3 (August 1990), 230-250.
15. G. J. Nutt, ''Collected Papers on Olympus'', University of Colorado, Department of Computer Science Technical Report CU-CS-518-91, February 1991.
16. C. M. Pancake, ''Where are we Headed?'', *Communications of the ACM 34*, 11 (November 1991), 53-64.

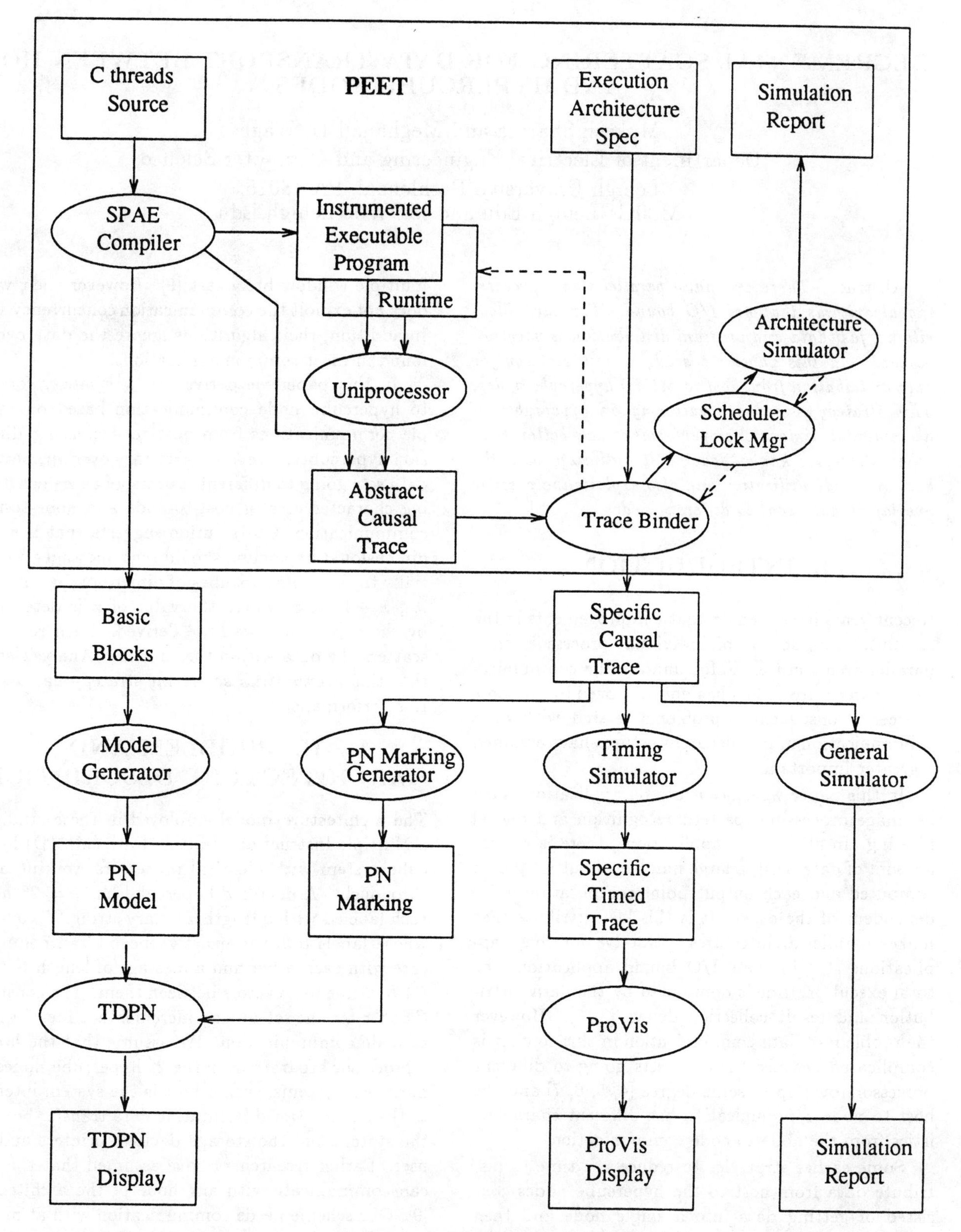


Figure 1: The PEET Tracing Facility

# DECREMENTAL SCATTERING FOR DATA TRANSPORT BETWEEN HOST AND HYPERCUBE NODES

Mukesh Sharma and Meghanad D. Wagh
Department of Electrical Engineering and Computer Science,
Lehigh University, Bethlehem, PA 18015.
MS0G@Lehigh.Edu and MDW0@Lehigh.Edu

Abstract –*There are many parallel signal processing algorithms that are I/O bound. For such algorithms, fast data and program distribution is very important. In this paper we study a new strategy for data distribution from host to MIMD hypercube nodes. This strategy, based on scattering on hypercubes of decremental dimensions, performs much better than other strategies known today. It optimizes both the host and node utilization and also exploits the possible overlap in data sent to different nodes.*

## I. INTRODUCTION

Recent years have seen dramatic improvements in the computational speeds of individual processors in a parallel architecture. Unfortunately the communication between processors has not improved to the same degree. Consequently, problems related with data communication in parallel architectures have assumed a greater importance.

In this paper we concentrate on applications such as image processing, pattern recognition and digital filtering. In all of these applications, there is a large amount of data used, a large number of output points computed and each output point computation is independent of the other. It is this last attribute that makes parallel architectures attractive for these applications [1]. In these I/O bound applications the total execution time is dominated by the data distribution and result collection delays [2, 3]. However the problem of data communication in this context is complicated because the data sets going to different processors overlap to some degree [4, 5, 6, 7] and the host to node communication has different characteristics from the node to node communication.

Some earlier strategies to reduce the time to distribute data from host to the hypercube nodes consisted of getting data into a single node and then broadcasting it from there [8]. This implies that while the broadcast is going on within the hypercube, the host is idle. Recently Prasad and Murty have given several new strategies to keep host busy concurrently with the node broadcast through scattering and multiple window broadcast [9]. However their work does not exploit the communication concurrency fully. In addition, their algorithms ignore the data overlap found in most common applications.

In this paper we derive a new strategy for host to hypercube node communication based on multiple communications from host to decreasing dimension hypercubes. We allow arbitrary overlaps between data sets going to different processors as well as differing characteristics of host-to-node and node-to-node communication. Our solution suggests that host of a dimension $d$ hypercube should send messages sequentially to the roots subcubes of dimensions $d-1$, $d-2$, ..., $x+1$, $x$, $x$, where the value of $x$ is determined by the expressions we have derived. Each root then scatters the data within the subcube. Analysis shows that this *decremental scattering* strategy has a superior performance.

## II. PROBLEM AND ARCHITECTURE DESCRIPTION

The architecture model employed in these studies is a Multiple Instruction Multiple Data (MIMD) hypercube system with identical processors working asynchronously. A degree $d$ hypercube has $p = 2^d$ nodes each labeled with a length $d$ binary string. Two nodes whose labels differ in exactly one bit can communicate with each other and a message of length N takes $\beta + N\tau$ time to go across between them. The constants $\beta$ and $\tau$ are the set-up and incremental times for node to node communication. We assume that the host is a processor separate from the $2^d$ hypercube nodes. In most real systems, such a host is the system interface to the outside world facilitating data transfer between the system and the storage devices, printers and the user. Earlier researchers have assumed that the host can communicate with any node of the architecture [9]. Our scheme needs communication with at most $d$ properly chosen processors. Since host has a greater load, length $N$ message transmission between the host and a node requires $\beta_h + N\tau_h$ time where $\beta_h$ and $\tau_h$ are the set-up and incremental times for host to node communication. To simplify the analysis, we assume that the different communication parameters are related by

$$(\beta_h/\beta) = \sigma \quad \text{and} \quad \tau_h = \tau. \tag{1}$$

Note that generally $\sigma > 1$ and $\tau_h = \tau$ because each depends on the transmission baud rate.

We consider applications in which the tasks in each processor are essentially independent. Examples of such applications arise in diverse applications. In digital filtering of a long sequence, each processor is required to compute one (or more) output points by taking inner product of a part of the input sequence with the filter sequence. The inner products being executed in distinct processors are independent of each other. Image processing through various two dimentional filters clearly fall in the same category. Pattern recognition problems can be solved by distributing the image within the processors and letting each search in its domain for the pattern independently. Note that even though the region along the border between two processors needs to be multiply assigned, there needs be no interaction between the neighboring processors.

In each of these applications, neighboring nodes share certain amount of data. We abstract this notion of data sharing as follows. Let $S$ denote the set of input points and $S_i, 0 \leq i < p$ the set of points to be transported to processor $i$. Clearly, $S = \cup_{i=0}^{p-1} S_i$. We assume that all sets $S_i$ are of equal size $M$, i.e., $|S_i| = M, \quad 0 \leq i < p$, and that the number of elements common to any pair of sets $S_i$ and to $S_{i+j}$ is independent of $i$, i.e.,

$$|S_i \cap S_{i+j}| = k_j, \quad 0 \leq i < p \text{ and } j = 1, 2, \ldots$$

For efficient data transport, the important parameters are the set size $M$, and the number of *new* elements in set $S_i$ beyond the collection of all earlier sets. We denote this number of extra elements by $\Delta$ and define it as:

$$\Delta = |\cup_{j=0}^{i} S_j| - |\cup_{j=0}^{i-1} S_j|.$$

The value of $\Delta$ is decided by the application and the algorithm. In a large number of Digital Signal Processing applications it is possible to predetermine the structure and details of the computations, allowing the user to detemine the data overlap. Many of the typical bilinear algorithms have as much as 70 to 80% redundancy in data sets required for various computations. If a data transport algorithm can exploit this redundancy, then the overall efficiency of the parallel computations will improve a great deal.

We calculate $\Delta$ in two sample cases. The first case, typical of digital filtering, is characterized by

$$|S_i \cap S_{i+j}| \supset |S_i \cap S_{i+j+1}|. \tag{2}$$

Here, $M > k_1 > k_2 \ldots$ and

$$\Delta = |\cup_{j=0}^{i} S_j| - |\cup_{j=0}^{i-1} S_j|$$

$$= |\cup_{j=0}^{i-1} S_j| + |S_i| - |[\cup_{j=0}^{i-1} S_j] \cap S_i| - |\cup_{j=0}^{i-1} S_j| \tag{3}$$

The last step is obtained by using the Principle of Inclusion and Exclusion [10]. Now distributing the Intersection over Union in the above equation and using (2), we get from (3),

$$\Delta = M - k_1.$$

In the second case, we assume $k_3 = 0$. Thus,

$$|[\cup_{j=0}^{i-1} S_j] \cap S_i| = |\cup_{j=0}^{i-1} [S_j \cap S_i]|$$

$$= |[S_{i-1} \cap S_i] \cup [S_{i-2} \cap S_i]| \tag{4}$$

Equation (4) used $k_3 = 0$. Applying the principle of Inclusion and Exclusion [10] again, we get

$$|[\cup_{j=0}^{i-1} S_j] \cap S_i| = k_1 + k_2 - k_2', \tag{5}$$

where $k_2'$ denotes the cardinality of set $S_i \cap S_{i-1} \cap S_{i-2}$. Finally, combining the results of (3) and (5) we get:

$$\Delta = M - k_1 - (k_2 - k_2').$$

## III. DATA DISTRIBUTION ALGORITHMS

Three methods of distributing data from host to all hypercube processors are available in the literature. The simplest method of required data transfer is Sequential Loading. In this method the host sends $M$ elements to each of the p processors sequentially. This therefore entails a cost of

$$T_1 = p(\beta_h + M\tau_h) \tag{6}$$

Clearly this method of *sequential loading* leaves much to be desired since its use of the available links is very sparse and consequently the communication cost very high.

In order to exploit the large number of links in the hypercube and the fact that communication between nodes is cheaper than that between host and nodes, Saad and Schultz have proposed [8] that the data be first downloaded from the host into a specific node $P_0$ of the hypercube and then scattered from there using the standard data scatter algorithms [8]. The time required to complete the data distribution by this *data scattering* method is given by

$$T_2 = \beta_h + Mp\tau_h + d\beta + M(p-1)\tau \tag{7}$$

The first two terms of the equation represent the time for host to node transfer of N elements while the remaining terms represent the time for scattering the data in the hypercube. This is a good method because during scatter process, the available links of the hypercube are used fairly efficiently.

However, in the *data scattering* method described in [8], the host is idle during the data scatter in hypercube. Prasad and Murthy have therefore suggested a strategy which calls selecting a suitable size subcube of dimension $x$ [9]. The host downloads the data for this subcube into a single node from where it is scattered using standard scatter algorithms. Concurrent with the scattering however, the host sequentially sends data to the remaining nodes of the hypercube. This *sequential/scattering* strategy completes the required data transmission in time given by

$$T_3 = \min_x\{\beta_h + M2^x\tau_h + \max\{(p-2^x)\beta_h + M(p-2^x)\tau_h, x\beta + M(2^x-1)\tau\}\} \quad (8)$$

The first two terms of the equation represent the time to transfer elements from the host to the root of the designated size $x$ subccube. The maximum gives the larger of the two times: scattering time and sequential host loading time. The minimum function indicates that one should choose $x$ appropriately to minimize the cost $T_3$. One may notice that when $x = d$, the degree of the hypercube, one gets the cost of the *data scatter* procedure. Thus, at all times, $T_3 \le T_2$.

There are two problems with the *sequential/scattering* strategy described above. Firstly, since x is an integer, the terms within the max function of (8) cannot be balanced very well resulting in idle nodes or host. Secondly, this method cannot exploit any data overlap because of the host loading individual data sets into many of the nodes.

Our strategy for host to hypercube node communication is based on partitioning the hypercube into progressively smaller degree $d-1, d-2, \ldots, x+1, x, x$ subcubes. We appropriately choose $x$, $(0 \le x < d)$ to minimize the overall communication cost. The host sends data sequentially to the designated roots of each of these subcubes. Each root then scatters the data within the subcube using a small modification of the standard algorithm [1] as follows. At the $i$-th step, $i = 0, 1, \ldots, t-1$, each node of the degree $t$ subcube whose label $q$ ends in $t-i$ zeros transfers appropriate amount of data to node $q + 2^{t-i-1}$. For example, when $t = 3$, at 0th step, node 0 sends data to node 4; at 1st step, nodes 0 and 4 send data to 2 and 6 respectively; and at the 2nd step, nodes 0, 2, 4, and 6 send data to 1, 3, 5 and 7 respectively. Assuming that the data sets being sent to consecutively labeled nodes have an overlap, this scattering scheme allows us to exploit that overlap to the maximum extent. Note that because of the overlap, the amount of data communicated at step $i$ (along each of the participating link) is $M + (2^{t-i-1} - 1)\Delta$.

The total data distribution time using this scheme is obtained by noting that when $\beta_h > \beta$, the last subcube will finish last. Thus the total time expended in this data propagation is the sum of the times spent by the host on all its sequential communications with the subcubes added with the time required to scatter the data in degree $x$ hypercube. This gives the total complexity as

$$T_4 = \min_x\{(d-x+1)\beta_h + (M-\Delta)(d-x+1)\tau_h + \Delta 2^d\tau_h + x\beta + x(M-\Delta)\tau + (2^x-1)\Delta\tau\} \quad (9)$$

By comparing (9) with (6)-(8) one can see the merit of the *Diminishing Scattering* strategy. Note that this strategy requires host to communicate with at most $d$ hypercube nodes, where $d$ is the hypercube degree. This is a feature lacking earlier strategies [9]. A small number of host connections is desirable for realistic implementation.

When the communication parameters are related as in (1), expression for $T_4$ takes the form

$$T_4 = \min_x\{((d-x+1)\sigma + x)\beta + (M-\Delta)(d+1)\tau + (2^d + 2^x - 1)\Delta\tau\} \quad (10)$$

Fig.s 1 and 2 show a comparison of the computational times of the *Diminishing Scattering*(**DS**) strategy with those obtained from earlier algorithms. In Fig. 1, the number of data points sent to each processor, $M = 100$ and $\beta = 800\mu$ and in Fig. 2, $M = 500$ and $\beta = 6500\mu$. In both simulations, $\sigma = 1.5$ and $\tau = 8\mu$. The overlap (defined as overlap $= M - \Delta$) between data sets going to consecutive hypercube nodes is varied between 0 and $M - 1$.

These figures show that the *Diminishing Scattering* strategy is substantially better than the earlier ones. It can exploit the data overlap to lower distribution time. Note that in digital filtering, where each output point is being evaluated by an independent processor, the data overlap is indeed $M - 1$, allowing a substantially lower transport time through this new strategy.

## IV. CONCLUSION

This paper describes a communication strategy to distribute data and program from host to hypercube

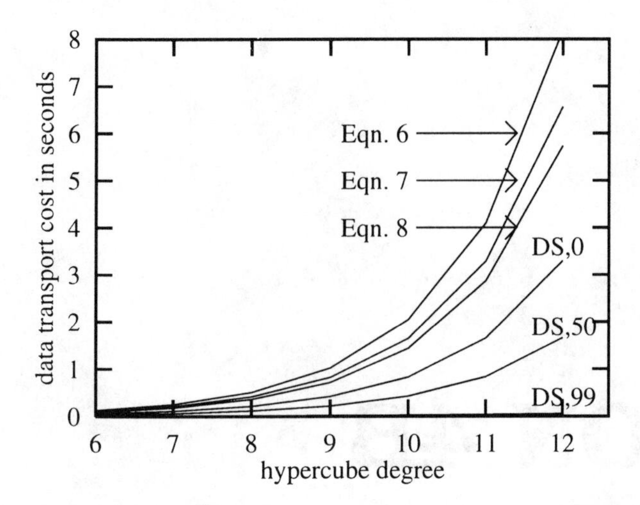


Fig. 1. Comparison of the *Diminishing Scattering* (DS) strategy with others when $M = 100$ and data overlap = 0, 50 and 99

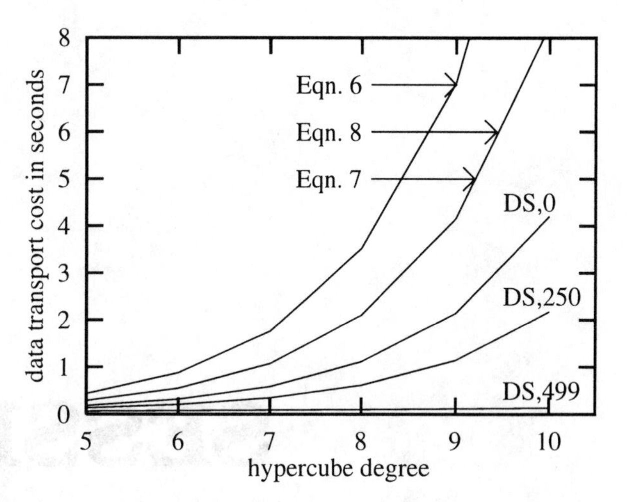


Fig. 2. Comparison of the *Diminishing Scattering* (DS) strategy with others when $M = 500$ and data overlap = 0, 250 and 499

nodes. The new strategy proposed is based on partitioning the hypercube into diminishing degree subcubes. The host sends data sequentially to the selected roots of these subcubes. Using a modified scattering technique described here, the roots then distribute the data into respective subcube nodes.

This strategy, called *Decremental Scattering*, uses subcubes of degree $d-1, d-2, \ldots, x+1, x, x$, where the integer $x$ is chosen appropriately to minimize the total communication time. The choice of $x$ takes into account the communication parameters in both host to node and node to node communication, the data set sizes as well as any possible overlap in the data sets.

Simulation results indicate that the *Decremental Scattering* strategy is highly superior to other strategies presented in the literature. This success of the strategy can be attributed to the fact that it maximizes the usage of all nodes and the host, as well as it meaningfully exploits the data set overlap. Thus, for widely varing communication conditions, this strategy consistently performs better. In applications such as digital filtering where the overlap is very large, this communication technique is highly recommended.

## References


[1] S. Ranka and S. Sahni, *Hypercube Algorithms with applications to image processing and Pattern recognition.* New York, NY: Springer-Verlag, 1990.

[2] E. Jensen, "The honywell experimental distributed processor -an overview," *Computer*, vol. 11, pp. 28–38, Jan. 1978.

[3] T. Agarwala, "Communication, computation and computer architecture," in *1977 Int. Commun. Conf. Rec.*, June 1977.

[4] R. E. Blahut, *Fast Algorithms for Digital Signal Processing.* Reading, MA.: Addision-Wesley, 1985.

[5] M. D. Wagh and H. Ganesh, "A new algorithm for the discrete cosine transform of arbitrary number of points," *IEEE Trans. on Computers*, vol. C-29, pp. 269–277, 1980.

[6] K. A. Doshi and P. Varman, "Optimal graph algorithms on a fixed-size linear array," *IEEE Trans. on Computers*, vol. C-36, pp. 460–470, Apr. 1987.

[7] M. Sharma, *A new algorithm for calculation of the discrete Hartley transform.* M.S. dissertation, Lehigh University, 1989.

[8] Y. Saad and M. H. Schultz, "Data communication in hypercubes," *J. Parallel and Distributed Comput.*, no. 6, pp. 115–135, 1989.

[9] V. V. R. Prasad and C. S. R. Murthy, "Downloading node programs/data into hypercubes," *Parallel Computing*, no. 17, pp. 633–642, 1991.

[10] K. P. Bogart, *Discrete Mathematics.* Lexington, MA.: D. C. Heath and Company, 1988.

# SESSION 4B

## CACHE/MEMORY MANAGEMENT

# Memory Reference Behavior of Compiler Optimized Programs on High Speed Architectures

**John W. C. Fu**

Intel Corporation
1900 Praire City Road,
Folsom, CA 95630
jfu@pcocd2.intel.com

**Janak H. Patel**

Center for Reliable and High-Performance Computing
University of Illinois at Urbana-Champaign
1308 W. Main Street, Urbana, IL 61801
patel@crhc.uiuc.edu

**Abstract**

*High speed architectures usually employ some form of parallelism or concurrency. Parallel or concurrent execution of a program not only increases the rate at which references are issued to the memory system but also changes the behavior of these references, relative to its serial-scalar execution.*

*This paper reports the variations in program memory reference behavior when automatically transformed by a compiler and executed on parallel and vector architectures. Using traces of the PERFECT benchmark set, executed on an Alliant FX/80 in a single scalar processor, single vector processor, scalar multiprocessor and vector multiprocessor modes, measurements are reported for issue rates, reference locality and data sharing.*

## 1. Introduction

High speed architectures usually employ some form of parallelism or concurrency. For example, a multiprocessor executes a program in parallel using multiple processors, while a vector processor overlaps multiple arithmetic operations using one or more pipelined arithmetic units. With respect to its serial-scalar execution, parallel and vector execution increases the rate at which references are issued to the memory subsystem and changes the behavior of these memory references. The variations in memory reference behavior of compiler optimized programs for parallel and vector execution is the focus of this paper.

Consider, the simple matrix multiple, $C = A \times B$. Figure 1 shows a typical triple nested loop form. In scalar execution, each iteration of the loop loads A(i,k), B(k,j) and C(i,j), performs the multiply and the add, and finally writes to C(i,j). Ignoring accesses needed to maintain the index values and the use of registers, to a cache or memory port, references may appear as follows:

$$a(1,1), b(1,1), c(1,1), c(1,1), a(1,2), b(2,1), c(1,1), c(1,1) \ldots c(n,n)$$

assuming a $n \times n$ matrix. There are two main differences when the loop is optimized for vector execution. First, an optimizing compiler restructures the program to execute efficiently on the chosen architecture. For example, on a machine that implements compound vector instructions such as dot-product, the compiler targets the loop for this form of execution. In this case the matrix multiply is transformed into a series of inner-products. Second, memory references, where possible, are made through vector instructions. A possible sequence of references to the cache or memory port is:

$$a(1,1), a(1,2), .., a(1,VR), b(1,1), b(2,1), .., b(1,VR), c(1,1), ..$$

where VR, is the vector register length. In the original scalar execution, access to each array is interleaved as the loop is iterated but in the vector version references are interleaved by the vector register length.

How memory references behave is an important consideration for building effective memory systems. Early studies concentrated on single processor scalar executions. Much of this work was specific to virtual memory systems [1] and cache memories [2]. More recently, studies have focused on parallel program executions on multiprocessors. Early characterization of parallel program executions were reported in [3]. With the increasing interest in multiprocessors, a number of studies have reported characterizations of sharing and the effect of cache coherence protocols on parallel program execution. Some measures used in [4, 5] are used in this paper.

This paper differs from previously published studies in two important aspects. First, the memory reference behavior of 4 "real" numerical programs, executed on 4 execution architectures are examined. Previous studies limited their studies to a single architecture. The architectures considered in this paper are single processor scalar (1S), single processor vector (1V), multiprocessor scalar (8S) and multiprocessor vector (8V). Second, previous studies of multiprocessor program execution used programs written specifically for the selected machine.

**Figure 1: Matrix Multiply**

```
DO I = 1, N
 DO J = 1, N
  DO K = 1, N

     C(I,J) = (CI,J) + A(I,K) * B(K,J)

  CONTINUE
```

The programs used in this paper are large numerical programs originally developed for a variety of different architectures. Each program is used "as is" and vectorized and parallelized by an optimizing compiler.

The base machine used in this study is the Alliant FX/80 [6]; a shared memory vector multiprocessor. Each program is executed and traced in 4 execution architectures or modes. The traces are examined and characterization measures reported for memory reference issue rates, reference localities and data sharing. A recent paper by [7] also used the Alliant FX/80 as the target machine but was concerned with characterizing memory references as a function of the problem size.

The results, presented in this paper, show that vector executions have a much higher issue rate than scalar execution but single processor vector and scalar-multiprocessor executions have a similar range of issue rates. Multiprocessor memory reference issue rates can be very high. Scalar executions show better memory reference locality than vector executions and the degree of data sharing, in parallel execution, is higher for scalar than vector.

As with any trace captured from a real machine, these traces are biased to the target machine; in this case the Alliant FX/80. Moreover, memory reference behavior on a complex architecture depend on a number of factors beyond the base machine organization, for example, the algorithm implemented and the quality of the compiler. These results should be interpreted as an *example* of how memory reference behavior can vary across these particular execution architectures. The results are important because while studies have shown memory reference behavior for single and multiprocessors with respect to specific mechanisms, we are not aware of a study that presents measurements across scalar and vector processing in single and multiprocessor modes.

Section 2 describes the Alliant FX/80, the tracing method and the traces. Section 3 looks at the memory reference issue rates or the memory bandwidth requirement for the execution modes. Section 4 reports locality and data reuse measures. How sharing occurs is an important consideration in multiprocessor memory design. Section 5 presents some results for data sharing in scalar and vector multiprocessor executions. Section 6 provides concluding remarks.

## 2. Method

This section briefly describes the Alliant vector multiprocessor system, the execution modes and how the memory traces were generated and collected.

### 2.1. Alliant Vector Multiprocessor

The Alliant vector multiprocessor consists of two processor types: the Interactive Processor (IP) and the Computation Element (CE). The IPs are primarily used for input/output processing and are not considered further in this paper. There are 8 CEs and these form the main computation resource. A CE consists of a scalar processor with a vector unit. A program can be compiled and executed in a combination of scalar, vector, single and multiprocessor modes. A program compiled for pure scalar execution uses only the scalar processor and its scalar registers. A vectorized program executes the vector instructions on the vector unit but all other instructions are executed on the scalar processor. A scalar program can be executed on a single processor or up to 8 scalar processors. Similarly, a vectorized program can be executed on a single CE or in vector-parallel mode on up to 8 CEs.

### 2.2. Address Traces

Program execution and tracing is by detailed simulation of the Alliant FX/80 system running on an Alliant FX/80. Tracing includes system and library calls and occurs before the shared cache system. The programs used, ADM, ARC2D, BDNA and DYFESM, are members of the PERFECT benchmark set [8]. Each program is compiled with the FX Fortran compiler which uses the VAST re-structurer. The execution modes are generated as follows. If the original program is compiled without optimization the mode is single processor scalar (1S). If the original program is parallelized it is 8 processor scalar (8S), if vectorization is used the mode is single processor vector (1V) and finally, if the program is parallelized and vectorized the mode is 8 processor vector (8V).

Since these programs can generate more than 2 billion references we statistically sample a program execution to obtain a *sampled trace* [9]. It is this sampled trace of memory references that is analyzed. Each sampled trace consists of about 40 samples of over 200,000 data references. Instruction references are not considered in this paper.

### 2.3. Static Reference Characteristics

Table 1 shows the memory space referenced by the traces[1] and the characteristics of the vector references. For vector executions, *prct vect* is the fraction of all references that are vector references. The *stride distance* and *vector length* distributions are the fraction of the vector references within the stride and vector length ranges. The vector length is the size of vector loads or stores and not the size of the vectors declared in the programs. The maximum vector load length in this case is 256 bytes. The memory space referenced by each trace

[1] Recall that these are sampled traces, so the size of the memory space is not the size of the execution memory space.

ranges from 9.2 KB for ADM to 804 KB for ARC2D. In all cases, single processor scalar execution has the smallest data size. The memory space referenced increases by the largest when a program is vectorized, and except for ARC2D, 8V execution references the most data blocks. This increase in data size is because program restructuring requires the compiler to generate temporary variables to remove data dependencies. For example, scalar expansion transforms a scalar variable into a vector of variables to remove dependencies and allow vectorization.

Parallelizing the programs (1V → 8V) reduces vectorization and vector lengths but increases vector stride distance. The most significant reduction in vectorization occurs in ADM where it is reduced by about 50%. Parallelization allocates vectors to multiple processors and this decreases the vector length. ADM has the most significant decrease in vector lengths; where parallelization reduces almost all vector lengths to be 64 bytes or less. Vector allocation also increases the vector stride distance. For example, in ARC2D, 54% of the vector accesses have a stride distance of 1 but after parallelization 11.5% of these vector accesses are allocated to processors with a stride of 64 bytes i.e. consecutive elements are allocated to consecutive processors[2].

### 2.4. Execution Mode Performance

Table 2 shows the execution performance of the 4 programs for the 4 execution modes. These results are for actual (and not emulated) execution of the programs on an Alliant FX/80. Each value is the average of 3 execution for each program. The relative speed up is the ratio of the cpu TIMEs. The sources were compiled as described above and timed using the routines supplied with the PERFECT benchmarks. These results are used to place the execution modes within a performance context and not as an evaluation of the Alliant or as a recommendation of one Alliant execution mode over another[3].

As expected, 8V and 1S show the highest and lowest performance respectively. However, the relative speed up of these two architectures show a surprising range, with a speed up of 38 for ARC2D but only a speed up of 4 for ADM. The execution time for a particular program decreases when it is vectorized. This is true for both single processor (1S → 1V) and multiprocessor (8S → 8V) vectorization. The relative speed up is significantly higher for multiprocessor than for single processor for 3 of the programs (ADM, ARC2D and BDNA).

The results for 1V and 8S are interesting to compare since these two modes are the most commonly considered architectures for computation speed up. ADM is the only program that shows scalar parallelization (8S) performing better than vectorization (1V). This suggests a vector processor is more desirable for these types of problems since the cost of a single scalar processor plus a vector unit is likely to be less than a system with 8 processors and the memory system less complex to implement. However, this result should be considered within the context of the Alliant, a machine designed for vector processing. Comparing the best vector processor design against the best scalar multiprocessor with the same cost range is more fair. Nevertheless, this result is interesting since intuition suggests that a multiprocessor must show significantly higher performance to justify the increased complexity for parallel execution.

**Table 1: Static Reference Characteristics.**

| ARCH | memory space referenced (KB) | | | |
|---|---|---|---|---|
| | ADM | DYFESM | ARC2D | BDNA |
| 1S | 17.9 | 9.2 | 640.0 | 49.7 |
| 1V | 18.6 | 9.3 | 804.3 | 55.5 |
| 8S | 17.9 | 9.2 | 750.7 | 50.2 |
| 8V | 19.2 | 9.6 | 764.3 | 57.7 |

| TRACE | ARCH. | prct vect. | stride distance | | | | vector length | |
|---|---|---|---|---|---|---|---|---|
| | | | 0 | 1 | 9-128 | >128 | 1-64 | 65-256 |
| ADM | 1V | 54.2 | 59.3 | 22.4 | 14.3 | 4.0 | 76.4 | 23.6 |
| | 8V | 23.6 | 33.7 | 17.3 | 36.3 | 12.8 | 99.0 | 1.0 |
| DYFESM | 1V | 61.3 | 40.2 | 37.4 | 17.2 | 5.3 | 70.4 | 29.4 |
| | 8V | 40.4 | 42.6 | 32.2 | 19.9 | 5.2 | 75.1 | 24.9 |
| ARC2D | 1V | 95.1 | 1.2 | 54.4 | 0.0 | 44.4 | 1.1 | 98.9 |
| | 8V | 83.4 | 0.3 | 38.9 | 11.5 | 49.3 | 5.9 | 94.1 |
| BDNA | 1V | 84.0 | 8.4 | 7.1 | 79.7 | 4.8 | 58.6 | 41.4 |
| | 8V | 72.0 | 16.3 | 7.1 | 73.5 | 5.1 | 62.9 | 37.1 |

**Table 2: Actual Execution Time for each Mode.**

| TRACES | ARCH | TIME (secs) | MFLOPS | relative speed up | | |
|---|---|---|---|---|---|---|
| | | | | 1S | 1V | 8S |
| ADM | 1S | 1463.1 | 0.34 | 1.00 | 0.77 | 0.44 |
| | 1V | 1128.8 | 0.45 | 1.30 | 1.00 | 0.57 |
| | 8S | 640.9 | 0.79 | 2.28 | 1.76 | 1.00 |
| | 8V | 353.8 | 1.42 | 4.14 | 3.19 | 1.81 |
| DYFESM | 1S | 1312.4 | 0.42 | 1.00 | 0.18 | 0.26 |
| | 1V | 241.5 | 2.29 | 5.44 | 1.00 | 1.39 |
| | 8S | 336.0 | 1.64 | 3.91 | 0.72 | 1.00 |
| | 8V | 201.3 | 2.74 | 6.52 | 1.20 | 1.67 |
| ARC2D | 1S | 8318.4 | 0.22 | 1.00 | 0.16 | 0.16 |
| | 1V | 1371.3 | 1.34 | 6.07 | 1.00 | 0.99 |
| | 8S | 1357.4 | 1.36 | 6.13 | 1.01 | 1.00 |
| | 8V | 218.6 | 8.43 | 38.05 | 6.27 | 6.21 |
| BDNA | 1S | 1072.0 | 0.84 | 1.00 | 0.46 | 0.79 |
| | 1V | 491.6 | 1.84 | 2.18 | 1.00 | 1.72 |
| | 8S | 846.7 | 1.07 | 1.27 | 0.58 | 1.00 |
| | 8V | 127.3 | 7.12 | 8.42 | 3.86 | 6.65 |

[2] This is because of the horizontal partitioning, dynamic allocation policy used by the Alliant. This is described in Section 5.

[3] Since, the Alliant FX/8 is designed as a vector multiprocessor this should clearly be the preferred execution mode. Alliant FX/80 performance results with all the PERFECT codes are available in [8].

## 3. Memory Reference Issue Rate

This section reports the data reference issue rate for the four execution modes while executing the benchmarks and assuming an idealized memory system, i.e. a processor does not stall for memory data. In calculating the issue rates each scalar reference is considered a single memory reference regardless of the data size being accessed. A vector memory instruction specifies multiple data elements. This is considered as multiple data accesses in consecutive cycles. For example, a load of $n$ elements is counted as $n$ consecutive references in $n$ consecutive cycles. This is the maximum rate memory references can be issued to a single ported memory system. In a multiprocessor, more than one processor can issue a reference within a particular cycle and the issue rate, to the memory, can exceed one reference per cycle. As processors issue references independently this is the maximum bandwidth the memory must supply.

Table 3 shows the mean and relative issue rates for the four modes of execution. The relative issue rate is the ratio of the issue rates for two architectures. If we assume each reference is for 8 bytes, the request bandwidth is $issue\ rate \times 8 \times 1/cycle\ time$. For example, for 1S, a memory bandwidth of 8.4 to 15.2 MB/secs is required if the cycle time is 200ns.

The lowest and highest bandwidth requirements are single scalar and multiple vector executions respectively. This is not surprising as they represent the two extremes of expected performance (Table 2). The issue rate for 8V is the only execution mode where the number of references issued per cycle exceeds 1. The issue rate of 8S does not exceed 1 though there are 8 processors executing highly parallel code such as ARC2D. For 8 processors this is surprisingly low but this does not preclude the possibility of multiple reference issues within a cycle during burst issue periods.

**Table 3: Memory Reference Issue Rates.**

| TRACES | ARCH | ISSUE RATE (ref/cycle) | relative issue rate | | |
|---|---|---|---|---|---|
| | | | 1S | 1V | 8S |
| ADM | 1S | 0.35 | 1.00 | 0.83 | 0.46 |
| | 1V | 0.42 | 1.20 | 1.00 | 0.55 |
| | 8S | 0.36 | 1.00 | 1.86 | 1.00 |
| | 8V | 0.77 | 2.20 | 1.83 | 2.26 |
| DYFESM | 1S | 0.38 | 1.00 | 0.79 | 0.48 |
| | 1V | 0.48 | 1.26 | 1.00 | 0.60 |
| | 8S | 0.80 | 2.10 | 1.67 | 1.00 |
| | 8V | 1.28 | 3.37 | 2.67 | 1.60 |
| ARC2D | 1S | 0.21 | 1.00 | 0.41 | 0.33 |
| | 1V | 0.51 | 2.43 | 1.00 | 0.81 |
| | 8S | 0.63 | 3.00 | 1.24 | 1.00 |
| | 8V | 3.85 | 18.33 | 7.55 | 6.11 |
| BDNA | 1S | 0.35 | 1.00 | 0.56 | 0.90 |
| | 1V | 0.63 | 1.80 | 1.00 | 1.62 |
| | 8S | 0.39 | 1.11 | 0.62 | 1.00 |
| | 8V | 1.89 | 5.40 | 3.00 | 4.85 |

As expected the issue rate of a vector execution is higher than the corresponding scalar execution. For example, ARC2D and BDNA have issue rates of 0.21 and 0.35 respectively for 1S. This increases to 0.51 and 0.63 references per cycle after vectorization. This increase in the rate of reference issue, due to vectorization, is less for ADM and DYFESM. These two both programs have a smaller fraction of vector references and shorter vector lengths (Table 1) than BDNA and ARC2D. The combination of limited number of vector accesses and short vectors is why the issue rates for 1V and 1S are similar.

Parallelizing a program results in a higher issue rate. This increase is higher for vector than scalar parallelization. Scalar parallelization (1S $\rightarrow$ 8S) increases the issue rate by a factor of 3 but vector parallelization (1V $\rightarrow$ 8V) can increase the issue rate by a factor of 7.5; almost the increase in the number of processors. This corroborates the conclusion in [10] that vector multiprocessors either need a very large number of parallel memory modules or some form of local memory.

The issue rates for 1V and 8S have a similar range and compatible with the performance results in Table 2. The average issue rate are 0.51 and 0.55 for 1V and 8S respectively. There is some correspondence between the relative issue rates given in Table 3 and the relative performance on the real machine given in Table 2 but in general these issues rates are not direct indicators of performance.

## 4. Memory Reference Locality

Locality measures are useful because they provide intuition in choosing cache and page parameters. For example, a large spatial measure may indicate that a cache block size needs to be large to capture spatial locality and a large temporal measure indicates that a large cache size is necessary because a large number of references occur between references to a particular block.

To measure the locality of the reference streams a "working set" method is used. Let $W_i$ and $W_{i+1}$ be two windows containing $N$ references each. Window $W_i$ is filled with $N$ references and the next $N$ references fills $W_{i+i}$. The references in the two windows are compared. References appearing in $W_{i+1}$ but not found in $W_i$ are references to *new blocks* in $W_{i+1}$. If $S(W_i)$ is the set of references in $W_i$ then the fraction of references that create new blocks in $W_{i+1}$, $B_{new} = \dfrac{S(W_i) - S(W_{i+1})}{N}$. This gives a rate at which new "blocks" are being created normalized to $N$. For an infinite cache this is the maximum miss rate over $N$ references.[4] The reuse rate for W(i+1) is, $B_{reuse} = \frac{reuse}{blocks}$ where *reuse* is a count of the references in $W_{i+1}$ that did not create a new block and *blocks* is the number of blocks in $W_i$ plus the blocks created in $W_{i+1}$. This is the amount of reuse normalized to the the number of blocks needed. Since, sampled traces are used, each window measure is averaged to get the sample average with the final measure being the average of all the sample averages. For the multiprocessor traces each processor only considers its own references and coherence effects are not considered. The final multiprocessor measure is the average over all the processors. A similar approach was used in [11].

Measurements for each execution mode are shown shown in Table 4 where $N = 10{,}000$ and the window size is 10,000 references. The results show that vectorization (1S $\rightarrow$ 1V and 8S $\rightarrow$ 8V) increases the rate at which new blocks ($B_{new}$) are created. This increase is larger for single processor where the new block creation rate for vector execution (1V) can be up to 4 times that of scalar execution (1S). This suggests that vector caches need to be larger than scalar caches to be effective. Similarly, parallelization (1S $\rightarrow$ 8S and 1V $\rightarrow$ 8V) increases the new block creation rate (except for BDNA 1V $\rightarrow$ 8V). This increase is expected since in parallel execution the data space and the references made to the data space are distributed across all the processors.

The reuse measure $B_{reuse}$ shows that scalar executions have much more reuse. However, recall in the matrix multiply (Figure 1) that scalar executions have more accesses to index variables which is likely to be the cause of these results. The lowest reuse is ARC2D executing on 8V. This program uses a vary large data set with about 90% vectorization.

## 5. Multiprocessor Executions

When multiple processors are used to execute a program in parallel the memory system must provide efficient access to data with minimal penalties for conflicts. This section presents measurements of the sharing characteristics of parallel scalar and vector executions. Cache memories are necessary to reduce the traffic to main memory these measures and are concerned with the effect of parallel execution on cache memories.

**Table 4: Memory Reference Locality Measures.**

| ARCH | ADM | | DYFESM | | ARC2D | | BDNA | |
|---|---|---|---|---|---|---|---|---|
| | $B_{new}$ | $B_{reuse}$ | $B_{new}$ | $B_{reuse}$ | $B_{new}$ | $B_{reuse}$ | $B_{new}$ | $B_{reuse}$ |
| 1S | 0.049 | 10.10 | 0.055 | 8.65 | 0.290 | 1.23 | 0.064 | 7.53 |
| 1V | 0.095 | 4.79 | 0.133 | 3.23 | 0.531 | 0.45 | 0.264 | 1.42 |
| 8S | 0.131 | 2.87 | 0.122 | 4.46 | 0.338 | 0.66 | 0.072 | 7.35 |
| 8V | 0.202 | 4.67 | 0.166 | 3.13 | 0.790 | 0.16 | 0.101 | 2.93 |

[4] Its the maximum since new references may have been referenced in $W_{i-1}$. This is sometimes referred to as the instantaneous miss ratio over $N$ references.

Program sharing characteristics depend not only on the algorithms implemented but also on how the program is parallelized and scheduled (mapped) onto the processors. Before the results are presented how loops are partitioned and scheduled on the Alliant FX/80 is briefly described.

### 5.1. Loop Partitioning and Scheduling

The source for program parallelism on the Alliant is the parallel or doall loop. There are two general considerations in loop scheduling: how loop iterations are partitioned and how the partitions are scheduled on the available processors. Loop iterations can be partitioned vertically or horizontally and scheduled statically or dynamically[5]. Loop iterations on the Alliant are horizontally partitioned and dynamically scheduled. In scalar execution, loop iterations are allocated to processors on a first-come-first-served basis. The Alliant FX/80 accelerates this process using a dedicated synchronization bus and registers. Vectorization requires additional considerations. Vectors are typically processed by strip-mining where the vectors are broken into strips of elements up to the length of the vector registers. For example, a loop with $N$ iterations is vectorized into a loop with $\lceil \frac{N}{VR} \rceil$ iterations where $VR$ is the length of the vector register. The FX Fortran compiler targets the inner loops for vectorization and outer loops for parallelization (outer-parallel-inner-vector execution) with the inner loop being strip-mined. Loop strips in a vector multiprocessor can be considered as vertical partitions with the block size being $VR$.

### 5.2. Data Sharing

Data sharing in parallel program execution is an important issue in the design of multiprocessor memory system. Access to shared data by processors can lead to resource conflicts, access serialization and degradation in performance. For example, shared read-write blocks in a system with private caches can suffer overhead in maintaining data coherence.

In cache designs, sharing can be classified as explicit (or true) or implicit (or false) sharing. Explicit sharing

[5] In horizontal partitioning consecutive loop iterations are scheduled on consecutive processors and vertical partitioning divides the iterations into "blocks" which are then assigned as a unit to a particular processor. Static scheduling assigns loops at compile time and the schedule is fixed throughout execution. This is the simplest approach but can result in load imbalance. Dynamic scheduling assigns loop iterations to processors at run time and can distribute the load across the processors but at higher synchronization costs and data migration can result.

**Table 5: Fraction of Data Shared**

| TRACE | type | 8-Processor-scalar (8S) | | | |
|---|---|---|---|---|---|
| | | processors sharing | | | |
| | | 1 | 2-4 | 5-7 | 8 |
| ADM | blocks | 0.070 | 0.494 | 0.357 | 0.079 |
| | references | 0.117 | 0.136 | 0.163 | 0.584 |
| DYFESM | blocks | 0.182 | 0.571 | 0.128 | 0.119 |
| | references | 0.045 | 0.097 | 0.093 | 0.765 |
| ARC2D | blocks | 0.351 | 0.605 | 0.043 | 0.001 |
| | references | 0.096 | 0.256 | 0.148 | 0.500 |
| BDNA | blocks | 0.490 | 0.322 | 0.147 | 0.041 |
| | references | 0.214 | 0.311 | 0.031 | 0.444 |

| TRACE | | 8-Processor-scalar (8S) | | | |
|---|---|---|---|---|---|
| ADM | blocks | 0.503 | 0.394 | 0.069 | 0.034 |
| | references | 0.446 | 0.157 | 0.095 | 0.302 |
| DYFESM | blocks | 0.440 | 0.484 | 0.022 | 0.053 |
| | references | 0.417 | 0.098 | 0.012 | 0.473 |
| ARC2D | blocks | 0.327 | 0.534 | 0.129 | 0.010 |
| | references | 0.196 | 0.372 | 0.319 | 0.113 |
| BDNA | blocks | 0.688 | 0.194 | 0.063 | 0.054 |
| | references | 0.451 | 0.144 | 0.179 | 0.225 |

occurs when two or more processors explicitly address the same data item. Implicit sharing occurs when two or more processors do not address the same data item but conflict because the data items being referenced are contained in the same cache block.

Table 5 shows the fraction of the memory blocks referenced accessed by a particular number of processors for 8S and 8V. The memory space is considered as a set of 8-byte blocks and the results are obtained by recording the accesses to the blocks. If $b$ is the fraction of blocks referenced by a single processor (private blocks) then $1-b$ is the fraction of blocks that are shared. For each fraction of the data space Table 5 also shows the fraction of the references made to the blocks.

The vector versions of the programs (8V) generally have a larger fraction of private data blocks than the corresponding scalar versions (8S), ranging from 70% for BDNA to 33% for ARC2D. The scalar versions have a higher percentage of shared block. For example, ADM has 50% of the blocks being shared for 8V but this increases to 93% for 8S.

It is interesting to compare the scalar results with the results reported in [4, 5]. The fraction of references to shared data areas, averaged over all traces, is reported as 0.27 to 0.36 for [4] and [5] respectively, compared with an average of almost 0.90 for these traces. Moreover, the fraction of data blocks shared in [5] averages about 0.33 but is about 0.70 for these traces. One possible explanation for these significant differences may be because of the compiler generated partitions and how loops are scheduled. The results report in [4, 5] use programs written specifically for the target machine and the data, presumably, is partitioned with minimal sharing. In contrast, the results reported here use a restructurer to optimize the PERFECT codes. Furthermore, the Alliant dynamically assigns partitions for processor efficiency and not to reduce sharing. Thus there is no guarantee that a data set accessed within a particular iteration by a particular processor will be accessed by the same processor in another iteration or another loop.

The vector sharing characteristics are closer to the results of [4, 5] where the average number of shared blocks is 0.50 and the fraction of shared references is 0.63. This increase in the private space in vector execution can be due to two possibilities. For parallel and vector execution a compiler must perform a number of program transformations to remove data dependencies. Some of these dependencies result in the use of private temporary variables. For example, scalar expansion and node splitting results in using additional data structures to allow vectorization and parallelization. Second, vectorizing the inner loop effectively blocks the inner iterations. However, data migration is still possible since the outer loop is dynamically assigned.

While the fraction of blocks shared by all 8 processors is low in both vector and scalar executions, the fraction of references to this shared space is surprisingly high. For example, in DYFESM only 12% and 5% of the data space is accessed by 8 processors for vector and scalar executions respectively, but 77% and 49% of the references are to the respective fraction of the memory space. This suggests significant data reuse but also a need to support data migration with low overhead data coherence. Table 6 shows how implicit sharing increases as a function of the block size. If 8 bytes is assumed to be the minimal data element size the increase in sharing at block sizes of 16, 64 and 256 bytes is due to implicit sharing.

As expected, when the block size increases implicit sharing also increases. This increase is higher for 8S than 8V. At 64 bytes, 74%-86% of the data space is being referenced by all 8 processors for ADM, DYFESM and AR2CD. This rapid increase in implicit sharing confirms that consecutive loop iterations are being allocated to

**Table 6: Data Shared as a Function of Block Size.**

| TRACE | 8-Processor-scalar (8S) | | | | | 8-Processor-Vector (8V) | | | | |
|---|---|---|---|---|---|---|---|---|---|---|
| | block | processors sharing | | | | block | processors sharing | | | |
| | size | 1 | 2-4 | 5-7 | 8 | | 1 | 2-4 | 5-7 | 8 |
| ADM | 8 | 0.07 | 0.49 | 0.36 | 0.08 | 8 | 0.50 | 0.39 | 0.07 | 0.03 |
| | 16 | 0.04 | 0.34 | 0.50 | 0.16 | 16 | 0.14 | 0.70 | 0.11 | 0.05 |
| | 64 | 0.03 | 0.06 | 0.05 | 0.86 | 64 | 0.16 | 0.07 | 0.07 | 0.71 |
| | 128 | 0.03 | 0.04 | 0.04 | 0.89 | 128 | 0.21 | 0.03 | 0.07 | 0.68 |
| DYFESM | 8 | 0.18 | 0.57 | 0.13 | 0.12 | 8 | 0.44 | 0.48 | 0.02 | 0.05 |
| | 16 | 0.03 | 0.52 | 0.25 | 0.20 | 16 | 0.33 | 0.57 | 0.04 | 0.06 |
| | 64 | 0.02 | 0.10 | 0.14 | 0.74 | 64 | 0.26 | 0.23 | 0.13 | 0.38 |
| | 256 | 0.02 | 0.06 | 0.09 | 0.83 | 256 | 0.13 | 0.23 | 0.10 | 0.54 |
| ARC2D | 8 | 0.35 | 0.61 | 0.04 | 0.00 | 8 | 0.33 | 0.53 | 0.13 | 0.01 |
| | 16 | 0.03 | 0.79 | 0.18 | 0.00 | 16 | 0.27 | 0.52 | 0.19 | 0.02 |
| | 64 | 0.02 | 0.69 | 0.12 | 0.80 | 64 | 0.27 | 0.22 | 0.07 | 0.45 |
| | 256 | 0.01 | 0.07 | 0.02 | 0.90 | 256 | 0.24 | 0.23 | 0.02 | 0.51 |
| BDNA | 8 | 0.49 | 0.32 | 0.15 | 0.04 | 8 | 0.69 | 0.19 | 0.06 | 0.06 |
| | 16 | 0.21 | 0.54 | 0.15 | 0.10 | 16 | 0.42 | 0.28 | 0.22 | 0.28 |
| | 64 | 0.08 | 0.27 | 0.16 | 0.49 | 64 | 0.31 | 0.25 | 0.06 | 0.38 |
| | 256 | 0.07 | 0.20 | 0.08 | 0.65 | 256 | 0.33 | 0.03 | 0.01 | 0.63 |

**Figure 3: Reference Metrics; Write Run, Read Run and Processor Locality.**

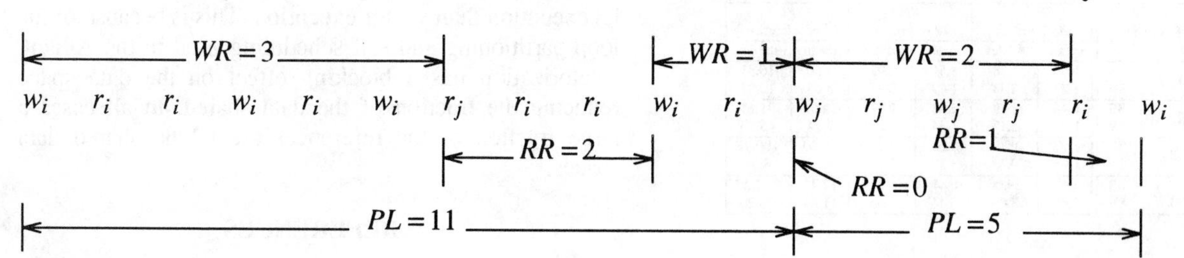


consecutive processors. The vector version retains a higher fraction of the memory space as private when the block size increases and confirms the blocking effect of parallel-vector execution. Even with a large block size of 256 bytes, 13%-33% of the data space remains private but in the scalar version less than 5% remain of the private space. This implies that for horizontally partitioned loops block sizes larger than the largest data item (8 bytes in this case) will result in large increases in implicit sharing.

### 5.3. Shared Data Access

The sequence of reads and writes to shared data is as important as how much of the data space is shared. In a system that employs private cache memories, references to read-write shared data may result in performance overheads because of coherence maintenance. When a processor writes to a shared block, other processors check their caches for block presence. If a copy of the block exists in more than one cache these blocks are either updated (an update protocol) or invalidated (an invalidate protocol). This section presents some characterization of shared data access related to coherence protocols for scalar and vector parallel executions. The metrics used are based on those used in [4, 5].

Consider, the sequence of references shown in Figure 3 where $R_i$ and $W_j$ are read and write references by processors $P_i$ and $P_j$ respectively. A write run (*WR*) is a count of the *number of writes* by a particular processor before an access by another processor. A write run begins with a write by $P_i$ and ends with a read or write by another processor $P_j$ where $i \neq j$. A long *WR* tends to favor an invalidate protocol as only the first write needs to be communicated to the other caches for invalidation with subsequent writes being local to the processor. A read or write by another processor terminates the write run because a bus access is necessary to invalidate or to source the cache block. Alternatively, a short write run tends to favor the update method because it avoids the continuous invalidation of blocks. However, the caches still have to be continuously updated, which can limit the cache access bandwidth for the processor.

A read run (*RR*) is a count of the *number of processors* that share a particular block after a processor write. A read run starts at the end of a write run and terminates at the start of the next write run. Read and write runs are disjoint. If a write run is terminated by a write then a read run of 0 processors is recorded. A large read run indicates a large number of processors sharing the data and for an invalidation protocol a large number of cache invalidations.

The last metric, processor locality (*PL*), is a count of the *number of references* to a particular block between writes by two different processors. The *PL* count starts at the write by processor $P_i$ and terminates at the next write by processor $P_j$ when the next *PL* count starts. The references in *PL* include the references in *WR* and *RR*. To a particular cache a write access causes a data block to be loaded into the cache which remains in the "owners" cache until written by another processor when it is invalidated (if an invalidation protocol is adopted). Thus, *PL* counts the number of references to the block before it is removed from the cache. A high *PL* value indicates that invalidation misses are unlikely to have a large effect on cache performance but a low value indicates that blocks are being ping-ponged between processors.

The mean and standard deviation for *WR*, *RR* and *PL* are given in Table 6. Vector execution (8V) generally results in short write runs but the difference is quite small. The longest write runs are for BDNA which generates more than twice the number of writes of the other programs. For BDNA/8S the standard deviation is high indicating that there are some very long and short write runs. The average write run for 8S and 8V are 2.14 and 1.52 respectively, as compared with 3.93 reported in [Egge87]. Because of the short write runs an update protocol may be more appropriate for these traces.

The scalar executions have longer read runs than the vector execution. The read sharing of shared data is highest in ARC2D for 8S. Recall in Table 6 that while 70% of the data space is shared by 2 or more processors about 90% of the references are to these areas. The size of ARC2D's read run suggest an update protocol but the low read run of the other traces indicate that an invalidate

**Table 7: Write and Read Run & Processor Locality**

| TRACES | | *WR* (# of writes) | | *RR* (# of procs) | | *PL* (# of references) | |
|---|---|---|---|---|---|---|---|
| | | mean | std dev | mean | std dev | mean | std dev |
| ADM | 8S | 1.56 | 3.38 | 1.51 | 2.02 | 9.60 | 16.81 |
| | 8V | 1.27 | 2.33 | 1.49 | 2.25 | 5.65 | 10.91 |
| DYFESM | 8S | 1.44 | 2.19 | 2.28 | 2.57 | 32.52 | 91.61 |
| | 8V | 1.07 | 2.47 | 1.44 | 1.88 | 7.14 | 45.73 |
| ARC2D | 8S | 1.07 | 0.38 | 5.00 | 3.22 | 50.60 | 100.98 |
| | 8V | 1.05 | 0.24 | 3.52 | 3.94 | 5.78 | 8.96 |
| BDNA | 8S | 4.47 | 30.06 | 0.41 | 0.59 | 17.43 | 133.29 |
| | 8V | 2.69 | 3.66 | 0.44 | 1.33 | 11.34 | 28.20 |

protocol will result in a small number of invalidations. The average read run length for 8S and 8V are 2.3 and 1.7 respectively as compared with 1.2. in [4].

The *PL* measure shows that vector executions are more likely to ping-pong blocks between the processors which is not as intuition would suggest. Since, 8S executions are horizontally partitioned and has the higher fraction of sharing, it would be expected that more data migration is likely to occur. A possible reason why *PL* is much higher for 8S than 8V is that shared variables such as index values are written by a single processor but interrogated by many processors. Since, scalar execution has more overhead in loop execution than vector execution this may account for the high *PL* value.

The general small values of both write run and read run do not conclusively argue for one protocol over another.

## 6. Conclusions

The execution of programs on different processor architectures lead to contrasting memory reference characteristics. This paper has presented memory reference characteristics from traces collected from single scalar (1S) and vector (1V) processor and 8 processor scalar (8S) and 8 vector processor (8V) executions. The paper reported measures for memory reference issue rate, data reuse and data sharing.

As expected 8V has the highest issue rate. For parallel and vectorizable programs the number of references issued per cycle increases by almost the factor the number of processors over that of the scalar and uniprocessor executions. This suggests that vector multiprocessor must use some form of local memory system. The issue rate of 1V and 8S are comparable.

The locality and reuse measures indicate that scalar and single processor executions will benefit more from cache memories than vector and parallel executions. The rate of reuse is generally much higher for single scalar executions. Vector executions show lower reuse but indicates that vectors accesses are likely to sweep caches and that either block sizes must be very large to contain multiple data elements or very small to reduce cache pollution. Vector caches need to be much larger than scalar caches.

Data sharing is significantly higher in scalar parallel execution than vector execution. This is because of the loop partitioning and self-scheduling used in the Alliant. Vectorization has a blocking effect on the data space reducing the fraction of the data shared. In all cases a large fraction of the references are to the shared data spaces.

## REFERENCES


[1] P. J. Denning, "The Working Set Model for Program Behavior," *Comm. of ACM*, vol. 11, no. 5, 1968.

[2] A. J. Smith, "Cache Memories," *ACM Computing Surveys*, vol. 14, no. 3, Sept. 1982, pp 473-530

[3] F. Darema-Rogers, G. F. Pfister and K. So, "Memory Access Patterns of Parallel Scientific Programs," *Proc. ACM Sigmetrics Conf.*, pp 215-225, May 1987, pp 46-58

[4] S. J. Eggers, "Simulation Analysis of Data Sharing in Shared Memory Multiprocessors," Report UCB/CSD89/501, University of California Berkeley, 1989.

[5] A. Agarwal and A. Gupta, "Memory Reference Characteristics of Multiprocessors Applications under MACH," *Proc. ACM Sigmetrics Conf.*, May 1988, pp 215-225.

[6] H. B. Lim and P. C. Yew, "Parallel Program Behavioral Study on a Shared Memory Multiprocessor," *Proc. Int'l Conference on Supercomputing*, 1991, pp 386-395.

[7] Alliant Computer Systems Corporation, FX/Series Product Summary, June 1985.

[8] M. Berry, et. al., "The Perfect Club Benchmarks: Effective Performance Evaluation of Supercomputers", *Int'l. Journal for Supercomputer Applications*, Fall 1989.

[9] J. W. C. Fu and J. H. Patel, "How to Simulate 100 Billion References Cheaply," CRHC Tech. Report, University of Illinois, 1991.

[10] "Vector Computer Memory Bank Contention," *IEEE Trans. Comp.*, vol C-36, pp 293-298, March 1987.

[11] A. L. Narasimha Reddy and P. Banerjee, "A Study of I/O Behavior of Perfect Benchmarks on a Multiprocessor," *Proc. 17th. Ann. Int'l Symp. on Comp. Arch.*, May 1990, pp 312-321.

# Iteration Partitioning for Resolving Stride Conflicts on Cache-Coherent Multiprocessors

Karen A. Tomko and Santosh G. Abraham *
Department of Electrical Engineering and Computer Science
University of Michigan
Ann Arbor, MI 48109-2122
email: karent@eecs.umich.edu

## Abstract

*We develop compile-time iteration partitioning techniques for private-cache shared-memory multiprocessors. Our techniques assign loop iterations to a set of processors so that cache coherency traffic due to interprocessor communication is minimized and load balance is maintained. In contrast to most previous research that has examined uniformly-generated dependences, we develop methods for non-uniform dependences that are generated by stride conflicts. Furthermore, we consider the effects of a long cache line size and minimize false coherency traffic. Our methods can handle conflicts between any two integer strides. We have conducted experiments on a 32-processor KSR-1 from Kendall Square Research which show 2x performance improvement using our partitioning algorithm over standard contiguous partitioning techniques.*

## 1 Introduction

In todays high performance multiprocessor systems, the remote memory latencies are over a hundred times greater than the cycle time of the processors [7]. Complex memory hierarchies have been developed to hide latency and provide data at sufficient rates to keep the processors busy. However, the memory hierarchy alone does not guarantee good performance. For example, a large parallel engineering application was found to spend forty percent of its time on cache stalls on a multiprocessor with a memory latency of less than 10 CPU cycles, a small latency by today's standards. This example demonstrates that it is essential for an application to make good use of the memory hierarchy in order to achieve efficient execution on high performance systems. The bandwidth requirements within the main loops of an application may be much greater than the bandwidth of the multiprocessor interconnection. Suitable partitioning techniques can greatly reduce communication requirements.

*The University of Michigan Center for Parallel Computing is partially funded by NSF grand CDA-92-14296 and this research was supported in part by ONR N00014-93-1-0163.

**doall** $i = 1, 200/5$
$a(i) = b(i) * b(5 * i)$
**enddo**
**doall** $i = 1, 200/5$
$b(i) = a(i) * a(5 * i)$
**enddo**

Figure 1: Stride conflict example

In this paper we introduce compile-time iteration partitioning techniques for shared-memory multiprocessors to minimize cache coherency traffic in loops with stride conflicts. A stride conflict occurs when an array is accessed with two different strides. For example, in Figure 1, the array B is accessed with a stride of 1 and a stride of 5 in the first loop and A is referenced with a stride of 5 and 1 in the second loop. Iteration i and iteration 5i of the first loop both reference the array element $b(5*i)$ and will require communication (cache-coherency traffic) if they are not executed by the same processor. Stride conflicts occur when array subscript expressions contain index coefficients (strides) greater than one or when an array is accessed by the same index variable in more than one dimension. In an empirical study [13], 15% of three-dimensional array references had subscript expressions with loop index coefficients greater than 1. Stride conflicts can be found in diverse applications such as mechanical CAE and computational chemistry [3].

Standard parallelization schemes such as contiguous or cyclic partitioning of the iteration space of parallel loops will generate cache coherency traffic in the presence of stride conflicts. However, in our *reference set partitioning* scheme, the iterations are assigned to processors so that the array references made by each processor are disjoint. As a result, cache coherence traffic is either completely eliminated (if the cache line size is one word) or is reduced to a negligible amount (otherwise). We also show how reference set partitioning can be used to reorder the iterations of a loop for better data locality.

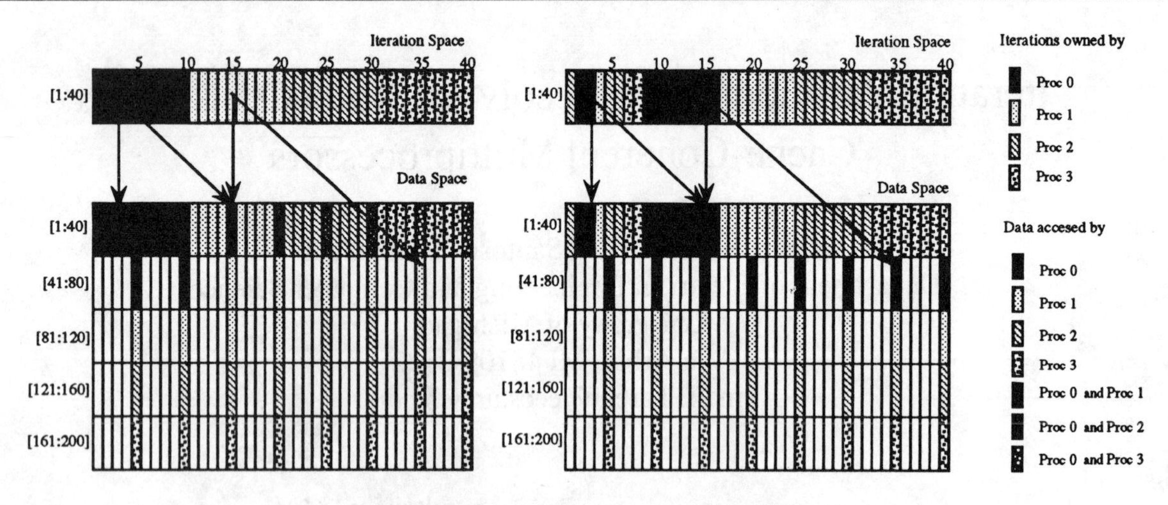


Figure 2: Stride conflict example partitioning

In Figure 2 we diagram contiguous partitioning and reference set partitioning (RSP) for a four processor system. Contiguous partitioning is a simple commonly used scheme which divides the iteration space of size N into p contiguous regions each of size N/p where p is the number of processor on which to run. The disadvantage of contiguous partitioning is that many of the array elements will be accessed by more than one processor causing coherency cache misses. In the figure, processor 0 reads 6 array elements that are written by the other processors. For example processor 0 will read $b(15)$ on iteration 3 and processor 1 writes to element $b(15)$ on iteration 15. The second diagram in the figure shows our reference set iteration partition with no cache coherency traffic. Our method identifies a region of the data space that is only accessed using the larger stride (41:200 in our example). Then, the RSP method partitions the iteration space that references this data contiguously among the processors. An iterative procedure assigns the remaining iterations to avoid sharing.

This paper is organized into 6 sections. Section 2 reviews related work, section 3 describes the program and machine models used in this paper, section 4 presents the details of our methods for iteration partitioning in the presence of stride conflicts, section 5 presents experimental results on the KSR-1, and section 6 summarizes our contributions.

## 2 Related Work

There is a large body of research on data allocation for communication reduction and locality optimization for memory hierarchies; we discuss the work most relevant to our own in this section.

Knobe, Lukas and Steele [9] present an algorithm for automatic allocation of array elements to processors for SIMD machines, based on array usage. They identify four types of conflicts that prevent communication-free parallelism-preserving data assignments to processors: stride, offset, scalar and cell conflicts. We assume techniques such as described in [9] can be used to identify stride conflicts and concentrate on resolution techniques.

Knobe and Natarajan [10] present an algorithm to minimize the interprocessor communication resulting from conflicts by dividing source programs into regions based on control flow and dependence information. Allocation requirements within a region are met and communication occurs between regions. Our objective is to partition scientific programs containing loop-level parallelism for current shared-memory parallel systems. We meet the conflicting allocation requirements by reducing the parallelism without incurring any communication costs.

Various heuristics to resolve conflicts have been developed. Gupta and Banerjee [6] have developed methods for automatic data partitioning for distributed memory MIMD multiprocessors. Kennedy and McKinley [8] use memory access cost estimates to guide transformations to increase data locality for parallel processors.

Offset conflicts arising from uniformly-generated dependences have been investigated both in the context of improving locality in sequential programs and in the context of reducing interprocessor communication in parallel programs. Gannon, Jalby and Gallivan [4] and Wolf and Lam [14] describe methods for improving data locality using program transformations. In contrast to our work, both analyses are restricted to uniformly generated dependences and cannot handle stride conflicts. Additionally, our work is useful in reducing cache coherency traffic in a parallel program as well as capacity misses in a sequential program.

Several researchers have developed methods for determining communication free partitions of index sets for subscript expressions having uniform dependencies [12], [11], [2]. Our work presents methods for communication free partitions of index sets with non-uniform dependencies. Abraham and Hudak [1] have developed automatic partitioning techniques for regular data-parallel loops with array accesses that have unit-coefficient linear subscripts. In contrast, this paper discusses automatic partitioning techniques for loops with stride conflicts. In [1], they develop iteration space partitions that satisfy load-balancing constraints and optimize cache-coherency traffic whereas we develop partitions that satisfy a zero communication constraint and minimize load imbalance.

Fang and Lu [3] develop a scheduling scheme that exploits reuse due to non-uniformly generated dependences. Neither our model or Fang and Lu's model is subsumed by the other. Our approach is similar to theirs in that we partition the iteration space into equivalence classes which access common array elements. For arrays that are accessed by two different strides we produce partitions that eliminate cache thrashing. We also address issues such as load balancing and multiple word cache line sizes in our partitioning scheme.

## 3 Machine and Program Model

We use a shared address space MIMD multiprocessor as our machine model and assume a memory hierarchy consisting of one or more levels of local cache, possibly a local memory, and one or more levels of remote memory where the latency to access each level of memory increases. Remote memory consists of the memory or cache storage local to other processors and communication between processors occurs when remote memory is accessed. The cache line sizes may be greater than one word. For section 5 we assume a uniprocessor system with one level of caching and main memory.

Our experiments were run on a 32-processor Kendall Square Research KSR-1 with the following characteristics. The KSR-1 is a cache-coherent shared-memory multiprocessor with a ring interconnection network. Each processor cell has a 256 KB primary instruction cache and a 256 KB primary data cache. They are referred to as the instruction subcache and data subcache respectively and have a cache subblock size of 64 bytes. The second level cache on each processor is 32 MB with a cache subblock size of 128 bytes. There is no main memory so a datum has no permanent home, it resides in the cache of the processor(s) that has(have) accessed it most recently. The access latencies for the KSR-1 memory hierarchy are given in Table 1.

We confine our analysis to loop nests of data-parallel application programs. The loop nests must meet the following constraints. The outermost loop is assumed to be a serial loop surrounding one or more parallel loops. The loops are not required to be perfectly nested. Subscript expressions are assumed to be of the form $k_j * i_j + c_j$ where $i_j$ is a loop index variable, and $k_j$ and $c_j$ are loop invariant integer expressions. Specifically, our partitioning techniques handle the case where there is at least one array with two such

Table 1: Memory access latencies for the KSR-1

| *Memory Component* | *Memory Access (Cycles)* |
|---|---|
| Subcache | 2 (sustainable rate = 1) |
| Local Cache (existing page) | 23.4 |
| Local Cache (new page) | 49.2 |
| Remote Cache (same ring) | 135-175 |
| Remote Cache (other ring) | 470-600(estimated) |

subscript expressions such that $k_1 \neq k_2$ (the case $k_1 = k_2$ has been handled by previous research). We assume that a loop index variable appears in only one dimension of each array subscript expressions so that each parallel loop can be partitioned independently. Thus our method extends to any number of loop iterations or array dimensions. The loop body may contain conditional statements, we assume that the body of a conditional will be executed. Additionally, we assume that computation time for each iteration of the loop is approximately the same.

## 4 Iteration Partitioning

In this section we will describe the reference set partitioning method. First we give a statement of the problem. Then we present the general algorithm for reference set partitioning followed by a simple example. We complete the section with a discussion of extensions to our method.

### 4.1 Problem Statement

The goal of our iteration assignment scheme is to partition the work of a loop nest amongst $p$ processors equitably and to minimize the interconnect traffic due to remote accesses. Let $S$ be the set of all possible values of the loop indices. We want to partition the indices into $p$ disjoint subsets such that each subset $s_j$ is a set of loop iterations which reference a unique set of array elements. We call the set of array elements the array reference set. The subsets $s_j$ of $S$ are a mathematical partition, meaning that $\bigcup s_j = S$ and $\bigcap s_j = \emptyset$. Our partition must satisfy the above constraints and minimize the objective function $\max_j(|s_j|)$ to ensure load balancing.

Given the number of processors, $p$; the number of iterations in the parallel loop, $n$; and the array subscript expressions for array references within the loop body, the following method will generate $p$ disjoint subsets $\{s_0, s_1, ..., s_{p-1}\}$ of iterations such that there is no data sharing when the subsets are assigned to processors $0, 1, ..., p-1$ respectively. Let $f_1 = k_1 i + c_1$ and $f_2 = k_2 i + c_2$ represent the array subscript expressions where $k_1, k_2, c_1$ and $c_2$ are loop invariant, integer expressions and $|k_1| \neq |k_2|$. We assume throughout this discussion without loss of generality that $|k_2| > |k_1|$. $f_1$ and $f_2$ are linear func-

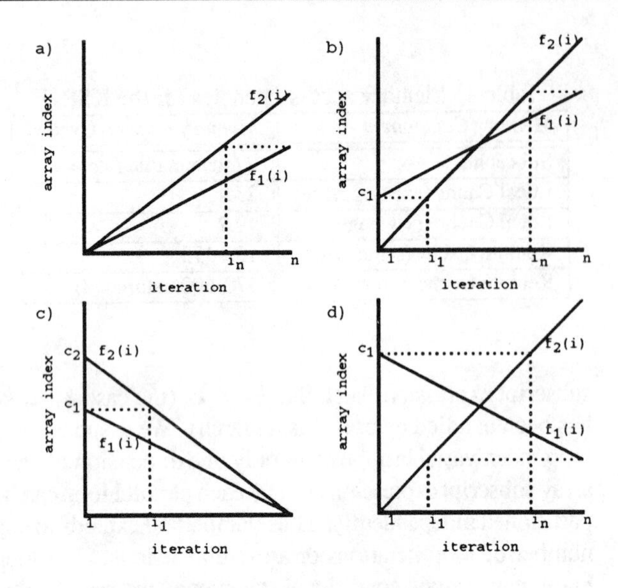


Figure 3: Data Access Functions

tions with integer domain $\{1 : n : 1\}$ and integer ranges $\{k_1 + c_1 : k_1 n + c_1 : k_1\}$ and $\{k_2 + c_2 : k_2 n + c_2 : k_2\}$. $f_1$ and $f_2$ can be thought of as real functions with domain $\{1 : n\}$ and ranges $\{k_1 + c_1 : k_1 n + c_1\}$ and $\{k_2 + c_2 : k_2 n + c_2\}$. Data sharing between iterations occurs when $f_1(i_1) = f_2(i_2)$. If the ranges of $f_1$ and $f_2$ do not overlap then there is no data sharing and contiguous partitioning can be used, we consider the case when the ranges overlap. The range of $f_2$ is obviously larger than that of $f_1$ thus there are one or two regions of data which are accessed only by array subscript expression $f_2$. Using the inverse function $f_2^{-1}$, we can get the iterations which access these data elements. Denote these iteration intervals as $\{1 : i_1\}$ and $\{i_n : n\}$. Some examples for $f_1$ and $f_2$ are given in Figure 3 with the iteration intervals marked. The iterations within these intervals have no data sharing amongst them so they can be partitioned equitably between the $p$ processors. Once each processor is assigned a range of iterations, an iterative algorithm is used to determine which other iterations to assign to each processor. In the following paragraphs we give the algorithm in greater detail and give an example.

### 4.2 Method

The reference set partitioning algorithm is given in Figure 4. We propose that a compiler can determine that a loop may be partitionable by the RSP method and insert the code in Figure 4. This code is executed at run time by each processor to determine its partition. This scheme is general and can handle cases where the strides $k_1$, $k_2$, and the effects of $c_1$, $c_2$ are not known at compile time. If the strides and offsets are known, the compiler can perform the partitioning and the only source of overhead is an additional level of loop nesting.

The first step in the RSP algorithm is to determine whether there is any data sharing between the data accessed by $f_1$ and $f_2$. This can be done with a standard dependence test. We use the interval test, which performs as well as a combination of the GCD test and Banerjee-Wolfe tests. If there is no data sharing then a contiguous partitioning scheme is used. Next we determine whether the strides, $k_1$ and $k_2$ have the same sign (Figure 2: a, b, and c). Assuming that the signs are the same, the loop is partitioned in two steps. We split the loop at the intersection point, $f_1(i_i) = f_2(i_i)$ or $i_i = \frac{c_1 - c_2}{k_2 - k_1}$, and partition the iterations $\{1 : i_i\}$ then partition iterations $\{i_i : n\}$. The loop must be split only if the intersection point lies withing the iteration space, i.e. $1 < i_i < n$. As stated in the previous section, there is at least one range of data, accessed by $f_2(1 : i_1)$ or $f_2(i_n : n)$, which is not accessed by $f_1$. Procedure PARTITION_SS uses this range, denoted $lb : ub$, as a starting point for partitioning. These iterations are partitioned in contiguous chunks amongst the processors. A processor is assigned iterations $\{low : high\}$. Given the data accessed by $f_1(low : high)$, the algorithm determines iterations $\{new\_low : new\_high\}$ such that $f_2(new\_low : new\_high) = f_1(low : high)$ and assigns these iteration to the same processor. At each step fewer iterations are assigned and the algorithm converges toward the intersection point. If $k_1$ and $k_2$ do not have the same sign (Figure 2: d) then we use a similar partitioning approach. However in this case data accessed by $f_1(1 : i_i)$ will be accessed by $f_2(i_i : n)$ and the partitioning algorithm oscillates around the intersection point where it eventually converges.

In the following paragraphs we document the correctness of the RSP algorithm by showing that the iteration sets are disjoint, $\bigcap s_j = \emptyset$; all iterations are assigned to a processor, $\bigcup s_j = \{1 : N\}$; and there is no data sharing between processors, $\bigcap f_{1,2}(s_j) = \emptyset$. We also show that the algorithm converges and discuss the maximum set size generated by the algorithm.

The algorithm is iterative so we show that at each step our criterion are met. Let $i$ be an iteration assigned to processor $pid$ such that $low \leq i \leq high$ at some step in the algorithm and the strides have the same sign. We show that the iteration sets are disjoint as follows. Iterations $low - 1$ and $high + 1$ may be assigned to other processors, assume that they are assigned to processors, $pid - 1$ and $pid + 1$ respectively. We show that the new iteration interval assigned to $pid$ at this step does not overlap with the new iterations intervals of $pid - 1$ or $pid + 1$. Assume that $direction =$ BACKWARD, $new\_low_{pid} = \lfloor \frac{k_1(low-1)+c_1-c_2}{k_2} \rfloor + 1 = new\_high_{pid-1} + 1$ and $new\_high = \lfloor \frac{k_1 high + c_1 - c_2}{k_2} \rfloor = new\_low_{pid+1} - 1$. So there is no overlap of the iteration intervals. Similarly there is no overlap if $direction =$

RSP $(k_1, k_2, c_1, c_2, 1, N)$
  $dependence \leftarrow$ INTERVAL_TEST$(k_1, k_2, c_1, c_2, 1, N)$
  **if** $(!dependence)$ **then return**
  **if** $(k_1 k_2 > 0)$ **then**
    $i_1 \leftarrow \lceil \frac{k_1+c_1-c_2}{k_2} \rceil - 1$
    PARTITION_SS $(k_1, k_2, c_1, c_2, 0, i_1, \text{FORWARD})$
    $i_n \leftarrow \lfloor \frac{k_1 N+c_1-c_2}{k_2} \rfloor$
    PARTITION_SS $(k_1, k_2, c_1, c_2, i_n, N, \text{BACKWARD})$
  **else if** $(k_1 k_2 < 0)$ **then**
    $i_1 \leftarrow \lceil \frac{k_1 N+c_1-c_2}{k_2} \rceil - 1$
    PARTITION_DS $(k_1, k_2, c_1, c_2, 0, i_1, \text{BACKWARD})$
    $i_n \leftarrow \lfloor \frac{k_1+c_1-c_2}{k_2} \rfloor$
    PARTITION_DS $(k_1, k_2, c_1, c_2, i_n, n, \text{FORWARD})$

PARTITION_SS $(k_1, k_2, c_1, c_2, lb, ub, direction)$
  $low \leftarrow lb + 1 + pid \cdot \lceil \frac{ub-lb}{p} \rceil$
  $high \leftarrow lb + 1 + (pid + 1) \cdot \lceil \frac{ub-lb}{p} \rceil - 1$
  $high \leftarrow \min(high, ub)$
  **do while** $(low \leq high)$ **and**
      $((\text{FORWARD}$ **and** $low \leq \min(n, i_i))$ **or**
      $(\text{BACKWARD}$ **and** $high \geq \max(1, i_i)))$
    ASSIGN_ITERATIONS $(pid, low : high)$
    **if** $(direction = \text{FORWARD})$ **then**
      $low \leftarrow \lceil \frac{k_1 low+c_1-c_2}{k_2} \rceil$
      $high \leftarrow \lceil \frac{k_1(high+1)+c_1-c_2}{k_2} \rceil - 1$
    **else if** $(direction = \text{BACKWARD})$ **then**
      $low \leftarrow \lfloor \frac{k_1(low-1)+c_1-c_2}{k_2} \rfloor + 1$
      $high \leftarrow \lfloor \frac{k_1 high+c_1-c_2}{k_2} \rfloor$

PARTITION_DS $(k_1, k_2, c_1, c_2, lb, ub, direction)$
  $low \leftarrow lb + 1 + pid \cdot \lceil \frac{ub-lb}{p} \rceil$
  $high \leftarrow lb + 1 + (pid + 1) \cdot \lceil \frac{ub-lb}{p} \rceil - 1$
  $high \leftarrow \min(high, ub)$
  **do while** $(low \leq high)$ **and**
      $((\text{FORWARD}$ **and** $low \leq i_i)$ **or**
      $(\text{BACKWARD}$ **and** $high \geq i_i))$
    ASSIGN_ITERATIONS $(pid, low : high)$
    $old\_high \leftarrow high$
    $old\_low \leftarrow low$
    **if** $(direction = \text{FORWARD})$ **then**
      $low \leftarrow \lceil \frac{k_1 old\_high+c_1-c_2}{k_2} \rceil$
      $high \leftarrow \lceil \frac{k_1(old\_low+1)+c_1-c_2}{k_2} \rceil - 1$
    **else if** $(direction = \text{BACKWARD})$ **then**
      $low \leftarrow \lfloor \frac{k_1(old\_high-1)+c_1-c_2}{k_2} \rfloor + 1$
      $high \leftarrow \lfloor \frac{k_1 old\_low+c_1-c_2}{k_2} \rfloor$
    $direction \leftarrow other(direction)$

Figure 4: Reference Set Partitioning Algorithm

FORWARD. Therefore within the iterations assigned during one step of the algorithm the iteration sets are disjoint.

All of the iterations $\{1 : n\}$ are assigned to a processor by the RSP algorithm. In the previous paragraph we have shown that there is no overlap between iterations assigned to different processors and there are no gaps at each step of the algorithm. We now show that there are no gaps between the iterations assigned at each step. Assume that $direction = \text{FORWARD}$ and let $old\_high_{p-1}$ be the highest iteration assigned on the previous step, $low_0$ and $high_{p-1}$ be the extremes of the iterations assigned at the current step of the algorithm, and $new\_low_0$ be the lowest iterations assigned in the next step of the algorithm. Assume that there is no gap between the previous iteration and the current iteration. Then there is no gap between the current iteration and the next because $new\_low_0 = \lceil \frac{k_1 low_0+c_1-c_2}{k_2} \rceil$, $high_{p-1} = \lceil \frac{k_1(old\_high_{p-1}+1)+c_1-c_2}{k_2} \rceil - 1$, and $old\_high_{p-1} + 1 = low_0$ so $new\_low_0 = high_{p-1} + 1$. Thus there are no gaps and no overlap between the iterations assigned in the current step and the iterations assigned in the next step of the RSP algorithm. Likewise, we can show that there are no iteration gaps when $direction = \text{BACKWARD}$ and when the signs of the strides are not the same.

There is no data sharing between processors with the RSP algorithm. Let $i$ be an iteration assigned to processor $pid$ such that $low \leq i \leq high$. Processor $pid$ accesses data addressed by $f_1(i)$ and $f_2(i)$. Iteration $j$ accesses data addressed by iteration $i$ when $f_2(j) = f_1(i)$, or when $j = \frac{k_1 i+c_1-c_2}{k_2}$. Iteration $j$ will be assigned to processor $pid$ in the next step of the algorithm because $\frac{k_1 low+c_1-c_2}{k_2} \leq j \leq \frac{k_1 high+c_1-c_2}{k_2}$ therefore there is no data sharing due to $f_1(i)$ access. Likewise, access $f_2(i)$ is accessed by iteration $m$ when $f_2(i) = f_1(m)$, or when $m = \frac{k_2 i+c_2-c1}{k_1}$. With a little algebra we see that $i = \frac{k_1 m+c_1-c_2}{k_2}$ and was thus assigned to processor $pid$ on the previous iteration. Therefore, iteration $i$ does not share any data with iterations owned by other processors. Similarly, we can show that this also holds for the case when $k_1$ and $k_2$ have different signs.

The algorithm converges in $log_{k_2/k_1} n$ steps if $f_1$ and $f_2$ intersect within the iteration space, otherwise we terminate the while loop when the iteration intervals reach the bounds of the iterations space. We know it will converge in $log_{k_2/k_1} n$ steps because the routine PARTITION_SS will iterate at most $log_{k_2/k_1} n$ times because we divide by $k_2/k_1$ at each step and there are at most n iterations to partition. Similarly, PARTITION_DS will converge in $log_{k_2/k_1} n$ steps.

As mentioned previously, we would like as equitable a partition as possible which does not share data. In the example partition in Figure 2, the processors are assigned 10, 9, 11 and 10 iterations each. This example shows how our partitioning method can provide nearly optimal load balancing. An optimally load balanced partition has a maximum set size of $\lceil n/p \rceil$. Our partitioning method will have a maximum set size less than $\lceil n/p \rceil + (k_2+1)log_{k_2/k_1}n$. We derive the upper bound as follows. For this analysis we assume the offsets are zero in order to simplify the mathematics. Let $\alpha = \lceil \frac{n - \lfloor \frac{k_1 n}{k_2} \rfloor}{p} \rceil$, which is the initial size of the iteration interval assigned to a processor. At each step of the algorithm an extra iteration can be assigned to a processor due to rounding. This gives us the series $\alpha + (\frac{k_1}{k_2}\alpha + 1) + (\frac{k_1}{k_2}(\frac{k_1}{k_2}\alpha + 1) + 1) + \cdots$. We can rewrite this as the summation $\sum_{j=1}^{m} [(\frac{k_1}{k_2})^j \alpha + 1 + (\frac{k_1}{k_2})^{j-1}] - 1$ where $m = log_{k_2/k_1}n$. This summation is bounded by the expression, $\alpha(\frac{1}{1-\frac{k_1}{k_2}}) + m + m(\frac{1}{1-\frac{k_1}{k_2}})$. If we substitute in the values for $\alpha$ and $m$ and make worst case assumptions regarding the values of $k_1$ and $k_2$ (i.e. $k_2 - k_1 = 1$) then we get the expression $\lceil n/p \rceil + (k_2+1)log_{k_2/k_1}n$. The maximum is greater than optimal by $(k_2+1)log_{k_2/k_1}n$ which is small with respect to $n$ when $k_1, k_2 \ll n$.

### 4.3 Example: Two Positive Strides

Consider a loop with bounds 1 and 40, data access functions $f_1 = i$ and $f_2 = 5i$. This is diagramed in Figure 3, graph a. The following steps will generate the partition given in Figure 2. Determine $i_n$ as follows. $f_2(i_n) = f_1(n)$ therefore $k_2 i_n + c_2 = k_1 n + c_1$ and $i_n = (k_1 n + c_1 - c_2)/k_2$ which in this example gives us $i_n = 8$. The iterations 9 to 40 have no data sharing and can be divided up amongst the processors. Assign iterations 9:16, 17:24, 25:32 and 33:40 to processors 0, 1, 2 and 3 respectively. Now consider a specific processor, it is responsible for data in the range $low : high$. We can iteratively determine the other iterations that must reside on this processor by plugging the bounds into the equation above which gives us, $high = (k_1 high + c_1 - c_2)/k_2$. We determine the new lower bound by using $low - 1$ in the same equation and adding one, $low = (k_1(low-1) + c_1 - c_2)/k_2 + 1$. Processor 2 in our example is assigned iterations {25:32, 5:6, 1} using this method iteratively. Assuming that the strides and offsets are known at compile time, a sophisticated compiler will generate the following code after partitioning using the RSP scheme.

```
i_n ← N/k_2
low ← i_n + 1 + pid · ⌈(N-i_n)/p⌉
high ← i_n + 1 + (pid + 1) · ⌈(N-i_n)/p⌉ - 1
high ← min(high, N)
do i ← 0 to ⌊log_{k_2} N⌋
    do j ← low to high
        a(j) ← b(j) · b(5j)
    low ← (low - 1)/k_2 + 1
    high ← high/k_2
```

### 4.4 Extensions

In this section we describe some extensions to the reference set partitioning method. We have adapted the algorithm for machines with a large cache line size. We have also adapted the algorithm to reorder iterations on a uniprocessor to improve locality. We also discuss how to partition cases not specifically handled by reference set partitioning.

The reference set partitioning iteratively assigns smaller and smaller ranges of iterations to each processor and the ranges may consist of a single iteration. On a processor with a large cache line size neighboring iterations may access data in the same cache line even though there is no true sharing of the data. If these iterations are assigned to distinct processors, false sharing is introduced. We have modified the algorithm to assign iterations in blocks so that false sharing will not occur. Let $l$ be the cache line size. The number of iterations that will access the same cache line is $l_{k_1} = \lceil l/k_1 \rceil$, and is referred to as the adjusted cache line size. To prevent false sharing our algorithm can easily be modified to force the iteration intervals to begin and end on adjusted cache line size boundaries. In Figure 2, if $l = 2$ and $k_1 = 1$ then $l_{k_1} = 2$ and the iterations are assigned in groups of two. The iteration sets generated by the modified RSP algorithm are { {1:2, 9:16}, {3:4, 17:24}, {5:6, 25:32}, {7:8, 33:40} }.

Partitioning based on adjusted cache line size eliminates false sharing at the cost of assigning some iterations that fall at the borders of iteration intervals to the 'wrong' processor. Thus we introduce some residual interprocessor communication. For instance, in the previous example, iterations 1 and 5 are assigned to distinct processors, 0 and 2, even though they both access b(5). The residual communication will result in write/read sharing of data and will affect at most $2p \cdot log_{k_2/k_1}n$ cache lines. The algorithm executes at most $2 \cdot log_{k_2/k_1}n$ steps and at each step there can be iteration intervals assigned to each of $p$ processors. Therefore there can be at most $2p \cdot log_{k_2/k_1}n$ iteration interval borders.

Reference set partitioning also assumes, for convergence, that $|k_1| \neq |k_2|$. If $k_1 = k_2$ then the dependence distance between the data accesses is constant (or uniform). Several researchers have provided partitioning algorithms which handle this case [12], [11], [2] and [1]. If $k_1 = -k_2$ the dependencies are not uniform and are thus not addressed in the papers mentioned above. There is however a simple solution for this case. In Figure 5 we give an example diagram for $k_1 = -k_2$. In the iteration interval $\{1 : i_1\}$ there is no data sharing so these iterations can be partitioned amongst processors. There is also no sharing within the interval $\{i_1 : i_i\}$ so it can also be partitioned. The interval $\{i_i : n\}$ is symmetric to $\{i_1 : i_i\}$ and the required partition can easily be derived from mirroring the partition of $\{i_1 : i_i\}$. For example, if $i_i - j$ is assigned to a particular

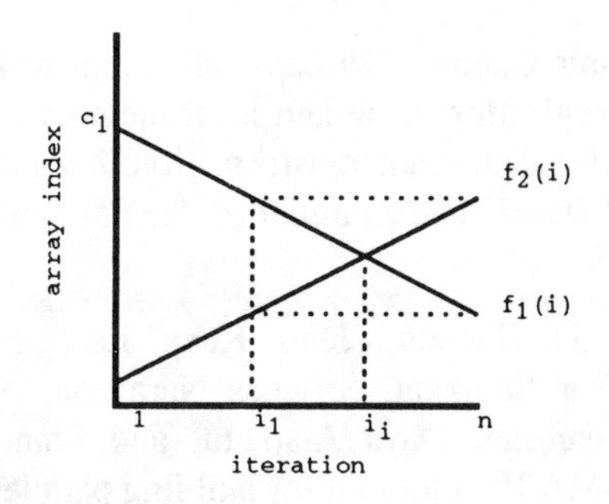


Figure 5: Data Access Functions: k1 = -k2

$i_n \leftarrow N/k_2$
$blks \leftarrow k_2N/C$
**do** $pid \leftarrow 1$ **to** $blks$
  $low \leftarrow i_n + 1 + pid \cdot \lceil \frac{N-i_n}{blks} \rceil$
  $high \leftarrow i_n + 1 + (pid+1) \cdot \lceil \frac{N-i_n}{blks} \rceil - 1$
  $high \leftarrow \min(high, N)$
  **do** $i \leftarrow 0$ **to** $\lfloor log_{k_2} N \rfloor$
    **do** $j \leftarrow low$ **to** $high$
      $A(j) \leftarrow B(j) \cdot B(5j)$
    $low \leftarrow (low-1)/k_2 + 1$
    $high \leftarrow high/k_2$

Figure 6: Iteration reordering for improved locality on uniprocessors

processor, then $i_i + j$ is also assigned to the same processor.

The index sets for partitioning data amongst processors in a multiprocessor can also be used to improve performance on a uniprocessor system. The partition sets contain the indices of iterations which access common array elements. If we execute all of the iterations of an index set, $s_i$, one after another then we reduce capacity cache misses. The reuse distance for an array element is the number of iterations between subsequent references to the element. Our goal is to reduce the reuse distance to eliminate capacity misses whenever possible. We have modified the reference set partitioning algorithm above by replacing $p$ with $\frac{k_2N}{C}$ where $C$ is cache size. Using iteration sets generated in this manner, we block the algorithm into iteration sets that fit entirely into the cache, and we can take advantage of the locality available within the set. In Figure 6 we reorder the iterations of our standard example.

# 5 Experimental Results

We ran two sets of experiments on the KSR-1, the results are given in Figures 7 and 8. Our experiments were run on hand restructured loops that were instrumented with timing calls. We used the same program for both tests, they differ only in the values of $k_1$, $k_2$, $c_1$ and $c_2$ that were used. The

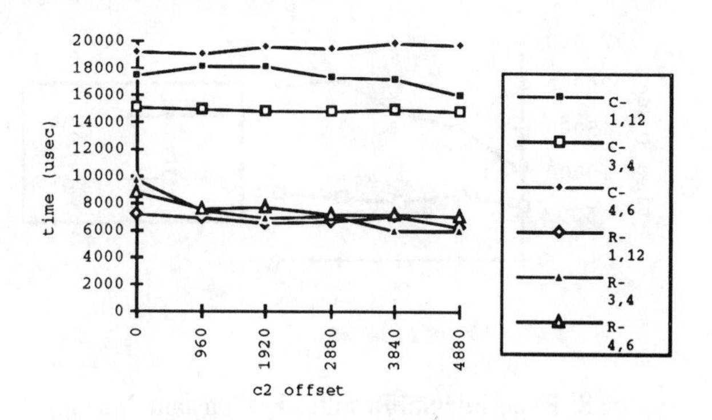


Figure 7: Experimental Results: $LCM(k_1, k_2) = 12$, Varying Offset

test program code is similar to that given in Figure 1. We compare our partioning algorithm to standard contiguous partitioning.

For the first test we varied the offset value, $c_2$ from 0 to 4880 and used three different $(k_1, k_2)$ pairs, $c_1$ was set to 0. We choose stride pairs such that $LCM(k_1, k_2)$ was a constant. For the second test we varied $LCM(k_1, k_2)$ from 4 to 20 by setting $k_1$ to 1 and varying $k_2$ from 4 to 20. The offsets were set to 0. These experiments were chosen because the amount of data sharing for a contiguous partition is proportional to $\frac{N}{LCM(k_1,k_2)}$. The data sharing for the reference set partition is negligible.

The results of the first experiment show that the $c_1$ and $c_2$ offsets do not affect the execution time very much, as expected. Additionally, the execution time for strides with the same LCM do not vary much. The results of the second experiment are less intuitive. We expect the amount of data sharing and the execution time to decrease as the LCM of the strides is increased, but we observe the opposite. This occurs because the KSR-1 has a large cache line size (Table 1) so multiple remote data are fetched together. When the stride is 16 or larger there is only one $f_2$ access per cache line and the curve starts to fall as the LCM increases.

In both experiments the execution time for reference set partitioning is constant with regard to the parameters we have varied. Execution times for RSP were less than half the execution time of contiguous partitioning in our tests.

# 6 Conclusion

Many researchers have studied partitioning of parallel do-loops containing uniform data dependencies. In this paper we have presented a compiler method for the partitioning of parallel do-loops with non-uniform data dependencies. The method presented generates a communication free partition for do-loops having conflicting data access strides. The methods are suited to improving data locality on uniprocessors as well as parallelization on multiprocessor systems. Our experimental results, on the KSR-1,

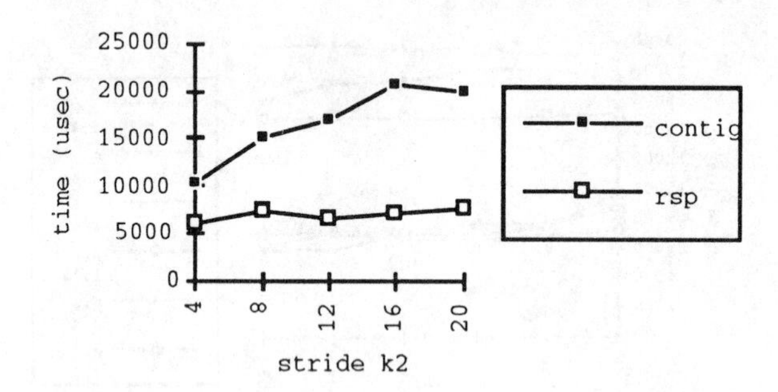


Figure 8: Experimental Results: $k_1$ Constant, Varying $k_2$

show that these methods can improve performance on such loops by a factor of two.

We have automated a subset of the partitioning methods outlined in this paper. Our compiler tool consists of a filter for detection of candidate loops and a restructurer to generate the partitioned versions of the loops. The current restructurer assumes that the data access functions have the form, $f_j = k_j i$. Given a program written in Fortran 77 we generate parallel Fortran for the KSR-1. We are integrating these software tools into the ASCOT (Architecture-Specific Compile-time Optimization Tool), a source to source restructurer being developed at the University of Michigan and built upon SIGMA II and the SIGMA Toolbox developed by Gannon et. al. [5].

The main limitation of our method is that it does not easily extend to more than two data access functions. If we increase the number of data access functions to just three the complexity increases dramatically. There can be sharing between $f_1$ and $f_2$, between $f_2$ and $f_3$, and between $f_1$ and $f_3$, which requires solving a system of equations at each step of the partitioning algorithm.

There are two open questions regarding this work. Can these methods be generalized to handle an arbitrary number of data access functions? Can these methods be generalized to handle a wider class of non-uniform data dependencies?

## References

[1] Santosh G. Abraham and David E. Hudak. Compile-time partitioning of iterative parallel loops to reduce cache coherence traffic. *IEEE Transactions on Parallel and Distributed Systems*, 2(3):318–328, July 1991.

[2] E. H. D'Hollander. Partitioning and labeling of index sets in do loops with constant dependence vectors. In *Proceedings of the International Conference on Parallel Processing*, pages 139–144, 1989.

[3] Jesse Fang and Mi Lu. An iteration partition approach for cache or local memory thrashing on parallel processing. In *Proceedings of the 4th Workshop on Languages and Compilers for Parallel Computing*, pages 313-327, 1991.

[4] Dennis Gannon, William Jalby, and Kyle Gallivan. Strategies for cache and local memory management by global program transformation. *Journal of Parallel and Distributed Computing*, 5(5):587–616, October 1988.

[5] Dennis Gannon, Jenq Kuen Lee, Bruce Shei, Sekhar Sarukkai, Srivinas Narayana, Neelakantan Sundaresan, Daya Atapattu, and François Bodin. SIGMA II: A tool kit for building parallelizing compilers and performance analysis systems. *To appear in* Elsevier, 1992.

[6] Manish Gupta and Prithviraj Banerjee. Demonstration of automatic data partitioning techniques for parallelizing compilers on multicomputers. *IEEE Transactions on Parallel and Distributed Systems*, 3(2):179–193, March 1992.

[7] Kendall Square Research Corporation, 170 Tracer Lane, Waltham, MA, 02154-1379. *KSR1 Principles of Operation*, 1991.

[8] Ken Kennedy and Kathryn S. McKinley. Optimizing for parallelism and data locality. Technical Report Rice COMP TR92-175, Rice University, Department of Computer Science, 1992.

[9] Kathleen Knobe, Joan D. Lukas, and Guy L. Steele, Jr. Data optimization: Allocation of arrays to reduce communication on SIMD machines. *Journal of Parallel and Distributed Computing*, 8:102–118, 1990.

[10] Kathleen Knobe and Venkataraman Natarajan. Data optimization: Minimizing residual interprocessor data motion on SIMD machines. In *Proceedings of the Third Symposium on the Frontiers of Massively Parallel Computation*, pages 416–423, 1990.

[11] Jih-Kwon Peir and Ron Cytron. Minimum distance: A method for partitioning recurrences for multiprocessors. *IEEE Transactions on Computers*, 38(8):1203–1211, August 1988.

[12] Weijia Shang and Jose A. B. Fortes. Independent partitioning of algorithms with uniform dependencies. In *Proceedings of the International Conference on Parallel Processing*, pages 26–33, 1987.

[13] Zhiyu Shen, Zhiyuan Li, and Pen-Chung Yew. An empirical study of fortran programs for parallelizing compilers. *IEEE Transactions on Parallel and Distributed Systems*, 1(3):356–364, 1990.

[14] Michael E. Wolf and Monica S. Lam. A data locality optimizing algorithm. In *Proceedings of the ACM SIGPLAN'91 Conference on Programming Language Design and Implementation*, pages 30–44, 1991.

# PERFORMANCE AND SCALABILITY ASPECTS OF DIRECTORY-BASED CACHE COHERENCE IN SHARED-MEMORY MULTIPROCESSORS *

Silvio Picano & David G. Meyer
School of Electrical Engineering
Purdue University
West Lafayette, IN 47907

Eugene D. Brooks III & Joseph E. Hoag
Massively Parallel Computing Initiative (MPCI)
Lawrence Livermore National Laboratory (LLNL)
Livermore, CA 94550

**Abstract** – *We present a study that accentuates the performance and scalability aspects of directory-based cache coherence in multiprocessor systems. Using a multiprocessor with a software-based coherence scheme, efficient implementations rely heavily on the programmer's ability to explicitly manage the memory system, which is typically handled by hardware support on other bus-based, shared memory multiprocessors. We describe a scalable, shared memory, cache coherent multiprocessor and present simulation results obtained on three parallel programs. This multiprocessor configuration exhibits high performance at no additional parallel programming cost.*

**Keywords:** directory-based cache coherence, parallel programming costs, Fortran D, multiprocessor simulation.

## 1 Introduction

General purpose machines, such as the BBN TC2000 [1], are achieving wide performance advantages (and cost advantages) over traditional mainframe technology. Yet these new parallel computers come with their own unique design and implementation problems with regards to shared and non-shared memory paradigms, scalability issues, parallel programming costs, and cache coherence support.

In the current parallel programming environment, programmers distinctly annotate programs with directives and instructions specifically aimed at a particular multiprocessor that will extract the desired performance. The resulting programs become difficult to understand because in addition to the underlying algorithm, details associated with the machine architecture are embedded in the programs.

In this paper, we analyze three parallel programs. We introduce a multiprocessor configuration with several important design features to enhance performance with little or no cost to an application programmer. We simulate the execution of the parallel programs on a multiprocessor simulator and show that high performance is achieved without additional programming costs and concerns.

*Work performed under the auspices of the U.S. Department of Energy by the Lawrence Livermore National Laboratory under contract No. W-7405-ENG-48.

## 2 The Parallel Programs

The parallel programs used in this study are described in the sections which follow.

### 2.1 A Network Simulation Program

The network simulation program [2] simulates and computes performance statistics of a proposed communications network for scalable multiprocessors. Simulated cpu ports interface to this network and communicate with other cpus via packet requests and responses. Multi-stage interconnection networks provide the communication vehicles for any number of simulated cpu and memory ports. The cpu and memory ports are connected to a request network, and a separate response network returns the requests back to the requesting cpu port.

### 2.2 An Iterative Relaxation Program

An iterative relaxation method is a stencil algorithm, which sets each element of a multi-dimensional grid array to the average of itself and its neighbors. The problem space is represented as a set of discrete elements, and at each iteration, each element is recalculated as a function of itself and its nearest neighbors. For our particular program, we average each element with its 8 nearest neighbors in a 2-dimensional, square grid. The domain decomposition is such that each processor will be responsible for the update of a fixed group of contiguous rows in the grid.

### 2.3 A Gaussian Elimination Program

Gaussian elimination is a method of solving a linear systems equation. The first part of the algorithm is referred to as the *reduction*, where the matrix is reduced to an upper triangular form. The second part of the algorithm, the *back-solve*, starts when the reduction is complete. Vector elements of the unknown solution are successively solved for and substituted into the remaining equations.

# 3 Performance Potential

## 3.1 Multiprocessor Simulation

A hardware enforced, cache coherent multiprocessor simulator system is described in [3, 4]. We detail the simulator's configuration of interest below:

1. RISC cpus with fully pipelined functional units, register scoreboarding, and compile-time branch prediction are used,
2. instruction pipeline stalls and flushes due to data hazards and wrong branch predictions are modeled exactly,
3. 2x2 switch nodes comprise a request and a response network,
4. each network is a worm-hole-routed, multi-buffered, multi-stage cube type with 64-bit data paths,
5. cache memory latency is 3 clock cycles with full pipelined access (1 access per clock cycle),
6. memory accesses and invalidations are faithfully modeled through the cache memories and networks,
7. a maximum of 5 outstanding shared memory requests per PE is allowed,
8. a *lock-up free*, write-back, shared data cache per PE (64 kilobyte, 2-way set associative, 16-byte line size) is used,
9. an invalidation-based protocol (fully documented in [4]) allows any number of read-only copies of a cache line and grants an exclusive copy of a cache line to a processor when a write operation occurs,
10. the presence flag technique [5] implemented at the memory controllers facilitates cache coherence, and
11. barrier synchronization is supported by software.

We do not model the effects of instruction cache misses or private data cache misses in the results presented below. Recall that 2x2 switch node elements comprise the networks. While a 4 PE system uses 4 network stages (i.e., 2 networks, each having 2 stages), a 32 PE system uses 10 network stages - which accounts for the increasing latency as the PE count increases. Invalidation distribution refers to the quantity of read/write sharing in the program. Before a write operation takes place, a processor must make sure that it has the only exclusive write access to this data. At the memory controller, invalidations are sent to every cache having a copy of this data before the write operation can continue.

## 3.2 Network Simulation Results

This version of the program ignores any logical, domain decomposition of the network simulation program's structure when scheduling parallel activity. We summarize some important run-time statistics in Table 1, and we

Table 1: Network Simulation Execution Statistics

| PEs | Cache Hit Ratio (%) | Latency (cycles) | Invalid. Dist. (%) | |
|---|---|---|---|---|
| | | | Width-1 | Width-2 |
| 2 | 85 | 10 | 96 | 4 |
| 4 | 89 | 13 | 93 | 6 |
| 8 | 92 | 15 | 92 | 6 |
| 16 | 92 | 19 | 92 | 7 |
| 32 | 90 | 28 | 91 | 7 |
| 64 | 88 | 33 | 89 | 8 |

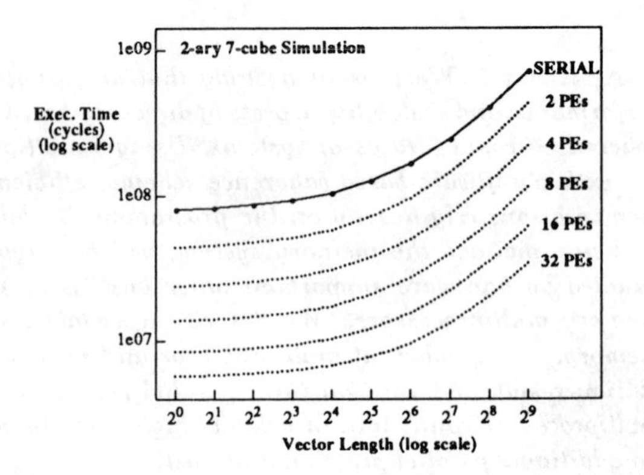


Figure 1: Network Simulation Execution Time

show execution time results in Figure 1, for 2-ary 7-cube network simulations. [1]

Analyzing the simulator's statistics more closely, we note that the shared data caches (64 kilobytes each) experience higher rates of both capacity misses and invalidation misses with respect to the rate of shared memory requests. For the 2-processor results, the capacity miss rate grows 1.08 times faster than the shared memory reference rate, while the invalidation miss rate grows 1.38 faster than the shared memory reference rate. In terms of total quantity of misses, capacity misses are 3 times greater than invalidation misses.

This situation reverses at higher processor counts. Analyzing the 32-processor results in particular, we note that capacity misses now grow 10 times slower than misses due to invalidations and 14 times slower than the shared memory reference rate. In absolute terms, capacity misses are only 5% of the total cache miss rate. Providing large, coherent caches is one key to achieving scalable performance for this application.

### 3.2.1 Iterative Relaxation Results

In Table 2, we summarize some of the important run-time statistics. In each category, the first column of numbers represents the statistics for the *128x128* problem size, while the second column of numbers represents the statistics for

[1] This problem size represents the operation of 7-stage networks having 2x2 switch nodes. [3]

Table 2: Iterative Relaxation Execution Statistics

| PEs | Cache Hit Ratio (%) | Latency (cycles) | Invalid. Dist. (%) | |
|---|---|---|---|---|
| | | | Width-1 | Width-2 |
| 2 | 96-93 | 10-10 | 95-98 | 5-1 |
| 4 | 96-93 | 12-13 | 93-98 | 6-2 |
| 8 | 98-93 | 17-15 | 83-97 | 15-1 |
| 16 | 98-95 | 25-19 | 75-96 | 22-2 |
| 32 | 96-95 | 30-28 | 71-98 | 25-1 |
| 64 | 90-97 | 33-33 | 53-82 | 42-17 |

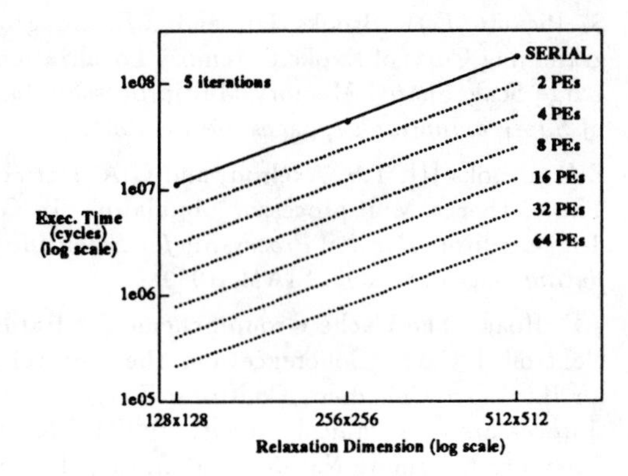


Figure 2: Iterative Relaxation Execution Time

Table 3: Gaussian Elimination Execution Statistics

| PEs | Cache Hit Ratio (%) | Latency (cycles) | Invalid. Dist. (%) | |
|---|---|---|---|---|
| | | | Width-1 | Width-2 |
| 2 | 51-50 | 16-15 | 99-99 | 1-0 |
| 4 | 51-50 | 20-19 | 99-99 | 1-0 |
| 8 | 52-51 | 24-22 | 99-99 | 1-0 |
| 16 | 56-52 | 30-26 | 97-99 | 2-0 |
| 32 | 62-54 | 38-31 | 97-99 | 3-1 |
| 64 | 71-58 | 50-38 | 95-99 | 5-1 |

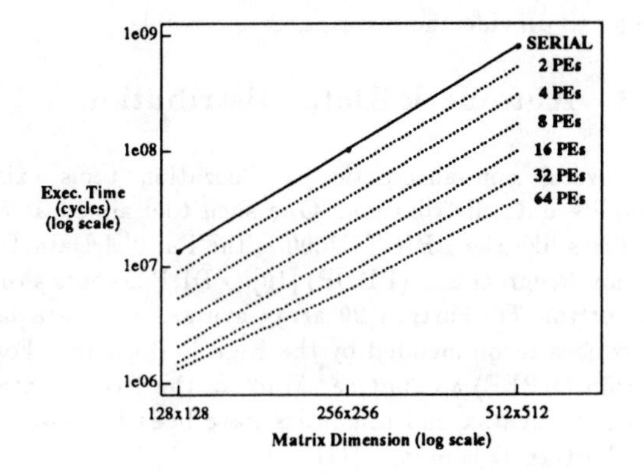


Figure 3: Gaussian Elimination Execution Time

the *512x512* problem size. We show the results of the simulations on this parallel program in Figure 2. In additional to the very high cache hit ratios, we see that width-1 and width-2 invalidation distributions combined make up over 95% of all invalidations across all numbers of processors. This distribution is a direct result of the decision to tile the 2-dimensional data in a blocked-row fashion to minimize invalidation requests (i.e., data sharing) from other PEs.

### 3.2.2 Gaussian Elimination Results

We summarize some of the important run-time parameters in Table 3, and we show serial and parallel execution time results in Figure 3. In each category, the first column of numbers represents the statistics for the *128x128* problem size, while the second column of numbers represents the statistics for the *512x512* problem size. From the table, we see some unexpected behavior in the cache hit ratio statistics. The increasing number of PEs actually increases the individual cache hit ratios, which is unlike many parallel applications. Investigating this further, we note that this program has a high quantity of barrier synchronization, which is actually implemented as an efficient software routine. The program's load imbalance and the execution time spent in barrier routines artificially inflates the cache hit ratio statistics. In terms of absolute performance, this program does not perform well because of the high level of invalidation traffic resulting from exclusively caching data that a PE will be unlikely to use again in a subsequent iteration. This performance degradation is evident through the increasing shared memory request latency at higher PE counts.

## 4 Related Work

### 4.1 Directory Limitations

As well documented in the literature [5, 6, 7], one major design constraint is the large amount of additional memory needed to implement the presence flag coherence technique. However, recent studies with Cerberus [6, 4] and Stanford's DASH project [7] show that *partial* directory states can reduce the memory requirements while maintaining a high level of desired performance. Work in [8] provides another method of reducing the directory requirements by using a combined hardware/software scheme to dynamically allocate directory state as shared data is referenced. Software-assisted cache coherence schemes [9] are alternatives to directory-based schemes. These schemes promote the use of compile-time reference marking combined with hardware-based coherence mechanisms.

### 4.2 Manual Data Distribution

Using simple and natural parallel program decompositions, each of the three programs required close to 8 parallel processors to regain the serial program's performance on the BBN TC2000 [10, 2]. [2] This performance degradation is due to 1) the disabling of the data caches, and 2) the frequent use of the remote shared memory. Program optimizations to selectively enable data caching and to explicitly place shared data in proper regions of the memory hierarchy resulted in significant increases in parallel performance. However, programming costs and complexity to perform these optimizations are significant.

The resulting source programs are difficult to understand if the reader has little knowledge of the BBN TC2000 multiprocessor. The code modifications are necessary to explicitly place data in the multiprocessor's memory hierarchy to provide the fastest access possible.

### 4.3 Automatic Data Distribution

Several semi-automatic parallelization tools exist to simplify data distribution. One such tool aimed at architectures like the BBN TC2000 is the Parallel Data Distribution Preprocessor (PDDP) [10]. PDDP accepts a dialect of Fortran 77, Fortran 90 array syntax, and data layout directives recommended by the High Performance Fortran Forum (HPFF) committee. Much of the PDDP data distribution syntax and semantics have been borrowed from the Fortran D language [11].

The Gauss elimination program has been rewritten using PDDP primitives and has been benchmarked on the BBN multiprocessor. The initial PDDP program introduced some performance problems due to hot spot accesses when all processors attempt to read the pivot row (stored in one local memory). Additional code is necessary to minimize the performance penalty of this hot spot. This new PDDP program yielded from 40-50% performance loss over the best efforts reported on the BBN machine. From this experience, we see that automatic data distribution tools have some side-effects which still complicate performance and programming concerns [10].

## 5 Summary and Open Issues

Parallel programs make critical demands on non-uniform memory access multiprocessors because of the high communication requirements they pose. The resulting performance is strongly dependent on data placement, movement, and allocation strategies. Using incremental program modifications, we obtain significant performance gains on the BBN TC2000 multiprocessor through the effective use of the memory hierarchy. While automatic data distribution models may alleviate some of the programming costs incurred, resulting performance may not approach the multiprocessor's capabilities.

We described a multiprocessor configuration that has several important design features to enhance multiprocessor performance. Detailed execution time results of the parallel programs on this multiprocessor showed high performance at no additional parallel programming cost. We also believe the multiprocessor configuration used in this study is a conservative one, and that more aggressive designs can be implemented with current technology.

[2] The BBN TC2000 has facilities allowing programmers to maintain cache coherence in their software.

## References


[1] BBN Advanced Computers Inc., Cambridge, MA. *Inside the TC2000*, 1989.

[2] S. Picano, E.D. Brooks III, and J.E. Hoag. Programming Costs of Explicit Memory Localization on a Large Scale Shared Memory Multiprocessor. In *Proc. of Supercomputing'91*, pages 36–45, 1991.

[3] E.D. Brooks III, T.S. Axelrod, and G.A. Darmohray. The Cerberus Multiprocessor Simulator. In G. Rodrigue, editor, *Parallel Processing for Scientific Computing*, pages 384–390. SIAM, 1989.

[4] J.E. Hoag. The Cache Group Scheme for Hardware-Controlled Cache Coherence and the General Need for Hardware Coherence Control in Large-Scale Multiprocessors. Technical Report UCRL-LR-106975, Lawrence Livermore National Laboratory, Livermore, CA, March 1991.

[5] L.M. Censier and P. Feautrier. A New Solution to Coherence Problems in Multicache Systems. *IEEE Transactions on Computers*, C-27(12):1112–1118, Dec. 1978.

[6] E.D. Brooks III and J.E. Hoag. A Scalable Coherent Cache System With Incomplete Directory State. In *Proc. of the International Conference on Parallel Processing*, pages 553–554, August 1990.

[7] A. Gupta, W.-D. Weber, and T. Mowry. Reducing Memory and Traffic Requirements for Scalable Directory-Based Cache Coherence Schemes. In *Proc. of the International Conference on Parallel Processing*, pages 312–321, August 1990.

[8] D.J. Lilja and P.C. Yew. Combining Hardware and Software Cache Coherence Strategies. In *ACM International Conference on Supercomputing*, 1991.

[9] S.L. Min and J.-L. Baer. Design and Analysis of a Scalable Cache Coherence Scheme Based on Clocks and Timestamps. *IEEE Transactions on Parallel and Distributed Systems*, 3(1):25–44, Jan. 1992.

[10] E.D. Brooks III, B.J. Heston, K.H. Warren, and L.J. Woods. The 1992 MPCI Yearly Report: Harnessing the Killer Micros. Technical Report UCRL-ID-107022-92, Lawrence Livermore National Laboratory, Livermore, CA, Aug. 1992.

[11] S. Hiranandani, K. Kennedy, and C. Tseng. Compiler Optimizations for Fortran D on MIMD Distributed-Memory Machines. In *Proc. of Supercomputing'92*, pages 86–100, 1992.

# Compiling for Hierarchical Shared Memory Multiprocessors

J. D. Martens and D. N. Jayasimha
The Ohio State University
Department of Computer and Information Science
Columbus, Ohio 43210-1277
martens@cis.ohio-state.edu, jayasim@cis.ohio-state.edu

Abstract—*Issues related to compiling for shared memory multiprocessors with multilevel memory hierarchies are discussed. In particular, a method using simple and regular sections for parallelizing nested iterative loops with constant dependence distances is described.*

## INTRODUCTION

In this paper we present a method for parallelizing iterative loops (e.g., `for` or `do` loops) for shared memory multiprocessors with multilevel memory hierarchies. By hierarchical shared memory multiprocessor architecture (HSMA), we mean a multiprocessor with a set of partially-shared memories such that a variable residing at level $i$ in the hierarchy could be shared by a larger number of processors than a variable at level $j$ ($j < i$). Architectures which fall into this class include the Tree-structured Hierarchical Memory Multiprocessor (THMM)[2], Cedar[5], Mabbs' Hierarchical Memory Structure[6], and Wei and Levy's multilevel hierarchy[11]. The use of local and hierarchical memory is not to be confused with the use of caches; in particular, we are not concerned with cache coherence.

Such a hierarchical memory exploits the notions of *partial sharing of variables* and the *degree of sharing* of a partially-shared variable (*PSV*) by mapping *PSVs* with a low degree of sharing to lower levels of the memory. Typically, the degree of sharing of a *PSV* changes over its lifetime—this information is used to move the *PSV* to the appropriate level of the hierarchy. Hierarchical memory architectures overcome the disadvantages of high latency and synchronization costs usually associated with large scale multiprocessors [2]. A number of researchers have proposed parallel algorithms which behave hierarchically and map well to hierarchical architectures[2, 8, 9, 10].

## COMPILING FOR HSMAs

### Background and Notation

We assume that the reader is familiar with standard parallelizing compiler terminology and techniques as discussed in [13]. Our work focuses on nested parallel loops in which subscripts in each dimension of the loop are of the form $i \pm k$, where $i$ is a loop index and $k$ is a constant; it is not necessary that the loop be perfectly nested. The *iteration space* is the set of all values taken on by loop indices, and is related to but distinct from the *data space*, which is the set of array locations written to (or *defined*) or read from (*used*) during the loop's execution. In order to execute a loop in parallel, we partition the iteration space into disjoint *tiles*; the union of all tiles is precisely the iteration space. Each tile is statically mapped to a particular PE. Associated with each tile is the set of data defined or used within the tile; this set is termed a *footprint*. The footprints are not necessarily disjoint—often footprints of neighboring tiles overlap.

The following notation and non-standard terminology is used throughout the rest of this paper:

- The number of PEs is $p$; we number the PEs $0 \ldots p-1$. $m$ refers to the number of a particular PE, i.e., $\mathrm{PE}_m$.

- The body of a loop consists of a sequence of one or more statements $S_i$; $use^j_m(L_i)$ is the set of data used by $\mathrm{PE}_m$ in loop $L_i$; $j$, if present, is an element of the vector of values taken on by induction variables of the outermost loops. $def^j_m$ is defined similarly.

- $lb^d$ and $ub^d$ are the lower and upper bounds of the $d^{th}$ dimension of a loop nest, respectively.

- $lo^d_m$ and $hi^d_m$ are the lower and upper bounds of the $d^{th}$ dimension of $\mathrm{PE}_m$'s tile, respectively.

- Two or more PEs are referred to as *neighbors* if the tiles mapped to the PEs are contiguous.

Communication and Synchronization Primitives. In order to write a parallel program or to parallelize a sequential program, support for communication and synchronization among PEs is needed. On a HSMA, a prerequisite for communication is that a variable be placed at a level in the hierarchy accessible by every PE needing access to the variable. We assume each shared variable has both *state* and *data* fields associated with it. As in [15], the state field is a generalization of the full/empty bit of the Denelcor HEP[4], being an arbitrary integer rather than just a single bit; the state is used for synchronization: producers set the state of a variable to a predetermined value when they write the datum, and consumers wait for the state to reach a particular value before reading the datum. The data field is simply the value of the variable. In order for two or more PEs to agree upon a location at which to communicate, we require a function called *commonLevel(P)*, which is a machine-dependent function that returns a memory unit efficiently accessible by all PEs in the list $P$. We also need language primitives for communication; we base ours on the atomic instruction of [15], with the following form: {address; condition; operation on state; operation on value}. The operational semantics is given in figure 1. Addresses are of the form *variable@memoryUnit*. Initially all data memory is uninitialized and all state values are zero.

```
while not condition do skip od
operation on address.state
operation on address.value
```

Figure 1: Operational semantics of synch. instruction.

Generation of Communication Statements. The producer sets the state of a shared variable to a loop-specific value[3], and the consumer waits for the state to reach the appropriate value. Currently our method does not distinguish between array elements shared between just two PEs and elements shared among more than two PEs; as a result, a PE producing a value writes it once for each consumer. This may result in the same value with the same state being written more than once to a particular location. The consumers do not change consumed variables' states, so barriers are needed between loops to ensure that values are consumed before being overwritten.

## The Method

Our method involves the following steps, which are later discussed in more depth. Detailed discussion and several examples (including matrix multiplication and five-point stencil) may be found in the full paper[7].

1) Parallelization of the loop nest.

2) Partitioning (or "tiling") of the iteration space.

3) Determining the use/def sets (the sets of data read and written, respectively) of the various PEs.

4) Determination of intersections between each PE's def set and other PEs' use sets.

5) Addition of communication and synchronization to satisfy dependences due to the intersections of disparate PEs' use and def sets.

Parallelization of the Loop Nest. The first step is to parallelize the loop nest. If the loops are perfectly nested, then unimodular transformations[12] may be applied to produce at least one level of parallelism as close as possible to the outermost level of nesting. Otherwise, we attempt to obtain outer loop parallelism either directly or via loop interchange. Failing this, we attempt to distributed the loop nest and apply unimodular techniques.

As an example, consider the program fragment and associated dependence information (produced by the Tiny program restructuring tool[14]) shown in figure 2. Tiny

```
2: for i := 1,n do
3:   a(i) := a(i)*b(i)
4:   c(i) := a(i-1)*2
2: endfor

flow dependence 3: --> 4:(<)  (1)
```

Figure 2: Example sequential loop.

indicates a flow dependence from $S_3$ to $S_4$. After loop distribution we parallelize the loop as in figure 3. There is still a flow dependence from $S_6$ to $S_9$, which indicates possible communication between PEs executing these statements.

Tile the Iteration Space, One Tile Per PE. Often it suffices to partition the iterations of the outermost `doall`; however, if the loop can be restructured for multiple adjacent levels of parallelism, then better load balancing and lower communication costs can often be obtained. Currently, a block partition is used for the tiling, but other partitions may be used in the future.

```
L1  5: doall i := 1,n do
    6:   a(i) := a(i)*b(i)
    5: endfor
L2  8: doall i := 1,n do
    9:   c(i) := a(i-1)*2
    8: endfor

flow dependence 6: --> 9
```

Figure 3: Example sequential loop after parallelization.

In the example of figure 3, the tile will be one-dimensional (a line segment). $PE_m$ executes iterations $lo_m \ldots hi_m$:

$$lo_m = hi_{m-1} + 1, \text{ where } hi_{-1} = lb - 1$$

$$hi_m = lb - 1 + \left\lceil \frac{(m+1)(ub - lb + 1)}{p} \right\rceil$$

In this example $lb = 1$ and $ub = n$.

Determine Footprints of Tiles. Two possible representations for footprints of tiles which allow efficient manipulation of data are simple sections and regular sections[1]. A regular section is a rectangular portion of an array, with each regular section boundary parallel to a coordinate axis. Simple sections are more flexible in that they also allow boundaries at $45^o$ angles to the axes. Since data represented by regular sections are easier to deal with from both expositional and computational points of view, the regular section abstraction will be assumed. In the loop class under consideration, it is possible to determine statically whether the increased complexity of simple sections would more closely match the shape of a footprint, in which case the compiler could generate code which uses simple rather than regular sections.[a]

In the example of figure 3, data used and defined by $PE_m$, i.e., $PE_m$'s footprint, is represented by the following regular section:

$$def_m(L1) = \{a(lo_m : hi_m)\}$$
$$use_m(L1) = \{a(lo_m : hi_m), b(lo_m : hi_m)\}$$
$$def_m(L2) = \{c(lo_m : hi_m)\}$$
$$use_m(L2) = \{a(lo_m - 1 : hi_m - 1)\}$$

Determine Intersections of Footprints. As explained above, each tile is statically mapped to a PE. Although the tiles are disjoint, the corresponding footprints may not be. The intersections of disparate footprints must be found. Sections are closed under intersection, i.e., the intersection of two regular (or simple) sections can be precisely represented as a regular (or simple) section. With $p$ PEs there will be $O(p^2)$ pairwise intersections (each PE determines the relevant intersections between its footprint and those of the other $p - 1$ PEs). Any array element that is within a nonempty intersection of sections must be mapped to the hierarchy. This is discussed in detail in the full paper[7].

Without static knowledge of $p$ and the loop bounds, a considerable portion of this calculation may have to occur at run-time. With a large number of PEs this is not desirable, so a less expensive technique is needed.

If dependence distances are available and these distances are small relative to the footprint sizes, then the problem can be simplified. In this case, all communication required for correct execution of a tile is among PEs with neighboring tiles. The condition under which we can assume this simplification is:

$$\bigwedge_{i \in \{1 \ldots d\}} \left(dd_i \le \frac{n_i}{\sqrt[d]{p}}\right) \qquad (1)$$

where $dd_i$ is the displacement of the dependence vector in the $i^{th}$ dimension (i.e., the dependence distance in the $i^{th}$ dimension), $d$ is the dimensionality of the tile, $n_i$ is the total number of iterations in the $i^{th}$ dimension, and $p$ is the number of PEs.

**Theorem:** Condition (1) holds iff all communication due to a particular dependence is among adjacent tiles.

Intuitively, condition (1) says that the magnitude of each component of a dependence vector is no more than the size of a tile in that dimension. For a proof, see [7].

Often not all of $dd$, $n_i$, and $p$ will be known at compile time, in which case the compiler will have to generate multiversion code: one version for "short" communication satisfying condition (1), and another version for cases in which condition (1) is not satisfied.

In figure 3, if $n \ge p$, then condition (1) is satisfied, so communication will be with PEs executing neighboring tiles. For $n < p$, high-numbered PEs will have no work, so communication among *active* PEs is still between neighboring tiles. Since this is a one-dimensional partition, the PEs with tiles neighboring those of $PE_m$ will be $PE_{m\pm1}$. The data communicated will thus be:

$$def_m(L1) \cap use_{m+1}(L2),$$

which is just the element $a(hi_m)$. So, after loop L1, $PE_m$ must write $a(hi)$ to $commonLevel(m, m+1)$. Performing a complementary analysis for $PE_m$ on L2, we see that $PE_m$ must read $a(lo_m - 1)$ from $commonLevel(m, m-1)$ prior to executing loop L2.

[a]Note that we do not lose parallelism by approximating footprints using sections. At worst, extra communication costs are incurred. This is because we restrict ourselves to `doall` parallelism, and we do not base our dependence analysis on sections.

Communication and Synchronization. Since the loops under consideration (after restructuring) are `doall` loops, there will be no loop-carried dependences. This simplifies matters considerably. At the beginning of a `doall`, we have an initialization phase in which all data needed by the PE for execution of its entire tile (i.e., all live data used within the tile that does not already reside within the PE) is obtained. Then the PE executes its tile. After the tile has been executed, all modified data that will be subsequently needed by other PEs is written to appropriate levels of the hierarchy; this is termed loop finalization.

The approximate code generated is given in figure 4 (for simplicity, boundary conditions and the ceiling function of the $lo_m, hi_m$ calculations have been omitted). Unless a memory unit is specified, all variable names refer to local addresses.

```
doall m := 0, p-1 do
  lo := 1 + m * n div p
  hi := (m + 1) * n div p
  for i := lo,hi do
    a(i) := a(i)*b(i)
  endfor
  Barrier
  {a(hi)@commonLevel(m, m+1); true; state := 1;
    value := a(hi)}  /* send a(hi) to PE m+1 */
  {a(lo-1)@commonLevel(m, m-1); state=1; nop;
    a(lo-1) := value}  /* get a(lo-1) from PE m-1 */
  for i := lo,hi do
    c(i) := a(i-1)*2
  endfor
endfor
```

Figure 4: Example of figure 3 with communication added.

## CONCLUSION

Several extensions to the work presented are being examined. One of these is to discover an inexpensive scheme for avoiding the extra communication that arises due to the fact that we restrict ourselves to pairwise intersections of footprints. A second improvement under investigation is a more sophisticated way of using state information to reduce the need for barriers. Finally, we are looking into some architectural improvements to simplify compilation and improve performance.

Acknowledgments. The authors would like to thank John Boyd, Himanshu Gupta, David Lutz, Loren Schwiebert, N. S. Sundar, and Yu-Chee Tseng for comments on an early draft of the full paper.

# References


[1] Paul Havlak and Ken Kennedy. An implementation of interprocedural bounded regular section analysis. *IEEE Transactions on Parallel and Distributed Systems*, 2(3):350–360, July 1991.

[2] D. N. Jayasimha. Partially shared variables and hierarchical shared memory multiprocessor architectures. In *11th Annual IEEE International Conference on Computers and Communications*, pages 63–71, April 1992.

[3] D. N. Jayasimha and J. D. Martens. Some architectural and compilation issues in the design of hierarchical shared memory multiprocessors. In *International Parallel Processing Symposium*, pages 567–572, March 1992.

[4] Harry F. Jordan. HEP architecture, programming, and performance. In Janusz S. Kowalik, editor, *Parallel MIMD Computation: HEP Supercomputer and Its Applications*, pages 1–40. The MIT Press, Cambridge, MA, 1985.

[5] J. Konicek et al. The organization of the cedar system. In *International Conference on Parallel Processing*, volume I, pages 49–56, August 1991.

[6] Stephen A. Mabbs and Kevin E. Forward. Optimizing the communication architecture of a hierarchical parallel processor. In *International Conference on Parallel Processing*, volume I, pages 516–520, August 1990.

[7] J. D. Martens and D. N. Jayasimha. Compiling for hierarchical shared memory multiprocessors. Technical report, Department of Computer and Information Science, The Ohio State University, 1993.

[8] Henk Meijer and Selim G. Akl. Optimal computation of prefix sums on a binary tree of processors. *International Journal of Parallel Programming*, 16(2):127–136, 1987.

[9] F. P. Preparata and J. Vuillemin. The cube connected cycles: a versatile network for parallel computation. *Communications of the ACM*, 24(5):300–309, May 1981.

[10] Peiyi Tang and Pen-Chung Yew. Software combining algorithms for distributing hot-spot addressing. *Journal of Parallel and Distributed Computing*, 10, 1990.

[11] Sizheng Wei and Saul Levy. Efficient hierarchical interconnection for multiprocessor systems. In *Supercomputing '92*, pages 708–717, 1992.

[12] Michael E. Wolf and Monica S. Lam. A loop transformation theory and an algorithm to maximize parallelism. *IEEE Transactions on Parallel and Distributed Systems*, 2(4):452–471, October 1991.

[13] Michael Wolfe. *Optimizing Supercompilers for Supercomputers*. MIT Press, Cambridge, MA, 1989.

[14] Michael Wolfe. The Tiny loop restructuring research tool. In *International Conference on Parallel Processing*, volume II, pages 46–53, August 1991.

[15] C.-Q. Zhu and P.-C. Yew. A scheme to enforce data dependence on large multiprocessor systems. *IEEE Transactions on Software Engineering*, pages 726–739, June 1987.

# SESSION 5B

## MAPPING/SCHEDULING

# Efficient Use of Dynamically Tagged Directories Through Compiler Analysis

Trung N. Nguyen* Zhiyuan Li* David J. Lilja†

*Department of Computer Science
†Department of Electrical Engineering
University of Minnesota
Minneapolis, MN 55455

## Abstract

Dynamically tagged directories have been recently proposed as a memory-efficient mechanism for maintaining cache coherence in large-scale shared-memory multiprocessors. In order to efficiently use these directories, the number of pointer operations must be minimized and pointers should be allocated as late as possible. If pointers are allocated too early, frequent pointer overflow will occur, which in turn may cause cache thrashing. Following the *delayed allocation marking* strategy, we present compiler algorithms to identify memory references that do not need to allocate a pointer. As a result, we reduce processor-memory network traffic and increase the data cache hit ratio, which will reduce the average memory latency. We demonstrate the effectiveness of this compiler optimization by implementing it in the Parafrase-2 parallelizing compiler.

*Key phrases* – cache coherence, dynamically tagged directories, optimizing compilers.

## 1 Introduction

A major problem with shared-memory multiprocessors is the latency of global memory accesses. In order to reduce this memory reference latency, private data caches are often used. Unfortunately, the presence of multiple private caches introduces the well-known cache coherence problem. Bus-based solutions to this problem [7, 10, 16, 18, 20] have limited scalability. By replacing the bus with a multistage interconnection network, the system can be made more scalable. Directory-based hardware schemes [1, 2, 4, 9, 19, 21] and compiler-assisted software schemes [6, 11, 14] are typically used in this environment to maintain coherence.

Software schemes perform static program analysis to predict temporal locality and potential cache incoherence. To maintain cache coherence, software schemes insert special instructions in a program to invalidate a cache line or to reset the whole cache. Despite their many advantages, software schemes are highly sensitive to the precision of data dependence analysis. The presence of aliases, procedure calls, and unknown symbolic terms can reduce the precision of the dependence analysis. Further, if the processors are scheduled dynamically, certain temporal locality is inherently undetectable at compile time.

Directory-based schemes, on the other hand, invalidate only those blocks that are actually stale because they can disambiguate memory references precisely at run time. Factors such as procedure calls and aliases pose no difficulty to these schemes. Unfortunately, directory-based schemes have two well-known limitations. First, they require interprocessor communication through the global memory, which is slow, and this communication increases the network traffic. Second, directories require a large amount of memory to store block-sharing information.

Several authors have recently proposed *dynamically tagged directories* [5, 8, 12, 13, 16], in which pointers are allocated only for data currently in a data cache. These directories maintain a cache of pointers in each memory module. Typically, each pointer consists of an address tag plus $n$ fields for storing $n$ different processor numbers. When a processor requests a block from the memory, the directory allocates a pointer to the block by setting the tag bits to the address of the requested block, and then setting one of the processor number fields to the identification number of the requesting processor. In this way, the directory can keep track of which processors have copies of which memory blocks. The dynamically tagged directories significantly reduce the number of pointers to cache blocks and thus the memory overhead, although they do not address the interprocessor communication problems.

These dynamically tagged coherence directories differ from each other primarily in the organization of the cache of pointers, and in the method used to create a free pointer when the directory overflows. For instance, the *tag cache* directory [16] associates $n$ pointers of $log_2p$ bits each (where $p$ is the number of processors in the system) with each address tag pointing to the first $n$ processors that request a copy of the block. If more than $n$ processors attempt to simultaneously share the same block, the directory overflows the additional pointers to a second-level directory cache that maintains an individual bit pointer for each of the $p$ processors. If this second-level directory cache overflows then some data block in the cache must be invalidated. The *LimitLESS* directory [5], on the other hand, generates a software interrupt to force a pointer to overflow to the memory.

A directory may also associate one address tag with a single processor pointer [12]. When more than one processor requests access to the same block, multiple pointers are allocated, each with the same tag value, but each pointing to a different processor. In a fully associative implementation of this directory, up to $p$ pointer entries can have the same address tag. In an $a$-way set associative implementation of the directory pointer cache, the number of processors that can simultaneously share a block is limited to $a$ since at most $a$ pointers can fit into a single set. When there are more requests for pointers than those available (i.e. when the directory overflows), an active pointer must be evicted from the directory. This eviction is accomplished by selecting a pointer and sending a message to the processor specified in the processor number field requesting that the processor invalidate its cached copy of the block specified in the address tag field. After the processor acknowledges the invalidation request, the pointer can be allocated to the block and processor combination of the current request. Any cache replacement policy, such as random or least-recently-used, can be used to select a pointer for eviction.

To summarize, there are two ways of handling directory overflow:

1. by invalidating a data cached block, or
2. by overflowing a pointer to memory.

The overflow-to-memory approach requires either a complicated interface between the memory and the directory or a rather lengthy software interrupt to handle the overflow. Furthermore, when a processor writes to a block which has multiple copies in different caches that must be invalidated to maintain coherence, the memory needs to be searched to locate pointers to those copies. Most schemes so far have chosen the simpler invalidating approach [8, 12, 13, 16].

The invalidating approach, however, has an important implication: a pointer should be allocated as late as possible. As we shall see in the next section, if pointers are allocated too early, then severe cache thrashing may occur because of frequent pointer overflow which in turn causes frequent invalidation. Such thrashing would consequently result in high network delay and high miss ratio. *Delayed allocation marking* [12], on the other hand, avoids this problem by not allocating a pointer at each reference to a block unless any further delay could cause cache incoherence.

In this paper, we demonstrate that delayed allocation marking can be effectively supported by compiler analysis. We have implemented two straightforward marking algorithms in a parallelizing compiler and have performed an experiment on the Perfect Club benchmark programs [3]. Both static and dynamic data are collected to show the effectiveness of the algorithms. In Section 2, we discuss the background for this work. In Section 3, we present our algorithms and their implementation. We describe the experiment and its results in Section 4. A summary and conclusions are given in Section 5.

## 2 Background

To demonstrate our compiler analysis, we consider the parallel execution of programs in the form of DOALL loops. We point out, however, that this analysis can be easily adapted to other parallelism models. A DOALL loop has no data dependences between different iterations and it terminates only when all iterations are completed. Due to the complexity of nested DOALL loops, we consider only singly nested DOALL loops. If several nested loops are parallelizable, we make the most beneficial one a DOALL loop, which usually is the outermost loop. Processors may be reassigned at the entry and the exit of a DOALL loop. We call such a point a ***boundary***. Under this model, the following sequence of events can turn a cache copy of a memory location, $M$, stale [6]:

$S_1$ : Processor $P_i$ accesses memory location $M$.

$S_2$ : Processors are reassigned (a boundary is crossed).

$S_3$ : Processor $P_j$ ($i \neq j$) writes to $M$.

```
doall i = l1, m1
   . = x                (s1)
   . = x                (s2)
    ...
   . = x                (sn)
end doall

x = .                   (t)

doall j = l2, m2
   . = x
end doall
```

Figure 1: A program segment example.

$S_4$ : Processors are reassigned (a boundary is crossed).

In event $S_1$, an initial copy of $M$ is stored in processor $P_i$'s local cache. After event $S_2$, other processors may access $M$. If one of them, say $P_j, j \neq i$, writes to $M$ (event $S_3$), then the cache copy of $M$ in $P_i$ turns stale. Since processors may be assigned dynamically, we must assume $P_i \neq P_j$. If $P_i$ may read $M$ again after event $S_4$, then the cached copy of $M$ in $P_i$ must be invalidated so the $P_i$ may not read the stale copy. In the old directory-based cache schemes, as soon as $P_i$ reads $M$ (event $S_1$), a pointer will be allocated in the directory to indicate that $P_i$ has a cached copy of $M$. When $S_3$ occurs, that copy is invalidated.

The pointer allocation strategy described above, however, does not suit dynamically tagged directories if the invalidating approach to directory overflow is used to handle pointer overflow. Take the program segment in Figure 1 for example. Scalar $x$ is referenced in statement $s1$ in every iteration of the DOALL loop with index $i$. Suppose $n$ processors execute this DOALL loop. If every processor allocates a pointer to its cached copy of $x$ as soon as $x$ is referenced, then $n$ pointers are needed in statement $s1$. Suppose the directory cache is, say, four-way set associative. At most four pointers may reside in the pointer cache simultaneously. All others must be evicted. Thus, all but four cached copies of $x$ must be invalidated. When $x$ is later referenced in statement $s2$, most processors will have to fetch $x$ from the memory again. This cycle of pointer overflow and cache invalidation continues for $x$ through the end of the DOALL loop.

On the other hand, in the same example, delayed allocation marking allocates a pointer only for the $x$ reference in statement $s_n$. All processors will have valid copies of $x$ until $s_n$ is executed. Of course, during the execution of $s_n$, some pointers will be evicted and some copies of $x$ invalidated. Here, however, the invalidation is justified, because $x$ will be rewritten after the DOALL loop, which may cause an incoherence.

The above example illustrates the essence of delayed allocation marking. To support this strategy, we assume that two types of memory references may be issued by an instruction. One forces a pointer allocation, while the other does not. We call the first type a *p-reference* and the second type an *n-reference*. Given a program, the compiler marks each memory reference as either a p-reference or an n-reference.

## 3 Compiler Algorithms

This section describes two compiler algorithms for marking p-references and n-references. We apply these algorithms to one routine at a time. With the assistance of interprocedural analysis, they can also be applied across different routines.

The first algorithm performs an *intraboundary analysis.* For each boundary (i.e. each processor reassignment point), it examines the program segments that are reachable from that boundary without crossing any other boundaries. It conservatively assumes that event $S_3$ may occur (*cf.* Section 2) after any other boundary is crossed. The second algorithm performs a *cross-boundary* analysis. This algorithm searches the program to determine whether the event sequence in Section 2 may occur for a particular variable. The cross-boundary analysis is more difficult to implement and more time-consuming to execute, but it gives better results by marking fewer p-references, and by further delaying the p-references.

### 3.1 Intraboundary Analysis

The intraboundary analysis handles one boundary at a time. For a given *starting* boundary, the intraboundary analysis examines the memory references that may occur before the program execution flows across any other boundary (an *ending* boundary). For each variable, it marks as p-references only the last memory references to the same variable before an ending boundary is crossed. All other (non-last) references can safely be marked as n-references without causing any coherence problem. Note that multiple ending boundaries exist only if the starting boundary (e.g. the boundary $B$ in Figure 2) begins

```
starting boundary, B:

          x = 5          n_references to x
          b = 4 + x      n_references to x, b
          if ( ) goto 11
          b = 1 - x      p_references to x, b

ending    boundary for B:

          DOALL 10 i=1,5

10        CONTINUE

new       boundary:

11        b = 2 - x      p_references to x, b

ending    boundary for B

          DOALL          ...
```

Figure 2: An example for boundaries and p/n-references.

a serial execution region. In Figure 2, the goto statement extends the serial region to statement 11, skipping the boundary at the exit of the first DOALL loop. As a result, $B$ has two ending boundaries. The p-references and the n-references in this serial region are marked in Figure 2.

The algorithm for the intraboundary analysis is shown in Figure 3. We assume that the compiler has the control flow graph for each routine of the given program. For each boundary, $B$, we extract a control flow subgraph, $F(B)$, such that $B$, the starting boundary, is in the entry node of $F(B)$ and each ending boundary is in an exit node of $F(B)$. This is done by function *make_IBC_graph* which makes a depth-first search over the control flow graph. Beginning at the node that contains $B$, every node that is reachable without crossing another (ending) boundary is included in $F(B)$. All the edges linking the nodes in $F(B)$ are also included in $F(B)$. We omit the details of this simple search.

Function *mark_last* (*cf.* Figure 3) traverses $F(B)$ and locates the p-references. This is a simple variation of the familiar iterative algorithm for computing reaching definitions. If there exists an execution path from $B$ to an ending boundary, say $C$, such that $r$ is the last memory reference to a certain variable, then $r$ is included in the *reach* set for $C$. In the algorithm in Figure 3, the reach sets for all ending boundaries (corresponding to $B$) are unioned to *mark_set*. After the algorithm iterates over all starting boundaries, *mark_set* contains all the references that should

```
intra_boundary_alg (flowgraph)
{
   mark_set = NULL;
   find_boundary ();
   for each boundary node, B, do
   {
      make_IBC_graph (B);

      mark_last (F(B));

      /* This call is not needed for
         intraboundary only analysis */
      summarize_scopes (F(B), B);

      for each ending boundary node, e, in F(B) do
         mark_set = mark_set ∪ e →reach;
   }
}
make_IBC_graph (B)
{
    add B to F(B) and to work_list;
    for each node, v, in work_list do
    {
       remove v from work_list;
       if ((v==B) or (v is not a boundary node)) then
          for each successor, u, of v do
          {
             if (u is not yet in F(B)) then
             {
                add u to F(B);
                add u to work_list;
             }
             add arc (v, u) to F(B);
          }
    }
}
mark_last (IBC_graph)
{
    for each node, v, in the IBC_graph do
    {
       v →KILL = all variables referenced in v;
       /* v →ref is the set of references in v */
       v →reach = v →ref;
    }
    change = true;
    while (change) do
    {
       change = false;
       for each node, v, in the IBC_graph do
       {
          oldreach = v →reach;
          v →reach = v →ref ⋃(u,v) u →reach;
          for each r in v →reach do
             if (r refers to a variable in v →KILL)
                remove r from v →reach;
          if (v →reach ≠ oldreach) then
             change = true;
       }
    }
}
```

Figure 3: The intraboundary algorithm

```
summarize_scopes (IBC_graph, B)
{
    for each ending boundary node, C,
        corresponding to B do
    {
        let s = scope (B, C);
        s →modset = all variables modified in s;
        s →readset = all variables read in s;
        s →lastref = C →reach;
    }
}
```

Figure 4: Pseudocode for summarizing_scopes

be marked as p-references by intraboundary analysis. Note that the function *summarize_scopes* is performed to prepare for cross-boundary analysis which will be addressed in the next section.

### 3.2 Cross-boundary Analysis

The p-references are marked conservatively by the intraboundary analysis. In the cross-boundary analysis, only the references involved in an event sequence that may cause cache incoherence are marked as p-references. Therefore, the cross-boundary analysis further reduces the number of p-references, but it takes extra compile time. Our algorithm for this analysis is based on the *scope* between a pair of boundaries which is defined as follows.

**Definition:** The *scope* between a starting boundary, $B$, and one of its ending boundaries, $C$, is the program code which is reachable from $B$ and extends to $C$. We denote this scope by *scope*$(B, C)$.

For each scope, we keep the set of its modified variables, the set of its used variables, and the reach set of the ending boundary which contains the last memory references in the scope. We collect such information during the intraboundary analysis by calling the function *summarize_scopes* (*cf.* Figure 3). The details of this function are given in Figure 4.

The algorithm for the cross-boundary analysis is shown in Figure 5. We maintain a *reachability matrix*, $R$, whose rows and columns are scopes. $R[v, u]$ is 1 if and only if scope $v$ may reach scope $u$. It is a simple matter to compute $R$ within $O(n^2)$ time, where $n$ is the number of nodes in the flow graph of the given routine. After $R$ is computed, the algorithm calls function *mark_ref* to locate memory references that may be involved in an event sequence that causes cache incoherence. This function has a time complexity of $O(N^3)$, where $N$ is the number of scopes. We consider this time complexity too high for a compiler. In our experiment, we modify the function so that, instead of iterating over triples of ($u$, $v$, $w$), we iterate over pairs of ($u$, $v$), which reduces the complexity to $O(N^2)$. Correspondingly, we only examine whether $r$ refers to a variable in $v \rightarrow modset$. If it is true, then we assume $r$ is involved in an event sequence that may cause cache incoherence.

## 4 Experiment

We implemented two passes that are invoked by Parafrase 2 [17]. The first pass, *showstat*, marks p-references and n-references using four different marking schemes (listed in the increasing order of capabilities):

1. *All:* marking all references as p-references, as in the conventional dynamically tagged directories.
2. *No Index:* marking references to index variables (i.e. iteration counters) of serial DO loops as n-references. (Such variables are not shared among processors and thus do not create cache incoherence.)
3. *Intra:* marking by the intraboundary analysis.
4. *Cross:* marking by the cross-boundary analysis.

The next pass, *gentrace*, inserts code into the original program, which is then compiled and run to generate a memory access trace [15]. A p-reference in the program results in a set of *p-accesses* (accesses that need to allocate a pointer) at run time. Likewise, an n-reference results in a set of *n-accesses* at run time. Different marking schemes result in different traces.

The marking schemes are compared by two measures: the *static counts* of p-references in each program and the *dynamic counts* of p-accesses in each trace. The dynamic counts show the ultimate difference at run time. The static counts, on the other hand, provide a broader picture of the pointer allocation requirement independent of any specific input data.

We use the Perfect Club benchmark programs [3] in our experiment. Table 1 shows, for each program, the number of DOALL loops found by Parafrase 2, the number of subroutine and function calls, the static and dynamic counts of the total references and

```
cross_boundary_alg (flowgraph)
{
    intra_boundary_alg (flowgraph);
    mark_list = NULL;
    for each scope v do
        for each scope u do
            R[v, u] = 0;
    compute reaching matrix R;
    mark_ref ();
}


mark_ref ()
{
    for all scope triples (u, v, w) do
    {
        if (R[u, v] == 1 and R[v, w] == 1) then
            for each reference, r, in u →lastref do
                if (r refers to a variable in both
                    v →modset and w →readset) then
                    mark_list = mark_list ∪ r;
    }
}
```

Figure 5: The cross-boundary algorithm

the percentage of the array references. Due to the lack of an exact array data dependence analysis in Parafrase 2, only scalar references are covered in our experiment. All array references will be marked as needing a pointer.

The static counts are gathered from all programs except css for which the compiler runs out of memory. The dynamic counts are gathered from nine programs for which we are able to reduce the traces to manageable sizes without altering the memory reference behavior. In order to reduce the traces, we modify the test programs to shorten the execution time. For most of the test programs, computations are repeated over time steps in the outermost serial loops. We only reduce the number of the time steps, so the memory reference behavior should remain the same. The details of the reduction are as follows: apsi, from 720 to 2; lwsi, from 100 to 1; mtsi, from 62 to 11; sdsi, from 1000 to 4; srsi, from 100 to 1; and tisi, from 40 to 15. For lgsi, the problem matrix size is reduced to 4x4x4x4 (from 8x8x8x8), and the number of sweeps is changed from 2 to 1. With tfsi, two meshes are removed from the input file, leaving just one.

The effects of different marking schemes on the static and the dynamic counts are shown in Figure 6 and Figure 7, where each bar represents the percentage of the static p-references (or the dynamic p-accesses) over the total references (or the total accesses). Note that for the three programs in Figure 6, only the static counts are available. As both figures show, the "No Index" scheme is insufficient for most of the Perfect programs. On the other hand, our two algorithms are able to reduce the static counts of pointer allocations for scalar references by approximately 50%. The dynamic counts also indicate that the majority of all run time scalar references do not need to allocate a pointer. Our algorithms contribute the largest fraction of the reduction of the dynamic counts for all programs except sdsi and tfsi.

Table 1: Programs' characteristics

| Prog. | Doalls | Calls | Static | | Dynamic | |
|---|---|---|---|---|---|---|
| | | | Total | %Array | Total | %Array |
| apsi | 72 | 247 | 7946 | 26.06 | 3.8M | 34.28 |
| lgsi | 31 | 133 | 3369 | 48.29 | 31.1M | 54.90 |
| lwsi | 16 | 62 | 2486 | 38.54 | 92.7M | 55.51 |
| mtsi | 45 | 98 | 3698 | 40.59 | 2.7M | 18.76 |
| nasi | 48 | 108 | 7429 | 30.42 | N/A | N/A |
| ocsi | 32 | 149 | 3126 | 25.91 | 17.2M | 28.32 |
| sdsi | 48 | 163 | 3234 | 38.71 | 6M | 32.57 |
| smsi | 26 | 87 | 7270 | 20.17 | N/A | N/A |
| srsi | 46 | 83 | 6018 | 33.18 | 59.6M | 30.49 |
| tfsi | 37 | 77 | 4461 | 32.71 | 72.8M | 42.26 |
| tisi | 22 | 45 | 1460 | 41.37 | 9.4M | 36.17 |
| wssi | 108 | 219 | 4636 | 35.42 | N/A | N/A |

For both the intraboundary analysis and the cross-boundary analysis, we have also considered the effect of aliasing (variables with different names that refer to the same memory location). Unfortunately, the alias analysis algorithm in Parafrase 2 does not provide global analysis. Hence, the aliasing information is very limited. For most of the test programs, the use of this limited alias analysis makes no difference in the counts. A few programs have counts that differ by less than 1 percent.

*In-lining* integrates the body of a subroutine in place of every call which thereby allows complete in-

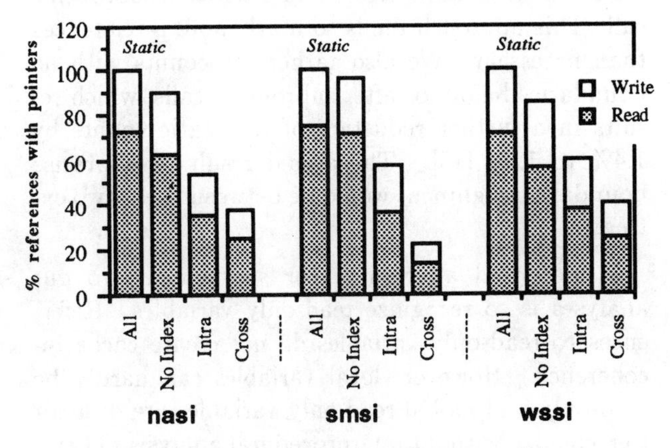


Figure 6: Percentage of p-references (1)

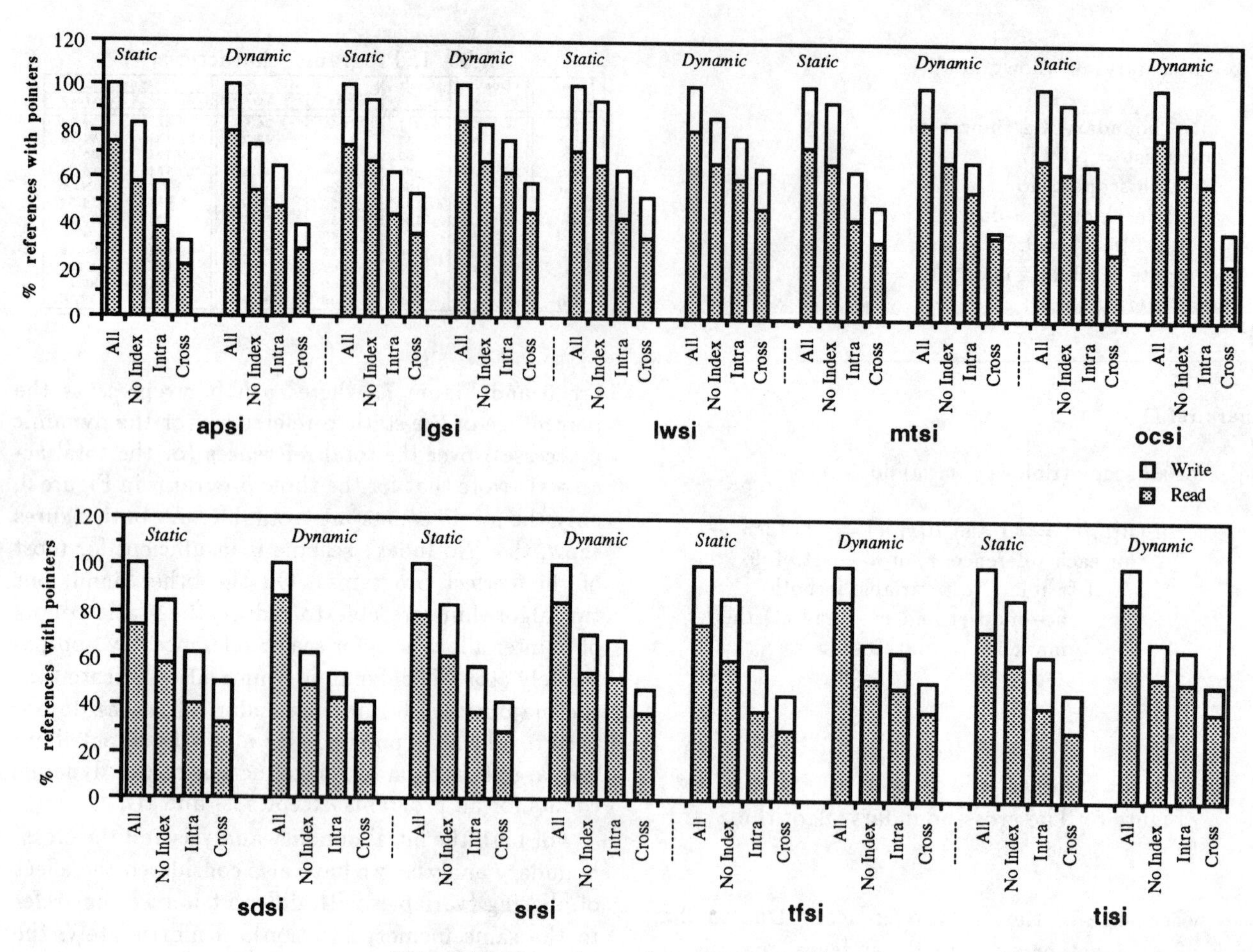


Figure 7: Percentage of p-references (2)

terprocedural analysis. Unfortunately, Parafrase 2 does not perform satisfactory in-lining for the programs in our experiment. Instead, we make conservative assumptions about subroutine calls in the data flow analysis. Moreover, we take the most conservative approach to the boundary assignment by assigning boundaries both before and after each subroutine call. This approach tends to mark more p-references than necessary. We also gather the counts with no boundaries before or after subroutine calls, which results in a further reduction of the static counts by 1.4% to 12% [15]. The actual result of a precise boundary assignment would lie between these two extremes.

A potential alternative (or supplement) to our analyses is to recognize read-only variables. References to read-only variables do not create cache incoherence. However, local variables can hardly be read-only and global read-only variables are difficult to recognize without interprocedural analysis. Therefore, we do not examine the effect of this possibility.

## 5 Conclusion

We have implemented two compiler algorithms to support delayed allocation marking on dynamically tagged directories for data caches. Our algorithms identify those memory references that do not need to allocate a pointer, and they delay the pointer allocation as much as possible. Currently, our implementation is limited to scalar references only. However, the results from the Perfect benchmark programs are very encouraging. Both the static counts from the source programs and the dynamic counts from run time memory traces indicate that the number of scalar memory references that need to allocate a pointer is significantly reduced by our algorithms. This reduction improves the program performance in two ways: First, we reduce the number of pointer operations and the contention at the coherence directories. Second, we reduce pointer overflow and thus reduce unnecessary cache invalidations. The net result of these changes should be a reduction in the average memory latency.

We are currently conducting simulations in which we feed our generated traces to a memory-processor simulator [12]. From the simulation results, we will be able to see more clearly the effect of our algorithms on the network traffic and the data cache miss ratio.

We have not extended our algorithms to array references because of the lack of an exact array data dependence analysis in the compiler, Parafrase 2, for our experiment. However, our data show that the number of array references is comparable to the number of scalar references. In our fufure work, we will examine the traces to study the effect of an ideal array data dependence analysis.

## Acknowledgement

This work is supported in part by the National Science Foundation, grant no. CCR-9210913 and CCR-9209458, by IBM Corporation, grant no. 340-659, by the Army Research Office, contract no. DAAL03-89-C-0038 with the University of Minnesota Army High Performance Computing Research Center, and by the Graduate College of the University of Minnesota.

## References

[1] A. Agarwal, R. Simoni, J. Hennessy, and M. Horowitz. An evaluation of directory schemes for cache coherence. In *Proc. 15th Annual International Symposium on Computer Architecture*, pages 280–289, June 1988.

[2] J. Archibald and J. Baer. An economical solution to the cache coherence problem. In *Proc. 11th Annual International Symposium on Computer Architecture*, pages 355–362, June 1984.

[3] M. Berry, D. Chen, P. Koss, D. Kuck, and S. Lo. The Perfect club benchmarks: Effective performance evaluation of supercomputers. Technical Report CSRD-827, University of Illinois, Urbana, IL, May 1989.

[4] L. M. Censier and P. Feautrier. A new solution to coherence problems in multicache systems. *IEEE Transactions on Computers*, C-27(12):1112–1118, December 1978.

[5] D. Chaiken, J. Kubiatowicz, and A. Agarwal. LimitLESS directories: A scalable cache coherence scheme. In *Proc. Sixth International Conference on Architectural Support for Programming Languages and Operating Systems*, pages 224–234, 1991.

[6] H. Cheong and A. V. Veidenbaum. Compiler-directed cache management in multiprocessors. *Computer*, 23(6):39–47, June 1990.

[7] J. R. Goodman. Using cache memory to reduce processor-memory traffic. In *Proc. 10th Annual International Symposium on Computer Architecture*, pages 124–131, June 1983.

[8] A. Gupta, W. Weber, and T. Mowry. Reducing memory and traffic requirements for scalable directory-based cache coherence schemes. In *Proc. 1990 International Conference on Parallel Processing, Vol. I: Architecture*, pages 312–321, August 1990.

[9] D. V. James, A. T. Laundrie, S. Gjessing, and G. S. Sohi. Scalable coherent interface. *Computer*, 23(6):74–77, June 1990.

[10] R. Katz, S. Eggers, D. A. Wood, C. Perkins, and R. G. Sheldon. Implementing a cache consistency protocol. In *Proc. 12th Annual International Symposium on Computer Architecture*, pages 276–283, June 1985.

[11] R. L. Lee, P. Yew, and D. J. Lawrie. Multiprocessor cache design considerations. In *Proc. 14th Annual International Symposium on Computer Architecture*, pages 253–262, June 1987.

[12] D. J. Lilja and P. Yew. Combining hardware and software cache coherence strategies. In *Proc. 1991 ACM International Conference on Supercomputing*, pages 274–283, 1991.

[13] W. Michael. A scalable coherence cache system with a dynamic pointing scheme. In *Proc. Supercomputing '92*, pages 358–367, 1992.

[14] S. L. Min and J. Baer. Design and analysis of a scalable cache coherence scheme based on clocks and timestamps. *IEEE Transactions on Parallel and Distributed Systems*, 3(1):25–44, January 1992.

[15] T. N. Nguyen, Z. Li, and D. Lilja. Compiler analysis for efficient use of dynamically tagged directories. Technical Report 93-036, AHPCRC, Univ. of Minnesota, 1993.

[16] B. W. O'Krafka and A. R. Newton. An empirical evaluation of two memory-efficient directory methods. In *Proc. 17th Annual International Symposium on Computer Architecture*, pages 138–147, June 1990.

[17] C. D. Polychronopoulos, M. B. Girkar, M. R. Haghighat, C. L. Lee, B. P. Leung, and D. A. Schouten. Parafrase-2: An environment for parallelizing, partitioning, synchronizing, and scheduling programs on multiprocessors. In *Proc. 1989 International Conference on Parallel Processing*, St. Charles, IL, August 1989.

[18] L. Rudolph and Z. Segall. Dynamic decentralized cache consistency schemes for MIMD parallel processors. In *Proc. 12th Annual International Symposium on Computer Architecture*, pages 340–347, June 1985.

[19] C. K. Tang. Cache design in the tightly coupled multiprocessor system. In *AFIPS Conf. Proc. Nat. Comput. Conf.*, pages 749–753, 1976.

[20] C. P. Thacker and L. C. Stewart. Firefly: A multiprocessor workstation. In *Proc. Second International Conference on Architectural Support for Programming Languages and Operating Systems*, pages 164–172, October 1987.

[21] W. C. Yen, D. W. L. Yen, and K. Fu. Data coherence problem in a multicache system. *IEEE Transaction on Computers*, C-34:56–65, January 1985.

# Trailblazing: A Hierarchical Approach to Percolation Scheduling*

Alexandru Nicolau and Steven Novack
Department of Information and Computer Science
University of California
Irvine, CA 92717

**Abstract:** Percolation Scheduling (PS) is a system for performing parallelizing transformations for the VLIW and superscalar computation models. PS has various useful properties, such as completeness with respect to local transformations, and appears to be an effective means of exploiting instruction level parallelism. However, compilers based on PS typically suffer from inefficiencies caused by the incremental application of PS transformations and significant code explosion. In this paper we present a non-incremental extension of PS that provides asymptotic efficiency improvements over normal PS. This approach dramatically reduces compilation time and achieves better parallelization by performing non-local transformations not feasible in PS. Simulation results comparing normal PS with our new technique are presented. The approach presented is adaptable to other global instruction level parallelization systems.

## 1 Introduction

Percolation Scheduling (PS)[14] was shown in to have various desirable properties, such as completeness with respect to local transformations[2], and appears to be an effective means of exploiting instruction level parallelism[12, 15, 18]. However, PS-based compilers [8, 5, 4] typically suffer from two efficiency problems: code explosion and linear operation moves. *Code explosion*[a] refers to the duplication of instructions and operations that may occur when parallelizing programs with conditional branches (e.g. when a conditional jump that used to follow some instructions is "moved" so that it precedes them, the instructions it moves past must be duplicated and inserted onto each of the conditional's branches). Code explosion always has the effect of increasing the number of operation moves and (often) increases program size. Sometimes, code explosion is desirable, as this may be the price of achieving speedups when an operation can move more on one path than on another; however, many code explosions are unnecessary in the sense that eventually all instructions that were duplicated could either be deleted or merged if the duplication does not result in increased parallelism.

In PS, *unnecessary* code explosion is a by-product of using strictly incremental PS transformations. For example, if an operation is located at a *join point*[b] and is independent of all instructions located in the body of the conditional (i.e. in the "then" or "else" branches), then the operation could be moved directly to the node preceding the conditional without visiting any node in the body of the conditional. The problem of *linear operation moves* refers to the fact that moving an operation from $A$ to $B$ using normal PS transformations requires a visit at each node on *every* control path from $A$ to $B$. The severity of these efficiency problems increases with the size of the region over which scheduling is performed, and since global scheduling regions are needed to effectively extract parallelism from many programs[12, 15], good solutions to these problems are critical.

The problems of unnecessary code explosions and linear operation moves could both be solved if moving an operation past multiple instructions could be done in constant time. Our solution is to structure the program graph's instructions and global information so that non-incremental moves can be made without visiting any instruction that is bypassed. We represent instruction level programs using the Hierarchical Task Graph (HTG) of Girkar and Polychronopoulos[10] and we define a technique called *Trailblazing* that exploits this structure by extending the PS core transformations to navigate through the HTG hierarchy. At the lowest level sub-graphs in the hierarchy, Trailblazing is able to perform the same fine-grained transformations as normal PS, while at higher levels, Trailblazing is able to move operations across large blocks of code in constant time, including loops (past which normal PS is unable to move operations at all). Trailblazing attempts to move operations at the "highest" possible level while moving operations to (or keeping them at) lower levels when needed to expose more parallelism. In this fashion, Trailblazing provides efficient code motion without sacrificing the "completeness" of normal PS and enables some code motion not possible in normal PS. Using the hierarchy of HTG's, Trailblazing is also able to provide a meaningful framework for making parallelism versus useful code explosion trade-offs.

Before describing our use of HTG's and Trailblazing, we briefly discuss related work (Section 2) and define some useful terminology (Section 3). Section 4 describes our adaptation of HTG's, and Section 5 defines the Trailblazing technique itself. Finally, Section 6, presents some simulation results.

## 2 Related Work

Code duplication[3] and lookahead windows[13] are both efficiency improving methods that have been proposed to reduce the compilation time and space complexity of fine-grained compilers. Both are orthogonal to, and could be used in conjunction with, the technique presented in this paper. [12] also presents a non-incremental version of PS, called Global Scheduling (GS), however GS requires incremental updates of global information for each move. Trailblazing could be incorporated into GS in order to decrease the overhead of maintaining global information. Various program graph representations, such as Dependence Flow Graphs[17], Program Dependence Graphs (PDG)[9], Static Single Assignment form[6], and Hierarchical Task Graphs (HTG's)[10], have been proposed for use in instruction level parallelization, and while each

*This work was supported in part by NSF grant CCR8704367 and ONR grant N0001486K0215.

[a] Although, we discuss it in the context of PS, code explosion is a problem common to any general approach to the parallelization of programs with conditionals.

[b] A *join point* is an instruction that has multiple predecessors. This instruction, and its successors, may be duplicated when paths are split. For example, the join point of an if-then-else block is the first instruction of the block which is executed after one of the conditional's branches (the "then" or "else") has executed.

may be useful for performing Trailblazing, HTG's seemed most natural for our purposes. Another non-incremental scheduling technique, Region Scheduling (RS)[11], provides non-incremental code movements on PDG's. RS is typically used at a coarser grain level than PS or Trailblazing and some transformations that can easily be performed by Trailblazing to extract fine-grain parallelism may be too expensive using RS when the program is not well structured.

## 3 Background and Terminology

In this section, we define some terms that will be used when discussing Trailblazing and PS. PS is a system for performing parallelizing transformations on the program graphs of the VLIW computation model.[c] A *program graph* is a directed graph wherein each node is a whole instruction and edges represent control flow. An *instruction* is a set of "RISC-type" operations possibly containing multiple conditional jumps that combine to yield a single control path. In the IBM VLIW model[7], it is natural to view each VLIW instruction as a tree consisting of a special *entry* operation that behaves as a NOP followed by a (possibly empty) binary tree of conditionals branches. Each *leaf* of the conditional tree is a pointer to a successor instruction. An empty conditional tree is just a single such leaf. Leaves in the conditional tree of an instruction are also referred to as "leaves of the instruction". Each functional operation is associated with one of the edges in the conditional tree, or the edge connecting the entry to the tree, and writes back its result only if the edge is on the path selected by the evaluation of the conditionals in the instruction. When discussing data dependencies, WRITES(OP) and READS(OP) refer to sets of registers[d] that are defined and used by OP, respectively. Similarly, WRITES(LEAF) and READS(LEAF) refer to the registers that are defined and used by operations and conditionals along the path from the entry of an instruction to LEAF. Other attributes are also associated with leaves, including INSTR(LEAF) which refers to the instruction that contains LEAF and SUCC(LEAF) which refers to the successor instruction pointed to by LEAF.

PS parallelizes program graphs by repeated application of a pair of transformations called MOVE-OP and MOVE-CJ that move an operation or a conditional jump up one instruction in the program graph while preserving the semantics of control and data flow. When an operation, OP, moves from a source instruction (denoted FROM) to a destination instruction (denoted TO), it exits the FROM instruction along one of the control paths coming into the entry of FROM and enters the TO instruction at a control path connecting a leaf of TO to FROM. Data dependencies can prevent this movement in two ways: OP can have a true data dependence on some operation in TO in which case we say that op can not be *injected* into TO at LEAF, and OP can be prevented from moving up out of FROM due to false dependencies, such as write-after-write and write-after-read, in which case we say that OP can not be *extracted* from FROM. All false dependencies on an operation, OP, can often be removed using a process called *renaming*[6] in which the destination of OP, and all of its uses, are replaced with a free register.

When an operation, OP, moves past a node, FROM, that has multiple predecessors, care must be taken to ensure that only the path along which OP is moved will be affected by the move. Trailblazing can accomplish this in two ways. The first, called *Node-isolation*, refers to copying FROM to FROM' so that the path along which the move occurs goes through FROM, while all other remaining paths go through FROM'. The second, called *oper-isolation*, avoids duplicating FROM (which may contain many operations other than OP) by inserting an empty instruction, NEW, before FROM. Then OP is moved from FROM to NEW and NEW is isolated using NODE-ISOLATE. OPER-ISOLATE is useful when splitting the entire node is either prohibitively expensive or not allowed (these cases do not occur for normal PS, but are relevant in the context of Trailblazing).

## 4 Hierarchical Task Graphs

An HTG is a directed acyclic graph containing five types of node: START and STOP nodes indicating the entry and exit of HTG's respectively, Simple nodes that (for our purposes) represent VLIW instructions, Compound nodes, representing sub-HTG's, which we use to represent if-then-else blocks, [e] and finally, Loop nodes that represent loops whose bodies are sub-HTG's[10]. In the remainder of this paper, *instruction* will refer exclusively to VLIW instructions and *node* will be used collectively for any type of HTG node (including simple nodes). Edges in the HTG represent control flow and with the exception of simple nodes (VLIW instructions), all nodes have at most one successor. As for VLIW instructions without conditionals, we define each non-simple node, B, to have a single "leaf" which points to its successor (denoted SUCC(B)) in the HTG. For any compound or loop node, B, WRITES(LEAF) is the set of registers defined anywhere in HTG(B) and READS(LEAF) is the set of registers that are both used in HTG(B) and live at the START node of HTB(B).[f] When moving an operation, OP, Trailblazing (see Section 5) can determine whether or not OP depends on any node in HTG(B) by comparing the READS and WRITES sets of OP with those of B. If no dependency exists, then OP can move non-incrementally across B without visiting any nodes in HTG(B). Since compound and loop nodes both act as "bridges" across their sub-HTG's, we refer to them collectively as *bridge nodes* or just *bridges.* Given a bridge B, HTG(B) is referred to as *the region bridged by B* or just *bridged region* if B is understood.

Storing global information, such as live register sets, at each node in the program would prevent truly non-incremental moves due to the need to update information at each node bypassed by a move. Trailblazing enables truly non-incremental moves by mapping global information to the HTG hierarchy in such a way as to allow an operation move across a bridge node to affect only the information stored at the bridge itself, and not at any node in the bridged region. For instance, global live register information is mapped to the hierarchy by partitioning it into "local" live information sets stored at each node,

[c] PS can also be used on other fine-grained computation models such as superscalar and super-pipelined machines.

[d] Registers may be actual or virtual depending on the implementation. For this paper, the distinction is unimportant.

[e] We consider an if-then-else structure to be the smallest region for which a conditional jump dominates all other nodes in the region and another node, containing only one successor, post-dominates all other nodes in the region.

[f] We may also refer to these sets as Writes(B) and Reads(B).

```
procedure trailblaze(op, from-leaf, barrier)
  interference ← Live(Instr(from-leaf)) - contribution(op)
  trail ← find-trail(op, from-leaf, barrier, interference)
  if trail = NULL then
    Suspend(op)
  else
    if | preds(Instr(from-leaf)) | > 1 then
      node-isolate(Instr(from-leaf),trail)
    end if
    – find-trail may do renaming by changing Writes(op) to
    – new-write. If so, then extract will leave behind the
    – copy operation "old-write ← new-write".
    extract(op, FIRST(trail))
    inject(op, LAST(trail))
    for each leaf on trail, FIRST to LAST do
      if | preds(Instr(leaf)) | > 1 then
        oper-isolate(op,Instr(leaf),trail)
      end if
      update-local-live(Instr(leaf))
    end do
  end if
end procedure
```

Figure 1: The Trailblaze transformation

N. If N is STOP(B)[g], then LOCAL-LIVE(N) = GEN(N) $\cup$ $\mathcal{L}$(SUCC(B)) $\cap$ WRITES(B); otherwise, LOCAL-LIVE(N) = GEN(N) $\cup$ $\bigcup_{\text{LEAF} \in \text{N}}$ LOCAL-LIVE(SUCC(LEAF)) − WRITES(LEAF). The function $\mathcal{L}$(N), which computes live information from LOCAL-LIVE sets, will be defined below, but first we should define the GEN set of a node N. Conceptually, GEN(N) is the set of live registers "generated" by N. For each VLIW instruction, N, X $\in$ GEN(N) if and only if X is used in N. For any bridge B, GEN(B) is the information generated in B itself, or flowing into B through STOP(B), that makes it to START(B) on some path without getting "killed". More formally, for any bridge B, X $\in$ GEN(B) if and only if one or both of the following two conditions hold: (1) X $\in$ GEN(N) for some N $\in$ HTG(B) and X is not defined on some path from START(B) to N, or (2) X $\in$ $\mathcal{L}$(SUCC(B)) and X is defined on some, but not all, paths from START(B) to STOP(B).

Global live register information for any node, N, is computed from LOCAL-LIVE sets by the recursive function $\mathcal{L}$(N). If N is in HTG(B) for some bridge B, then $\mathcal{L}$(N) = LOCAL-LIVE(N)$\cup$($\mathcal{L}$(SUCC(B))−WRITES(B)); otherwise, $\mathcal{L}$(N) = LOCAL-LIVE(N). A correctness proof of this approach as well as its generalization to a broad class of global information is provided in [16].

## 5 Trailblazing PS (TiPS)

In this section we describe how Trailblazing PS (TiPS) moves operations in HTG's. The main difference between TiPS and normal PS is that the MOVE-OP transformation of normal PS is replaced with the TRAILBLAZE transformation in TiPS. Conditionals are still moved in TiPS using the normal PS transformation, MOVE-CJ.[h] The TRAILBLAZE transformation shown in Figure 1 is responsible for trying to move the non-conditional operation, OP, as far forward as possible along the "best" (according to some evaluation function) path from its current position to a specified "barrier" node. The barrier node is used to ensure that OP does not travel "too far" according to the particular scheduling rules being used. For instance, the Global Resource-constrained Percolation (GRiP)[15] scheduling guidance rules employed by our compiler schedule nodes in the program graph in a top-down fashion, so the barrier node is usually chosen to be the node currently being scheduled, since trying to move an operation higher than this would be futile.

The first step of the TRAILBLAZE transformation is to call the FIND-TRAIL function which returns a linked list of leaves, called the TRAIL, that characterizes the path of instructions and bridges that will be visited when moving the operation, OP.[i] The first leaf on TRAIL is the leaf from which OP will be extracted and the last leaf on TRAIL is the destination leaf where OP will be injected. The nodes of the intermediate leaves on TRAIL may have to be isolated (see Section 3) in order to preserve the semantics of control flow and will need to have their LOCAL-LIVE sets updated to reflect the movement of OP. When searching for a TRAIL, if FIND-TRAIL encounters a node that has multiple predecessors, the order in which paths will be tried is determined by such factors as predicted path probabilities and user-supplied assertions.

On any given path, the TRAIL for OP can be terminated by data dependencies, resource constraints, or artificial limitations such as speculative scheduling strategies, or cost vs. performance trade-offs. It should be noted that one of the advantages of our approach is that these issues are completely isolated from the transformations proper, and thus, our TiPS transformation system is orthogonal to the particular scheduling heuristics used, which will vary from machine to machine.

True data dependencies, indicated by a non-empty intersection of READS(OP) and the WRITES set of the leaf being considered, will always terminate the TRAIL. For example, Figure 2 shows a sequence of 3 nodes in an HTG. Rectangles represent instructions, ovals represent non-simple nodes, and triangles represent single-entry, multiple-exit sub-graphs containing instructions, compound nodes, and/or loop nodes. The numbers in dark circles are used to label nodes referred to in the text. If we apply TRAILBLAZE to "A = X + B" in node 1, then one possible TRAIL would be the nodes, 1 through 5, in succession, with no definitions of X or B occurring in nodes 3 or 5, and the search being terminated by the definition of X in node 6. Notice that when creating TRAIL (and when subsequently moving OP), no node in the arbitrarily large region bridged by node 3 is visited, so TiPS is able to bypass the entire region in constant time, with only a single visit at node 3. In contrast, normal PS would require OP to incrementally traverse each path through the region, causing unnecessary code explosion at each join-point encountered. When FIND-TRAIL encounters a true data dependence at a bridge node, as happens in the example when trying to move OP across node 4 (since X $\in$ WRITES(4)), FIND-TRAIL "descends" into the bridged region and continues from the STOP node. When FIND-TRAIL reaches the START node of a region bridged by the bridge B, the search continues from B (as in node 2).

False dependencies are detected by FIND-TRAIL by

[g] Read "the STOP node of the region bridged by B".

[h] Although, conditionals are not allowed to move into or out of bridged regions since doing so would cause the region to have multiple successors, which is undefined in HTG's.

[i] Recall that each leaf of a node indicates a different continuation of the node, therefore the leaves of nodes rather than the nodes themselves are the true representation for a path of nodes

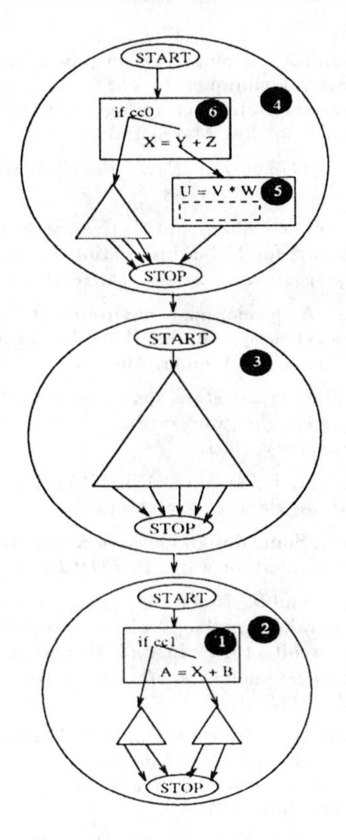


Figure 2: Finding a TRAIL.

maintaining a set of registers, called the INTERFERENCE set, that contains all registers that are live[j] or defined at some leaf on TRAIL. While computing TRAIL, if OP writes to a register contained in INTERFERENCE, then there is a false dependence on OP and the TRAIL will be terminated if the dependence can not be removed by renaming (e.g. if OP is a STORE or |INTERFERENCE| = number of registers).

If no trail is found, then the operation is currently unmovable and will be suspended until it can be moved; otherwise, the operation is *extracted* from its current position and *injected* into its new destination. For example, in Figure 2, the operation will be extracted from its position in node 1 and injected into node 5 at the position indicated by the dashed rectangle. When entering or leaving a region bridged by B (as opposed to crossing B), information associated with B must be updated. For example, in Figure 2, when moving the operation from node 1 to node 5, A will be added to Writes(4) and if no other definitions of A exists in the region bridged by node 2, A will be removed from WRITES(2). READS sets and possibly some LOCAL-LIVE information may also need to be updated.

When FIND-TRAIL encounters a conditional branch, as is the case in node 1 of the example, it is necessary to decide whether or not to allow the operation to move above the conditional. Making this decision is called *speculative scheduling* and is done by TiPS using a predicate called SPECULATION-WORTHWHILE which attempts to weigh the relative benefit and cost of the decision. For example, when enough resources are available, speculative scheduling is always worthwhile, but when there is a lack of resources, it is essential to ensure that potentially useless operations do not compete for resources with operations that are always useful. Again, while TiPS allows the introduction of such "guidance" heuristics, it is not tied to a particular decision which will have to vary with the specific architecture.

[j] Contribution(op), referred to in Figure 1, is the set of registers that have their last use in op, and therefore are live only because of op, and should not be included in Interference.

| code | TiPS moves | TiPS SU | PS moves | PS SU | compile time (PS/TiPS) |
|---|---|---|---|---|---|
| CF | 622 | 4.11 | 18230 | 4.38 | 9.96 |
| DOT | 514 | 3.20 | 15761 | 3.16 | 14.18 |
| crale | 1318 | 5.40 | 17312 | 4.33 | 13.56 |
| unriems | 3364 | 7.72 | 71654 | 7.27 | 4.34 |
| cmp | 506 | 2.34 | 3379 | 2.34 | 6.46 |
| queens | 2115 | 3.23 | 23731 | 3.09 | 11.68 |
| AVG | 1407 | 4.33 | 25011 | 4.10 | 10.03 |

Table 1: TiPS vs. PS: Compile Time and Performance

One important class of cost vs. performance trade-offs is whether or not to allow useful code duplication. Unnecessary code duplication is eliminated by bypassing bridges containing if-then-else structures. Bridge nodes also facilitate managing useful code duplication. When FIND-TRAIL descends into a bridged region that contains an if-then-else structure, a predicate called EXPLOSION-WORTHWHILE is used to make a decision about whether or not to allow the operation to split the join-point of the conditional (via OPER-ISOLATE or NODE-ISOLATE). For the example in Figure 2, we assume that EXPLOSION-WORTHWHILE indicates that it would be worthwhile to apply OPER-ISOLATE at the STOP node of node 4, thus allowing TRAILBLAZE to proceed to node 5; if the operation is then actually moved, oper-isolation will be performed in order to leave a copy of the operation on the incoming edges of the STOP node that were not part of the TRAIL. Since the compound node encapsulates all of the information about the if-then-else structure, this decision can be made based on global criteria such as how much duplication has already occurred for this if-then-else block, the expected cost of isolation (which can be significant when doing node-isolate on a bridge node), and the expected benefit of allowing the operation to move along only one branch of the conditional.

## 6 Results

In this section, we describe how Trailblazing PS (TiPS) and HTG's are integrated into the scheduler of the UCI VLIW compiler and provide simulation results comparing the performance of this enhanced scheduler with a version that uses normal PS. To switch the scheduler from normal PS to TiPS, we simply use the TRAILBLAZE transformation in place of the (normal PS) MOVE-OP transformation. HTG's are integrated by adding a pre-scheduling phase that structures the program's CFG into an HTG and a post-scheduling phase that converts the HTG back into a CFG (i.e. compound and loop nodes are replaced by the sub-HTG's that they represent and START and STOP nodes are deleted). A post-pass phase is then used to perform local transformations that were previously not allowed due to structural limitations of HTG's (e.g. conditional jumps can not move out of bridged regions).

Table 1 shows simulation results comparing the performance of a version of our compiler that uses TiPS with a version that uses normal PS. In both cases, the compiler is targeted for a VLIW architecture with 8 homogeneous functional units. COMPILE TIME RATIO is the ratio of normal PS compilation time to TiPS compilation time.

Compilation time includes all of the time spent compacting the code, including the conversion to and from HTG's for the TiPS compiler. SU is the speed-up defined as the ratio of sequential to parallel cycles observed while simulating the benchmarks before and after compaction. MOVES refers to the total number of transformations performed: MOVE-OP plus MOVE-CJ (PS) or TRAILBLAZE plus MOVE-CJ (TiPS).[k] CF and DOT in the table refer to two kernels from the Livermore Loops, Casual Fortran (CF) (kernel 15) and Discrete Ordinates Transport (DOT) (kernel 20). Crale and unriems are both modules from two large computational fluid dynamics programs of the same names. Cmp is the Unix command to compare two files. Queens is the Eight Queens problem taken from the Stanford Benchmarks. Some of the smaller loops in these kernels were arbitrarily unrolled two or three iterations to expose some amount of code to parallelize, but the loop restructuring that we would normally use for Perfect Pipelining[1, 15] was disabled in order to isolate the effects of Trailblazing. It should be emphasized that for most of these benchmarks, scheduling more iterations and performing Perfect Pipelining would significantly and similarly improve speedup for both PS and TiPS, while the differences in compilation time between TiPS and PS would become even more dramatic. Therefore, these results represent conservative estimates of the improvements obtainable using TiPS.

The results in Table 1 shows that TiPS is able to provide a 10 fold improvement in compilation time without degrading the performance of the parallelized code. The main reason for this is that, using normal PS, the worst case cost of moving operations across if-then-else blocks is exponential in the number of conditionals in the block, whereas using TiPS, if-then-else blocks are placed in compound nodes and operations that are able to move across the if-then-else blocks will instead move across the compound nodes, without visiting any nodes in the if-then-else blocks. Another factor that contributes to the compile time improvement is that TiPS decreases the overhead of visiting each instruction by performing injection and extraction only at the end-points of what would otherwise be a sequence of incremental moves using normal PS. For instance, even for the unriems benchmark, which is predominately straight-line code (only 1% of the operations are conditional jumps), TiPS is able to achieve a more than 4 fold improvement in scheduling time while actually improving speedup. The non-incremental moves provided by TiPS can improve the performance of the parallelized code by enabling code motions that are not feasible in normal PS. For the benchmarks in Table 1, non-incremental moves improve performance by allowing operations to move across loops (an impossible transformation in normal PS) and across instructions which have no available resources (impossible for normal PS when resource constraints are integrated into the scheduling process). For one benchmark, the speedup produced by TiPS is less than that produced by PS (by about 6%) due to cost vs. performance trade-offs and MOVE-CJ restrictions, but for most programs (and on average), TiPS *improves* speedup while significantly decreasing compilation time, with commensurate decreases in the size of the compacted code.

[k] The number of moves are presented in addition to the compilation time to provide a clearer picture of the effects of Trailblazing.

## References


[1] A. Aiken and A. Nicolau. Perfect pipelining: A new loop parallelization technique. In *1988 European Symposium on Programming*. Springer Verlag Lecture Notes in Computer Science no. 300, March 1988.

[2] A. S. Aiken. *Compaction-Based Parallelization*. PhD thesis, Cornell University, 1988.

[3] D. Bernstein, D. Cohen, and H. Krawczyk. Code duplication: An assist for global instruction scheduling. In *24th Annual Symposium on Microarchitecture*, 1991.

[4] G. Bockle. A development environment for fine-grained parallelism extraction. In *3rd Workshop on Compilers for Parallel Computers*, Vienna, Austria, 1992.

[5] M. Breternitz. *Architecture Synthesis of High-performance applications-specific processors*. PhD thesis, Carnegie-Mellon University, 1991.

[6] R. Cytron and J. Ferrante. What's in a name? In *1987 Int'l. Conf. on Parallel Processing*.

[7] K. Ebcioglu. Some design ideas for a vliw architecture for sequential-natured software. In *IFIP Proceedings*, 1988.

[8] K. Ebcioglu and T. Nakatani. A new compilation technique for parallelizing loops with unpredictable branches on a vliw architecture. In *2nd Workshop on Programming Languages and Compilers for Parallel Computing*, Urbana, IL, 1989.

[9] J. Ferrante, K. Ottenstein, and J. Warren. The program dependence graph and its use in optimization. *ACM Transactions on Programming Languages and Systems*, 9(3):319–349, July 1987.

[10] M. Girkar and C. Polychronopoulos. Automatic extraction of functional parallelism from ordinary programs. *IEEE Transactions on Parallel and Distributed Systems*, 3(2):166–178, March 1992.

[11] R. Gupta and M. Soffa. Region scheduling: An approach for detecting and redistributing parallelism. *IEEE Transactions on Software Engineering*, 16(4):pp. 421–431, April 1990.

[12] S. Moon and K. Ebcioglu. An efficient resource constrained global scheduling technique for superscalar and vliw processors. Technical report, IBM, 1992.

[13] T. Nakatani and K. Ebcioglu. Using a lookahead window in a compaction-based parallelizing compiler. In *23rd Annual Int'l. Symp. on Microarchitecture*, 1990.

[14] A. Nicolau. Uniform parallelism exploitation in ordinary programs. In *1985 Int'l. Conf. on Parallel Processing*.

[15] S. Novack and A. Nicolau. An efficient global resource constrained technique for exploiting instruction level parallelism. In *1992 Int'l. Conf. on Parallel Processing*.

[16] S. Novack and A. Nicolau. Trailblazing: A hierarchical approach to percolation scheduling. Technical Report TR-92-56, University of California at Irvine, 1992.

[17] K. Pingali, M. Beck, R. Johnson, M. Moudgill, and P. Stodghill. Dependence flow graphs: an algebraic approach to program dependencies. In *Advances in Languages and Compilers for Parallel Processing*. The MIT Press, 1991.

[18] R. Potasman. *Percolation-Based Compiling for Evaluation of Parallelism and Hardware Design Trade-Offs*. PhD thesis, University of California at Irvine, 1991.

# Contention-Free 2D-Mesh Cluster Allocation in Hypercubes*

*Stephen W. Turner, Lionel M. Ni,* and *Betty H.C. Cheng*

Department of Computer Science
Michigan State University
East Lansing, Michigan 48824-1027
{turner,ni,chengb}@cps.msu.edu

**Abstract**

Traditionally, each job in a hypercube multiprocessor is allocated with a subcube so that communication interference among jobs may be avoided. Although the hypercube is a powerful processor topology, the 2D mesh is a more popular application topology. This paper presents a 2D-mesh cluster allocation strategy for hypercubes. The proposed auxiliary free list processor allocation strategy can efficiently allocate 2D-mesh clusters without size constraints, can reduce average job turnaround time compared with that based on subcube allocation strategies, and can guarantee no communication interference among allocated clusters when the underlying hypercube implements deadlock-free E-cube routing. The proposed auxiliary free list strategy can be easily implemented on hypercube multicomputers to increase processor utilization.

## 1 Introduction

The problem of subcube allocation has been studied extensively to maximize processor utilization and minimize system fragmentation in hypercubes. Several strategies have been proposed and implemented for subcube allocation, including the buddy strategy [1], the gray code (GC) strategy [2], the modified buddy strategy [3], and the free list strategy [4]. Of these approaches, only the free list strategy has been shown to perform optimally, since it provides perfect subcube recognition.

For hypercube machines, such as the nCUBE-2 and the newly announced nCUBE-3, the restriction of allocating subcubes causes low processor utilization. Although the hypercube is a powerful network topology [5], 2-D and 3-D meshes are more popular application topologies. For example, grid domain decomposition for solving partial differential equations is an application that can easily be implemented on 2-D and 3-D meshes. In addition, 2-D and 3-D meshes are more efficient at allocating exactly the number of processors requested. For example, if the optimal number of processors for a task is 600, then the smallest subcube that can be allocated is 1024 processors, resulting in a waste of 424 processors, while a 2-D mesh may allocate a $20 \times 30$ cluster.

Consider the 4-dimensional cube shown in Figure 1, in which one job is allocated a $2 \times 5$ mesh, and another job is allocated a $2 \times 3$ mesh. With a restriction to subcube allocation, both jobs cannot be simultaneously executed, even though the total number of processors, 16, is sufficient. Without the subcube restriction, both clusters may be allocated in the 4-cube, as shown in Figure 1. A closer look reveals that communication from node 0100 to node 1010 in the $2 \times 5$ cluster will cause link contention with communication between nodes 0110 and 0010 in the $2 \times 3$ cluster, if the popular deadlock-free E-cube routing is used [6]. This contention results in ***intercluster communication interference***, which is desirable to avoid. Many known processor allocation strategies have been developed to guarantee contention-free cluster allocation, such as those subcube allocation strategies for hypercubes, the one used in Intel Touchstone (2D mesh topology), and the one used in CM-5 (fat tree topology).

It is well-known that the hypercube may embed a $k$-dimensional mesh with dilation 1 [7]. The *auxiliary free list method*, presented here, addresses the problem of low processor utilization by allowing 2-D mesh allocation that is contention-free and uses existing deadlock-free E-cube routing. In addition, it improves performance over the free-list method by decreasing job turnaround time and increasing processor utilization. This paper only deals with 2-D mesh allocation. For applications not requiring a 2-D mesh, the closest 2-D mesh is allocated. However, intracluster communication interference may occur.

The remainder of this paper is organized as follows: Section 2 discusses the auxiliary free list allocation and deallocation algorithms. Section 3 presents a brief analysis of the algorithms associated with the auxiliary free list, as well as some performance results

*This work was supported in part by the NSF grants CCR-9209873, CDA-9121641, and MIP-9204066.

relevant to the study. Section 4 gives concluding remarks, as well as directions for future work.

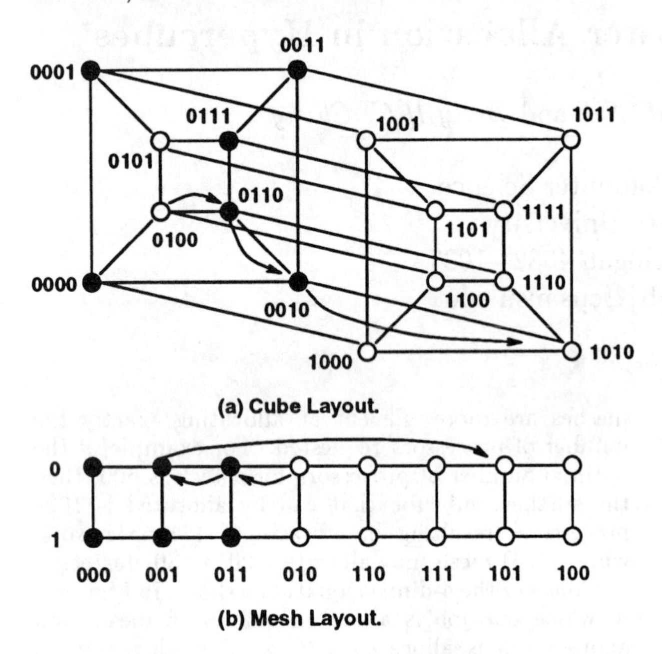


Figure 1. Communication interference.

# 2 Contention-Free Allocation

Let $Q_n$ denote an $n$-dimensional hypercube of $2^n$ nodes, in which the address of nodes and subcubes are represented by an $n$-bit string $\alpha = a_{n-1}a_{n-2}\ldots a_0$, $a_i \in \{0, 1, *\}$, with bit $a_i$ corresponding to dimension $i$ and '$*$' representing the "don't care" symbol. To embed a 2-dimensional mesh in a hypercube, its node addresses may be represented by the 3-tuple $(Q_k, X, Y)$, where $Q_k$ is the $k$-cube required to embed an $X \times Y$ 2-D mesh with dilation 1 ($k = \lceil \log_2 X \rceil + \lceil \log_2 Y \rceil$).[1] This embedding produces $2^k - XY$ *leftover* nodes.

The basic idea of the proposed auxiliary free list is to supplement existing allocation methods by partitioning the leftover nodes into subcubes, which are then stored in the auxiliary free list. Leftover nodes are partitioned into subcubes so that clusters may be guaranteed to be contention-free. In this paper, the auxiliary free list is used to supplement the free list algorithm [4], although it may be used with any of the common subcube allocation methods.

The original free list algorithm maintains a list of currently available subcubes for each dimension $i$ ($0 \leq i \leq n$). When an incoming request requires a subcube of dimension $k$, the free list algorithm can determine, in $O(n)$ time (for a hypercube $Q_n$), whether a $k$-cube can be obtained. There are three main issues to consider with auxiliary free list allocation: (1) how to partition leftover nodes, (2) allocation using the auxiliary free list, and (3) deallocation using the auxiliary free list. Subcubes stored in the auxiliary free list are referred to as *leftover*, while subcubes stored in the free list are referred to as *free*. Both leftover and free subcubes may be partitioned to allocate 2-D meshes.

[1] If a dilation 1 embedding is not necessary, then $k = \lceil \log_2 XY \rceil$.

## 2.1 Partitioning Leftover Subcubes

The central idea of the auxiliary free list algorithm is its partitioning algorithm (Figure 2). For a dilation-1 2-D mesh embedding, assume that $k = \lceil \log_2 X \rceil + \lceil \log_2 Y \rceil$, that $X$ and $Y$ are not powers of 2, and that $Y \geq X$. In a mesh layout of the subcube, let $G_X$ and $G_Y$ be the set of gray codes used to address the $X$ and $Y$ dimensions, respectively. The first $X$ and $Y$ gray codes of $G_X$ and $G_Y$ are used to address the embedded mesh in the $X$ and $Y$ dimensions, respectively. Let $\Delta = 2^{\lceil \log_2 X \rceil} - X$, and $\Gamma = 2^{\lceil \log_2 Y \rceil} - Y$. The following additional information is used by the partitioning algorithm: $S_x = \{x_0, x_1, \ldots x_j\}$ ($\sum_0^j x_i = X$ and $x_i$ are unique powers of 2), $S_y = \{y_0, y_1, \ldots y_k\}$ ($\sum_0^k y_i = Y$ and $y_i$ are unique powers of 2), and $S_x$ and $S_y$ are sorted in ascending order. $S_\delta = \{d_0, d_1, \ldots d_\ell\}$ ($\sum_0^\ell d_i = \Delta$ and $d_i$ are unique powers of 2), and $S_\gamma = \{g_0, g_1, \ldots g_m\}$ ($\sum_0^m g_i = \Gamma$ and $g_i$ are unique powers of 2), and $S_\delta$ and $S_\gamma$ are sorted in ascending order.

There are two cases for the partitioning algorithm: (1) $(\Delta \times Y) > (\Gamma \times X)$; (2) $(\Gamma \times X) > (\Delta \times Y)$. For brevity, the algorithm is presented as handling only case (1). Here, **replace**$(i, j)$ is the function that replaces the least significant $j$ bits in the $i$th binary reflected gray code (BRGC) with '$*$', and **insert**$(a, i)$ is the function that inserts string $a$ into the dimension $i$ auxiliary free list. $L_o$ refers to the meshes ownership list, and $\bullet$ is the string concatenation operator. "Add $a$ to $L_o$" is to add the string $a$ (representing a cube address) to the ownership list of the 2-D mesh. The ownership list is maintained to expedite deallocation of the 2-D mesh, and is explained further in Section 2.3.

The results of the partitioning algorithm are illustrated in Figure 3, in which a $5 \times 5$ mesh has been embedded in a 6-cube. The cube is pictured in a mesh-layout form.

The partitioning algorithm takes advantage of a property that enables the algorithm to easily compute cube addresses by using an address from some gray code subset and replacing the $i$ least significant bits with '$*$' [8]. For example, in the 4-bit BRGC, the sequence of 4 codes $\{1010, 1011, 1001, 1000\}$ has "10" as its two most significant bits. The cube address of this sequence is therefore $10**$. The values in the sets $S_x$, $S_y$, $S_\delta$, and $S_\gamma$ are all powers of two because they represent placeholders for the $X$ component or $Y$ component of cube addresses.

As Figure 3 illustrates, the leftover subcubes

**Algorithm:** *Partitioning.*

**Input:** a subcube of dimension $k$.

**Output:** A set of leftover subcubes.

**Procedure:**

```
begin
  i_x = X
  For i = 0 to ℓ do begin
    c_x = replace(i_x, log2 δ_i);
    c_y = replace(i_x, ⌈log2 Y⌉);
    a = c_x • c_y;
    insert(a, (log2 δ_i + ⌈log2 Y⌉));
    Add a to L_o;
    i_x = i_x + δ_i;
  end
  i_x = 0;
  For i = 0 to k do begin
    i_y = Y;
    For j = 0 to m do begin
      c_x = replace(g_{i_x}, log2 x_i);
      c_y = replace(g_{i_y}, log2 γ_j);
      a = c_x • c_y;
      insert(a, (log2 x_i + log2 γ_j));
      Add a to L_o;
      i_y = i_y + γ_j;
    end
    i_x = i_x + x_i;
  end
end
```

Figure 2. The partitioning algorithm.

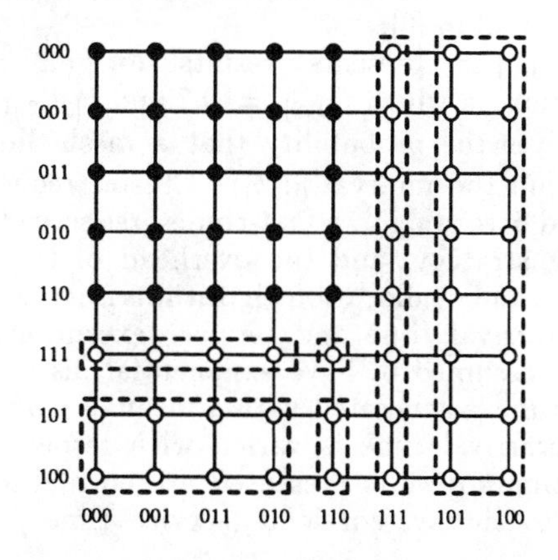


Figure 3. Allocation of a 5 × 5 mesh.

generated by the partitioning algorithm are disjoint, and many pairs of cubes are adjacent. To keep them in this form, they are kept separate from the normal free list, as the free list deallocation algorithm would rearrange this format.

Intuitively, it can be seen that using the partitioning algorithm can increase system utilization: the previous example of a 20 × 30 mesh leaves 424 extra nodes that would not be allocated by the original free list algorithm. The partitioning algorithm will generate an 8-cube, a 7-cube, a 5-cube, and a 3-cube as available leftover cubes.

Theorem 1 states that the auxiliary free list algorithm will allow multiple 2-D meshes to be allocated within one subcube without communication interference. Due to space limitations, the proof is omitted, but it may be found in [8].

**Theorem 1** *The leftover cubes created by the 2-D mesh partitioning algorithm suffer no communication interference from the 2-D mesh cluster, and the 2-D mesh cluster suffers no communication interference from the leftover cubes.*

## 2.2 Allocation of 2-D Meshes

The auxiliary free list is intended to supplement, rather than replace, the free list algorithm. Therefore, cluster allocation using the auxiliary free list is a simple decision procedure, in which the auxiliary free list is checked first for an incoming cluster request. In the algorithm below, assume that **exist**$(A, i)$ is a boolean function that determines whether a subcube of dimension $i$ exists in list $A$. The function **get**$(A, i)$ obtains a dimension $i$ subcube from list $A$. **Partition**$(Q)$ is a call to the partitioning algorithm on a subcube $Q$, and **Mark**$(Q, X, Y)$ marks an $X \times Y$ mesh region of $Q$ as allocated. Here, $A$ always refers to the auxiliary free list. Again note that a leftover cube may be partitioned to produce new leftover cubes.

The auxiliary free list allocation algorithm checks only dimensions $k$ and $k+1$ of the auxiliary free list. This algorithm may be modified to check for the existence (in the auxiliary free list) of any higher dimension cubes from $k+2$ to $n$. However, checking more than one higher dimension could result in greater overhead in managing the ownership list of the 2-D mesh cluster. For example, if a new request requires a 4-cube, and an 8-cube is the next highest auxiliary free list cube available, then the 8-cube would be decomposed into a 7-cube, a 6-cube, a 5-cube, and two 4-cubes, each of which would then be added to the original ownership list, as well as to the auxiliary free list. If many similar requests were to occur, then a specific 2-D mesh cluster's ownership list could grow to be very long. Therefore, the algorithm is restricted to checking only the size $(k+1)$ auxiliary free list. This tradeoff between higher list-management overhead and greater flexibility of the auxiliary free list will be the topic of future investigations.

## 2.3 Deallocation of 2-D Meshes

When a 2-D mesh request is deallocated, the mesh must first be partitioned into a set of subcubes, allowing it to be readily coalesced with the leftover

```
Algorithm: Auxiliary Free List Allocation.
Input: An X × Y mesh request.
Output: Allocated X × Y mesh.
Procedure:
    begin
      If exist(A, k);
      then begin
        Q = get(A, k);
        Partition(Q);
        Mark(Q, X, Y);
      else begin
        If exists(A, k + 1);
        then begin
          Q = Decompose(A, k + 1);
          Partition(Q);
          Mark(Q, X, Y);
        end
        else
          Perform the free list algorithm;
      end
    end
```

Figure 4. Auxiliary free list allocation.

subcubes generated when it was originally allocated. This algorithm is similar to the partitioning algorithm in that it partitions the area composing the 2-D mesh into several subcubes, which may then be coalesced with the original leftover subcubes created by the partitioning algorithm.

Leftover subcubes originally created by the partitioning algorithm are linked together by an ownership list that allows the deallocation procedure to quickly check (using a linked-list traversal) whether these leftover cubes are free or allocated. In the event that all of the leftover cubes are currently free when the 2-D mesh is deallocated, coalescing will restore the original $k$-cube. In this case, the free-list deallocation algorithm [4] is performed on the $k$-cube. Otherwise, coalescing will produce several disjoint subcubes, which may be added to the original free list. In any event, the ownership list is destroyed when the 2-D mesh is deallocated. Subcubes that are still allocated will then be deallocated using the free list deallocation algorithm when they become free. Note that this process will eventually restore the original $k$-cube when all of the leftover subcubes become free. See [8] for complete details of the deallocation algorithms.

# 3 Analysis and Performance

For a complete analysis of the complexity of the original free list algorithms, refer to [4]. Both the allocation and deallocation of 2-D mesh clusters do not add significant overhead to the original free list algorithm. For a requested subcube of dimension $k$ in a hypercube of dimension $n$ ($k \leq n$), auxiliary free list allocation adds $O(k^2)$ complexity to the free list algorithm algorithm, which is $O(n)$. More importantly, auxiliary free list deallocation adds no more than $O(k^3)$ complexity to the free list deallocation algorithm, which is normally at least $O(n^3)$. The existence of the auxiliary free list deallocation algorithm makes it often unnecessary to perform the free list deallocation algorithm, resulting in a greater time savings. Therefore, the methods introduced in this paper do not introduce significant time complexity to the free list algorithms. A more complete analysis of the algorithms involved in the auxiliary free list method is available in [8].

Extensive simulations have been conducted to compare the performance of the free list method with the auxiliary free list method when allocating 2-D mesh requests. In our investigations, 2-D mesh requests were generated and allocated using the free list and auxiliary free list algorithms on a hypercube of dimension 10 (1024 processors). The $X$ and $Y$ dimensions of incoming requests were varied under numerous distributions. Simulation results depend on the workload distribution, including such factors as the mesh size, the mesh geometry (rectangular or square), and the job's service time. With this in mind, many simulations were run, including: square meshes, in which the $X$ dimension was randomly generated on [1,16] and [1,32] uniform distributions; rectangular meshes, in which the $X$ and $Y$ dimensions were randomly generated on [1,16] and [1,32] random distributions; and numerous interval distributions [9] for rectangular meshes. The interval distributions were used to simulate bipartite distributions, in which there are many small requests coupled with many very large requests. The results for all of these simulations may be found in [8].

This paper presents results for an interval distribution, in which $P_{[1,8]} = 0.7$ and $P_{[25,32]} = 0.3$ ($P_{[a,b]}$ gives the probability that a mesh dimension falls within the interval $[a, b]$). Cluster requests are processed according to a first-come-first-serve (FCFS) queueing strategy, and the overhead of the cluster allocation and deallocation algorithms is ignored. The job interarrival time and service (execution) times are both assumed to have exponential distributions. Service time is generated with a mean of 4.0 seconds, and interarrival time is varied with respect to the service time. Results presented are normalized with respect to the system load (service time / arrival time).

For different simulations, random number streams were generated with the same seed values to guarantee the same input values. The interarrival time, service time, $X$ dimension, and $Y$ dimension, as well as a probability generator used by the interval distribution, are all independent. Simulation runs were generated with a 95 % confidence interval.

Figure 5 plots the average job turnaround time (JTT) versus the system load. The graph shows that, for heavy loads, the average job turnaround time using the auxiliary free list is lower than that of the free list.

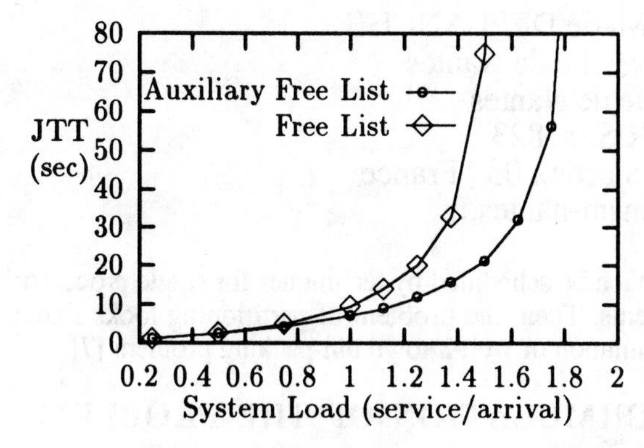


Figure 5. JTT vs. system load.

Figure 6 plots the system utilization versus the system load. Although performance is similar for light loads, the auxiliary free list gives better processor utilization under high loads. Here, the auxiliary free list shows approximately a 7% increase in system utilization over the free list.

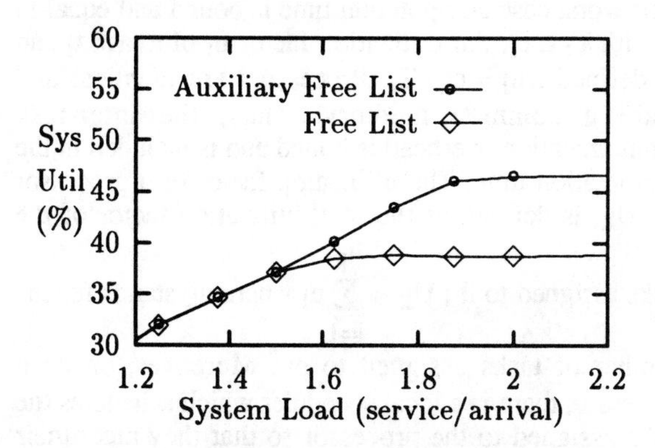


Figure 6. System utilization vs. system load.

It is also possible to gain greater increases in system utilization, as well as significant decreases in turnaround time. For example, a constant load application was simulated, in which 50% of mesh requests were $4 \times 6$ and the other 50 % were $4 \times 2$ meshes. For this distribution, the auxiliary free list method obtained 100% utilization, while the free list method obtained only 50% utilization. In addition, the system load at which turnaround time started to increase exponentially was 4.0 for the auxiliary free list, while it was 2.0 for the free list (that is, the auxiliary free list performed twice as well as the free list alone).

## 4 Conclusions

We have developed a new cluster allocation algorithm that decreases average job turnaround time and increases processor utilization in a hypercube multiprocessor. Using our strategy, allocated clusters suffer no communication interference from other allocated clusters, yet they are not required to be a perfect subcube. The auxiliary free list strategy, when used in combination with the free list strategy, has same or better performance for both job turnaround time and system utilization for 2-D mesh cluster requests.

While the results presented here are promising, we have extensive plans for future work. The algorithms are still being updated, and we intend to explore different versions of the partitioning algorithm to explore various tradeoffs. For example, the auxiliary free list allocation algorithm is purposely not as flexible as the free list algorithm in that it cannot check for subcubes of dimension higher than $k + 1$ when it attempts to allocate a $k$-cube. We will study the performance effects of removing this restriction. We also intend to study the effects of partitioning to minimize the number of leftover subcubes versus partitioning to find the largest possible leftover subcubes. Other future work will be to simulate applications having different cluster sizes and shapes.

## References

1. K. C. Knowlton, "A fast storage allocator," *Communications of the ACM*, vol. 8, pp. 623–625, October 1965.
2. M.-S. Chen and K. G. Shin, "Processor allocation in an n-cube multiprocessor using gray codes," *IEEE Transactions on Computers*, vol. C-36, pp. 1396–1407, December 1987.
3. A. Al-Dhelaan and B. Bose, "A new strategy for processor allocation in an n-cube multiprocessor," in *Proceedings of the International Phoenix Conference on Computers and Communications*, March 1989.
4. J. Kim, C. R. Das, and W. Lin, "A top-down processor allocation scheme for hypercube computers," *IEEE Transactions on Parallel and Distributed Systems*, vol. 2, no. 1, pp. 20–30, 1991.
5. Y. Saad and M. H. Schultz, "Topological Properties of Hypercubes," *IEEE Transactions on Computers*, vol. C-37, pp. 867–872, July 1988.
6. L. M. Ni and P. K. McKinley, "A survey of wormhole routing techniques in direct networks," *IEEE Computer*, vol. 26, pp. 62–76, Feb. 1993.
7. M. Y. Chan and F. Y. L. Chin, "On embedding rectangular grids in hypercubes," *IEEE Transactions on Computers*, vol. C-37, pp. 1285–1288, October 1988.
8. S. W. Turner, L. M. Ni, and B. H. C. Cheng, "Contention-free 2D-mesh cluster allocation in hypercubes," Tech. Rep. MSU-CPS-ACS-76, Michigan State University, Department of Computer Science, East Lansing, Michigan 48824, November 1992.
9. Y. Zhu, "Efficient processor allocation strategies for mesh-connected parallel computers," *Journal of Parallel and Distributed Computing*, vol. 16, pp. 328–337, 1992.

# A TASK ALLOCATION ALGORITHM IN A MULTIPROCESSOR REAL-TIME SYSTEM

Jean-Pierre BEAUVAIS & Anne-Marie DEPLANCHE
Ecole Centrale de Nantes/Université de Nantes
Laboratoire d'Automatique de Nantes
Unité associée au C.N.R.S. n°823
1 rue de la Noë , 44072 NANTES cedex 03 , France
E-mail : beauvais@lan01.ensm-nantes.fr

Abstract -- *This paper describes a heuristic algorithm in $O(n^2)$ which assigns a set of real-time tasks to a set of processing sites. The main objective that we try to achieve is to balance the load over the different processors of the system to improve the safety of the system as well as checking the scheduling conditions that must be satisfied in a real-time context and taking into account given allocation constraints.*

## INTRODUCTION

The proliferation of distributed computing systems has prompted widespread investigation into the problem of how best to distribute these tasks most effectively among multiple processors. This problem of assigning tasks to processors has been formulated as a constrained optimization problem. The constrained equations or inequations allow to describe system attributes such as limited memory capacity, task redundancy, task precedence, network topology etc... The function to optimize is usually a cost function that depends on the objective of the allocation.

In this paper, we will consider safety-critical periodic tasks which must meet their deadline and we assume an "a priori" knowledge of the task characteristics i.e. the model of tasks is a deterministic one. Moreover, these tasks must respect allocation constraints.

As a rule, allocation problems tend to be computationally intensive and belong to the well-known class of NP-complete problems. This has generated two types of resolution algorithms : the exact algorithms (graph theoretical [2] - [13], mathematical approaches [4] - [9], enumerative methods [12]) searching an optimal allocation and the approximate algorithms (employing techniques such as simulated annealing [14], clustering [3] - [10], partitioning [6], bin-packing[1] - [5] ...) searching a sub-optimal allocation.

Our method to treat the considered problem of allocation is a heuristic one. We have employed here the partitioning method which seeks to partition the task set into groups and to assign the groups to distinct processors. The group of tasks assigned to a processor can then be scheduled by techniques for single processor systems. Then, the problem of partitioning looks like a formulation of well-known bin-packing problem [7].

## FORMULATION OF THE PROBLEM

Software and hardware configurations - hypothesis

The basic data of the problem we are considering consist of a set $P = \{p_1, p_2, ..., p_m\}$ of m processors and of a set $T = \{t_1, t_2, ..., t_n\}$ of n safety-critical tasks where n is greater than m. Tasks are periodic with a period $T_i$ (the deadline corresponding to the end of the period) and their worst case computation time is bound and equal to $c_i$ with $T_i \geq c_i$. An utilization factor $u_i$ of a task $t_i$ can be defined : $u_i = c_i / T_i$. Processors are identical and share a common memory. Thus, the inter-task communication overhead is bound and is included in the computation time. The utilization factor of a processor $p_k$, $U_k$, is defined as the total utilization factor of the tasks assigned to it ; $U_k = \sum_{i=1}^{n_k} u_i$ where $n_k$ stands for the number of tasks assigned to $p_k$. Moreover, on each processor, there is a local scheduler which schedules the tasks assigned to the processor so that they meet their deadlines. We have chosen as scheduling policy, the earliest deadline one which assigns to a task a dynamic priority according to the proximity of its deadline. It is optimal in its class and owns a necessary and sufficient scheduling condition :

$$\forall\, k=1 \ldots m\ , U_k = \sum_{i=1}^{n_k} u_i \leq 1 \quad [11]$$

Definition of the problem

In this paper, the problem is to find an assignment of periodic safety-critical tasks to identical processors in order to obtain a balance between the processor utilization factors considered as the processor loads. The allocation must respect the next requirements :

1. at a time, a task is active once even if its code

can be duplicated on several processors to insure standby redundancy.

2. there is no migration of tasks during execution.

3. a task can only be assigned to some processors. These processors define the feasible processor set of the task. This allocation constraint represents the fact that tasks require sometimes special resources (peripheral devices ...) that exist on certain processors.

An allocation can be defined by an assignment matrix X (nxm) of assignment variables $X_{ik}$ such as :
$X_{ik} = 1$ if $t_i$ is assigned to $p_k$
$X_{ik} = 0$ otherwise

For all the tasks, it is possible to represent the feasible processor set by a feasible allocation matrix R (nxm) of variables $R_{ik}$ that verify :
$R_{ik} = 1$ if $t_i$ can be assigned to $p_k$
$R_{ik} = 0$ otherwise

The allocation problem is then to find, if it exists, an allocation matrix X that verify the next problem :
minimize

$$\mathrm{cost}(X) = \sum_{k=1}^{m} U_k^2 = \sum_{k=1}^{m} \left(\sum_{i=1}^{n} u_i\, x_{ik}\right)^2 \text{ [8]} \qquad (1)$$

with the constraints

$$\forall\, k=1,\dots,m \qquad U_k = \sum_{i=1}^{n} u_i\, x_{ik} \le 1 \qquad (2)$$

$$\forall\, i=1,\dots,n,\ \forall\, k=1,\dots,m \quad x_{ik} \le R_{ik} \qquad (3)$$

$$\forall\, i=1,\dots,n,\ \forall\, k=1,\dots,m \quad \sum_{k=1}^{m} x_{ik} = 1 \qquad (4)$$

## A HEURISTIC APPROACH

The search space of such a problem (size of the order of $m^n$) makes the research of an optimal solution intractable and we propose a partitioning heuristic algorithm which operates like a bin-packing heuristic. In the same way as you take packets and try to put them into bins in accordance with bin capacities, our algorithm consists in taking the tasks to be allocated, one by one, according to a specified order and then, putting them into one of their feasible processors, provided its utilization factor does not exceed its scheduling limit.

In order to guarantee effective execution, our algorithm allocates tasks to processors one by one without backtracking. That is why the order in which tasks are considered for allocation is a critical factor.

### Task selection

The task selection employs three metrics per task :

1. the utilization factor , $u_i$, that represents the "size" of the task $t_i$.

2. the dynamic feasible assignment degree, $\mu_i$, which is defined by the number of processors to which task $t_i$ can be assigned. This metric is a dynamic one and is initialized with the cardinality of the feasible processor set of the task $t_i$, i.e. $\sum_{k=1}^{m} R_{ik}$. Then, $\mu_i$ decreases since, throughout the allocation process, some processors which belonged to the feasible processor set of the task $t_i$ must be deleted from this set : they can no more schedule the task $t_i$. If the dynamic feasible assignment degree of a task is one, it has no freedom at all and must be assigned to the only site that remains in its feasible processor set.

3. the dynamic load degree, $\beta_i$, which is defined as the sum of the current load of the processors in the feasible processor set of task $t_i$.

These metrics give rise to two different versions of our algorithm depending on we favour the utilization required by the tasks or their freedom degree :

- Case 1 : the utilization factor is the primary consideration. Tasks are considered in decreasing order of utilization. In the event of a tie, the task with the smallest dynamic feasible assignment degree is chosen (tasks with fewer site choices are allocated earlier). In case of another tie, the task with the largest load degree is selected (as its candidate sites are more heavily loaded).
- Case 2 : the dynamic feasible assignment degree is the primary consideration. Tasks are considered in increasing order of degree. The evaluation order for the last metrics is : maximum utilization factor and maximum load degree.

### Processor selection

After determining the task to be considered next, up to two metrics will be used to guide the selection of a processor among those in the feasible processor set of task $t_i$. These two metrics are :

1. the current utilization factor or load $U_k$ of processor $p_k$ ; this is the sum of the utilization factors of the tasks already assigned to it (possibly including the initial load of the site). It indicates the way the site is busy i.e. the current filling of the equivalent bin.

2. the potential load $P_k$ of processor $p_k$, which is

defined as the sum of the utilization factors of the unassigned tasks that have $p_k$ in their feasible processor set. A processor whose potential load is great is in a position to be chosen by many and "large" tasks.

In order to attempt to reach the load-balancing objective, the processor selection is executed by considering the utilization factors in increasing order among the feasible processors for the selected task. In the event of a tie, the processor with the minimum potential load is chosen. It is important to notice that not all metrics have to be systematically calculated for every task allocation.

## PERFORMANCE ANALYSIS

The two versions of the algorithm have been implemented with the C™ language on a Sun™ IPX Sparc station under UNIX™.

The performance criteria

We have chosen to analyze the reliability and the optimality of the algorithm.

The reliability represents the capacity of the algorithm to find an allocation, either optimal or sub-optimal, when it exists. Reliability =

$$\frac{\text{number of allocations found by the heuristic}}{\text{number of feasible configurations}}$$

The reliability is the best one when it is 100%.

The optimality represents the capacity of the algorithm to find an allocation whose cost is as near as possible from the minimum cost. We have evaluated the average, on a certain number of configurations, of the relative difference between the cost of the allocation produced by the heuristic and the minimum cost . Optimality

$$= \text{average}\left(\frac{\text{heuristic cost - minimum cost}}{\text{minimum cost}}\right).$$ The optimality is the best one when it is 0%.

To measure the heuristic performances, as we have defined them, it is necessary to run an exact algorithm which produces an optimal allocation. It is a great drawback and limitation because, when the size of the data of the problem increases, the execution time of the exhaustive research becomes prohibitive. For example, for m = 5 and 6, we could not run the exhaustive search and the results, not given here, give a comparison of the two versions of the algorithm.

The choice of the samples

A particular care has been given to choose the configuration samples to be tested. We have generated a number of configurations by varying the number of processors m, the global load coefficient f and the number of tasks n in the configuration. The global load coefficient of the system is defined as the ratio of the total load of the system and the maximum admissible load of the system : $f = \sum_{i=1}^{n} u_i / m$. f measures the severity of the system load.

Each sample, defined by a m value chosen in {2,3,4,5,6} is composed of about 1000 configurations, f varying from 0.2 to 1 and n varying, for example, from 3 to 12 for m=2 and from 7 to 20 for m=6. The allocation constraints are drawn randomly.

The results

We have generated two test groups. The first one is composed of configurations without allocation constraints and the two algorithm versions become only one. The second one takes into account allocation constraints. The two versions are named further h1 (priority to $u_i$) and h2 (priority to $\mu_i$).

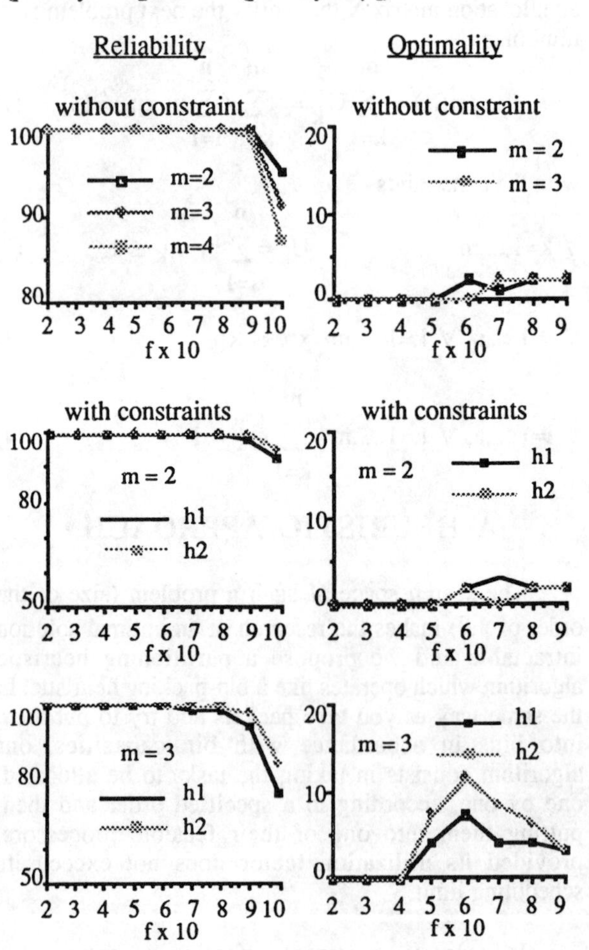

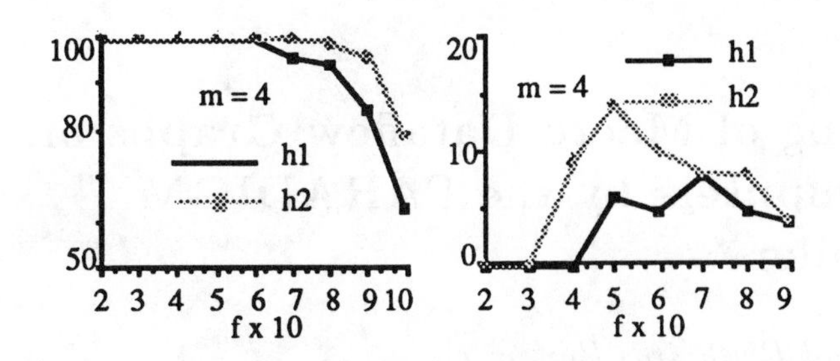


The results show that the reliability of the algorithm is very good. It decreases when we approach the global saturation as fast as the number of processors is great.

The second version of the algorithm is always better in reliability terms. From this point of view, it is preferable to consider first rather the allocation constraints of the tasks than their utilization factor. Nevertheless, for the limit utilization cases, the conjunction of the two versions of the algorithm improves interestingly these percentages since some failure cases are not connected as shown in the next table. Furthermore, supplementary measures no developped here show that the configurations for which the algorithm fails have tasks whose utilization factors are greater than the average and less dispersed.

Global reliability

| | without const. | with constraints | | |
|---|---|---|---|---|
| m | | h1 | h2 | h1+h2 |
| 2 | 98% | 97% | 98% | 99% |
| 3 | 97% | 90% | 94% | 96% |
| 4 | 95% | 83% | 92% | 94% |

The optimality of the algorithm is quite good : in the case without constraint, it is globally equal to 1% to 2% while in the constrained case, it varies globally from 1% to 6.5% on the different samples.

## CONCLUSION

Our interest is moving now towards taking into account many additional considerations in our model. Our aim is to integrate precedence constraints that define an order between the execution of the tasks and also to consider the communication costs so that the algorithm could be applied to distributed systems. Another further main line of our research is to consider fault-tolerance systems and thus to add redundancy constraints to the tasks.

## REFERENCES


[1] J.A. Bannister, K.S. Trivedi, "Task allocation in fault-tolerant distributed systems", *Acta Informatica*, (Dec, 1983), pp. 261-281.

[2] S.H. Bokhari, "Dual Processor Scheduling with Dynamic Reassignment", *IEEE Transactions on software engineering*, (Jul, 1979), pp. 341-349.

[3] N.S. Bowen, C.N. Nikolaou, A. Ghafoor, "On the Assignment Problem of Arbitrary Process Systems to Heterogeneous Distributed Computer Systems", *IEEE Transactions on Computers*, (Mar, 1992).

[4] W.W. Chu, "Optimal File Allocation in a Multiple Computer System", *IEEE Transactions on Computers*, (Oct, 1969), pp. 885-889.

[5] M.R. Garey, R.L. Graham, J.D. Ulman, "Worst-case Analysis of memory allocation algorithms", *4th Annual Symp. on the Theory of Computing*, (1972), pp. 143-150.

[6] T. Hamada, C.K. Cheng, P.M. Chan, "An efficient Multilevel Placement Technique Using Hierarchical Partitioning", *IEEE Transactions on Circuits and Systems*, (1992), pp. 432-439.

[7] D.S. Johnson "Fast Algorithms for Bin-Packing", *Journal of Computer and System Sciences*, (1974), pp. 272-314.

[8] H. Lu, M.J. Carey, *Load-Balanced Task Allocation in Locally Distributed Computer Systems*, Computer Sciences Dpt, University of Wisconsin-Madison, Report #633, (Feb, 1986), 23 pp.

[9] P.R. Ma, E.Y. Lee, M. Tsuchiya, "A Task Allocation Model for Distributed Computing Systems", *IEEE transactions on Computers*,, (Jan, 1982), pp. 41-47.

[10] K. Ramamritham, "Allocation and Scheduling of Complex Periodic Tasks", *10 th Int. Conf. on Distributed Computing Systems*, (1990).

[11] O. Serlin, "Scheduling of Time Critical Processes", *Spring Joint Computers Conf.*, (1972), pp. 925-932

[12] C.C. Shen, W.H. Tsai, "A Graph Matching Approach to Optimal Task Assignment in Distributed Computing System Using a Minmax Criterion", *IEEE Transactions on Computers*, (Mar, 1985).

[13] H.S. Stone, "Critical Load Factors in Two-Processor Distributed Systems", *IEEE Transactions on Software Engineering*, (Mar, 1978), pp. 254-258.

[14] K.W. Trindell, A. Burns, A.J. Wellings, "Allocating Hard Real-Time Task : A NP-Hard Problem Made Easy", *Real-Time Systems*, (Jun, 1992).

# Processor Allocation and Scheduling of Macro Dataflow Graphs on Distributed Memory Multicomputers by the PARADIGM Compiler *

*Shankar Ramaswamy and Prithviraj Banerjee*

Center for Reliable and High Performance Computing
University of Illinois at Urbana-Champaign
Urbana, IL 61801
E-mail : {shankar,banerjee}@crhc.uiuc.edu

**Abstract :** *Functional or Control parallelism is an effective way to increase speedups in Multicomputers. Programs for these machines are represented by Macro Dataflow Graphs (MDGs) for the purpose of functional parallelism analysis and exploitation. Algorithms for allocation and scheduling of MDGs have been discussed along with some analysis of their optimality. These algorithms attempt to minimize the execution time of any given MDG through exploitation of functional parallelism. Our preliminary results show their effectiveness over naive algorithms.*

**Keywords :** *Macro Dataflow Graphs, Distributed Memory Multicomputers, Allocation and Scheduling, Parallelizing Compilers, Optimization.*

## 1 Introduction

Distributed Memory Multicomputers offer significant advantages over shared memory multiprocessors in terms of cost and scalability. Unfortunately, writing efficient software for them is an extremely laborious process for users. The PARADIGM compiler project at Illinois is aimed at devising a parallelizing compiler for distributed memory multicomputers that will accept sequential FORTRAN 77 programs as input. The completed PARADIGM compiler will automatically:

- Generate data partitioning specifications [1].
- Partition computations and generate communication for data parallel programs [2].
- Exploit high level communication primitives [3].
- Exploit functional parallelism [4, 5].

This paper deals with the last aspect of the PARADIGM compiler, viz., Functional parallelism exploitation. Functional or Inter loop parallelism has long been recognized as a technique to further increase speedups beyond Data or Intra loop parallelism. The two techniques are orthogonal to each other in that the application of one does not preclude the use of the other. In the context of functional parallelism, the notion of Macro Dataflow Graphs (MDG) has been proposed and used by researchers, for example, by Prasanna and Agarwal in [6]. They are a coarse grain representation of programs; nodes are typically loops or nested loops, and edges define precedence constraints among these loops. The weight of a node is its computation cost and the weight of an edges is the cost of communicating data between the nodes involved. Exploiting functional parallelism involves defining the degree of data parallelism for each node in the MDG and scheduling the execution of the nodes in a manner which allows the program to be executed in the shortest time.

Any compiler that wants to exploit functional parallelism therefore, has to:

- Identify and construct the MDG of the given program automatically. Girkar and Polychronopoulos have dealt with this aspect in [7].
- Allocate processors to each node and schedule the MDG in such a way as to minimize its execution time.

Our focus has been entirely on the second step in compilation outlined above. Obtaining an optimal solution to this step has been shown to be NP-complete by Lenstra and Kan in [8]. The two major approaches to the approximate solution of this problem have been; first, a bottom up approach like those used by Sarkar in [9], and Gerasoulis and Yang in [10, 11] and, second, a top down approach like the ones used by Prasanna and Agarwal in [6], Belkhale and Banerjee in [4, 5] and in this paper. A bottom up approach considers the MDG to be made up of lightweight nodes and coalesces these nodes together to form larger nodes during its construction of a schedule. Top down approaches start with the assumption of heavyweight nodes in the MDG and break them down during the process of constructing an optimal schedule. Top down methods take a more global view of the problem than the bottom up approaches, thus performing better.

We first look at the computation and communication cost models used in our work. A few definitions and assumptions are then stated, followed by a discussion of the scheduling algorithm and its optimality. Next we describe our allocation algorithms and present some preliminary results obtained using our algorithms. Finally, we discuss the the implications of our work and the future directions we will be taking.

*This research was supported in part by the Office of Naval Research under Contract N00014-91J-1096, and in part by the National Aeronautics and Space Administration under Contract NASA NAG 1-613.

# 2 Cost Models

## 2.1 Computation Model

For the computation model, we use Amdahl's laws, i.e., the execution time of a node ($T_i$) as a function of $p_i$ is given by:

$$T_i(p_i) = (\alpha_i + \frac{1-\alpha_i}{p_i}) \cdot \tau_i \quad (1)$$

where $\tau_i$ is the execution time of the node on a single processor and $\alpha_i$ is the fraction of the node that has to be executed serially.

## 2.2 Communication Cost Model

A many to many communication of $L_{ij}$ bytes involving $p_i$ sending and $p_j$ receiving processors is assumed to consist of $p_i \cdot p_j$ messages of $\frac{L_{ij}}{p_i \cdot p_j}$ bytes each, i.e. each of the $p_i$ sending processors sends a message to each of the $p_j$ receiving processors, all messages being of equal size. There are three cost components to this transfer, $t_1$ representing the startup and processing costs for $p_j$ messages at the sending processors, $t_2$ representing the network cost for a message, and, $t_3$ representing the startup and processing costs for $p_i$ messages at the receiving processors. We can write expressions for these components as:

$$t_1 = p_j \cdot (t_s + t_p \cdot \frac{L_{ij}}{p_i \cdot p_j}) \quad (2)$$

$$t_2 = t_n \cdot \frac{L_{ij}}{p_i \cdot p_j} \quad (3)$$

$$t_3 = p_i \cdot (t_s + t_p \cdot \frac{L_{ij}}{p_i \cdot p_j}) \quad (4)$$

where $t_s$ denotes the startup cost required for a message at a processor, $t_p$ denotes the processing cost per message byte at a processor and $t_n$ denotes the network cost per message byte.

The total communication cost $C_{ij}$ is a sum of these components and can be written down as:

$$C_{ij} = (p_i + p_j) \cdot t_s + L_{ij} \cdot (\frac{1}{p_i} + \frac{1}{p_j}) \cdot t_p + L_{ij} \cdot \frac{1}{p_i \cdot p_j} \cdot t_n \quad (5)$$

# 3 Assumptions

Listed below are a few assumptions made for our algorithms and their analysis.

**Assumption 1** *MDG specification*
The $i$th node in the MDG is specified by a pair $(\alpha_i, \tau_i)$ and the edge joining nodes $i$ and $j$ is specified by the length of the message transferred $L_{ij}$. Gupta and Banerjee have shown the estimation of computation and communication costs to be feasible in [1]. Our assumption therefore says that these estimated costs will fit the cost models we outlined earlier.

**Assumption 2** *Communication cost parameters*
The values of $t_s$, $t_p$, and $t_n$ which are required to compute communication costs can be determined for a particular architecture and are, therefore, available to our algorithms. This is a realistic assumption and has been made by numerous researchers in some form or the other.

**Assumption 3** *Computation to Communication Ratio*
For the theoretical analysis that follows, we assume that the weight of any node will be at least an order of magnitude higher than the weight of any edge. This means we can neglect communication costs for the analysis.

**Assumption 4** *Control Dependence*
Like many other researchers, we have considered only data dependences between nodes in the MDG. Control dependences are assumed to exist only within nodes and not across them. This is a reasonable assumption considering the nodes in the MDG represent sufficiently large computation.

# 4 Definitions

We present below the definitions of two terms that are used in the proofs of the theorems that follow in Section 5.

**Definition 1** *Critical Path Weight*
The critical path is defined to be the path in the MDG whose nodes have the largest aggregate weight when at most $k$ processors can be used by any node. This weight is called the critical path weight and is denoted by $C_k$.

**Definition 2** *Average Finish Time*
The average finish time ($A_k$) of an MDG with $n$ nodes, when each node can use at most $k$ processors is given by:

$$A_k = \frac{1}{p} \cdot \sum_{i=1}^{n} T_i(p_i) \cdot p_i \quad (6)$$

where $p$ is the maximum number of processors available at any given time in the given multicomputer system, $p_i$ is the number of processors used by the $i$th node ($1 \leq p_i \leq k$), and $T_i(p_i)$ is the execution time of the node on $p_i$ processors.

If the computation cost model is used in the expression above, we have:

$$A_k = \frac{1}{p} \cdot \sum_{i=1}^{n} \tau_i \cdot (1 + \alpha_i \cdot (p_i - 1)) \quad (7)$$

# 5 Scheduling Algorithm

## 5.1 PSA

Our allocation heuristics use a scheduling algorithm called the Prioritized Scheduling Algorithm (PSA) whose steps are outlined below:

**Step 1** Do a Topological Sort of the MDG with the directions of all edges reversed (all other parameters remaining the same). The ordering produced decides the priority of each node.

**Step 2** The next node to be processed is the one with the highest priority and whose precedence constraints have been met at the current time. Schedule the selected node at a time when its processor requirement is met. Repeat this step until no node is left unexecuted.

**Step 3** The earliest time at which all the nodes finish executing is the execution time of the MDG.

Similar scheduling algorithms are described for example, by Liu in [12], by Garey, Graham and Johnson in [13], by Wang and Cheng in [14], and by Turek, Wolf and Yu in [15]. While these researchers have published results similar to Theorem 1 below, the assumptions used by them have been limiting. Our result is more general in its assumptions and implications. In fact, as shown in Corollaries 2 and 3, our result matches those obtained by earlier researchers for the specific assumptions they use.

### 5.2 Optimality of PSA

**Theorem 1** *If each task uses at most $x$ processors in a $p$ processor system, the finish time obtained using the PSA ($T_{psa}$) is bounded by a factor of the optimum finish time ($T_{opt}$) as given by the following equation:*

$$T_{psa} \leq 2 + \frac{x-2}{p-x+1} * T_{opt} \tag{8}$$

*under the assumptions made in Section 3.*

**Proof** If each task uses at most $x$ processors, then a maximum of $x-1$ processors will be forced to be idle, waiting for the $x$th processor to be free to start the next waiting task. This should mean at least $p-(x-1)$ or $p-x+1$ processors are busy at all times, if there is enough work for them. There not being enough work for $p-x+1$ processors means all other unexecuted tasks are dependent on one or more of the currently executing tasks. There are two cases in which this could happen:

**Case 1** Due to poor scheduling by the PSA in spite of there being enough work (The optimal algorithm is able to keep all processors busy). By poor scheduling we mean the PSA has made some wrong choices as compared to the optimal algorithm and therefore, put off working on some of the tasks which precede all other unexecuted tasks. In this situation, the worst case time for which the PSA will be in such a situation is the maximum length of any chain of tasks from start to finish. This is the critical path weight ($C_x$) of the graph when no task uses more than $x$ processors. Therefore, we can guarantee that at least $p-x+1$ processors will be busy for at least $T_{psa}-C_x$ units of time and at least 1 processor will be busy for $C_x$ units. This means the area of useful work done is given by $(p-x+1)*(T_{psa}-C_x)+1*C_x$ in the worst case. The area covered by the optimum is $T_{opt}*p$. Since we have a worst case for the PSA algorithm, we have:

$$(p-x+1)*(T_{psa}-C_x)+1*C_x \leq p*T_{opt} \tag{9}$$

**Case 2** Due to poor scheduling by the PSA, but the optimal algorithm may not have enough work either due to properties of the task graph. In this situation, if the optimal algorithm does not have enough work, the PSA could have work since it could put off working on some tasks. This means the idle time of the PSA for such situations is bounded by that of the optimal algorithm. However, due to poor scheduling, i.e. making the wrong choices, the PSA could have idleness where the optimal algorithm does not. Again, the duration of this situation, by the same argument as above, is bounded by $C_x$. This means the area $(p-x+1)*(T_{psa}-C_x)+1*C_x$ does include some idle time in the first term, but that is less than or same as the idle time in the term $p*T_{opt}$. Therefore, we can write:

$$(p-x+1)*(T_{psa}-C_x)+1*C_x \leq p*T_{opt} \tag{10}$$

It can be noted that both the cases above give us the same relationship between $T_{psa}$, $T_{opt}$ and $C_x$. Further, $T_{opt}$ must be at least $C_x$, i.e. $T_{opt} \geq C_x$. Therefore, using this fact and either Equation 9 or 10, we have:

$$(p-x+1)*T_{psa} \leq (2p-x)*T_{opt}$$

$$\Rightarrow T_{psa} \leq \frac{2p-x}{p-x+1} * T_{opt} \tag{11}$$

After some re-arrangement of terms, we obtain:

$$T_{psa} \leq 2 + \frac{x-2}{p-x+1} * T_{opt} \tag{12}$$

which is the required result □.

Some special cases of $p$ and $x$ are considered to show the validity of our result.

**Corollary 1** $x = p = 1$

This is the uniprocessor case.

$$T_{psa} = T_{opt} \tag{13}$$

which is precisely the case.

**Corollary 2** $x = 1$

Each task uses only one processor.

$$T_{psa} \leq (2 - \frac{1}{p}) * T_{opt} \tag{14}$$

this result matches those in [13] and [12].

**Corollary 3** $x = \frac{p}{2} + 1$

Each task can use approximately half as many processors available.

$$T_{psa} \leq (3 - \frac{2}{p}) * T_{opt} \tag{15}$$

this result matches those in [15] and [14].

As we have seen in Theorem 1, the PSA algorithm is within a factor of the optimum, the factor being dependent on $x$. It may be construed at first sight that a good choice of $x$ is one which minimizes this factor. There is, however another aspect to the choice of $x$. This aspect being the influence $x$ has on the time $T_{opt}$. Theorem 2 below takes into account this aspect and provides a method of choosing $x$ optimally.

**Theorem 2** *The optimum value of the bound on the number of processors to be used by any node in a given MDG with $n$ nodes is given by that value of $x$ which minimizes the expression:*

$$\frac{2 \cdot p}{x} + \frac{(x-2) \cdot p}{(p-x+1) \cdot x} \tag{16}$$

*under the assumptions of Section 3.*

**Proof** Consider the perfect schedule time ($T_{opt}$) for a given processor allocation with a bound $x$ being placed on the number of processors used by any node ($1 \leq p_i \leq x, i = 1, n$). It is limited only by the average finish time ($A_x$) and the critical path weight $C_x$.

$$T_{opt} = \max(A_x, C_x) \tag{17}$$

If the bound is changed to $p$, and the corresponding value of the perfect schedule time is $T_{opt}^{min}$, we can see that:

$$T_{opt}^{min} = \max(A_p, C_p) \tag{18}$$

It must be noted that for any given problem, $T_{opt}^{min}$ represents the theoretically best possible finish time.

We now consider the relationship between $T_{opt}$ and $T_{opt}^{min}$. In imposing a bound on the number of processors used by any node, we can reduce the number of processors used by a node from $p$ to $x$ in the worst case. From the computation model, we can see that this can blow up the computation costs of the node by a factor no greater than $\frac{p}{x}$. Since the quantities $C_x$ and $C_p$ represent the critical path weights when each node in the MDG uses at most $x$ and $p$ processors respectively, we can write:

$$C_x \leq \frac{p}{x} C_p \tag{19}$$

From the definition of the average in Equation 7, we can see that it decreases as the number of processors used by the nodes of the MDG are either decreased or kept the same. Since imposing the bound means either decreasing or keeping the number of processors used by a node, we have:

$$A_x \leq A_p \Rightarrow A_x \leq \frac{p}{x} \cdot A_p \tag{20}$$

since $\frac{p}{x} \geq 1$.

From Equations 19, 20, 17 and 18 we can write:

$$T_{opt} \leq \frac{p}{x} \cdot T_{opt}^{min} \tag{21}$$

Using Theorem 1, we can establish the relationship between $T_{psa}$ and $T_{opt}^{min}$ as:

$$T_{psa} \leq (2 + \frac{x-2}{p-x+1}) \cdot \frac{p}{x} \cdot T_{opt}^{min} \tag{22}$$

From this expression, we can see that the optimum choice of $x$ is the value that minimizes:

$$\frac{2 \cdot p}{x} + \frac{(x-2) \cdot p}{(p-x+1) \cdot x} \tag{23}$$

which is the required result □.

# 6 Allocation Algorithms

The allocation algorithms described below are all heuristics built around the PSA.

## 6.1 Heuristic Change Critical Tasks (CCT)

**Step 1** Allocate one processor to each node in the MDG and schedule using the PSA. Denote the finish time obtained to be $ft_{last}$. Also mark all nodes as changeable.

**Step 2a** Find and mark all critical path nodes, i.e. all nodes which lie on at least one critical path of the MDG for the current processor allocation.

**Step 2b** Find the critical path node which offers the biggest change per processor increase and is marked changeable. Increase the node's processor allocation by one. Compute the finish time using PSA, call it $ft_{current}$.

**Step 3** The following cases exist with respect to the difference $ft_{last} - ft_{current}$ :

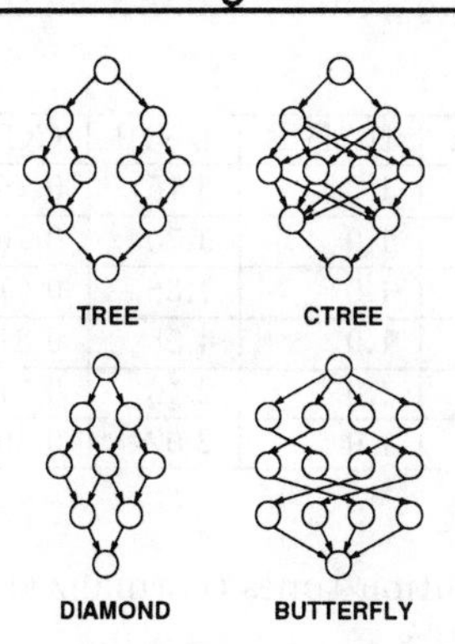


Figure 1: Some regular MDGs used as benchmarks

$ft_{last} - ft_{current} > 0$ which means increasing the processor allocation for the node is a good move. Therefore accept the change and mark all node using less than $p$ processors changeable.

$ft_{last} - ft_{current} \leq 0$ which means increasing the processor allocation for the node is a bad move. Mark the node unchangeable and negate the increase in processor allocation done in Step 2b.

**Step 4** Check if there is any node that is marked changeable, if so, go to Step 2a else go to Step 5.

**Step 5** Schedule using the PSA and compute the final finish time.

## 6.2 Heuristic Optimal Bound Change Critical Tasks (OBCCT)

This heuristic uses the theory developed in Theorem 2 to come up with bounds on the maximum number of processors used by any node $x$. The value of the bound $x$ is used in Step 3.a to mark all nodes using less than $x$ processors changeable instead of the value of $p$ processors used for CCT. All other steps for this heuristic are identical to the ones in CCT.

## 6.3 Other Heuristics

In addition to the heuristics described above, we also implemented two other heuristics called NAIVE, which allocates $p$ processors to each node in the MDG and UNIP, which allocates only one processor per node in the MDG. These were used for comparison with our heuristics. NAIVE corresponds to what most data parallelizing compilers use currently.

# 7 Results

The heuristics proposed above were implemented and tried out on some synthetic MDGs. The aim of this exercise was to try and see the effectiveness of the heuristics over naive schemes for these graphs. We chose synthetic graphs since we did not have any MDGs generated from real benchmarks at this stage. The synthetic MDGs are varied and complex in structure. Four of them, Diamond, Tree, Ctree, and Butterfly are regular in structure. We have shown the shape of these graphs in Figure 1. For our experiments we used larger versions of these graphs. In addition, we also used two randomly generated graphs called Random1 and Random2 to make sure our methods worked well for

| Benchmark | NAIVE | UNIP | CCT | OBCCT |
|---|---|---|---|---|
| Random1 | 1.0 | 3.15 | 0.75 | 0.76 |
| Random2 | 1.0 | 3.70 | 0.76 | 0.79 |
| Butterfly | 1.0 | 1.86 | 0.67 | 0.68 |
| Diamond | 1.0 | 4.55 | 0.81 | 0.84 |
| Ctree | 1.0 | 2.43 | 0.77 | 0.79 |
| Tree | 1.0 | 2.02 | 0.89 | 0.90 |

Table 1: Execution times (normalized to NAIVE) for various heuristics

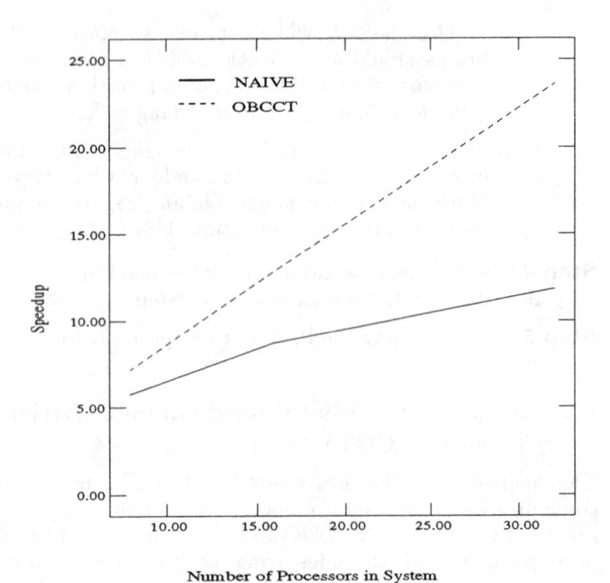


Figure 2: Speedup Curves for OBCCT and NAIVE

irregular MDGs. A pair $(\alpha, \tau)$ was generated randomly for each node in the MDG. To check is there were benefits to using our algorithms even if $\alpha$ is low (nodes have almost linear speedups), we purposely kept it between 1 and 10 percent. In addition a communication size $L_{ij}$ was generated randomly for each pair of nodes $i$ and $j$ that need to communicate.

The results obtained are shown in Table 1. From this table, it can be observed that both our allocation algorithms give an improvement over the NAIVE heuristic or the UNIP heuristic. The quality of improvement for the OBCCT algorithm is only slightly inferior to that of the CCT algorithm. Its worst case complexity (not derived due to lack of space), however, is lower than the CCT algorithm. This would make it favorable to use the OBCCT algorithm in practice.

Our algorithms are more effective as compared to the NAIVE heuristic when the number of processors in the system is increased. This effect is clear from Figure 2 showing the speedup curves for OBCCT and NAIVE for the Butterfly benchmark. This shows the OBCCT has a linearly increasing profile whereas the NAIVE heuristic is tending towards a saturating value.

# 8 Conclusions

Practically feasible algorithms have been presented for the allocation and scheduling of MDGs. Important theoretical results have been derived for with respect to the optimality of the scheduling algorithm used. Results obtained on synthetic benchmark graphs show the potential of our methods even when the nodes of the MDG are highly parallel.

Our future work is aimed at including communication costs in our theoretical analysis. Another area of interest is the development of optimal allocation algorithms to replace the heuristics we are currently using. After this has been done we propose to test our algorithms on real benchmarks.

# References

[1] M. Gupta and P. Banerjee, "Demonstration of Automatic Data Partitioning Techniques for Parallelizing Compilers on Multicomputers," *IEEE Transactions on Parallel and Distributed Computing*, pp. 179–193, March 1992.

[2] E. Su, D. Palermo, and P. Banerjee, "Automating Parallelization of Regular Computations for Distributed Memory Machines in the PARADIGM compiler," in *Proceedings of the International Conference on Parallel Processing*, 1993.

[3] M. Gupta and P. Banerjee, "A Methodology for High-level Synthesis of Communication on Multicomputers," in *Proceedings of the International Conference on Supercomputing*, 1992.

[4] K. P. Belkhale and P. Banerjee, "Approximate Algorithms for the Partitionable Independent Task Scheduling Problem," in *Proceedings of the International Conference on Parallel Processing*, pp. 72–75, 1990.

[5] K. P. Belkhale and P. Banerjee, "A Scheduling Algorithm for Parallelizable Dependent Tasks," in *Proceedings of the International Parallel Processing Symposium*, pp. 500–506, 1991.

[6] G. N. S. Prasanna and A. Agarwal, "Compile-time Techniques for Processor Allocation in Macro Dataflow Graphs for Multiprocessors," in *Proceedings of the International Conference on Parallel Processing*, pp. 279–283, 1992.

[7] M. Girkar and C. D. Polychronopoulos, "Automatic Extraction of Functional Parallelism from Ordinary Programs," *IEEE Transactions on Parallel and Distributed Computing*, pp. 166–178, March 1992.

[8] J. K. Lenstra and A. H. G. R. Kan, "Complexity of Scheduling under Precedence Constraints," *Operations Research*, pp. 22–35, January 1978.

[9] V. Sarkar, *Partitioning and Scheduling Parallel Programs for Multiprocessors*. MIT Press, 1989.

[10] T. Yang and A. Gerasoulis, "A Fast Static Scheduling Algorithm for DAGs on an Unbounded Number of Processors," in *Proceedings of IEEE Supercomputing 91*, pp. 633–642, 1991.

[11] T. Yang and A. Gerasoulis, "A Parallel Programming Tool for Scheduling on Distributed Memory Multiprocessors," in *Proceedings of the Scalable High Performance Computing Conference*, pp. 350–357, 1992.

[12] C. L. Liu, *Elements of Discrete Mathematics*. McGraw-Hill Book Company, 1986.

[13] M. R. Garey, R. L. Graham, and D. S. Johnson, "Performance Guarantees for Scheduling Algorithms," *Operations Research*, pp. 3–21, January 1978.

[14] Q. Wang and K. H. Cheng, "A Heuristic of Scheduling Parallel Tasks and Its Analysis," *SIAM Journal on Computing*, pp. 281–294, April 1992.

[15] J. W. Turek, J. and P. Yu, "Approximate Algorithms for Scheduling Parallelizable Tasks," in *Proceedings of the 4th Annual Symposium on Parallel Algorithms and Architectures*, pp. 323–332, 1992.

# SESSION 6B

## COMPILER (II)

# LOCALITY AND LOOP SCHEDULING ON NUMA MULTIPROCESSORS

Hui Li, Sudarsan Tandri, Michael Stumm, and Kenneth C. Sevcik
Computer Systems Research Institute
University of Toronto
Toronto ON M5S 1A4
CANADA

## Abstract

*An important issue in the parallel execution of loops is how to partition and schedule the loops onto the available processors. While most existing dynamic scheduling algorithms manage load imbalances well, they fail to take locality into account and therefore perform poorly on parallel systems with non-uniform memory access times. In this paper, we propose a new loop scheduling algorithm, Locality-based Dynamic Scheduling (LDS), that exploits locality, and dynamically balances the load.*
**Key Words**: Locality, Loop Scheduling, NUMA Multiprocessors, Data Partitioning, Locality-based Dynamic Scheduling.

## 1 Introduction

Loops are a major source of parallelism for todays parallelizing compilers. An important issue in the parallel execution of loops is how to partition and schedule the loops onto the available processors. A number of algorithms have been proposed for this purpose. For example, *static scheduling* algorithms such as block, cyclic, and block-cyclic scheduling, partition the loop into fixed-sized chunks and distribute the chunks evenly across processors statically at the beginning of the computation. *Dynamic scheduling* algorithms, on the other hand, assign the loop partitions at run time, depending on the speed and progress of the processors. For example, self scheduling [5] partitions the loop into fixed size chunks, which are conceptually organized in a single system-wide queue, and each processor obtains a new chunk from the queue when it has completed its previous chunk. More recent proposals, such as guided self scheduling (GSS) [10], factoring [4], and trapezoid [13], vary the size of the chunks; they start with large chunks in order to reduce the overhead in accessing the central queue and then progressively use smaller chunks in order to maintain good load balance. These scheduling algorithms are described in detail in Section 2.

All of the loop scheduling algorithms listed above assume a shared memory architecture with uniform memory access (UMA) costs, and hence need not take data locality into consideration. However, many of the more modern, especially scalable, shared memory multiprocessors have non-uniform memory access (NUMA) cost; *i.e.*, the cost of accessing memory increases with the distance between the accessing processor and the target memory. Examples of multiprocessors with non-uniform memory access costs include DASH [6], Hector [14], BBN [12], RP3 [7], and Cedar [8]. In these systems, data locality is important for good application performance, and the loop scheduling algorithms should take this into account.

In this paper, we introduce a new loop scheduling algorithm that takes data locality into consideration, and compare its performance with other well known scheduling algorithms. The next section describes some of the existing loop scheduling algorithms. Section 3 describes data locality and why it is an important factor that cannot be neglected in loop scheduling algorithms. In particular, we argue that data locality is important even in systems with hardware-based cache coherence and in cache-only-memory architectures (COMA), such as the KSR [1]. The locality based dynamic scheduling (LDS) algorithm we propose in this paper is described in Section 4, and compared against the affinity scheduling algorithm developed at the University of Rochester, the only other loop scheduling algorithm we are aware of that also takes memory access locality into consideration. In Section 5, the results of experiments comparing LDS to the other scheduling algorithms are described.

# 2 Scheduling Algorithms

## 2.1 Static Scheduling

Static scheduling algorithms, such as block scheduling, cyclic scheduling, and block cyclic scheduling, assign a fixed number of loop iterations to each processor.

**Block scheduling** divides the loop into blocks of $\lceil N/P \rceil$ iterations, where $N$ is the number of iterations and $P$ is the number of processors. Each processor is assigned a separate block. If the amount of computation performed by each iteration differs, then block scheduling can perform poorly because of load imbalance. For example, in an iteration space where the amount of computation per iteration increases linearly, the first few blocks will entail very little computation while the latter ones will involve much more. The first few processors will therefore finish their computations early and have to wait for others to complete, resulting in poor speedup.

**Cyclic scheduling** assigns loop iterations to processors in a cyclic order, so that processor $p$ will execute the iterations $p, p+P, p+2P, \ldots$, where $P$ is again the number of processors executing the loop. In contrast to block scheduling, cyclic scheduling obtains better load balance for triangular iteration spaces and other iteration spaces where the amount of computation increases/decreases linearly with the iterations.

**Block cyclic scheduling** is a compromise between block scheduling and cyclic scheduling. This algorithm assigns blocks of a fixed size to processors in a round robin fashion. If the block size is equal to one, then block-cyclic scheduling degenerates to cyclic scheduling and if the block size is $\lceil N/P \rceil$, then block-cyclic scheduling is same as block scheduling. Hence, block cyclic scheduling forms a continuum between block and cyclic scheduling algorithms.

The static algorithms ignore the fact that the amount of computation performed per iteration may differ, or that it cannot always be determined a priori (for example, the amount of computation could be dependent on the data). Moreover, the speed of each processor may also differ because of multitasking interference. Therefore static scheduling often suffers from load imbalance, resulting in poor speedup.

## 2.2 Dynamic Scheduling

If the imbalance becomes large, then it is necessary to dynamically adjust the work assigned to each processor at run-time in order to balance the load. This is done by grouping together one or more iterations to form subtasks which are dynamically allocated and executed by the processors. The subtasks need not be of fixed granularity, and in fact the granularity could vary dynamically. If the granularity is very large ($N/P$ iterations per subtask) then we effectively have block scheduling. On the other hand, if the granularity is very small, then the data structure controlling the subtasks could become a bottleneck because of the number of accesses each processor must perform to this structure. Self scheduling, guided self scheduling, factoring, and the trapezoid method are examples of dynamic scheduling algorithms.

# 3 Locality and Scheduling

**Self scheduling** partitions the loops into subtasks containing one or more iterations [5]. Each processor then continuously allocates and executes one subtask at a time until no subtasks are left for processing. If the number of iterations per subtask is fixed and greater than one, then this scheduling strategy is generally referred to as ***fixed-size chunking*** [5].

With fixed-size chunking it can be difficult to choose the correct granularity. Small granularity increases the overhead of accessing the data structure controlling the subtasks that still need to be executed. Larger granularity can lead to load imbalances when the last set of subtasks is being executed. Hence in the algorithms that follow, the size of the subtasks is dynamically adjusted with the progress of the computation.

**Guided self scheduling(GSS)** uses a subtask granularity of $\lceil n/P \rceil$ iterations, where $n$ is the total number of remaining iterations [10]. With this algorithm, the subtasks are composed of a large number of iterations at the start of the computation, then progressively fewer until the size is one. In this scheme, there will be at least $P-1$ subtasks consisting of only one iteration, and each will be executed independently. When the execution time of the iterations differ, it is possible that an early subtask could be so large that it does not complete by the time all other subtasks have completed [4]; this load imbalance problem is addressed by the factoring algorithm.

**Factoring** is similar to GSS in that the size of the subtask decreases as the computation progresses, but it assigns $\lceil n/(2P) \rceil$ iterations to $P$ consecutive subtasks, where $n$ is equal to the number of remaining iterations at the beginning of these allocations [4]. Hence $P$ consecutive subtasks will be of the same size, before the granularity is decreased. If the variance of the amount of computation performed by each iter-

| Scheme | No. of Iterations = 500 and P = 4 |
|---|---|
| GSS | 125 94 71 53 40 30 22 17 12 9 7 5 4 3 2 2 1 1 1 1 |
| Factoring | 63 63 63 63 31 31 31 31 16 16 16 16 8 8 8 8 4 4 4 4 2 2 2 2 1 1 1 1 |
| Trapezoid | 62 58 54 50 46 42 38 34 30 26 22 18 14 8 |
| LDS | 63 55 48 42 37 32 28 25 22 19 17 14 13 11 10 8 7 7 6 5 4 4 3 3 3 2 2 2 1 1 1 1 1 1 1 1 |

Table 1: Subtask Sizes for dynamic scheduling algorithms. The GSS, factoring and trapezoid algorithms are described in Section 2. LDS is described in Section 4.

| Arch. | Cache | Local Mem. | Remote Mem. |
|---|---|---|---|
| Hector | 1 | 10 | 24 |
| DASH | 1 | 22 | 61 |
| RP3 | 1 | 10 | 15 |

Table 2: Latency for memory read operation in processor clocks

ation is large, then factoring performs better than GSS [4].

The **Trapezoid** method also assigns a decreasing number of iterations to subtasks and thus is a variation of GSS. In this case, however, the subtask size decreases linearly instead of exponentially [13]. The total number of iterations, $N$, is partitioned into $S = \lceil 2N/(f+1) \rceil$ subtasks, where $f = \lfloor N/(2P) \rfloor$ is the size of the first task. Consecutive subtasks differ by $\lfloor (f-1)/(S-1) \rfloor$ iterations.

For comparison, the subtask sizes employed by the dynamic scheduling algorithms GSS, factoring, and the trapezoid method for a problem size with 500 iterations executing on four processors is given in Table 1.

In NUMA systems, managing data locality is important due to the increased cost of accessing remote memory. Table 2 shows the difference between remote memory access costs and local memory access costs for different architectures configured with 64 processors. On these systems having most of the accessed data local to the accessing processor can be a major factor in improving performance [2, 3, 11].

In parallelizing a loop, it is important to consider the partitioning of both the data space and the loop iteration space, and how both are mapped onto the processors. For good performance, it is essential that the loop partitions and scheduling match the data partitions. Best performance is achieved when all data required by a loop partition is local to the processor on which the partition is scheduled. A mismatch in the scheduling of the loop partitions and the data partitions can have a heavy performance penalty on NUMA multiprocessors, as will be shown in Section 5.

Consider for example the simple loop shown in Figure 1. The iteration space and the data space are two dimensional. Because the inner loop $j$ is sequential, the data space must also be partitioned row-wise and implicitly the iteration space must also partitioned row-wise. Because of the simple reference pattern, the loop and data partitions match. If the loop partition $i = 0$ is scheduled on the processor which has the row $A[0][*]$, then all the accesses to $A$ are local. Otherwise all the accesses would be non-local. We call the static scheduling algorithm *Block-D* if both the data partitioning and the loop scheduling occur in blocks. (Analogously we use the terms *Cyclic-D* and *Block-cyclic-D* if both data partitioning and the loop scheduling match.)

```
parallel_for(i=0; i < N; i++)
        for(j=0; j< N; j++)
                A[i][j] = ...
```

Figure 1: Simple Program

In general, dynamic scheduling algorithms can achieve good load balance, but at the cost of decreased locality in data accesses, since each subtask may be scheduled on any of the processors regardless of the location of the data it must access. The cost of an average memory access would increase on systems with non-uniform memory access cost. This can lead to a decrease in performance, not only because of increased latencies to access the remote data, but also because of increase in network traffic and congestion.

It is interesting to note that, while cache memory helps reduce the effects of non-uniform memory access costs, it does not eliminate them entirely. In practice, even on a multiprocessor with hardware

```
for(k=0; k < N-1; k++) {
        d0 = A[k][k];
        parallel_for(j= k+1; j< N; j++) {
                A[j][k] /= d0;
                d1 = A[j][k];
                for ( i = k+1; i < N; i++)
                        A[j][i] -= d1*A[k][i];
        }
}
```

Figure 2: LU Decomposition

cache consistency (such as the DASH multiprocessor), the average memory response time can be reduced by exploiting memory locality. We illustrate this using LU decomposition as an example. The core of the LU decomposition code is shown in Figure 2. It consists of an outer sequential loop and a parallel loop. In the innermost loop, one of the rows of the matrix $A$ is modified based on the pivot row $k$. Consider the execution of the parallel loop $j = 5$ running on processor $P1$, and trace the computation. $P1$ accesses the elements of the fifth row and modifies it, causing a valid copy of this row to be in the cache of processor $P1$ and the copy in the memory to become invalid (assuming a write-back cache). If in the next invocation of the parallel loop, loop $j = 5$ is executed on another processor (say $P7$), then the main memory needs to be updated with the values still cached on $P1$, and the values in $P1$ need to be invalidated when $P7$ modifies them. Thus the validation and invalidation traffic on the network can become excessive if there is no locality. Similar effects are possible in cache-only memory architectures (COMA), such as the KSR.

Some multiprocessors, such as RP3 and BBN Butterfly/TC-2000 allow remote memory to be accessed in an interleaved and/or randomized manner. By distributing the memory accesses more evenly across the processors, the number of hot-spots at the memories and in the network is reduced. Thus the NUMA machine behaves like an UMA machine with one cost (close to the maximal one) for accessing the memory. While randomized access has the possible advantage of reducing memory and network hotspots, it also has the disadvantage of not exploiting locality. Because a single processor can execute using only local memory, the speedup of applications that exclusively access shared data in this randomized manner will be poor. The execution time of the application would be better, if data locality were exploited.

# 4 Locality-based Dynamic Scheduling

In this section, we propose a new scheduling algorithm, Locality-based Dynamic Scheduling (LDS), that addresses both locality and load balancing. LDS (see Figure 3) is based on the following principles:

1. The data space is partitioned to reside on $P$ processors. Typical data partitions are block, cyclic, and block-cyclic. Often, the partition chosen is constrained by rest of the computation.

2. Each processor, when it is ready to execute the next subtask, computes the size of this subtask. The size can be chosen as in any of the dynamic scheduling algorithms as a function of the number of remaining iterations and the number of processors. In our experiments, we set the subtask size to $\lceil n/(2P) \rceil$ (where $n$ is the number of remaining unscheduled iterations and $P$ is the number of processors). This creates subtasks about half as large as those by GSS in order to avoid overly large initial subtask sizes (see Table 1.

3. Once the subtask size has been determined, the processor must decide which iterations to execute as part of its subtask. The dynamic scheduling algorithms sequentially take iterations from the loop iteration space; that is the first subtask of size $S1$ includes iterations $1, 2, \ldots, S1$, the second subtask of size $S2$ includes iterations $S1 + 1, \ldots, S1 + S2$, and so on. In LDS, on the other hand, the iterations are chosen such that locality is maximized. For example, if the data distribution is cyclic, and the processor $p$ has to execute a subtask of $S1$ iterations, then it executes the iterations $p+P, p+2P, \ldots, p+P*S1$. If the data distribution is block then the subtask would include iterations $p*B+1, p*B+2, \ldots, p*B+S1$, where $B$ is the block size. If all the scheduled local iterations are completed, iterations are acquired from the processor with the most unscheduled iterations.

The LDS algorithm is related to the *affinity scheduling* algorithm (AFS) proposed by Markatos and LeBlanc in that AFS also takes locality into account [9]. AFS divides the iterations of a loop into block partitions of $\lceil N/P \rceil$ iterations, where $N$ is the

1. Determine the subtask size $S = \lceil n/(2P) \rceil$ based on the total number of unscheduled iterations ($n$) and the number of processors ($P$).
2. If the processor has $r > 0$ locally assigned, unscheduled iterations, then the subtask includes $\min(r, S)$ of those iterations. Otherwise, if $r = 0$ then $\min(r_{max}, S)$ iterations are acquired from the processor with the most unscheduled iterations, where $r_{max}$ is the maximum number of unscheduled iterations on that processor.
3. Execute the subtask.
4. Repeat 1–3 until $n = 0$.

Figure 3: Locality-based Dynamic Scheduling Algorithm

total number of iterations and $P$ is the number of processors, and assigns each partition to a different processor. When a processor becomes free, it takes the next subtask of $1/k$ iterations from its local partition, where $k$ is a parameter of the algorithm that is chosen statically between 2 and $P$. Once the entire local partition has been executed, the processor determines the processor with the most remaining iterations, and takes fraction $\lceil 1/P \rceil$ of them. The implementation of AFS uses $P$ local locks to protect the local partitions and a global lock to protect the data structure indicating the processor with the most remaining iterations.

LDS is different from AFS in the following ways:

- AFS assumes that data is copied into local storage when first accessed. This can be done by hardware on machines with cache coherence or by the operating system. AFS does not utilize the information of data placement. Instead, it assumes that data accessed in iteration $j$ is likely to be adjacent to the data used in iteration $j+1$, and thus partitions the iteration space into blocks and assigns a block to each processor. Therefore, memory locality can be exploited only when the data is also partitioned and distributed in blocks. LDS, on the other hand, takes data placement into account, by always having the processor first execute those iterations which have the data local to the processor. For this reason, LDS can easily accommodate other data partitioning methods, such as cyclic or block-cyclic.

| | Do not consider Locality | Consider Locality |
|---|---|---|
| Static | Block<br>Cyclic<br>Block-cyclic | Block-D<br>Cyclic-D<br>Block-cyclic-D |
| Dynamic | GSS<br>Self<br>Factoring<br>Trapezoid | AFS<br>LDS |

Table 3: Comparing the various scheduling algorithms

- In AFS, each processor independently schedules iterations from its local partition using a parameter $k$. The best value for $k$ will depend on the application and may be difficult to choose. If $k$ is small, then a processor with an exceptionally large proportion of the workload assigned to it could make the size of the first subtask too large so that later dynamic scheduling will not be able to adjust for the load imbalance. The maximum load imbalance in a loop with a linear iteration space in AFS is $\frac{N(P-k)}{P(P-1)k} + 1$ iterations. When $k$ approaches $P$, on the other hand, the worst-case load-imbalance approaches that of GSS, but at the cost of an increase in the number of synchronization operations by a factor of $P$, since the total number of lock operations performed by AFS is $O(kP \log \frac{N}{kP}) + O(P \log \frac{N}{P^2})$ [9]. In LDS, the worst-case load-imbalance will be one iteration and the number of synchronization operations will be $O(P \log(N))$.
- AFS uses one way to determine the number of iterations to be taken from a local partition and another way to determine the number of iterations to be taken from other partitions. LDS uses the same algorithm for the both.

Table 3 gives a comparative classification of the scheduling algorithms we have discussed. The differences between AFS and LDS in particular are summarized in Table 4. Differences in performance are shown in the next section.

# 5 Experimental Results

In this section, we present performance results from experiments involving four benchmark programs: Matrix Multiplication, LU Decomposition, Successive

| | AFS | LDS |
|---|---|---|
| Locality | block only | any data dist. |
| Lock Ops | $O(kP\log\frac{N}{kP})+$ $O(P\log\frac{N}{P^2})$ | $O(P\log(N))$ |
| Max Imbal. | $\frac{N(P-k)}{P(P-1)k}+1$ iters | one iter. |

Table 4: Comparison of AFS and LDS

Over Relaxation, and Transitive Closure.

- Matrix Multiplication: The regular matrix multiplication i-j-k algorithm is parallelized at the outer $i$ loop, multiplying $400 \times 400$ matrix of double precision numbers.
- LU Decomposition: This algorithm has a sequential outer loop, a parallel loop and an inner most sequential loop. A matrix of $400 \times 400$ double precision numbers is partitioned by row. The computation of LU decomposition is skewed in that the lower rows must be recalculated more frequently than the upper rows; *i.e.*, row 1 is calculated only once, where as elements of row $N-1$ are processed in $N-1$ iterations.
- Successive Over Relaxation: SOR is similar to LU decomposition in that it has a sequential outer loop and a parallel inner loop. Because each processor must access all the neighboring elements of the element being computed, locality plays a major role in obtaining good performance.
- Transitive Closure: Transitive closure has a loop structure similar to LU decomposition. Unlike the previous algorithms, the sequential inner most loop may or may not be executed, and hence the computation is dynamic. The input values determine the variation of iteration execution time. High variance can cause load imbalance. A matrix size of $800 \times 800$ integers is processed.

The experiments were performed on Hector, a scalable shared memory multiprocessor [14, 11]. Hector consists of sets of processor-memory pairs connected together by buses, several buses connected together by local rings, and several local rings connected together by a global ring (see Figure 4). Hector provides a single global physical address space; each memory module contains one portion of the global memory. Access time to memory is a function of memory hierarchy.

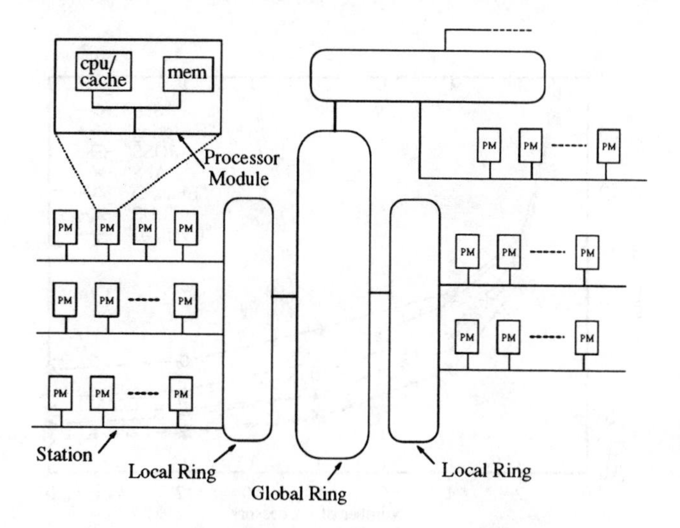


Figure 4: General Architecture of Hector

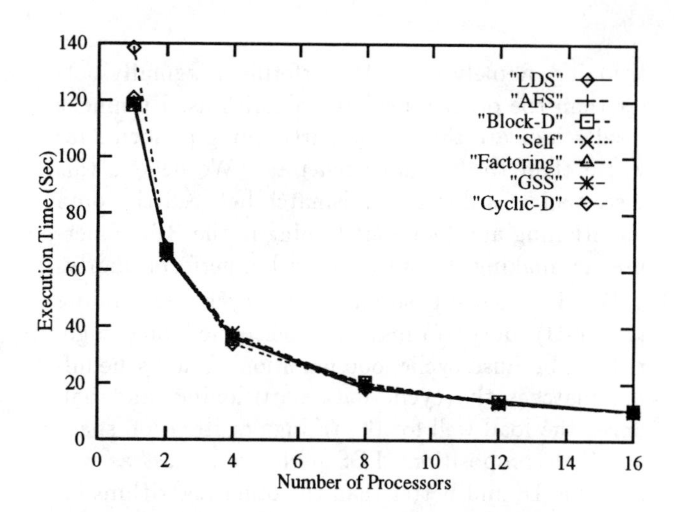


Figure 5: Execution Times for Matrix Multiply

Figures 5-8 show the response times of the four applications listed above when run with the different scheduling policies.

In matrix multiplication, all processes must access all of the matrix $B$. Assuming $B$ does not fit in the cache, thus the overhead of accessing $B$ will dominate the total overhead of accessing the data elements. Good locality in accessing the elements of matrices $A$ and $C$ is automatically achieved through the caching. For this reason, and because the load in this computation is well balanced, all of the scheduling algorithms perform equally well, as shown in Figure 5. Our results for matrix multiplication differ from those of a similar experiment performed on the RP3 by Hummel *et al.* [4]. In our case, static par-

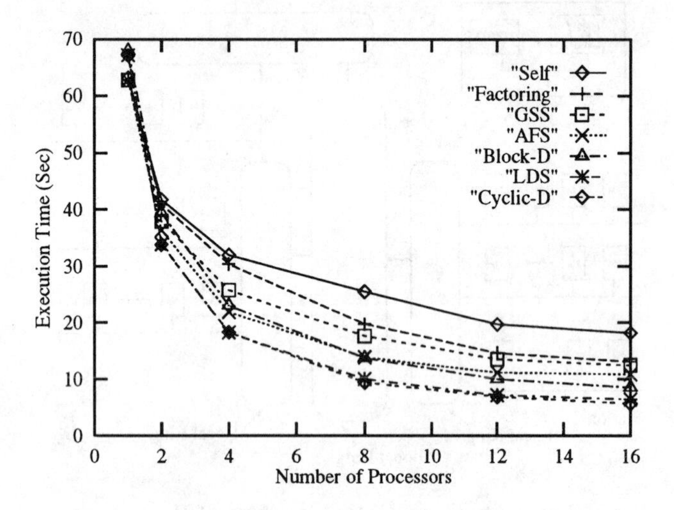


Figure 6: Execution Times for LU Decomposition

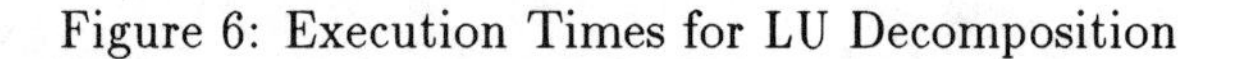

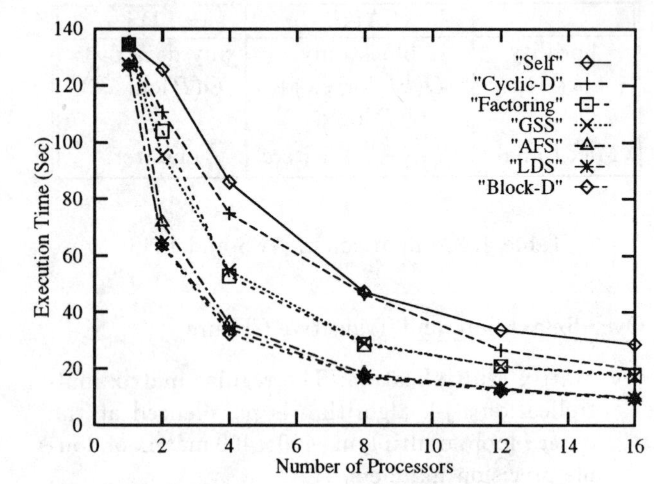


Figure 7: Execution Times for Successive Over Relaxation

titioning, namely cyclic-D, performs marginally better than the other scheduling algorithms. Hummel's results indicate that static partitioning performs far worse than the dynamic schemes. We believe this discrepancy is due to a mismatch between the data partitioning and loop partitioning in the RP3 experiments, making the static algorithm perform poorly.

For LU decomposition, static cyclic scheduling (cyclic-D) outperforms all other scheduling algorithms, because cyclic loop partitioning and scheduling matches the cyclic data partitioning, and balances the load well for the triangular iteration space of LU decomposition. LDS performs almost as well as cyclic-D, and better than the other algorithms for the same reasons. The execution time of the program using LDS is slightly higher than that for cyclic-D because of the run-time scheduling overhead. The results are shown in Figure 6.

For SOR, the static scheduling algorithm that matches the data partitioning, namely block-D, performs best, and the best results are obtained when the data and iteration spaces are partitioned into blocks. Again, LDS performs almost as well as block-D because its iteration space is effectively also partitioned into blocks, given the block distribution of data. In this case affinity scheduling (AFS) performs equally as well because of block partitioning. From Figure 7, it is interesting to note that the performance of the other dynamic algorithms is substantially worse than Block-D, LDS, and AFS because of the lack of data locality.

The transitive closure experiment was chosen as a representative of computationally imbalanced iteration spaces. In this case, LDS outperforms all of the other scheduling algorithms, because it is able to dynamically balance the load, and yet exploit data locality. Block-D performs almost as well, because in this case the variance in the amount of computation of the blocks is not very large. One would expect that AFS could perform as well or better than block-D scheduling, but our results indicate that the overhead AFS incurs for locking will increase quadratically with the number of processors. For example, with $P = 16$, the number of lock operations performed will be about 431, with most locking occurring towards the end of the computation. The deterioration of AFS's performance as the number of processors increases is visible in Figure 8.

Our results indicate that no fixed scheduling algorithm will perform satisfactorily for all applications without taking data locality into account. The static algorithms perform better if the iteration partitions match the data partitioning. With the exception of LDS, the dynamic algorithms are all similar.

# 6 Conclusions

In this paper we have argued that data locality is an important factor to consider in partitioning and scheduling loops. While most existing dynamic scheduling algorithms manage load imbalances well, they fail to take locality into account and therefore perform poorly on parallel systems with non-uniform memory access times. We have presented a new scheduling algorithm, LDS, that is dynamic, yet takes

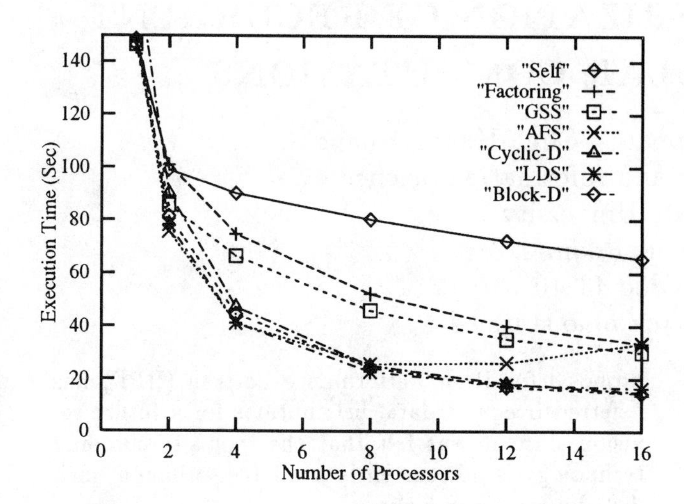


Figure 8: Execution Times for Transitive Closure

locality into account. We have presented experimental results that indicate:

- on NUMA systems, scheduling algorithms do not perform well over a variety of applications if they do not take locality into account;
- no single scheduling algorithm performed best across all applications considered;
- of all the dynamic scheduling algorithms, LDS performed best for the applications considered; and
- unless large load imbalances exist (that is, variance in loop execution times are high), appropriate static algorithms outperform the dynamic scheduling algorithms.

We are currently working on extending LDS to handle irregular data distributions.

## References


[1] Gordon Bell. Ultracomputers: A teraflop before its time. *CACM*, 35(8):27–47, August 1992.

[2] William J. Bolosky, Michael L.Scott, Robert P. Fitzgerald, Robert J. Fowler, and Alan L. Cox. NUMA policies and their relation to memory architecture. In *ASPLOS-IV Proccedings*, pages 212–221, April 1991.

[3] Stephen Curran and Michael Stumm. A comparison of basic CPU scheduling algorithms for multiprocessor Unix. *Computing Systems*, 3(4), Fall 1990.

[4] Susan F. Hummel, Edith Schonberg, and Lawrence E. Flynn. Factoring: A method for scheduling parallel loops. *CACM*, 35(8):90–101, August 1992.

[5] Clyde P. Kruskal and Alan Weiss. Allocating independent subtasks on parallel processors. *IEEE Transactions on Software Eng.*, SE-11(10):1001–1016, October 1985.

[6] Daniel Lenoski, James Laudon, Kourosh Gharachorloo, Wolf-Dietrich Weber, Anoop Gupta, John Hennessy, Mark Horowitz, and Monica S. Lam. The Stanford DASH multiprocessor. *IEEE Computer*, 25(3):63–79, March 1992.

[7] Jack G. Lipovski and Miroslaw Malek. *Parallel Computing: Theory and Comparisons*, Appendix C: The RP3. John Wiley and Sons, 1987.

[8] Jack G. Lipovski and Miroslaw Malek. *Parallel Computing: Theory and Comparisons*, Appendix D: Cedar. John Wiley and Sons, 1987.

[9] Evangelos P. Markatos and Thomas J. LeBlanc. Using processor affinity in loop scheduling on shared-memory multiprocessors. In *Supercomputing 92*, pages 104–113, November 1992.

[10] Constantine Polychronopoulos and David Kuck. Guided self scheduling: A practical scheduling scheme for parallel computers. *IEEE Transactions on Computers*, C-36(12):1425–1439, December 1987.

[11] Michael Stumm, Zvonko G. Vranesic, Ron White, Ron Unrau, and Keith Farkas. Experiences with the Hector multiprocessor. In *International Parallel Processing Symposium*, April 1993.

[12] Arthur Trew and Greg Wilson. *Past, Present, Parallel: A Survey of Available Parallel Computer Systems.* Springer-Verlag, 1991.

[13] Ten H. Tzen and Lionel M. Ni. Dynamic loop-scheduling for shared memory multiprocessors. In *Proceedings International Conference on Parallel Processing*, pages 247–250, August 1991.

[14] Zvonko G. Vranesic, Michael Stumm, David M. Lewis, and Ron White. Hector: A hierarchically structured shared memory multiprocessor. *IEEE Computer*, 24(1):72–79, January 1991.

# COMPILE-TIME CHARACTERIZATION OF RECURRENT PATTERNS IN IRREGULAR COMPUTATIONS

Kalluri Eswar P. Sadayappan Chua-Huang Huang
Department of Computer and Information Science
The Ohio State University
2036 Neil Avenue Room 228
Columbus Ohio 43210
{eswar,saday,chh}@cis.ohio-state.edu

Abstract – *Many engineering applications use irregular array access functions. Compile-time characterization of the computation structure of such applications is infeasible, making the automatic generation of efficient communication difficult. This paper considers the Inspector-Executor (IE) compilation model to handle such applications. In this model, a run-time inspection of the computation is followed by the actual execution. This paper discusses the issue of automatic compile-time identification of sections of a program that have recurrent computational patterns so that the cost of run-time inspection is amortizable over many executions. An algorithm is presented and an illustrative example is given.*

## 1 Introduction

Massively parallel distributed-memory multiprocessors have the potential to accelerate compute-intensive scientific and engineering applications, but their widespread use is currently impeded by the complexity of writing large parallel programs using the explicit message-passing paradigm. One approach that is currently planned by most vendors of parallel machines is a shared-space programming model along with directives for user specification of regular data distributions. It is then the compiler's responsibility to use the data distribution information to automatically generate the communication statements needed in the parallel program [4, 7]. Much research has already been done in compiler algorithms to generate efficient communication for programs with regular data distribution patterns (for example, block, cyclic, and block-cyclic distributions) and regular array access functions (for example, affine functions of loop indices)[5, 6, 8].

Many engineering applications, however, use irregular array access functions. Compile-time characterization of the computation structure of such applications is infeasible, making the automatic generation of efficient communication difficult. The current proposal for High Performance Fortran (HPF) has deferred irregular data distributions for a future revision since it was felt that the requisite compiler technology is not currently available to handle such distributions efficiently.

This paper addresses an important issue pertaining to efficient parallel implementation of scientific and engineering computations with irregular array accesses. A significant property of many unstructured engineering applications is that although the computational patterns are irregular, these irregular patterns often recur a number of times during the course of execution [3, 9]. For example, with circuit simulation, irregular array access patterns arise due to the irregular interconnectivity of the physical circuit modeled. However, the circuit structure does not change from one simulated time step to another. Therefore the non-zero structure of the sparse matrices used also does not change, so that many of the computations that operate on these matrices are characterized by an irregular albeit repetitive computational pattern. Such a section of the application can be analyzed once at run time (by essentially emulating the execution of the section) to determine an efficient communication schedule for it, and the cost of such an analysis can be amortized over the repeated uses of the generated schedule in repeated executions of that section. This model of compiling a section of a program is an example of Inspector-Executor compilation where the inspector corresponds to the code generated to perform the analysis, and the executor corresponds to the code generated to actually execute the section of the program. In this context, the recognition of sections of programs with repetitive computation patterns is clearly very important.

PARTI [2] is a system that has used the Inspector-Executor (IE) compilation model. The basis of the system is a set of primitives, calls to which can be embedded within a distributed-memory parallel program. The primitives can be used to determine the communication required and to carry out the communication required for arrays used in a parallel (DOALL) loop. An extension of Fortran 77, ARF, has also been designed that allows the programmer to specify the data distribution and the parallel loops. The ARF compiler [10] then generates the inspector and executor with embedded calls to the PARTI primitives. The programmer is responsible for identifying the sections of the program for which an IE compilation model is to be used.

This paper addresses the issue of automatic compile-time identification of sections of a program that exhibit recurring computational patterns, for which the IE compilation model is attractive. Such sections will, in general, be referred to as IE-sections. Although the actual computation structure is determinable only at run time, whether or not the computation structure will repeat is often determinable at compile-time itself. The framework that will be presented in this paper for the identification of IE-sections is general in that it includes sections that contain loops with inter-iteration dependences (DOACROSS loops).

The rest of the paper is organized as follows. Section 2 motivates the IE compilation model by using a small example. Section 3 elaborates on IE-sections and presents the framework and algorithm used to identify a particular type of IE-section. Section 4 provides concluding remarks.

# 2 IE Compilation

This section presents a small example of IE compilation and then explains the basic IE compilation model.

## 2.1 Inspector and Executor Example

Consider the program segment shown in Figure 1. It performs a computation similar to a Gauss-Jacobi iterative solution of a sparse linear system of equations.

```
real x(n), y(n), a(maxcols,n)
integer ncols(n), cols(maxcols,n)

do l = 1,lmax
   do i = 1,n
      do j = 1,ncols(i)
         y(i) = y(i) + a(j,i)*x(cols(j,i))
      end do
   end do
   do i = 1,n
      x(i) = f(y(i))
   end do
end do
```

Figure 1: IE compilation example

Assume that the arrays are to be distributed among the processors such that `x(i)`, `y(i)`, `ncols(i)`, `a(*,i)`, and `cols(*,i)` are all on the same processor, for each `i`. Assume also that the two `i` loops have been designated as parallel loops, with an implicit barrier between them and that the `l` loop is designated as a sequential loop. It can then be seen that the only array access that can possibly require communication between the processors is `x(cols(j,i))`. The values of `x` that need to be communicated between the processors depend on the integer arrays `cols` and `ncols`. Their values are input-dependent, so that it is impossible for a compiler to group together the communication of the values of `x` required for a group of elements of `y`. This can potentially cause poor performance because of the high startup costs involved in each communication between processors in a distributed-memory multiprocessor.

It can, however, be seen in this program segment that the values of `x` that need to be communicated remain the same in different iterations of the `l` loop, because the array `cols` remains unchanged and the subscripts for `cols` do not depend on `l`. The program segment can therefore be compiled so that at run time each processor emulates one iteration of the `l` loop, determining the values it needs to send and receive before each iteration of the `l` loop. This pass would be the **inspector** portion of the code generated and it would collect the communication information in certain auxiliary data structures. The other portion, the **executor**, would be the body of the actual `l` loop, where the necessary communication would be done by each processor before each iteration. If `lmax` is sufficiently big, the cost of the one-time analysis can be amortized over the `l` loop's iterations.

**IE compilation** is informally defined as compiling a section of a program to produce an inspector for preprocessing its computation structure and an executor for its actual execution. In general, an execution of the inspector might be required immediately prior to each execution of the executor. However, for computations such as the one in Figure 2, the inspector need only be executed once, before the first execution of the executor. A program section such as the body of the `l` loop in Figure 1 will be called an **IE-section**.

An outline of the code that might be generated by IE compilation to run on each processor for this example is shown in Figure 2. The set `iters(p)` refers to the iterations of the i loop that processor p is responsible for.

```
{ Inspector }
Mimic one iteration of the l loop to
determine values of x to be communicated
do l = 1,lmax
   { Executor }
   Using information generated by inspector,
   send relevant locally mapped x values and
   receive needed nonlocally mapped x values
   do i ∈ iters(p)
      ...
   end do
   do i ∈ iters(p)
      ...
   end do
end do
```

Figure 2: Example IE code outline for processor p

The inspector portion of the code generated can, in general, perform more work than just figuring out the communication required. For example, it might determine how the IE-section is to be parallelized and/or how the arrays are to be distributed for minimizing communication. The work to be done in the inspector should also, in general, be split among the processors in order to minimize the inspector overhead. The framework to be presented in Section 3 allows the automatic identification of IE-sections by a compile-time analysis of the program, independent of the specific kind of inspector generated.

## 2.2 Basic IE compilation

In this section, we describe more precisely the form of the source program we assume, and the general characteristics of any inspector. We consider a program containing only structured statement constructs (sequencing, if-then-else, single-entry-single-exit loops) and having only scalar and array variables. For simplicity, we also assume that there are no subprograms. A section in a program is defined to be a single-entry-single-exit region in the program.

The basic IE compilation model is illustrated in Figure 3. A section $\Sigma$ in the program is compiled into two code segments, its inspector $\Sigma_I$, and its executor $\Sigma_E$, with the executor immediately following the inspector. The inspector performs a preprocessing of the execution of the section in order to determine some information that will be useful for the actual execution performed by the executor. Informally, the

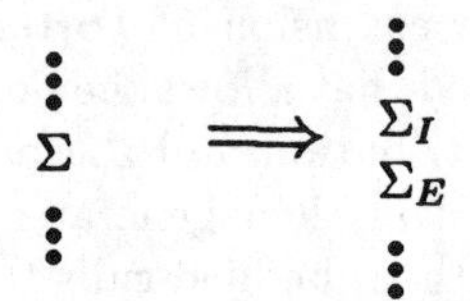


Figure 3: Basic IE compilation

inspector should be produced in such a way that it analyzes the structure of the execution of the section and not the actual execution. In other words, the inspector should emulate the execution of the section only to be able to determine what statements are executed and what variables are computed.

Consider a particular execution of a section $\Sigma$. Since, in general, $\Sigma$ can contain loops, each statement in $\Sigma$ may have multiple execution instances in this particular execution of $\Sigma$. Each statement instance reads and/or writes certain variable locations.

Based on the above, it is possible to talk about a **statement instance dependence graph** for this execution of $\Sigma$, in which each statement instance is a vertex and there is a directed edge from a vertex to another if the latter cannot execute unless the former has executed earlier. This may be the case because the first vertex writes a value into a location that is read by the second vertex (a write-read dependence), or the first vertex reads a value in a location that is overwritten by the second vertex (a read-write dependence), or the first vertex writes a value into a location that is overwritten by the second vertex (a write-write dependence). The directed edge is labeled by the set of variable locations that cause the dependence.

Without making specific assumptions about the properties of the inspector to be generated by the IE compilation scheme, the following general characteristic is assumed: an inspector $\Sigma_I$ for a section $\Sigma$ will be such that it can effectively be implemented as a computation that first determines the statement instance dependence graph of the execution of $\Sigma$ and then uses only the information in the statement instance dependence graph in its execution.

# 3 IE-sections

IE-sections have been informally defined to be sections that are amenable to the IE compilation model. In the case of basic IE compilation, any section is amenable to it. Sections 3.1 and 3.2 define two types of IE-sections along with the IE compilation models they are amenable to. Section 3.3 contains an algorithm for the identification of one of the types of IE-sections and section 3.4 illustrates the algorithm by tracing through an example. Section 3.5 considers how the specification of nonpartitionable regions can allow more more sections in a program to become IE-sections.

### 3.1 Context-free IE-sections

The first type of IE-section considered is defined using the notion of a statement instance dependence graph.

**Definition:** If the statement instance dependence graph of a section $\Sigma$ is identical in any execution of $\Sigma$, then $\Sigma$ is said to be a context-free IE-section, or an IE/C-section.

```
                      { beginning of program }
                      first_time(Σ) = true
                      ⋮
⋮                     if first_time(Σ) then
Σ        ⟹               Σ_I
⋮                         first_time(Σ) = false
                      end if
                      Σ_E
                      ⋮
```

Figure 4: IE/C compilation

Since the inspector in an IE compilation does not compute anything that cannot be computed from the statement instance dependence graph, and since this graph is identical in all executions of an IE/C-section, it can be compiled in such a way as to have its inspector execute only once, before the first execution of the section. This can be achieved by having a boolean variable `first_time(Σ)` associated with the section that tells whether $\Sigma$ has been previously executed. This method of IE compilation is called IE/C-compilation and is illustrated in Figure 4.

The invariance of the statement instance dependence graph implies that the set of statement instances in each execution of $\Sigma$ is the same and that the dependence structure between the statement instances is the same. This guarantees that regardless of how the data and the statement instances are partitioned among the processors, the same computation takes place in each execution of $\Sigma$ which means that $\Sigma$ is amenable to IE/C compilation thus allowing the amortization of the cost of its inspector over all its executions.

The definition of an IE/C-section cannot be used directly to test if a section is an IE/C-section at compile time since the statement instance dependence graph is only determinable at run time. However, it is possible to state conditions on the section to be tested, that are sufficient to conclude that its statement instance dependence graph will be invariant. These conditions are presented in Proposition 1.

**Proposition 1:** The following conditions on a section $\Sigma$ are, together, sufficient to guarantee that its statement instance dependence graph will be invariant in different executions:

(a) The sequence of statement instances executed in $\Sigma$ is invariant across different executions of $\Sigma$.

(b) For each array access `A(s1,s2,...)` in $\Sigma$, for each subscript expression $s_i$, the sequence of values taken by it in all the execution instances of the statement in which it occurs is invariant in different executions of $\Sigma$.

**Proof:** Condition (a) guarantees the invariance of the nodes in the statement instance dependence graph. Condition (b) guarantees the invariance of the dependences between the nodes because the dependences are determined by the variable locations that each statement instance accesses. The set of variable locations accessed by a statement instance is determined by the values of the subscript expressions of array accesses in it. Since condition (a) is true, each statement has a fixed sequence of instances in each execution of $\Sigma$. Each subscript expression in the statement thus has a sequence of values corresponding to the sequence of statement instances. If each subscript expression in the statement has the property that its sequence of values is the same in each execution of $\Sigma$, it must be the case that the set of variable locations accessed by each statement instance is also invariant in different executions of $\Sigma$, making the statement instance dependence graph invariant.

The conditions in Proposition 1 are formalized using a function `SI/C(Σ,v,S)` (SI is for Sequence-Invariant) which is defined for each variable (scalar or array) `v` used (i.e. read) in each statement `S` in the section $\Sigma$. The possible values for `SI/C` are `true` and `false`. `SI/C` is defined as follows.

**Definition:** If the sequence of values of variable `v` used during the execution instances of statement `S` remains invariant for any possible execution of $\Sigma$, then `SI/C(Σ,v,S)` is `true`, and is `false` otherwise.

Proposition 2 uses the function `SI/C` to give conditions that, if true, are sufficient to conclude the conditions of Proposition 1.

**Proposition 2:** A section $\Sigma$ is an IE/C-section if the following conditions are both true:

(a) For the condition P of a control statement S in $\Sigma$, for every variable v used in P, SI/C($\Sigma$,v,S) = **true**.

(b) For every non-control statement S in $\Sigma$, for every variable v used in an array subscript expression in S, SI/C($\Sigma$,v,S) = **true**.

**Proof:** If the sequence of values taken by each variable used in the condition of any control statement is invariant for different executions of $\Sigma$, the sequence of decisions made at any control statement in $\Sigma$ is also invariant, which means that the sequence of statement instances executed will be invariant. Therefore, condition (a) of Proposition 2 implies condition (a) of Proposition 1. Condition (b) of Proposition 2 implies condition (b) of Proposition 1 since the invariance of the sequence of values of all variables used in subscript expressions implies the invariance of the sequence of values of the expressions. From Proposition 1, $\Sigma$ has an invariant statement instance dependence graph, which means $\Sigma$ is an IE/C-section.

## 3.2 Loop-body IE-sections

A section's statement instance dependence graph must remain invariant in *all* its executions in order for it to be an IE/C-section because the inspector code will be executed only once irrespective of the number of invocations of the executor. However, if a section is such that its statement instance dependence graph remains invariant in a subsequence of the sequence of all its executions, it can still be compiled so that the inspector code is executed only once before each such subsequence of executor invocations, with the information produced by the inspector being used by the executor invocations in that subsequence.

A situation of great practical importance where this often arises is within loops. A section which is the body of a loop in the program may have the above property, with the subsequence of executions corresponding to one execution of the loop enclosing it. This type of IE-section is defined as follows.

**Definition:** A section $\beta$ which is the body of a loop L is said to be a loop-body IE-section or an IE/L-section if, for any execution of L, the statement instance dependence graph of $\beta$ is identical in any execution of $\beta$ within that execution of L.

An IE/L-section can be compiled so that the inspector of the section can be moved outside the loop, leaving only the executor of the section within the loop. This method of compilation, which was used in the example in Section 2.1, is called IE/L compilation and is illustrated in Figure 5.

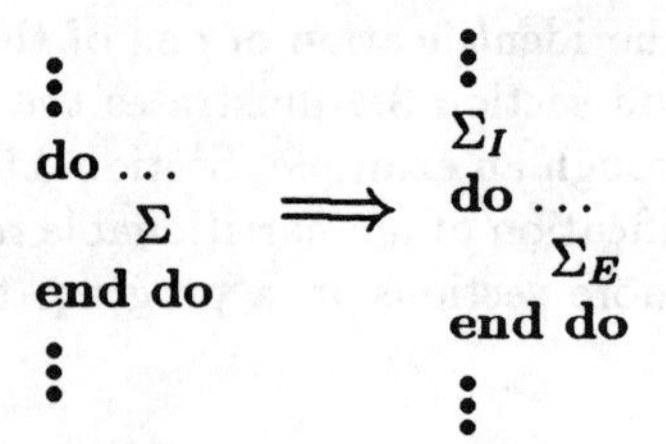


Figure 5: IE/L compilation

## 3.3 Compile-time Identification of IE/L-sections

The framework introduced in Section 3.1 can be used to develop an algorithm for the compile-time testing of whether a given loop body is an IE/L-section.

Let a section $\beta$ be the body of a loop L. A function SI/L($\beta$,v,S), which is closely related to the function SI/C, is used to characterize IE/L-sections. It is defined as follows.

**Definition:** Consider one execution of L, i.e., the execution of the loop body for *all* iterations of that invocation of L. SI/L($\beta$,v,S) is **true** if the sequence of values of v used during the execution instances of statement S remains invariant in the different executions of $\beta$ in *that* execution of L, and is **false** otherwise.

The function SI/L can be used to state Proposition 2′, analogous to Proposition 2 in Section 3.1.

**Proposition 2′:** A loop-body $\beta$ is an IE/L-section if the following conditions are both true:

(a) For the condition P of any control statement S in $\beta$, for every variable v used in P, SI/L($\beta$,v,S) = **true**.

(b) For every non-control statement S in $\beta$, for every variable v used in an array subscript expression in S, SI/L($\beta$,v,S) = **true**.

Compile-time testing of whether a loop-section $\beta$ is an IE/L-section can be performed by computing SI/L($\beta$,v,S) for all pairs (v,S), and then checking the conditions of Proposition 2′. If they are found to be true, it can be asserted that $\beta$ is an IE/L-section. If either of the conditions is found to be false, the compiler can conservatively assume that the section is not an IE/L-section.

Several definitions from the area of compiler design [1] will be used in the following for the computation of SI/L.

**Definition:** A flowgraph G is a triple G = (V,E,s), where (V,E) is a directed graph, s $\in$ V is the initial vertex, and there is a path from s to every other vertex of G.

The control structure of a program can be modeled using a flowgraph, with vertices representing statements, and the edges representing possible control transfers between statements.

**Definition:** A statement S that assigns a value to a scalar variable v or an element of an array variable v is called a definition of v.

**Definition:** A definition S′ of a variable v is said to reach a statement S if there is a path in the flowgraph from S′ to S such that none of the statements on the path (excluding S′ and S) is a definition of v.

**Definition:** A statement $S_2$ is said to be control dependent on a statement $S_1$ if the execution of $S_1$ determines whether or not $S_2$ is to be executed. $S_1$ is said to be a control predecessor of $S_2$.

An algorithm that computes a conservative approximation to SI/L is given in Figure 6. It is conservative in the sense that it may compute a value of **false** when the actual value is **true**. It extends the range of the SI/L function to include a value **dontknow** that is used temporarily.

The basic idea of the algorithm is that the sequence of values of a variable's use in a statement in $\beta$ may be different in different executions of $\beta$ if some values may be defined before the first execution and others during the executions of $\beta$. This is reflected in the initialization of SI/L done in step 1. The value **dontknow** is used to initialize all uses for which it cannot be concluded that the SI/L is **false**. Step 2 repeatedly propagates **false** values of SI/L until the values of SI/L stabilize. If a definition of a variable has a use for which the SI/L value has been set to **false**, that use may have different sequences of values in different iterations of the loop. This means that the sequence of values at any use which this definition may reach may not be identical in different iterations of L. This is why step 2 considers the set RD. If a use in a control predecessor of a statement has its SI/L value set to **false**, the sequence of decisions made at the control statement may be different in different iterations of L, which means the sequence of values of all uses in the controlled statement may not be invariant across different iterations of L just because the number of executions of the statement may be different. Hence the need to look at the set of statements in CP. When the SI/L values do not change, step 3 changes all the **dontknow** values to **true**. This is because any SI/L that is actually **false** will be set to **false** in step 2 by the propagation mechanism.

```
{ β is the body of a loop L }
Step 1
for each variable v used in a statement S in β do
    SI/L(β,v,S) :=
        false, if v has definitions reaching S
                 from both outside and inside L
        dontknow, otherwise

Step 2
repeat
    for each v and S such that SI/L(β,v,S) =
            dontknow do
        RD := set of definitions of v inside L reaching S
        CP := set of control predecessors of S inside L
        SI/L(β,v,S) :=
            false, if any use in any statement in
                   RD or CP has a false SI/L value
until no SI/L value changes

Step 3
for each v and S such that
        SI/L(β,v,S) = dontknow do
    SI/L(β,v,S) := true
```

Figure 6: SI/L computation algorithm

### 3.4 Example

Figure 7 shows an example program used to illustrate the SI/L computation algorithm in Figure 6. Let $\beta_1$ refer to the section which is the body of the outermost **while** loop (call it $L_1$), i.e., $\beta_1$ contains statements 3 thru 19. Let $\beta_2$ refer to the section which is the body of the next inner loop (call it $L_2$), i.e., $\beta_2$ contains the statements 7 thru 18.

Table 1 shows a trace of the execution of the SI/L computation algorithm for $\beta_1$. There is one row for each use of a variable within $\beta_1$. The RD column gives the definitions within $\beta_1$ that reach each use. Similarly, the CP column gives the control predecessors within $\beta_1$ for the statement containing each use. The $s_0$ column shows the initial values of SI/L($\beta_1$,v,S) assigned by step 1 of the algorithm. The use of p in statement 8 has its SI/L value set to **false** (F) because one of the definitions of p reaching it, statement 1, is outside $L_1$ while another reaching definition, statement 19, is inside $L_1$. SI/L($\beta_1$,c,12) is set to **false** for a similar reason. The rest of the uses have their SI/L values set to **dontknow** (D).

The columns $s_1$, $s_2$, and $s_3$ contain the values of SI/L after the first, second, and third iterations of the **repeat** loop in step 2. Assume that the **for** loop in step 2 considers the ($\beta$,v,S) triples in the order that they are given in Table 1. In the first iteration of the **repeat** loop, the value of SI/L($\beta_1$,r,7) remains **dontknow** because the value of SI/L($\beta_1$,q,9)

```
       real b(100,100), c(100), d(100), p, q, r
       integer a(1000), i, j, k

1      read p, a, b, c
2      do while (c(2) .gt. 1.0e-5)
3         r = 1.1
4         i = 1
5         k = 2
6         do while (i .lt. 10)
7            d(100) = 57.3 * r
8            q = arcsin(p)
9            r = q * 3
10           j = 1
11           do while (j .le. i)
12              c(j) = c(j) + b(a(k),j) * q
13              j = j + 2
             end do
14           if (k .ge. 5) then
15              c(1) = 2.0
             else
16              c(100) = 2.0
             end if
17           k = 2 * k
18           i = i + 1
          end do
19        p = p + 0.01
       end do
```

Figure 7: SI/L computation example program

is **dontknow**. Similarly, the value of SI/L($\beta_1$,j,12) remains **dontknow** because all uses in the statements 10, 13, and 11 (RD ∪ CP) have a **dontknow** SI/L value. The value of SI/L($\beta_1$,q,9) becomes **false** because the use p in statement 8 has a **false** SI/L value, and statement 8 is in the RD set of ($\beta_1$,q,9). SI/L($\beta_1$,q,12) is similarly set to **false**. In the second iteration of the **repeat** loop, the **false** value in SI/L($\beta_1$,q,9) propagates to make SI/L($\beta_1$,r,7) **false**. All other **dontknow** values remain unchanged. Finally, since no SI/L value changes in the third iteration of the **repeat** loop, the algorithm proceeds to step 3 where all the **dontknow** values are changed to **true** (T). The new values are shown in the SI/L$_4$ column. This column also shows the values of SI/L which need to be checked according to Proposition 2′ by circling them. It can be seen that all the circled values are **true**, which means that $\beta_1$ is an IE/L-section.

Table 2 illustrates how the SI/L computation algorithm operates when testing the section $\beta_2$. After the initialization of SI/L in step 1, the first iteration

Table 1: Trace of SI/L computation for $\beta_1$

| ($\beta$,v,S) | RD | CP | $s_0$ | $s_1$ | $s_2$ | $s_3$ | $s_4$ |
|---|---|---|---|---|---|---|---|
| ($\beta_1$,i,6) | 4,18 | – | D | D | D | D | (T) |
| ($\beta_1$,r,7) | 3,9 | 6 | D | D | F | F | F |
| ($\beta_1$,p,8) | 19 | 6 | F | F | F | F | F |
| ($\beta_1$,q,9) | 8 | 6 | D | F | F | F | F |
| ($\beta_1$,j,11) | 10,13 | 6 | D | D | D | D | (T) |
| ($\beta_1$,i,11) | 4,18 | 6 | D | D | D | D | (T) |
| ($\beta_1$,j,12) | 10,13 | 11 | D | D | D | D | (T) |
| ($\beta_1$,c,12) | 12,15,16 | 11 | F | F | F | F | F |
| ($\beta_1$,b,12) | – | 11 | D | D | D | D | T |
| ($\beta_1$,a,12) | – | 11 | D | D | D | D | (T) |
| ($\beta_1$,k,12) | 5,17 | 11 | D | D | D | D | (T) |
| ($\beta_1$,q,12) | 8 | 11 | D | F | F | F | F |
| ($\beta_1$,j,13) | 10,13 | 11 | D | D | D | D | T |
| ($\beta_1$,k,14) | 5,17 | 6 | D | D | D | D | (T) |
| ($\beta_1$,k,17) | 5,17 | 6 | D | D | D | D | T |
| ($\beta_1$,i,18) | 4,18 | 6 | D | D | D | D | T |
| ($\beta_1$,p,19) | 19 | – | F | F | F | F | F |

of the **repeat** loop in step 2 makes several **dontknow** values **false** mainly because several statements are control dependent on statement 11 where i is used whose SI/L value SI/L($\beta_2$,i,11) has been set to **false** in step 1. The SI/L value for ($\beta_2$,j,11) becomes **false** in the next iteration of the **repeat** loop, and no changes occur after that. After step 4, the resulting SI/L values are those shown in the SI/L$_4$ column. The circled values in that column are those that need to be all **true** for $\beta_2$ to be an IE/L-section. Therefore, it can be concluded that $\beta_2$ is not an IE/L-section.

### 3.5 Nonpartitionable Regions

Consider the program segment shown in Figure 8 which is part of some loop body being tested for be-

```
      :
   if (c .gt. 5) then
      a(i) = x + y
   else
      a(i) = x - y
   end if
      :
```

Figure 8: Example program segment

Table 2: Trace of SI/L computation for $\beta_2$

| ($\beta$,v,S) | RD | CP | $s_0$ | $s_1$ | $s_2$ | $s_3$ | $s_4$ |
|---|---|---|---|---|---|---|---|
| ($\beta_2$,r,7) | 9 | - | F | F | F | F | F |
| ($\beta_2$,p,8) | - | - | D | D | D | D | T |
| ($\beta_2$,q,9) | 8 | - | D | D | D | D | T |
| ($\beta_2$,j,11) | 10,13 | - | D | D | F | F | (F) |
| ($\beta_2$,i,11) | - | - | F | F | F | F | (F) |
| ($\beta_2$,j,12) | 10,13 | 11 | D | F | F | F | (F) |
| ($\beta_2$,c,12) | 12,15,16 | 11 | F | F | F | F | F |
| ($\beta_2$,b,12) | - | 11 | D | F | F | F | F |
| ($\beta_2$,a,12) | - | 11 | D | F | F | F | (F) |
| ($\beta_2$,k,12) | 17 | 11 | F | F | F | F | (F) |
| ($\beta_2$,q,12) | 8 | 11 | D | F | F | F | F |
| ($\beta_2$,j,13) | 10,13 | 11 | D | F | F | F | F |
| ($\beta_2$,k,14) | 17 | - | F | F | F | F | (F) |
| ($\beta_2$,k,17) | 17 | - | F | F | F | F | F |
| ($\beta_2$,i,18) | - | - | F | F | F | F | F |

ing an IE/L-section. Suppose that the use of c in the **if** statement has been determined to have a **false** SI/L value. In other words, there is a possibility of different decisions being made at the **if** statement in different iterations of the loop. This automatically disqualifies the section from being an IE/L-section. However, it can be seen in this program that the same location will be written into and the same values used in either clause of the **if** statement. The characterization of IE/L-sections presented above requires the invariance of the statement instance dependence graph. The granularity of the smallest unit in the program that may execute on a single processor has been assumed to be at the statement level. If the entire **if** statement is considered as a single unit which executes on the same processor (i.e., a region whose computation cannot be partitioned among the processors), the fact that the use c in the condition has a **false** SI/L-value is not problematic as long as it can be guaranteed that the variable locations written and read by the **if** statement are the same for different iterations of the candidate loop.

Many real applications contain loops whose bodies will not be IE/L-sections. However, with an appropriate specification of nonpartitionable regions, the region instance dependence graph for the body of a loop could become invariant across iterations of the loop, thus allowing IE/L compilation to be used.

# 4 Conclusion

IE compilation, which is a compilation strategy suitable for some programs with irregular, input-dependent computation structures was illustrated using an example. IE-sections, which are sections of a program that are amenable to IE compilation were then considered. Two types of IE-sections, context-free IE-sections and loop-body IE-sections were defined. A framework for the compile-time identification of IE-sections was defined. An algorithm for the compile-time identification of loop-body IE-sections was presented and illustrated using an example. The use of nonpartitionable regions in increasing the number of IE-sections was introduced.

# References


[1] A. Aho, R. Sethi, and J. Ullman, *Compilers. Principles, Techniques, and Tools*, Addison Wesley, 1986.

[2] H. Berryman, J. Saltz, and J. Scroggs, "Execution time support for adaptive scientific algorithms on distributed memory machines," *Concurrency: Practice and Experience*, Vol. 3, pp. 159-178, 1991.

[3] R. Das, et al., "The design and implementation of a parallel unstructured Euler solver using software primitives," ICASE Report No. 92-12, Institute for Computer Applications in Science and Engineering, 1992.

[4] G. Fox, et al., "Fortran D Language Specification," Technical Report Rice COMP TR90-141, Department of Computer Science, Rice University, 1990.

[5] M. Gupta and P. Banerjee, "A methodology for high-level synthesis of communication on multicomputers," *Sixth ACM International Conference on Supercomputing*, 1991.

[6] S. K. S. Gupta, et al., "On the generation of efficient data communication for distributed-memory machines," *Proceedings of the International Computing Symposium*, Vol. 1, pp. 504-513, 1992.

[7] High Performance Fortran Forum, *High Performance Fortran Language Specification*, Version 1.0, Draft, 1993.

[8] S. Hiranandani, K. Kennedy, and C. Tseng, "Compiler support for machine-independent parallel programming in Fortran D," In *Compilers and Runtime Software for Scalable Multiprocessors* (J. Saltz and P. Mehrotra, editors), Elsevier, 1991.

[9] P. Sadayappan and V. Visvanathan, "Circuit simulation on shared-memory multiprocessors," *IEEE Trans. Comput.*, Vol. 37, pp. 1634-1642, 1988.

[10] J. Wu, J. Saltz, H. Berryman, and S. Hiranandani, "Distributed memory compiler design for sparse problems," ICASE Report No. 91-13, Institute for Computer Applications in Science and Engineering, 1991.

# Investigating Properties of Code Transformations †

Deborah Whitfield
Department of Computer Science
Slippery Rock University
Slippery Rock PA 16057

Mary Lou Soffa
Department of Computer Science
University of Pittsburgh
Pittsburgh, PA 15260

Abstract -- *Creating highly optimized code for parallel machines is a difficult task, as the efficiency of the transformed code depends on the types of transformations applied, where they are applied, the order of their application, the architecture and underlying scheduler. This paper describes the utilization of an automatic transformer generator to experimentally investigate properties of transformations, including the cost, expected benefits, application frequency, and interaction frequency. Experimental results for these properties are presented. The properties of transformations are used to develop guidelines for ordering a set of transformations to reveal potential parallelism that exists in code.*

## INTRODUCTION

With the advent of numerous parallel architectures, deciding what code transformations to apply to parallelize source code and the order in which to apply them has become quite complex. Unlike traditional code optimizations, parallelizing transformations must be performed on program code to fully exploit the parallel architecture. The performance of the parallelized code is very much dependent on both the underlying architecture and the scheduler that is used to schedule the parallel events on the processors. Given an architecture, applying a transformation with one scheduler can enhance the performance of the parallel system but would degrade the performance under another scheduler [7]. Also, the order of application may have a significant impact on the performance of the transformed code, as transformations can interact with one another by creating or destroying the potential for further transformations. The cost and benefit of transformations also vary considerably. Thus, for a parallel system, the decisions as to what transformations to apply, where to apply them, and the order in which to apply them are important decisions that greatly impact on the performance of the parallel execution of the program. In order to make such decisions, properties of transformations must be understood. Some of these properties include the frequency that conditions for transformations occur in code, the frequency that transformations enable or disable others, and the cost and benefit of applying transformations.

† Partially supported by National Science Foundation Grant CCR-9109989 to the University of Pittsburgh.

An approach that has been used to address some of these issues is to manually implement a series of transformations in the construction of a parallelizing tool[1,2,6,10]. However, actually implementing a transformer is a time consuming process. Once such a transformer is developed, it is difficult to add or remove transformations. Thus, current systems suffer from having too few transformations implemented, decreasing their value as experimental tools. Another approach that has recently been advocated uses a specification language, Gospel, to specify transformations, and a transformer generator, Genesis, to automatically produce a transformer from specifications [8].

In this paper, we demonstrate a utility of the Gospel/Genesis system in experimentally investigating general characteristics of both traditional and parallelizing transformations. Both types of transformations are included in our study, as code produced for parallel systems can benefit from the application of both types of code transformations and they can affect one another. A set of traditional and parallelizing transformations were specified in Gospel and transformers were automatically produced using Genesis. With these transformers and a set of programs, we experimentally explored the frequency that the potential for applying transformations exists, the frequency that they interact with one another, and the cost and expected benefit of the transformations. From these experiments we identified transformations that are frequently applicable, frequently interact with other transformations, and have an extreme cost/benefit ratio. Using properties identified from the experimentation and from formally determinable properties, we developed a set of guidelines for application orders for the selected transformations.

## METHODOLOGY

The methodology used in this work to explore transformation properties involves using the specification technique of Gospel (1) to formally identify interactions that can potentially occur between transformations, and (2) to specify a set of transformations that are input to Genesis to automatically produce transformers that implement the transformations. These transformers are used to experimentally investigate properties of interest. We first present a brief overview of the Gospel/Genesis system[8,9].

In order to specify a transformation in Gospel, preconditions are first specified that describe the textual and global data dependencies that must be present in order to correctly apply a transformation*. The actual code modifications are specified using a set of primitive actions that move, modify, add and delete code. A set of post conditions that describes the conditions that existed after application of a transformation is also part of the specification. The pre- and post conditions are used to formally determine the interactions that can potentially occur between two transformations.

The pre-condition and action specifications in Gospel are used by Genesis to automatically generate transformers. Genesis produces transformers that employ computed data dependences and use an extended intermediate representation of a source program that maintains loop control structures and array references from the source program. Thus, the system is source code independent and can be used for any language that can be represented by the intermediate code format (e.g., FORTRAN, Pascal, C). The design of Genesis allows the user to specify whether the transformation should be applied at all valid program points or should be applied under the user's direction. The extended intermediate code allows the user to interact at the source level for loop transformations that are typically applied for parallel systems. The Genesis tool permits the experimental investigation of the cost and expected benefit of transformations by adding code that approximates the cost of applying transformations and their potential benefits. Such information can be used to determine the merits of including the particular transformation in a parallelizing compiler.

In our experiments, ten source programs were used to investigate eleven specified transformations**. Transformers were produced for Constant Folding (CFO), Copy Propagation (CPP), Constant Propagation (CTP), Dead Code Elimination (DCE), Invariant Code Motion (ICM), Loop Interchanging (INX), Loop Unrolling (LUR), Loop Fusion (FUS), Loop Circulation (CRC), Bumping (BMP), and Parallelization (PAR) [5,10].

## FREQUENCY AND INTERACTION

The first sets of experiments investigated the frequency that the potential for applying transformations occurs and the frequency with which the transformations interact. Interactions may occur by one transformation enabling the application of another transformation that previously could not be applied, (i.e., creating a potential application point) or one transformation disabling a transformation that previously was applicable (i.e., invalidating conditions that existed for another transformation). In order to determine the frequency of interactions, the potential for transformations to interact was used to guide the experimentation[9]. This information was obtained from formal studies in which proofs were developed showing the effect of an application of a transformation on another transformation.

Experimenting with applications of the transformations reveals that theoretically possible interactions do occur. The applicability of the 11 transformations was first established for each program. Next, one transformation was applied and the applicability of transformations was again established, showing the enabling and disabling of the transformation. Table 1 displays the interactions that occur, where the Applied (App) column displays the number of application points for each transformation that was found in 10 programs. The numbers of enabling and disabling interactions that were found are shown as enabling/disabling; a 0 indicates that no interactions were found even though theoretically they occur, whereas a dash indicates that no interaction occurred because it is theoretically impossible. For example, the entry for the [LUR, FUS] cell displays "1/7" which means that of the 49 application points found for LUR, only one instance of FUS was enabled, but 7 instances were disabled after LUR was applied. Since CPP and ICM† did not did not occur frequently in the ten source programs, they were eliminated from the table.

Table 1: Enabling and Disabling Interactions

| | App | DCE | CTP | CFO | LUR | FUS | INX |
|---|---|---|---|---|---|---|---|
| DCE | 34 | 0/- | -/2 | -/- | -/- | 0/- | 0/- |
| CTP | 97 | 13/- | -/- | 5/- | 41/- | 0/- | 0/- |
| CFO | 5 | -/- | 41/- | -/- | 0/- | 0/- | -/- |
| LUR | 49 | 0/- | 10/- | 0/- | /- | 1/7 | 2/6 |
| FUS | 11 | -/- | -/5 | -/- | -/1 | 1/0 | 0/6 |
| INX | 13 | -/- | -/- | -/- | -/- | 5/4 | 4/0 |

It is important that if one transformation is of more value than another, that the correct order be used to ensure the application of the more important transformation. These interaction frequencies are used later in order to include the act of enabling and disabling transformations as part of the benefit measure.

*As data flow conditions used in traditional optimizations can be expressed using data dependency notation, only data dependency conditions are used.

**A total of 21 transformations have been specified and implemented using Gospel and Genesis.

† It should be noted that ICM does not occur frequently in our experiments due to our extended form of intermediate code that does not include an address calculation.

## COST AND EXPECTED BENEFITS

The transformers were also utilized to experimentally investigate the cost and benefits of applying transformations. The code of the transformers produced by Genesis was enhanced to include a cost and benefit estimate of the implemented transformations. An estimate of the cost of a transformation is based on the number of statement elements and data dependences that are accessed.

All cost assessments are considered equal in the amount of work performed. These relative costs were verified using actual execution timings. The cost and average timing of the applied transformations in terms of microseconds are given as Cost/Timing for several transformations: CTP: 11/1.0, DCE: 12/1.8, CFO: 27/2.6, LUR: 23/3.8, INX: 54/17.2, CPP: 434/65.4, FUS: 485/81.4. By applying a curve fitting algorithm to determine how close the data matches the theoretical curve, the coefficient of regression results in 0.968 for a straight line indicating that the cost is linear to the machine timings. Hence, the cost assignments are an accurate measure of the expense involved in finding and applying transformations, and these estimated costs may be used to determine the relative cost of a transformation.

Benefit measures are based on the amount of execution time a transformation potentially saves. For the traditional transformations (Constant Folding, Constant Propagation, Copy Propagation, Dead Code Elimination, and Invariant Code Motion), the benefit is typically one unit for the removal of an operation. If the particular statement being optimized is within a loop(s), then the benefit is multiplied by the number of times the statement is expected to execute.

Experimentation was performed to determine the cost and expected benefit of several transformations. The cost/benefit ratios for several transformations follow, where the cost is given over the benefit: CFO: 18/1, CTP: 13/1, CPP: 68/1, DCE: 142/315, LUR: 23/251, FUS: 811/64, INX: 75/232. There are a number of uses of this information, including deciding whether a transformation should be included in a parallelizing compiler, and determining methods to improve the efficiency of a transformer by examining the expensive transformations for alternative methods of specification.

Another use of cost benefit information is to enumerate those transformation pairs with high interaction frequency (Table 1) and add or subtract the benefit to reflect the enabling or disabling interaction, respectively. This adjusted benefit is used for ordering transformations.

The cost/benefit information is also useful to change the definition of a transformation so that it is less expensive but may not find all application points. One specification of CTP handles multiple definitions of a variable and propagates the constant to the use if all of the constant definitions that reach the use are the same. However, an alternative specification of CTP allows only one definition to reach the use. Obviously, the second specification requires less checking and is less costly. Our experiments show that all occurrences of constant propagation had one definition that reached the use. Thus, the specification for CTP should be simplified for only one definition. Also, the cost/benefit measure can determine those transformations that do not merit application as they are expensive and their expected benefits are low.

## ORDERING TRANSFORMATIONS

Using Gospel and the Genesis tool, a number of different approaches can be used to develop guidelines for ordering transformations. Two guidelines for ordering transformations based on the intended use of the transformer follow: 1) Order the transformations based on only the cost/benefit measures. These guidelines would produce an efficient transformer that may be used in environments where compilation time is a factor, 2) Use the interaction of transformations to develop guidelines for parallelizing code. Instead of trying all possible combinations, these guidelines would aid the user in finding the potential parallelism in code by indicating what transformations might be enabled or disabled as a result of an application of another transformation.

### Using Cost and Benefit to Order Transformations

The cost and expected benefit measures determined in the experimentation were first used to develop guidelines for ordering transformations. These measures can be used to produce 3 different orderings as given in Table 2. The first ordering is based strictly on the cost and the second is computed using cost/benefit. The third ordering uses adjusted benefits for the transformations in the computation of the cost/benefit ratio. The adjusted benefit of FUS is -32 as FUS disables other transformations and thus the cost/benefit is not defined.

Table 2: Three Orderings

| Cost | | Cost/Benefit | | Cost/Adjusted Benefit | |
|---|---|---|---|---|---|
| CTP | 12.8 | LUR | 0.10 | CTP | 0.10 |
| CFO | 17.6 | INX | 0.33 | LUR | 0.12 |
| LUR | 23 | DCE | 0.45 | INX | 0.25 |
| CPP | 68.2 | FUS | 12.6 | DCE | 0.45 |
| INX | 74.8 | CTP | 12.8 | CFO | 1.9 |
| DCE | 142 | CFO | 17.6 | CPP | 68.2 |
| FUS | 811 | CPP | 68.2 | FUS | undef |

The most accurate ordering is given in the third column where all factors are taken into consideration (i.e., benefits include the enabling and disabling of transformations). This column also suggests that relative to the other transformations FUS and CPP are magnitudes more costly and hence serious consideration should be given before deciding to perform these transformations. Of course, for a given ordering the adjusted benefit may change (i.e., if FUS is ordered after INX, then the cost/benefit ratio would remain 12.6). By simply examining the cost versus the benefit of a transformations, several suggestions can be made to enhance the performance of the transformer. In most cases, the implementation of CPP is too costly, but the implementation of INX is well-worth the effort. Also, the possible benefits of FUS for the particular architecture need to be examined before drawing conclusions as to its merit.

Using Interactions to Order Transformations

Information about the interaction of transformations can also be used to develop ordering guidelines. For traditional optimizations, applying as many optimizations as possible typically results in more efficient code. For parallel transformations, the more opportunities for applying transformations, the better chance of exploiting parallelism. For both of these reasons, the criterion used for developing guidelines for ordering transformations is based on finding as many potential application points as possible. This allows the user to find the best code. Thus, our ordering guidelines are based on two rules: 1. If transformation A can enable transformation B, then order A before B -- <A, B>. 2. If transformation A can disable transformation B, then order A after B -- <B, A>. However, using these two rules to develop guidelines introduces some conflicts for which decisions need to be made before a complete ordering can be determined.

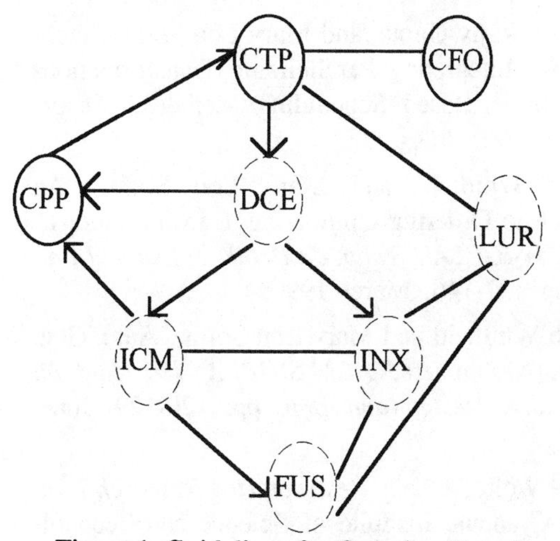

Figure 1. Guidelines for Ordering Transformations

Figure 1 shows the interactions among transformations. Some interactions may create conflicts, disabling the establishment of a definite ordering. Notice that the graph is cyclic as the conflicts that arise are not resolved. The arrows between the transformations represent the order in which the transformations should be applied; the lines without an arrow indicate bi-directional lines. The dashed circles (i.e., DCE, ICM, INX, FUS, and LUR) indicate that the transformation should be applied repeatedly (i.e., it enables itself).

In order to resolve the conflicts, we use several criteria that incorporate experimental results to develop guidelines for the order in which to apply transformations: 1. The frequency of enabling and disabling interactions that occur. 2. The cost of the transformations versus the benefit. 3. The effect a transformation has on the cost and benefit of other transformations.

An Example of Applying the Ordering Guidelines

The formal results, based only on the interactive capability of transformations shown in Figure 1, are coupled with the above three criteria to develop guidelines for ordering the application of transformations. These criteria were used jointly to resolve the conflicts: (CPP, ICM), (ICM, FUS), (INX, ICM), (LUR, FUS), (INX, FUS), (INX, LU R), (CTP, CFO), (CPP, CTP, DCE), and (CTP, LUR).

The first criterion utilizes the interaction frequency data that appears in Table 1 to help resolve some of the conflicts. The conflict between LUR and FUS occurs because LUR both enables and disables FUS. In practice the disabling does occur; however the enabling case was not found. The application of <LUR,FUS> produced six application points, whereas the reverse order produced seven. As these results suggest that there is not much difference between the orderings, we order LUR after FUS to allow for the maximum number of applications. Similarly, the resolution of (INX,FUS) involves examining four interactions. The application order <INX, FUS> produced 11 application points whereas the application order <FUS, INX> produced four. Hence, the application of FUS should follow the application of INX. Also, the conflict between INX and LUR is resolved using this method as <INX, LUR>. Hence, the three transformations are ordered in this example as <INX, FUS, LUR>. Likewise, the conflict between CTP and LUR is resolved as <CTP, LUR> since CTP enables LUR 41 times. This ordering would permit the user to find more potential parallelism. The conflict between CFO and CTP is not easily resolved even though CFO enables CTP more often than CTP enables CFO. In fact, there were many occurrences of CTP enabling CFO which enables CTP which enables LUR. Thus, the two transformations should be iteratively applied to reveal potential parallelism to the user. As far as the three-way conflict (CTP,CPP,DCE) is concerned, these interactions did not occur in the experimentation. Hence, the order <CTP,DCE,CPP> is sufficient.

The second criterion utilizes the cost and expected benefit of the transformations to resolve conflicts. When a decision needs to be made between two transformations that interact in such a way as to decrease the number of application points, then the more cost-effective transformations should be favored. This technique is used to resolve three of the conflicts: (CTP, CPP, DCE), (INX, FUS), (LUR, FUS). Since CPP is not a cost-effective transformation, it is applied last in the three-way conflict (i.e., <CTP, DCE, CPP>). Similarly, FUS is not a cost-effective transformation so in this example it is applied after both INX and LUR. The cost benefit ratios of LUR and INX are too close to determine a complete ordering between them.

A particular ordering of transformations may increase or decrease the cost of a particular transformation as compared to another ordering. For example, in one experiment the order <LUR,FUS> was applied, and in another experiment <FUS,LUR> was applied. The total cost of applying LUR before FUS was 94, whereas the application of LUR after FUS produced a cost of 71 for LUR. An interesting situation arises between LUR and INX. INX may reduce both the cost and the benefit of LUR as seen in the experimentation. The reduction in cost results if the loop to be unrolled is an outer loop of two tightly nested loops and is moved inside via interchanging. Thus, the interchanged loop structure is not copied. Similarly, a reduction in benefit occurs since fewer statements have been unrolled.

Applying the above criteria results in several decisions. The first decision is that ICM need not be applied as its application is not effective at the extended intermediate code level. Secondly, the conflicts between FUS, LUR, and INX should be resolved by applying INX, followed by LUR, followed by FUS. This ordering was chosen due to the extreme cost of applying FUS and the non-existent expected benefit. If these transformations were applied on an architecture that made FUS more beneficial, then this ordering would be reconsidered. Thirdly, the conflicts between CFO, CTP and LUR should be resolved by iterating the application of CTP and CFO and then applying LUR. These resolutions to the conflicts produce the optimization order: Iterate(CTP, CFO), DCE, CPP, INX, LUR, FUS

## CONCLUSION

Although automatic generation has been used in the development of peephole optimizers for compilers[3,4], the automatic generation of code transformations has not been exploited in parallelizing compilers. We demonstrated the utility of such a tool in experimenting with transformations and show that Genesis is a valuable tool that allows the user to experimentally determine properties of transformations that can be used to improve the performance of the tools for parallelizing programs. The experimental results presented are not definitive but should be used as an indication of the properties of the transformations considered. Although the experiments in this work were limited to a particular set of transformations, the tool can be used to investigate the same properties of other transformations as well as other properties.

## REFERENCES

[1] F.E. Allen, et al, "A Framework for Determining Useful Parallelism," *Proceedings of Supercomputing '88*, pp. 207-215, St. Malo, France, Feb. 1988.

[2] Vasanth Balasundaram, et al, "The ParaScope Editor," *Proceedings of Supercomputing '89*, pp. 540-549, Reno, Nevada.

[3] Christopher W. Fraser and Alan L. Wendt, "Automatic Generation of Fast Optimizing Code Generators," *SIGPLAN '88 Conference on Prog. Lang. Design and Imp.*, pp. 79-84, June 1988.

[4] Robert Giegerich, "Auto. Gen. of Machine Specific Code Optimizer," *9th ACM Symp. on Princ. of Prog. Lang.* pp. 75-81 Jan. 1982.

[5] David A. Padua and Michael J. Wolfe, "Advanced Compiler Optimizations for Supercomputers," *Communications of the ACM* vol. 29, no. 12, pp. 1184-1201, Dec. 1986.

[6] D. Polychronopoulos et al, "Parafrase-2," *1989 International Conf. on Par. Proc.* pp. 39-48, St. Charles, Illinois, Aug. 1989.

[7] Tia Watts, Rajiv Gupta, and Mary Lou Soffa, "Techniques for Integrating Parallelizing Transformations and Compiler-Based Scheduling Methods," *Supercomputing '92*, Nov. 1992.

[8] Deborah Whitfield and Mary Lou Soffa, "An Approach to Ordering Optimizing Transformations," *2nd ACM SIGPLAN Symp. on Princ. & Prac. of Par. Prog.* pp. 137-146, March 1990.

[9] Deborah Whitfield and Mary Lou Soffa, "Auto. Gen. of Global Optimizers," *ACM SIGPLAN '91 Conf. on Prog. Lang. Design and Imp.* pp. 120-129, June, 1991.

[10] Michael Wolfe, *A Loop Restructuring Research Tool*, Oregon Graduate Institute of Science and Technology, 1989.

# Compiling Efficient Programs for Tightly-Coupled Distributed Memory Computers*

PeiZong Lee, Institute of Information Science, Academia Sinica, Taipei, Taiwan, R.O.C.
Tzung-Bow Tsai, Dept. of EE, Chung-Cheng Univ., Taiwan, R.O.C.

## Abstract

In this paper, we present a systematic method for compiling programs on distributed memory parallel computers. First, we derive a dynamic programming algorithm for data distribution. Second, we use data-dependence information for pipelining data. Jacobi's iterative algorithm and the Gauss elimination algorithm for linear systems are used to illustrate our method.

## Introduction

It is our goal in this paper to present a systematic method for compiling programs on distributed memory parallel computers. First, we want to derive a dynamic programming algorithm for data distribution. Previously, Li and Chen [7], Gupta and Banerjee [3] formulated the component alignment problem from the whole source program. The data distribution schema they derived may result in a larger communication overhead. Unlike their methods, we deal with each nested Do-loop independently. Data distribution schema between two nested Do-loops may be different and may require some data communication between them. A dynamic programming algorithm can compute the minimum cost order of data distribution schema for executing a sequence of nested Do-loops in distributed memory computers.

Second, we want to use data-dependence information for pipelining data. In many cases, the method of using the component alignment algorithm for distributing data is not sufficient to provide enough information for generating efficient communication operations. However, if compilers can detect all data-dependence vectors, it is possible to distribute data according to the iterative space. It also allows compilers to generate efficient communication codes for pipelining data.

## Background

In this paper, we are concerned with distributed memory systems. The abstract target machine we adopt is a $q$-D grid of $N_1 \times N_2 \times \cdots \times N_q$ processors, where D stands for dimensional and $q$ is less than or equal to the deepest level of the Do-loop program. A processor on the $q$-D grid is represented by the tuple $(p_1, p_2, \ldots, p_q)$, where $0 \le p_i \le N_i - 1$ for $1 \le i \le q$. Such a topology can be embedded into almost any distributed memory system. For example, the $q$-D grid can be embedded into a hypercube computer using a binary reflected Gray code.

The parallel program generated from a sequential program for a grid corresponds to the SPMD (Single Program Multiple Data) model, in which each processor executes the same program but operates on distinct data items [1, 3, 4, 5, 8]. More precisely, in general, a source program has sequential parts (which must be executed sequentially) and concurrent parts (which can be executed concurrently). Each processor will execute the sequential parts individually; while all processors will execute the concurrent parts altogether by using message passing communication primitives. In practice, scalar variables and small data arrays used in the program are replicated on all processors in order to reduce communication costs; while large data arrays are partitioned and distributed among processors.

Gupta and Banerjee [3] proposed a data distribution function, which can map each array dimension to a unique dimension of the processor grid. In addition, the data distribution function can specify the method of partition to be "contiguous" or "cyclic" or "contiguous-cyclic". Lee and Tsai [6] have generalized data distribution functions for 1-D and 2-D data arrays by additionally allowing distributed data to be indexed increasingly or decreasingly, and allowing the distributions of different data dimensions to be dependent or independent.

Li and Chen noticed that a data distribution scheme must be given before analyzing communication costs [8]. However, to examine whether a data distribution scheme is good or not really depends on which commu-

*This work was partially supported by the NSC under Grant NSC 81-0408-E-001-505 and NSC 82-0408-E-001-016.

| Primitive | Cost on Hypercube |
|---|---|
| Transfer($m$) | $O(m)$ |
| Shift($m$) | $O(m)$ |
| OneToManyMulticast($m$, $seq$) | $O(m * \log num(seq))$ |
| Reduction($m$, $seq$) | $O(m * \log num(seq))$ |
| AffineTransform($m$, $seq$) | $O(m * \log num(seq))$ |
| Scatter($m$, $seq$) | $O(m * num(seq))$ |
| Gather($m$, $seq$) | $O(m * num(seq))$ |
| ManyToManyMulticast($m$, $seq$) | $O(m * num(seq))$ |

Table 1: Costs of communication primitives.

nication primitives are involved. Gupta and Banerjee suggested the following two steps to break this cyclical dependency. First, assume $N_1 = N_2 = \cdots = N_q$, then a data distribution scheme can be determined by using the component-alignment algorithm. Second, formulate the total execution time including the computation time and the communication time; then the values of $N_1$, $N_2$, ..., $N_q$ can be determined by requiring the optimal total execution time.

Table 1 shows the costs of some interesting communication primitives on the hypercube computer which are borrowed from [2] and [8]. The parameter $m$ denotes the message size in words, $seq$ is a sequence of identifiers representing the processors in various dimensions over which the collective communication primitive is carried out. The function $num$ applied to such a sequence simply returns the total number of processors involved.

## Distributing Data Using the Component Alignment Algorithm

In this section, we show an example of how to distribute data arrays using the component alignment algorithm. Given a program, we first construct a component affinity graph from the source program [7]. It is a directed weighted graph, whose nodes represent dimensions (components) of arrays and whose edges specify affinity relations between nodes. Two dimensions of arrays are said to have an affinity relation, if the difference of the two subscripts of these two dimensions is a constant value. The weight with an edge is equal to the communication cost and is necessary if two dimensions of arrays are distributed along different dimensions of the processor grid. The direction of an edge specifies the direction of the data communication according to the owner computes rule.

The component alignment problem is defined as partitioning the node set of the component affinity graph into $q$ ($q$ is the dimension of the abstract target grid) disjointed subsets so that the total weight of edges across nodes in different subsets is minimized, with the restriction that no two nodes corresponding to the same array are in the same subset [3] [7]. Although the component alignment problem is NP-complete, Li and Chen have proposed an efficient heuristic algorithm based on applying the optimal matching procedure to a bipartite graph constructed from the nodes in the two data arrays [7]. In this paper, we assume that the abstract target grid is 2-dimensional, therefore, $q$ is equal to 2.

Consider the following Jacobi's iterative algorithm for linear systems $A_{m \times m} X_m = B_m$.

```
1  DO 10 k = 1, MAX_ITERATION
2    DO 6 i = 1, m
3      V(i) = 0.0
4      DO 6 j = 1, m
5        V(i) = V(i) + A(i,j) * X(j)
6    CONTINUE
7    DO 9 i = 1, m
8      X(i) = X(i) + (B(i) - V(i)) / A(i,i)
9    CONTINUE
10 CONTINUE
```

There is an iterative loop from line 1 to line 10, whose body is from line 2 to line 9. Fig. 1 shows the corresponding component affinity graph. The edge weights of the graph are as follows.

$$\begin{aligned} c_1 &= \text{ManyToManyMulticast}(\tfrac{m^2}{N}, N) \text{ (line 5)} \\ c_2 &= \text{ManyToManyMulticast}(\tfrac{m}{N_1}, N_1) \\ &\quad + \text{OneToManyMulticast}(m, N_2) \text{ (line 5)} \\ c_3 &= N_1 * \text{OneToManyMulticast}(\tfrac{m}{N_1}, N_2) \text{ (line 8)} \\ c_4 &= N_1 * \text{OneToManyMulticast}(\tfrac{m}{N_1}, N_2) \text{ (line 8)} \end{aligned}$$

Along with each term, we indicate the line number in the program to which the affinity relation appears. The data size for array $A$ is $m^2$, the data size for arrays $V$, $B$, and $X$ each is $m$. The total number of processors is denoted by $N$, while $N_1$ and $N_2$ refer to the number of processors along which various array dimensions are initially assumed to be distributed. Note that $c_2$ is greater than $c_4$. Therefore, applying the component alignment algorithm on this graph, we get the following disjointed sets of dimensions: set 1 includes $A_1$ and $V$; set 2 includes $A_2$, $B$, and $X$. These two sets are mapped to dimensions 1 and 2, respectively, of the processor grid.

Next, we determine the partition strategy. As the iteration space is rectangular, the partition strategies for all array dimensions are "contiguous".

We now determine the value of $N_1$ and $N_2$. The total execution time including both the computation time and the communication time of an iteration from line 2 to line 9 is formulated as follows.

$$\begin{aligned} \text{Time} &= 2 * \tfrac{m^2}{N_1 * N_2} * t_f + \text{Reduction}(\tfrac{m}{N_1}, N_2) \text{ (line 5)} \\ &+ \ 3 * \tfrac{m}{N_2} * t_f + N_1 * \text{OneToManyMulticast}(\tfrac{m}{N_1}, N_2) \end{aligned}$$

(or $N_1 *$ Transfer$(\frac{m}{N_1})$ if $N_2 = 1$) (line 8)
+ OneToManyMulticast$(\frac{m}{N_2}, N_1)$ (line 5 and line 8) (communication cost because of the loop-carried dependence of $X$).

We assume that the average time of a floating point operation is $t_f$ and the average time of transferring a word is $t_c$. The optimal execution time can be obtained by substituting all possible $N_1$ and $N_2$ into the formula, where $N = N_1 * N_2$. Table 2 shows the execution time on three processor grids. The costs of communication primitives are based on Table 1. The case when $N_1 = 1$ and $N_2 = N$ is better than the other two cases for the computation time. However, this distribution scheme cannot be satisfied, as it requires more communication time than the other two cases. Therefore, we will show a better one in the next section.

## A Dynamic Programming Algorithm for Data Distribution

Consider Jacobi's iterative algorithm again. The body of the iterative loop contains two Do-loops: $L_1$ and $L_2$. $L_1$ is from line 2 to line 6; $L_2$ is from line 7 to line 9. Suppose that we compute $L_1$ and $L_2$ separately using the component alignment algorithm. Then the total execution time for computing an iteration should include not only the execution time for $L_1$ and $L_2$ but also the communication time for changing data layouts for $L_1$ to $L_2$ and the communication time because of the loop-carried dependence, see Fig. 2.

Fig. 3 shows the component alignment for $L_1$ and $L_2$. In $L_1$, $A_1$ and $V$ and $B$ are mapped to dimension 1 of the processor grid; $A_2$ and $X$ are mapped to dimension 2. In $L_2$, $A_1$ and $V$ and $B$ and $X$ are mapped to dimension 1; $A_2$ is mapped to dimension 2. Suppose that the execution time for $L_1$ is $\text{Time}_1$ and for $L_2$ it is $\text{Time}_2$; the communication time for changing data layouts for $L_1$ to $L_2$ is $\text{CTime}_1$ and the communication time because of the loop-carried dependence is $\text{CTime}_2$. Then,

$$\begin{aligned}
\text{Time}_1 &= 2 * \frac{m^2}{N_1 * N_2} * t_f \\
&\quad + \text{Reduction}(\tfrac{m}{N_1}, N_2) \text{ (line 5)} \\
\text{Time}_2 &= 3 * \tfrac{m}{N_1} * t_f \text{ (line 8)} \\
\text{CTime}_1 &= 0 \text{ (in this algorithm)} \\
\text{CTime}_2 &= \text{ManyToManyMulticast}(\tfrac{m}{N_1}, N_1) \\
&\quad + \text{OneToManyMulticast}(m, N_2) \\
&\quad \text{(loop-carried dependence of } X \\
&\quad \text{from line 8 to line 5).}
\end{aligned}$$

Note that, it is not necessary for sending or receiving data among processors for changing data layouts for $L_1$ to $L_2$ in this algorithm, because array $X$ is not modified outside $L_2$ and array $X$ will be immediately modified in $L_2$. The optimal execution time for $L_1$ is $2 * \frac{m^2}{N} * t_f$ when $N_1 = N$ and $N_2 = 1$. The optimal execution time for $L_2$ is $3 * \frac{m}{N} * t_f$ when $N_1 = N$ and $N_2 = 1$. The communication cost for updating array $X$ because of the loop-carried dependence is $m * t_c$. Therefore, the total execution time for computing an iteration is $(2 * \frac{m^2}{N} + 3 * \frac{m}{N}) * t_f + m * t_c$, which is better than the one using a different data distribution scheme in the last section.

Since this is our preferred implementation, we briefly describe it in below. Table 3 shows the data layouts of the corresponding parallel Jacobi's iterative algorithm for implementing linear systems $A_{4\times4} X_4 = B_4$ on a four-processor linear array. The $i$th row of data array $A$ and the $i$th elements of data arrays $V$, $B$, and $X$ are stored in processor $i - 1$. $V(i)$ and $X(i)$ are computed in processor $i - 1$. After computing a new version of data array $X$, it broadcasts to all processors for further computation.

In general, a program contains $s$ Do-loops or an iterative loop contains $s$ Do-loops, and the problem of finding the optimal execution time or the minimum cost order of data distribution schema can be obtained by the following dynamic programming algorithm. Let $L_1, L_2, \ldots, L_s$ be $s$ Do-loops in sequence in the program. Let $M_{i,j}$ be the cost of computing the sequence of loops $L_i, L_{i+1}, \ldots, L_{i+j-1}$ using the component-alignment algorithm, and $P_{i,j}$ be the distribution scheme, for $1 \le i \le s$ and $1 \le j \le s - i + 1$. Define $T_{i,j}$ to be the minimum cost of computing the sequence of loops $L_1, L_2, \ldots, L_{i+j-1}$ with the restriction that the final data distribution scheme after computing $T_{i,j}$ is $P_{i,j}$. Clearly, $T_{1,i}$ is equal to $M_{1,i}$.

**Algorithm 1.** A dynamic programming algorithm for computing the minimum cost order of data distribution schema of executing a sequence of $s$ Do-loops on the distributed memory computer.

Input: $M_{i,j}$ and $P_{i,j}$, where $1 \le i \le s$ and $1 \le j \le s - i + 1$.

Output: The minimum cost of executing $s$ Do-loops on the distributed memory computer.

1. **for** $i := 2$ to $s$ **do**
2. **for** $j := 1$ to $s - i + 1$ **do**
3. $T_{i,j} := \text{MIN}_{1 \le k < i}(T_{i-k,k} + M_{i,j} + cost(P_{i-k,k}, P_{i,j}))$ ;
4. $Minimum\_Cost := \text{MIN}_{1 \le k \le s}(T_{s-k+1,k} + loop\_carried\_dependence(T_{s-k+1,k}))$ .

Note that, $cost(P_{i-k,k}, P_{i,j})$ returns the communication cost of changing data layouts from $P_{i-k,k}$ to $P_{i,j}$. $loop\_carried\_dependence(T_{s-k+1,k})$ returns the communication cost incurred by the loop-carried dependence, if a sequence of distribution schema is used for computing $T_{s-k+1,k}$.

## Using Data-Dependence Information for Pipelining Data

Consider the following Gauss elimination algorithm for linear systems $A_{m \times m} X_m = B_m$.

```
1     {* Matrix triangularization.  *}
2 DO 8 k = 1, m
3    DO 8 i = k + 1, m
4       L(i,k) = A(i,k) / A(k,k)
5       B(i) = B(i) - L(i,k) * B(k)
6       DO 8 j = k + 1, m
7          A(i,j) = A(i,j) - L(i,k) * A(k,j)
8 CONTINUE
9     {* Triangular linear system UX = Y. *}
10 DO 12 i = m, 1, -1
11    V(i) = 0.0
12 CONTINUE
13 DO 17 j = m, 1, -1
14    X(j) = (B(j) - V(j)) / A(j,j)
15    DO 17 i = j - 1, 1, -1
16       V(i) = V(i) + A(i,j) * X(j)
17 CONTINUE
```

Fig. 4 shows the corresponding component affinity graph and the suggested component alignment. Although the program fragment from line 2 to line 8 prefers using a 2-D processor grid, the program fragment from line 13 to line 17 prefers to use a processor ring. In order to achieve a better load balance among processors, a processor ring is used. In addition, data arrays are partitioned along the first dimension. Because the index space includes an oblique pyramid and a triangle, "cyclical" data distribution schema will be used.

A naive compiler may generate a lot of OneToMany-Multicast operations for broadcasting $B(k)$ in line 5, $A(k,j)$ in line 7, and $X(j)$ in line 16 to all processors in the ring for each distinct $k$ and $j$. It will certainly incurs excessive communication overhead.

In effect, data communication for $B(k)$, $A(k,j)$, and $X(j)$ is due to the loop-carried dependence. A better method for generating efficient codes relies on the data-dependence information of each data token. For instance, line 5 is in the body of a two-nested loop, token $B(k)$ was generated in index $(k-1,k)^t$ and is used in indices $(k,0)^t + i(0,1)^t$, for all $k+1 \leq i \leq m$. The data-dependence vector corresponding to token $B(k)$ is $(0,1)^t$. As we want to map index $(k,i)^t$ to be executed in the virtual processor $i$, the index-processor mapping is $(0,1)$. $(0,1)$ will map the data-dependence vector $(0,1)^t$ to 1, which means that $B(k)$ will be used in the neighboring processor in the next consecutive step. Therefore, $B(k)$ can be arranged by pipelining to the neighboring processor instead of broadcasting to the neighboring processor. The detailed data-dependence information of each data token and the suggested index-processor mapping can be seen in Table 4. Fig. 5 shows the parallel program in a processor. In the program, OneToManyMulticast operations are substituted by Shift operations (send and receive operations).

## Conclusions

We have developed a dynamic programming algorithm for distributing data which can compute the minimum cost order of data distribution schema for executing a sequence of nested Do-loops in distributed memory computers. Then, we considered the improvement of communication time by pipelining data. Especially, the data-dependence information, which can be used to map iterations to be executed in specific processors, also can provide enough information for pipelining data.

## Acknowledgements

The authors would like to thank the anonymous referees for their valuable comments on this paper.

# References

[1] V. Balasundaram, G. Fox, K. Kennedy, and U. Kremer. A static performance estimator to guide data partitioning decisions. In *PPoPP*, pages 213–223, Williamsburg, VA, Apr. 1991. ACM.

[2] M. Gupta and P. Banerjee. Compile-time estimation of communication costs on multicomputers. In *IPPS*, pages 179–193, Beverly Hills, CA, Mar. 1992. IEEE.

[3] M. Gupta and P. Banerjee. Demonstration of automatic data partitioning techniques for parallelizing compilers on multicomputers. *IEEE TPDS*, 3(2):179–193, Mar. 1992.

[4] S. Hiranandani, K. Kennedy, and C-W. Tseng. Compiling Fortran D for MIMD distributed-memory machines. *Communications of the ACM*, 35(8):66–80, Aug. 1992.

[5] K. Ikudome, G. C. Fox, A. Kolawa, and J. W. Flower. An automatic and symbolic parallelization system for distributed memory parallel computers. In *Fifth Distributed Memory Comput. Conf.*, pages 1105–1114, Charleston, SC, Apr. 1990. IEEE.

[6] P.-Z. Lee and T. B. Tsai. Compiling efficient programs for tightly-coupled distributed memory computers. Technical Report TR-93-004, Institute of Information Science, Academia Sinica, April 1993.

[7] J. Li and M. Chen. Index domain alignment: Minimizing cost of cross-referencing between distributed arrays. In *Frontiers90: 3rd Symp. Frontiers Massively Parallel Comput.*, pages 424–433, College Park, MD, Oct. 1990. IEEE.

[8] J. Li and M. Chen. Compiling communication-efficient problems for massively parallel machines. *IEEE TPDS*, 2(3):361–376, July 1991.

| $N_1 \times N_2$ | Computation Time | Communication Time |
|---|---|---|
| $N_1 = 1$ <br> $N_2 = N$ | $(2 * \frac{m^2}{N} + 3 * \frac{m}{N}) * t_f$ | $(2 * m * \log N) * t_c$ |
| $N_1 = N$ <br> $N_2 = 1$ | $(2 * \frac{m^2}{N} + 3 * m) * t_f$ | $(m + m * \log N) * t_c$ |
| $N_1 = \sqrt{N}$ <br> $N_2 = \sqrt{N}$ | $(2 * \frac{m^2}{N} + 3 * \frac{m}{\sqrt{N}}) * t_f$ | $(\frac{1}{2} * m * \log N) *$ <br> $(\frac{2}{\sqrt{N}} + 1) * t_c$ |

Table 2: Computation time and communication time on three processor grids.

| | |
|---|---|
| processor 0 | $A_{11}$ $A_{12}$ $A_{13}$ $A_{14}$ $V_1$ $B_1$ $X_1$ $(X_1$ $X_2$ $X_3$ $X_4)$ |
| processor 1 | $A_{21}$ $A_{22}$ $A_{23}$ $A_{24}$ $V_2$ $B_2$ $X_2$ $(X_1$ $X_2$ $X_3$ $X_4)$ |
| processor 2 | $A_{31}$ $A_{32}$ $A_{33}$ $A_{34}$ $V_3$ $B_3$ $X_3$ $(X_1$ $X_2$ $X_3$ $X_4)$ |
| processor 3 | $A_{41}$ $A_{42}$ $A_{43}$ $A_{44}$ $V_4$ $B_4$ $X_4$ $(X_1$ $X_2$ $X_3$ $X_4)$ |

Table 3: Data layouts of the parallel Jacobi's iterative algorithm for implementing linear systems $A_{4\times4}\ X_4 = B_4$ on a four-processor linear array.

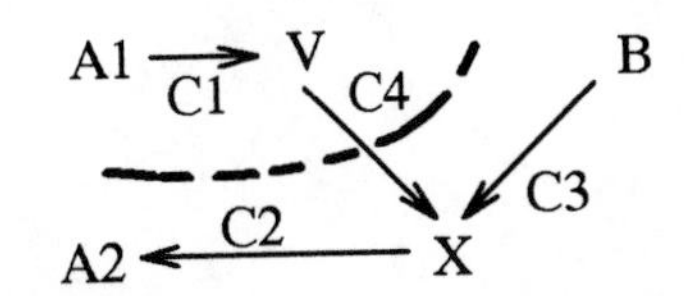


Figure 1: Component affinity graph of Jacobi's iterative algorithm.

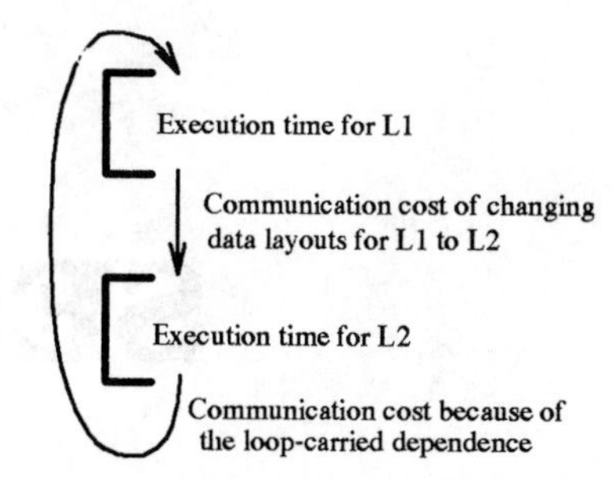


Figure 2: The total execution time for computing two Do-loops in an iteration.

| token | line | used in indices | virtual-PE mapping | dependence-vector mapping | used in PEs |
|---|---|---|---|---|---|
| $B(i)$ | 5 | $(0,i)^t + k(1,0)^t$ | $(0,1)((k,i)^t = i$ | $(0,1)(1,0)^t = 0$ | $(i-1) \bmod N$ |
| $B(k)$ | 5 | $(k,0)^t + i(0,1)^t$ | $(0,1)((k,i)^t = i$ | $(0,1)(0,1)^t = 1$ | all PEs |
| $A(i,j)$ | 7 | $(0,i,j)^t + k(1,0,0)^t$ | $(0,1,0)(k,i,j)^t = i$ | $(0,1,0)(1,0,0)^t = 0$ | $(i-1) \bmod N$ |
| $L(i,k)$ | 7 | $(k,i,0)^t + j(0,0,1)^t$ | $(0,1,0)(k,i,j)^t = i$ | $(0,1,0)(0,0,1)^t = 0$ | $(i-1) \bmod N$ |
| $A(k,j)$ | 7 | $(k,0,j)^t + i(0,1,0)^t$ | $(0,1,0)(k,i,j)^t = i$ | $(0,1,0)(0,1,0)^t = 1$ | all PEs |
| $V(i)$ | 16 | $(0,i)^t + j(1,0)^t$ | $(0,1)(j,i)^t = i$ | $(0,1)(1,0)^t = 0$ | $(i-1) \bmod N$ |
| $X(j)$ | 16 | $(j,0)^t + i(0,1)^t$ | $(0,1)(j,i)^t = i$ | $(0,1)(0,1)^t = 1$ | all PEs |

Table 4: Data-dependence information of each data token and the suggested index-processor mapping of the Gauss elimination algorithm.

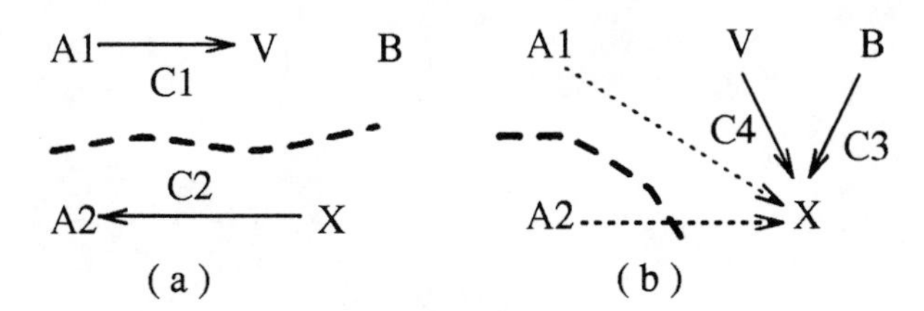


Figure 3: Component alignment for (a) lines 2–6 and (b) lines 7–9.

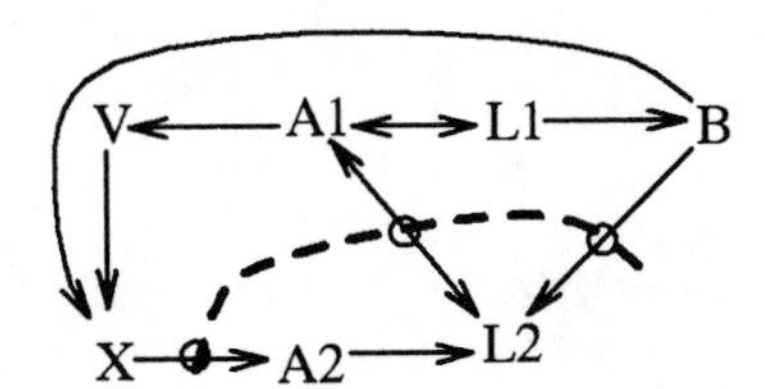


Figure 4: Component affinity graph and the suggested component alignment of the Gauss elimination algorithm.

```
1   {* Let m be the problem size, N be the number  *}
2   {*      of processors, and block = m / N. *}
3   REAL A(block, m), L(block, m), X(block), B(block)
4   REAL V(block), Apipeline(m), Xpipeline, Bpipeline
5   me = who_am_i() {* Return current processor's ID. *}
6               {* Matrix triangularization.  *}
7   do 15  k = 1, me
8     receive_from_left( Apipeline(k..m), Bpipeline )
9     send_to_right( Apipeline(k..m), Bpipeline )
10    do 15  i = 1, block
11      L(i, k) = A(i, k) / Apipeline(k)
12      B(i) = B(i) - L(i, k) * Bpipeline
13      do 15  j = (k + 1),  m
14        A(i, j) = A(i, j) - L(i, k) * Apipeline(j)
15  continue
16  do 34  k = (me + 1), m, N
17    pivot = ceiling(k / N)
18    send_to_right( A(pivot, k..m), B(pivot) )
19    do 24  i = (pivot + 1),  block
20      L(i, k) = A(i, k) / A(pivot, k)
21      B(i) = B(i) - L(i, k) * B(pivot)
22      do 24  j = (k + 1), m
23        A(i, j) = A(i, j) - L(i, k) * A(pivot, j)
24  continue
25  receive_from_left( A(pivot, k..m), B(pivot) )
26  do 34  k1  = (k + 1), min(m, (k + N - 1))
27    receive_from_left( Apipeline(k1..m), Bpipeline )
28    send_to_right( Apipeline(k1..m), Bpipeline )
29    do 34  i = (pivot + 1), block
30      L(i, k1) = A(i, k1) / Apipeline(k1)
31      B(i) = B(i) - L(i, k1) * Bpipeline
32      do 34  j = (k1 + 1), m
33        A(i, j) = A(i, j) - L(i, k1) * Bpipeline
34  continue
35              {* Triangular linear system UX = Y. *}
36  do 38  i = block, 1, -1
37    V(i) = 0.0
38  continue
39  do 44   j = m, ((m - N + 1) + (me + 1)), -1
40    receive_from_right( Xpipeline )
41    send_to_left( Xpipeline )
42    do 44  i = block, 1, -1
43      V(i) = V(i) + A(i, j) * Xpipeline
44  continue
45  do 58  j = (m - N + me + 1), 1, -N
46    pivot = ceiling(j / N)
47    X(pivot) = (B(pivot) - V(pivot)) / A(pivot, j)
48    send_to_left( X(pivot) )
49    do 51 i = (pivot - 1), 1 -1
50      V(i) = V(i) + A(i, j) * X(pivot)
51    continue
52    receive_from_right( X(pivot) )
53    do 58  j1 = (j - 1), max(1, (j - N + 1)), -1
54      receive_from_right( Xpipeline )
55      send_to_left( Xpipeline )
56      do 58  i = (pivot - 1), 1, -1
57        V(i) = V(i) + A(i, j1) * Xpipeline
58  continue
```

Figure 5: Generated parallel codes for the Gauss elimination algorithm.

# SESSION 7B

## SIMD/DATA PARALLEL

# Solving Dynamic and Irregular Problems on SIMD Architectures with Runtime Support

WEI SHU AND MIN-YOU WU
Department of Computer Science
State University of New York, Buffalo, NY 14260

*Abstract —One of the essential problems in parallel computing is: can SIMD machines handle asynchronous problems? This is a difficult, unsolved problem because of the mismatch between asynchronous problems and SIMD architectures. We propose a solution to let SIMD machines handle general asynchronous problems. Our approach is to implement a runtime support system which can run MIMD-like software on SIMD hardware. Substantial performance has been obtained on CM-2 and CM-5.*

## 1. Introduction

### 1.1. Can SIMD machines handle asynchronous problems?

The current parallel supercomputers have been developed along two major architectures, the SIMD (Single Instruction Multiple Data) architecture and the MIMD (Multiple Instruction Multiple Data) architecture. The SIMD architecture consists of a central control unit and many processing units. Only one instruction can execute at a time and every processor executes the same instruction. Advantages of a SIMD machine include its simple architecture which makes the machine potentially inexpensive, and its synchronous control structure which makes programming easy and communication overhead low [18]. The MIMD architecture is based on the duplication of control units for each individual processor. Different processors can execute different instructions at the same time. It is more flexible for different problem structures and can be applied to general applications. However, the complex control structure of MIMD architecture makes the machine expensive and the system overhead large. It is also difficult to program on an MIMD machine.

Application problems can be classified into three categories: synchronous, loosely synchronous, and asynchronous [6].

- The synchronous problems have a uniform problem structure. In each time step, every processor executes the same operation over different data, resulting in a naturally balanced load.
- The loosely synchronous problems can be structured iteratively with two phases, the computation phase and the synchronization phase. In the synchronization phase, processors exchange information and synchronize with each other.
- The asynchronous problems have no synchronous structure. Processors may communicate with each other at any time. The computation structure could be very irregular and the load imbalanced.

The synchronous problems can be naturally implemented on a SIMD machine and the loosely synchronous problems on an MIMD machine. Implementation of the loosely synchronous problems on SIMD machines is not easy; computation load must be balanced and the load balance activity is essentially irregular.

Solving the asynchronous problems is more difficult. First, a direct implementation on MIMD machines is nontrivial. The user must handle the synchronization and load balance issues at the same time, which could be extremely difficult for some application problems. In general, a runtime support system, such as LINDA [1], reactive kernel [15], or chare kernel [17], is necessary for solving asynchronous problems. Implementation of the asynchronous problems on SIMD machines is even more difficult because it needs a runtime support system and the support system itself is asynchronous. In particular, the support system must arrange the code in such a way that all processors execute the same instruction at the same time. We summarize the above discussion in Table 1.

Various application problems require different programming methodologies. Two essential programming methodologies are array-based and thread-based. The problem domain of most synchronous applications can be naturally mapped onto an array, resulting in the array-based programming methodology. The solution to asynchronous problems naturally demands the thread-based programming methodology, in which threads are individually executed and where information exchange can happen at any time.

### 1.2. Let SIMD machines handle asynchronous problems

To make a SIMD machine serve as a general purpose machine, we must be able to solve asynchronous problems in addition to solving synchronous and loosely synchronous problems. The major difficulties in executing asynchronous applications on SIMD machines are:

- the gap between the synchronous machines and asynchronous applications; and

Table 1: Implementation of Problems on MIMD and SIMD Machines

| | Synchronous | Loosely synchronous | Asynchronous |
|---|---|---|---|
| MIMD | easy | natural | need runtime support |
| SIMD | natural | difficult | difficult, need runtime support |

- the gap between the array processors and thread-based programming.

One solution, named the application-oriented approach, lets the user fill the gap between application problems and architectures. With this approach, the user must study each problem and look for a specific method to solve it [2, 19, 20, 22]. The region growing algorithm is an asynchronous, irregular problem and difficult to run on SIMD machines [22]. The other two implementations, the Mandelbrot Set algorithm [19] and the Molecular Dynamics algorithm [2, 20], solved a class of problems, which contains a parallelizable outer loop but has an inner loop for which the number of iterations varies between different iterations of the outer loop.

An alternative to the application-oriented approach is the system-oriented approach, which provides a system support to run MIMD-like software on SIMD hardware. The system-oriented approach is superior to the application-oriented approach because of two reasons:

- the system is usually more knowledgeable about the architecture details, as well as its dynamic states;
- it is more efficient to develop a sophisticated solution in system, instead of writing similar code repeatedly in the user programs.

The system-oriented approach can be carried out in two levels: instruction-level and thread-level. Both of them share the same underlying idea: if one were to treat a program as data, and a SIMD machine could interpret the data, just like a machine-language instruction interpreted by the machine's instruction cycle, then an MIMD-like program could efficiently execute on the SIMD machine [8]. The instruction-level approach implements this idea directly: the instructions are interpreted in parallel across all of the processors by control signals emanating from the central control unit [3, 4, 12, 14, 23]. The major constraint of this approach is that the central control unit has to cycle through almost the entire instruction set for each instruction execution because each processor may execute different instructions. A common method to reduce the average number of instructions emanated in each execution cycle is to perform *global or*'s to determine whether the instruction is needed by any processor [4, 23]. It was believed that a program would not diverge quickly and only a few instructions need to be emanated [4]. However, later research found that the programs diverged fast and the *global or* solution did not deliver much benefit [23]. It might be necessary to insert barrier synchronizations at some points to limit the degree of divergence for certain applications. Having a barrier at the end of each WHERE statement is a good idea [3]. Furthermore, the applications implemented in these systems are non-communicating [23] or has a barrier at the end of the program [4]. However, many applications have dependences and proper synchronization must be inserted to ensure correct execution sequence. The synchronization could suspend a large number of processors. Finally, this approach is unable to balance load between processors and unlikely to produce good performance for general applications.

We propose a thread-based model for a runtime system which can support loosely synchronous and asynchronous problems on SIMD machines. The thread-level implementation offers great flexibility. Compared to the instruction-level approach, it has at least two advantages:

- the execution order of threads can be exchanged to avoid processor suspension; and
- system overhead will not be overwhelming since granularity can be controlled with the thread-based approach.

The runtime support system is named as *Process kernel* or *P kernel*. The P kernel is thread-based as we assign computation at the thread-level. The P kernel is able to handle the bookkeeping, scheduling, load balancing, and message management, as well as to make these low-level tasks transparent to users.

### 1.3. Other related research

Perhaps the work that is closest to our approach is Graphinators and early work on combinators [8, 11, 7] and additional work on the interpretation of Prolog and FLAT GHC [10, 13]. The Graphinators implementation is an MIMD simulator on SIMD machines. It was achieved by having each SIMD processor repeatedly cycle through the entire set of possible "instructions". The Graphinator model suffered because of its fine granularity. Our work is distinguished from the work in terms of granularity control.

Our model is similar to the Large-Grain Data Flow (LGDF) model [9]. It is a model of computation that combines sequential programming with dataflow-like program activation. The LGDF model was implemented on shared memory machines. It can be implemented on MIMD distributed memory machines as well [24, 17]. Now we show that the model can be implemented on a SIMD distributed memory machine too.

## 2. The P Kernel Approach — Computation Model and Language

The computation model for the P kernel is a message-driven, nonpreemptive, thread-based model. Here, a parallel computation will be viewed as a collection of *processes*, each of which in turn consists of a set of threads, named as *atomic computations*.

Processes communicate with each other via *messages*. Each atomic computation is then the result of processing a message. During its execution, it can create new processes or generate new messages. All atomic computations of the same process share one *common data area*. Thus, a *process* $P_k$ consists of a set of *atomic computations* $A_{k_i}$ and one *common data area* $D_k$:

$$P_k = \{D_k, A_{k_1}, A_{k_2}, ..., A_{k_n}\}, n \geq 1$$

Once a process has been scheduled to a processor, all of its atomic computations are executed on the same processor. There is no presumed sequence to indicate which atomic computation will be carried out first. Instead, it depends on the order of arrival of messages.

With this thread-based model, the P kernel is a runtime support system on a SIMD machine built to manipulate and schedule processes, as well as messages. A program written in the P kernel language consists mainly of a collection of process definitions and subroutine definitions. A process definition includes a process name preceded by the keyword **process**, and followed by the process body, as shown below.

```
process ProcName {
    <Common Data Area Declarations>
    entry LABEL1: (message msg1) <Code1>
    entry LABEL2: (message msg2) <Code2>
    ... }
```

Here, bold-face letters denote the keywords of the language. The process body, which is enclosed in braces, consists of declarations of private variables that constitute the common data area of the process, followed by a group of atomic computation definitions. Each atomic computation definition starts with a keyword **entry** and its label, followed by a declaration of the corresponding message and arbitrary user code. One of the process definitions must be the *main* process. The first entry point in the main process is the place the user program starts. The user can write a program with the P kernel language, deal with the creation of processes, and sending messages between them. For details of the computation model and language, refer to [17].

We illustrate how to write a program in the P kernel language using the N-queen problem as an example in Figure 1. The algorithm used here attempts to place queens on the board one row at a time if the particular position is valid. Once a queen is placed on the board, the other positions in its row, column, and diagonals, will be marked invalid for any further queen placement. The atomic computation QueenInit in the MAIN process creates $N$ processes of type SUBQUEEN, each with an empty board and one candidate queen in a column of the first row. There are two types of atomic computations in process SUBQUEEN: ParallelQueen and ResponseQueen, and a common data area consisting of solutionCount and responseCount. Each atomic computation ParallelQueen receives a message that represents the current placement of queens, and a position for the next queen to be placed. Following the

```
Process MAIN
 { int solutionCount = 0; responseCount = 0;
 entry QueenInit: (message MSG1()) { int k;
   read N from input
   for (k = 1, N) {responseCount = N;
     OsCreateProc(SUBQUEEN,ParallelQueen,
         MSG2(1,k,empty board)) }
   }
 entry ResponseQueen: (message MSG3(m)) {
   solutionCount=solutionCount+m; responseCount--;
   if (responseCount==0) {
     print "# of solutions =", solutionCount;
     OsKillProc() }
   }
 }

Process SUBQUEEN
 { int solutionCount = 0; responseCount = 0;
 entry ParallelQueen:(message MSG2(i,j,board)){int k;
   invalidate row i, column j, and diagonals of (i,j)
   for (k = 1, N) {
     if (position (i+1,k) is marked valid) {
       if ((N-i) is larger than the grainsize)
           OsCreateProc(SUBQUEEN,ParallelQueen,
                     MSG2(i+1,k,board));
       else
           OsCreateProc(SEQQUEEN,SequentialQueen,
                     MSG2(i+1,k,board));
       responseCount++;
       }
     };
   if (responseCount==0) {
     OsSendMsg(ParentProcID(),ResponseQueen,
         MSG3(solutionCount));
     OsKillProc() }
   }
 entry ResponseQueen: (message MSG3(m)) {
   solutionCount=solutionCount+m; responseCount--;
   if (responseCount==0) {
     OsSendMsg(ParentProcID(),ResponseQueen,
         MSG3(solutionCount));
     OsKillProc() }
   }
 }

Process SEQQUEEN {
 entry SequentialQueen: (message MSG2(i,j,board)) {
   int k, count;
  call sequential routine, recursively generating all valid
     configurations
  OsSendMsg(ParentProcID(),ResponseQueen,
     MSG3(count));
  OsKillProc() }
  }
 }
```

Figure 1: The N-queen program.

invalidation processing, it creates new SUBQUEEN or SEQQUEEN processes by placing one queen in every valid position in the next row. The atomic computation ResponseQueen in processes SUBQUEEN and MAIN counts the total number of successful queen configurations. It can be triggered any number of times until there is no more response expected. The atomic computation SequentialQueen is invoked when the rest of rows are to be manipulated sequentially. This is how granularity can be controlled. In this example, there are two process definitions besides that of process MAIN. Atomic computations that share the same common data area should be in a single process, such as ParallelQueen and ResponseQueen. In general, only the atomic computations that are logically coherent and share the same common data area should be in the same process.

## 3. Design and Implementation

The main loop of the P kernel system is shown in Figure 2. It starts with a system phase which includes placing processes, transferring data messages, and selecting atomic computations to execute. It is followed by a user program phase to execute the selected atomic computation. The iteration will continue until all the computations are completed. The P kernel software consists of three main modules: computation selection, communication, and memory management, described as follows.

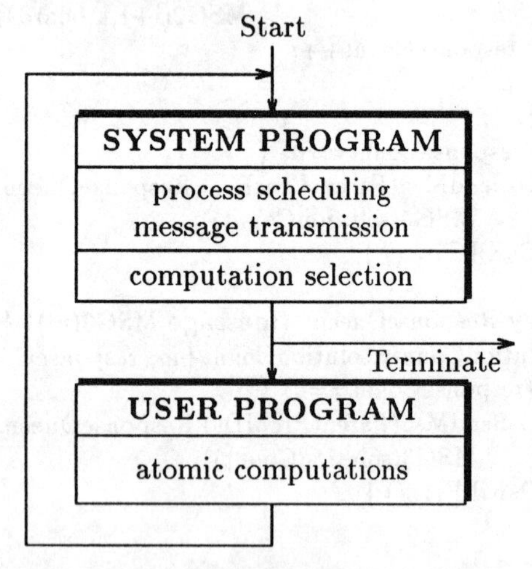


Figure 2: Flow chart of the P kernel system

### 3.1. Computation selection

A fundamental difference between the MIMD and SIMD systems is the degree of synchronization required. In an MIMD system, different processors can execute different threads of code, but not in a SIMD system. When the P kernel system is implemented on an MIMD machine, which atomic computation to be executed next is an individual decision of each processor. Whereas, due to the lock-step synchronization in a SIMD machine, the same issue turns to be a global agreement to decide which atomic computation to execute next.

Let's assume that there are $K$ atomic computation types, corresponding to the atomic computation definitions, represented by $a_0, a_1, ..., a_{K-1}$. During the life time of execution, the total number of atomic computations executed is far more than $K$. These atomic computations are dynamically distributed among processors. We define function $num(i,p,t)$ to record the number of atomic computation with type $a_i$ at processor $p$ in iteration $t$. In the system phase at iteration $t$, a computation selection function $\mathcal{F}$ is applied to generate a number $k$, where $k = \mathcal{F}(t)$ and $0 \leq k < K$. Followed by the user program phase at the same iteration, processor $p$ will be active if there is at least one atomic computation available with the selected type $a_k$. Let the function $act(i,t)$ count the number of active processors at iteration $t$ if the atomic computation type $a_i$ is selected,

$$act(i,t) = | \{ \ p \mid num(i,p,t) > 0, \ 0 \leq p < N \} |$$

where $N$ is the number of processors.

We present three computation selection algorithms here. The first one, $\mathcal{F}_{cyc}$, is a simple algorithm.

**Algorithm I: Cyclic algorithm.** Basically, it repeatedly cycles through all atomic computation types.

$$\mathcal{F}_{cyc}(t) = \min\{ \ i \mid act((i \bmod K), t) > 0, \\ \mathcal{F}_{cyc}(t-1) < i \leq \mathcal{F}_{cyc}(t-1) + K\} \bmod K$$

where $t \geq 1$ and $\mathcal{F}_{cyc}(0) = -1$. Here, it is not always necessary to carry out $K$ reductions to compute $act(i,t)$, since as long as the first nonzero $act(i,t)$ is found, the value of function $\mathcal{F}_{cyc}$ is determined.

To complete the computation in the shortest time, the number of iterations has to be minimized. Maximizing the processor utilization at each iteration is one of possible heuristics. If $act(k_1,t)$ is 100, $act(k_2,t)$ is 900, a computation selection function $\mathcal{F}(t)$ generating $k_2$ is intuitively better, leading to an immediate good processor utilization. An *auction* algorithm, $\mathcal{F}_{auc}$ is proposed based on this observation:

**Algorithm II: Auction algorithm.** For each atomic computation $i$, calculate $act(i,t)$ at iteration $t$. Then, the atomic computation with the maximum value of $act(i,t)$ is chosen to execute next:

$$\mathcal{F}_{auc}(t) = \min\{ \ j \mid act(j,t) = \max_{0 \leq i < K} act(i,t), \\ 0 \leq j < K\}$$

The cyclic algorithm is nonadaptive in the sense that the selection is made almost independent of the distribution of atomic computations. In this way, it could be the case that a few processors are executing one atomic computation type while many processors are waiting for execution of the other atomic computation types. The auction algorithm is runtime adaptive. It will maximize utilization in most cases. An adaptive algorithm is more sophisticated in general, however, experimental results show that in most cases, the cyclic algorithm performs better than the auction algorithm. It has been observed that when an auction algorithm is applied, at the near end of execution, the parallelism becomes smaller and smaller, and the program takes a long time to finish. This low parallelism phenomenon degrades performance seriously, which is characterized as the *tailing effect*.

We propose an improved adaptive algorithm to overcome the tailing effect. To retain the advantage of the auction algorithm, we intend to maximize the processor utilization as long as there is a large pool of atomic computations available. On the other hand, when the available parallelism falls down to a certain degree, we aim at exploiting large parallelism by assigning priorities to different atomic computation types. An atomic computation whose execution increases the parallelism gets a higher priority, and vice versa.

**Algorithm III: Priority auction algorithm.** For simplicity, we assume that the atomic computations $a_0, a_1, \ldots, a_{K-1}$ have been presorted according to their priorities, $a_0$ with highest priority and $a_{K-1}$ with the lowest one. Use $m = cN$ as a gauge of available parallelism, where $c$ is a constant and $N$ is the number of processors.

$$\mathcal{F}_{pri}(t) = \begin{cases} \min\{\, j \mid act(j,t) = \max_{0 \le i < K} act(i,t), \\ \qquad 0 \le j < K\}, \\ \qquad\qquad \text{if } act(j,t) > m \\ \min\{\, j \mid act(j,t) > 0, 0 \le j < K\}, \\ \qquad\qquad \text{otherwise} \end{cases}$$

Here, the constant $c$ is set to be 0.5. That is, if more than half of processors are active, the auction algorithm is used to maximize the processor utilization; otherwise, the priority is predominantly considered in favor of parallelism increase and the tailing effect prevention.

This algorithm can constantly provide better performance than that provided by the cyclic algorithm. Table 2 shows the performance for different computation selection algorithms with the 12-queen problem on a 4K-processor CM-2. All timing data presented in this paper are CM elapsed time.

Table 2: Execution time of the 12-queen Problem

| Cyclic | Auction | Priority Auction |
|---|---|---|
| 203 Seconds | 220 Seconds | 196 Seconds |

### 3.2. Communication

There are two kinds of messages to be transferred. One is the *data* message, which is specifically addressed to the existing process. The other kind of messages are *process* messages, which represent the newly generated processes.

**Transfer of data messages.** Assume each processor initially holds $d_0(p)$ data messages to be sent out at the end of a computation phase. Because of the SIMD characteristics, only one message can be transferred each time. Thus, the message transfer step must be repeated at least $D_0$ times to complete, where

$$D_0 = \max_{0 \le p < N} d_0(p).$$

The real situation is even more complicated. During each time of the message transfer, a collision may occur when two or more messages from different processors have the same destination processor. Therefore, we need to prevent the message loss due to the collision. Let $dest(p)$ be the destination of a message from processor $p$ and $src(q)$ be the source from which processor $q$ is going to receive a message, respectively. There is a collision if two processors $p_1$ and $p_2$ are sending messages to the same processor $q$, so that $dest(p_1) = dest(p_2) = q$. The processor $q$ can receive only one of them, say from $p_1$, by assigning $src(q) = p_1$. Thus, only processor $p_1$ can successfully deliver its message to the destination. The processors $p_i$ with $p_i \neq src(dest(p_i))$ have to wait for the next time to compete again. After the first transfer of data messages, there may still be some unsent messages since some processor $p$ has more than one message to send ($d_0(p) > 1$), or not every processor is able to send out its message during the first transfer due to collisions. Hence, at the $k$th transfer

$$d_k(p) = \begin{cases} d_{k-1}(p) - 1 & \text{if } p = src(dest(p)) \\ d_{k-1}(p) & \text{otherwise} \end{cases}$$

$$D_k = \max_{0 \le p < N} d_k(p), k \ge 1$$

As long as $D_k > 0$, we need to continue on with $D_k$, $D_{k+1}$, ... , $D_m$, such that $D_m = 0$.

Notice that for later transfers, it is most likely that only a few processors are actively sending messages, resulting in a low utilization in the SIMD system. To avoid such a case, we do not require all messages to be transferred. Instead, we attempt to send out only a majority of data messages. The residual messages are buffered and will be sent again during the next iteration. Because of the atomic execution model, the processor that fail to send messages will not be stalled. Instead, a processor can continue execution as long as there are some messages in its waiting queue.

**Process placement.** The handling of process messages is almost the same as that of data messages, except that we need to assign a destination processor ID to each process message. The assignment is called *process placement*. Two placement algorithms have been implemented in the P kernel, the Random Placement algorithm and the Extended Scatter algorithm.

**Random Placement algorithm.** It is a simple, moderate performance algorithm:

$$dest(p) = random() \bmod N$$

where $N$ is the number of processors. Once $dest(p)$ has been assigned, we can follow the same procedure as in the transfer of data messages. However, the two kinds of messages are different from each other in that the $dest(p)$ of a data message is fixed, whereas that of a process message can be varied. To take this advantage, we are able to reschedule the process message if the destination processor cannot accept the newly generated process because of some resource constraint or collision. Here, the rescheduling is simply a task to assign another random number as the destination processor ID. We can repeat this rescheduling until all process messages are assigned the destination processor IDs. However, it is practically a better choice that we only offer one or two chances for rescheduling, instead of repeating it until satisfaction. The unsuccessfully scheduled process messages are buffered similar to the residual data messages, waiting for the next communication phase.

**Extended Scatter algorithm.** This algorithm extends the scatter algorithm to problems with unpredictable structures [25]. It is based on the *generation order* of the processes in each processor. That is, processes can be labeled as they are generated. The algorithm schedules the $k$th generated process at processor $p$ to processor $scat(p, k)$,

$$scat(p, k) = (scat(p, k-1) + 1) \bmod N$$

where $scat(p, 0) = p$ and $N$ is the number of processors. This scatter function was designed for a small number of processors [25]. For a larger number of processors, such as 4K processors in CM-2, it spreads processes rather slowly. Thus, the scatter function is modified as follows:

$$scat(p, k) = (scat(p, k-1) + sp(k) + 1) \bmod N$$

where $scat(p, 0) = p$ and $sp(k)$ is a function to add extra hops for fast spread. The initial value of $sp(0)$ is set to be close to $0.1N$. Its successive values can be decreased such as $sp(k) = \gamma \cdot sp(k-1)$, where $\gamma$ is a decay factor, chosen to be 0.95 in our experiment. The impact of $sp(k)$ vanishes as $k$ gets larger, since processes should have been evenly spread during that time.

Table 3 compares the two algorithms. The Extended Scatter algorithm can distribute load more evenly than the Random Placement algorithm.

Table 3: Execution time of the 12-queen Problem

| Random Placement | Extended Scatter |
|---|---|
| 211 Seconds | 196 Seconds |

### 3.3. Memory management

Most SIMD machines are massively parallel processors. In such a system, any one of the thousands of processors could easily run out of memory, resulting in a system failure. A memory management provides features to improve the system robustness. When the available memory space on a specific processor becomes tight, we should restrict the new resource consumption, or release memory space by moving out some unprocessed processes to other processors.

In the P kernel system implemented on SIMD machines, we define three different states to measure the current usage of memory space at processor $p$. In the *nearly-full* state, we need to limit the new resource consumption since the available memory space is getting tight. It is accomplished by preventing newly generated process messages from being scheduled on. In the *full* state, incoming data messages are buffered at the original processors and waiting for change of the destination processor's state. In the *emergency* state, several actions can be taken before we declare its failure. One is to clear up all residual data messages and process messages, if there are any. Another is to redistribute the unprocessed process messages that have been previously placed for this processor previously.

## 4. Performance

The first version of the P kernel was running on a Connection Machine CM-2 in March, 1991. The P kernel is currently running on a CM-2 and a CM-5. The CM-2, at Rice University, has two partitions, each with 4K processors. Each processor has 32K bytes memory. We use one partition for testing performance. The CM-5, at Syracuse University, has 32 processors. Each processor has 32M bytes memory. The CM-5 is an MIMD/SIMD machine. We use the SIMD mode of CM-5 machines for our research simply because it is the only coarse grain SIMD machine currently available.

The P kernel is written in CM Fortran. It is compiled using the Slicewise 1.2 compiler on CM-2 and the Sparc 2.1, Beta 0.1 compiler on CM-5. Although Fortran is not the best language for system programs. it can be easily ported on other SIMD machine, such as Maspar MP-1. The problems with a Fortran implementation are lack of pointers and block FORALLs. First, lack of the pointer data structure in CM Fortran makes implementation of queues very difficult. Currently, we use the multiple dimension array with indirect addressing to implement queues. Hence, not only is the indirect addressing extremely slow, but accessing different addresses in CM-2 costs much more [4]. The second problem with current CM Fortran implementation is that it provides only single-statement FORALL and no procedure can be called from the FORALL body. This restriction makes the coding of the P kernel inefficient. High Performance Fortran (HPF) provides pointers as well as a block FORALL construct with PURE functions [5], which are eagerly desired features. Once these features are available, the P kernel can be implemented much easily and more efficiently. We have not found a single C-based language supported between different SIMD machines. TMC's old C* and the new C*, as well as Maspar's C are all quite different. This is why we are still hesitating to rewrite our system with C.

The total execution time of a P kernel program consists of two parts, the time to execute the system program, $T_{sys}$, and the time to execute the user program, $T_{usr}$. The *system efficiency* is defined as follows:

$$\mu_{sys} = \frac{T_{usr}}{T} = \frac{T_{usr}}{T_{usr} + T_{sys}},$$

where $T$ is the total execution time. The P kernel executes in a loosely synchronous fashion and can be divided into iterations. Each iteration consists of a system program phase and a user program phase. For each iteration, the execution time of the system program varies between 250 and 350 milliseconds on CM-2 and between 60 and 70 milliseconds on CM-5, respectively. The system efficiency depends on the ratio of the system overhead and grainsize of atomic computations. Table 4 shows that the system efficiency increases with the problem size (as well as the grainsize). High efficiency results from the high ratio of granularity and system overhead.

Table 4: System Efficiencies ($\mu_{sys}$) for the N-queen Problem on CM-5

| 11-queen | 12-queen | 13-queen |
|---|---|---|
| 97.6% | 99.3% | 99.9% |

In the system phase, every processor participates in the global actions. On the other hand, in the user program phase, not all processors are involved in the execution of the selected atomic computation. In each iteration $t$, the ratio of the number of participating processors ($N_{active}(t)$) to the total number of processors ($N$) is defined as *utilization* $u(t)$:

$$u(t) = \frac{N_{active}(t)}{N}$$

The *utilization efficiency* is defined as follows:

$$\mu_{util} = \frac{\sum_{t=1}^{m} u(t)T(t)}{\sum_{t=1}^{m} T(t)} = \frac{\sum_{t=1}^{m} T(t)N_{active}(t)}{N\sum_{t=1}^{m} T(t)}$$

where $T(t)$ is the execution time of $t$th iteration and $\sum_{t=1}^{m} T(t) = T_{usr}$. The utilization efficiency depends on the computation selection strategy and the load balancing scheme. Table 5 shows the utilization efficiencies for different problem sizes with the priority auction algorithm and extended scatter scheduling. In this example, granularity is large enough so that the extended scatter algorithm can produce optimal performance. A more sophisticated scheduling algorithm cannot improve performance since it only reduces communication overhead which is negligible in this case. Combining the priority auction algorithm and scatter scheduling, the system can reach high utilization efficiency.

Table 5: Utilization Efficiency ($\mu_{util}$) for the N-queen Problem on CM-5

| 11-queen | 12-queen | 13-queen |
|---|---|---|
| 76.4% | 81.6% | 84.6% |

In irregular problems, the grainsizes of atomic computations may vary substantially. The grainsize variation heavily depends on how irregular the problem is and how the program is partitioned. We use an index, called *busyness*, to measure the grainsize variation of atomic computation for each iteration:

$$b(t) = \frac{\sum_{k=1}^{N_{active}(t)} T_k(t)}{T(t)N_{active}(t)}$$

where $T_k(t)$ is the busy time of the $k$th participated processor at iteration $t$. The *busyness efficiency* is defined as:

$$\mu_{busy} = \frac{\sum_{t=1}^{m} b(t)T(t)N_{active}(t)}{\sum_{t=1}^{m} T(t)N_{active}(t)}$$

$$= \frac{\sum_{t=1}^{m}\sum_{k=1}^{N_{active}(t)} T_k(t)}{\sum_{t=1}^{m} T(t)N_{active}(t)}$$

Table 6 shows the busyness efficiencies for different problem sizes.

Table 6: Busyness Efficiency ($\mu_{busy}$) for the N-queen Problem on CM-5

| 11-queen | 12-queen | 13-queen |
|---|---|---|
| 54.2% | 56.0% | 57.2% |

Now we define the overall *efficiency* as follows:

$$\mu = \mu_{sys} * \mu_{util} * \mu_{busy}$$

Taking the N-queen problem as an example, the overall efficiencies on CM-2 and CM-5 are shown in Table 7. The low efficiency on CM-2 is due to the fact that the problem runs on too many processors and the fine grain machine suffers from heavy communication. The CM-5, using fewer but more powerful processors, achieved good performance.

Table 7: The Efficiencies ($\mu$) for the N-queen Problem

| CM-2 | | | CM-5 | | |
|---|---|---|---|---|---|
| 11-queen | 12-queen | 13-queen | 11-queen | 12-queen | 13-queen |
| 4.8% | 11.7% | 14.2% | 40.4% | 45.4% | 48.4% |

The second example is a loosely synchronous problem, the GROMOS Molecular Dynamics program [21, 20]. The test data is the bovine superoxide dismutase molecule (*SOD*), which has 6968 atoms [16]. The cutoff radius is predefined to 8 Å, 12 Å, and 16 Å. The overall efficiencies on CM-2 and CM-5 are shown in Table 8. Because of the small test data set, only 1K processors in CM-2 are used. On 1K processors in CM-2, this program achieved 102 MFLOPS for the cutoff radius of 16 Å. It is reasonable to assume that the performance results for a 1K system will scale up to a larger system when the work per processor remains constant.

Table 8: The Efficiencies ($\mu$) for Molecular Dynamics

| CM-2 | | | CM-5 | | |
|---|---|---|---|---|---|
| 8 Å | 12 Å | 16 Å | 8 Å | 12 Å | 16 Å |
| 24.6% | 32.2% | 42.3% | 28.9% | 47.0% | 60.8% |

## 5. Concluding Remarks

Experimental results have shown that the P kernel is able to balance load fairly well on SIMD machines for asynchronous application problems. System overhead can be reduced to a minimum with granularity control. Considering both the low cost of machines and ease of programming, this approach becomes much more attractive.

Performance is not the focus point in the current research stage. Instead, this work is mainly proof-of-concept based. The motivation is twofold: to prove whether a SIMD machine can handle asynchronous application problems, serving as a general-purpose machine; and to study the feasibility of providing a truly portable parallel programming environment between SIMD and MIMD machines.

**Acknowledgments**

We are very grateful to Reinhard Hanxleden for providing the GROMOS program, and Terry Clark for providing the SOD data. We also thank Alan Karp, Guy Steele, and Jerry Roth for their comments, and Ravikanth Ganesan for carefully proofreading the manuscript.

This research was partially supported by NSF grants CCR-9109114 and CCR-8809615. Use of the Connection Machine was provided by the Center for Research on Parallel Computation under NSF Cooperative Agreement No. CCR-8809615 with support from Keck Foundation and Thinking Machines Corporation. The performance data was gathered on the CM-5 at NPAC, Syracuse University.

## References

[1] N. Carriero and D. Gelernter. Linda in context. *Commun. ACM*, 32(4):444–458, April 1989.

[2] T. W. Clark, R. v. Hanxleden, K. Kennedy, C. Koelbel, and L. R. Scott. Evaluating parallel languages for molecular dynamics computations. In *Scalable High Performance Computing Conference*, April 1992.

[3] R. J. Collins. Multiple instruction multiple data emulation on the Connection Machine. Technical Report CSD-910004, Dept. of Computer Science, Univ. of California, February 1991.

[4] H. G. Dietz and W. E. Cohen. A massively parallel MIMD implemented by SIMD hardware. Technical Report TR-EE 92-4, School of Electrical Engineering, Purdue Univ., February 1992.

[5] High Performance Fortran Forum. High performance fortran language specification. Technical Report Version 1.0, January 1993.

[6] G.C. Fox. The architecture of problems and portable parallel software systems. Technical Report SCCS-78b, Syracuse University, 1991.

[7] W. D. Hillis and G. L. Steele. Data parallel algorithms. *Comm. of the ACM*, 29(12):1170–1183, December 1986.

[8] Paul Hudak and Eric Mohr. Graphinators and the duality of SIMD and MIMD. In *The 1988 ACM Conf. on Lisp and Functional Programming*, July 1988.

[9] Robert G. Babb II and David C. DiNucci. Design and implementation of parallel programs with large-grain data flow. In Leah H. Jamieson, Dennis B. Gannon, and Robert J. Douglass, editors, *The Characteristics of Parallel Algorithms*, pages 335–349. MIT Press, 1987.

[10] P. Kacsuk and A. Bale. DAP prolog: A set-oriented approach to prolog. *The Computer Journal*, 30(5):393–403, 1987.

[11] B. C. Kuszmaul. Simulating applicative architechtures on the connection machine. Master's thesis, MIT, 1986.

[12] M. S. Littman and C. D. Metcalf. An exploration of asynchronous data-parallelism. Technical Report YALEU/DCS/TR-684, Dept. of Computer Science, Yale University, October 1988.

[13] M. Nilsson and H. Tanaka. Massively parallel implementation of flat GHC on the Connection Machine. In *The Int. Conf. on Fifth Generation Computer Systems*, pages 1031–1039, 1988.

[14] M. Nilsson and H. Tanaka. MIMD execution by SIMD computers. *Journal of Information Processing*, 13(1), 1990.

[15] J. Seizovic. The reactive kernel. Technical Report Caltech-CS-TR-99-10, Computer Science, California Institute of Technology, 1988.

[16] J. Shen and J. A. McCammon. Molecular dynamics simulation of superoxide interacting with superoxide dismutase. *Chemical Physics*, 158:191–198, 1991.

[17] W. Shu and L. V. Kale. Chare Kernel — a runtime support system for parallel computations. *Journal of Parallel and Distributed Computing*, 11(3):198–211, March 1991.

[18] Thinking Machines Corp. *CM Fortran Reference Manual*, version 5.2-0.6 edition, September 1989.

[19] S. Tomboulian and M. Pappas. Indirect addressing and load balancing for faster solution to mandelbrot set on simd architectures. In *The 3rd Symposium on the Frontiers of Massively Computation*, pages 443–450, October 1990.

[20] Reinhard v. Hanxleden and Ken Kennedy. Relaxing SIMD control flow constraints using loop transformations. Technical Report CRPC-TR92207, Rice University, April 1992.

[21] W. F. van Gunsteren and H. J. C. Berendsen. GROMOS: GROningen MOlecular Simulation software. Technical report, Laboratory of Physical Chemistry, University of Groningen, The Netherlands, 1988.

[22] Marc Willebeek-LeMair and Anthony P. Reeves. Solving nonuniform problems on SIMD computers: Case study on region growing. *Journal of Parallel and Distributed Computing*, 8(2):135–149, February 1990.

[23] P. Wilsey, D. Hensgen, N. Abu-Ghazaleh, C. Slusher, and D. Hollinden. The concurrent execution of non-communicating programs on SIMD processors. In *The Fourth Symposium on the Frontiers of Massively Parallel Computation*, October 1992.

[24] M. Y. Wu and D. D. Gajski. Hypertool: A programming aid for message-passing systems. *IEEE Trans. Parallel and Distributed Systems*, 1(3):330–343, July 1990.

[25] M.Y. Wu and W. Shu. Scatter scheduling for problems with unpredictable structures. In *The Sixth Distributed Memory Computing Conference*, pages 137–143, April 1991.

# Evaluation of Data Distribution Patterns in Distributed-Memory Machines*

*Edgar T. Kalns, Hong Xu* and *Lionel M. Ni*

Department of Computer Science
Michigan State University
East Lansing, MI 48824-1027
{kalns,xuh,ni}@cps.msu.edu

**Abstract**

Determining an appropriate data distribution among different memories is critical to the performance of data-parallel programs on distributed-memory machines. By analyzing the computational load of data arrays and the communication complexity of various data movement operations in a program, this paper suggests a first-order cost model for determining a small set of appropriate data distribution patterns among many possible choices. A new data distribution specification, namely CYBLOCK, is proposed to enhance the expressiveness of data distribution specifications being proposed in High Performance Fortran. Cost analysis of two case studies: a linear system solver and a Purdue-set benchmark loop, are used to illustrate the proposed evaluation method. The model correctly predicts the relative performance of the case studies when implemented with various regular data distributions on an nCUBE-2 multicomputer.

## 1 Introduction

Distributed-memory machines, such as the Intel Paragon, TMC CM-5, nCUBE-2/3, and even workstation clusters, have demonstrated their high performance computing capability in solving so-called grand-challenge problems. However, the traditional message-passing programming model for these machines based on separate name spaces is tedious, time consuming, and error-prone for programmers. In contrast, the data-parallel or SPMD programming model based on a single name space provides an easier and familiar programming style for users. Due to their non-uniform memory access times, determining an appropriate data decomposition among different memories is critical to the performance of data-parallel programs on such distributed-memory machines. The data decomposition problem involves *data distribution*, which deals with how data arrays should be distributed, and *data alignment*, which deals with how data arrays should be aligned with respect to one another. The goal of data decomposition is to maximize the system performance by balancing the computational load and by minimizing remote memory accesses (or communication messages).

In order to provide high-level language support for data-parallel programming, several data-parallel Fortran extensions have been proposed, such as Fortran D [1] and Vienna Fortran [2]. In an effort to standardize data parallel Fortran programming, HPF (High Performance Fortran) is being proposed as a standard by the High Performance Fortran Forum led by Rice University [3] for distributed-memory machines.

An essential part of these data parallel Fortran extensions is the specification, through compiler directives, the distribution and alignment of data arrays. Typically, multiple data arrays are aligned to a *template* which is distributed among processors. The languages, however, do not provide the programmer any guidance in selecting a decomposition. Given the large number of distribution and alignment possibilities, it is difficult for programmers to select an appropriate data decomposition which maximizes program performance. The problem of finding optimal data storage patterns has been shown to be NP-complete [4]. Furthermore, the data alignment problem has also been shown to be NP-complete [5]. Consequently, several heuristics have been proposed to automatically determine an appropriate data decomposition scheme. A survey of some of these heuristics and their limitations and drawbacks can be found in [4]. However, due to their limited success, the programmer currently has no choice but to use a brute-force approach to select an appropriate decomposition, an impractical choice for programs with more than a few data arrays.

Focusing on two salient factors: computational load of a program and its data movement operations, this paper presents a simple model for assisting an HPF programmer in reducing the number of distribution candidates to a select few. We do not address

*This work was supported in part by the NSF grants CDA-9121641 and MIP-9204066.

the problem of data alignment in this paper. In contrast to other approaches which focus primarily on the data parallel Fortran code, our model focuses on both the HPF code and the compiler-generated message-passing code. Focusing on the latter code has particular merit. There are a number of optimizations such as overlapping of computation and communication, pipelining, and combining communication messages, which the compiler can exploit to generate more efficient code [6]. The choice of data distribution potentially enhances a compiler's ability to utilize such optimizations. Additionally, target architecture-specific features, such as communication latency, will greatly influence a node program's performance. For these reasons, our model addresses the computational load of data array elements and analyzes communication cost for both the HPF and message-passing node programs.

Section 2 reviews various data distribution methods proposed in HPF. A new and useful data distribution method is proposed to enhance the expressiveness of data distribution in HPF. Section 3 describes the computation cost model based on data array assignment statements. Section 4 describes the communication cost model. The tile model is applied to HPF code as a method for determining which dimensions of a data array ought to be distributed based on data reference patterns along dimensions. Sections 5 and 6 consider two case studies: solving a system of linear equations with Gaussian Elimination and Backward Substitution, and a Purdue-set benchmark loop. Section 7 concludes the paper.

## 2 Data Distribution Specification

**BLOCK** and **CYCLIC** distributions comprise two essential distributions, known as *regular* distributions, by HPF [3], Fortran D [1], and Vienna Fortran [2]. With the **BLOCK** specification, contiguous, evenly-sized segments of an array dimension are distributed to each processor. With the **CYCLIC** specification, elements of a dimension of an array are assigned to each processor in a round-robin fashion. These distributions are extended to the **BLOCK**($b$) and **CYCLIC**($c$) distributions in HPF [3]. With the **BLOCK**($b$) distribution, each processor is assigned a contiguous block of size $b$ while the **CYCLIC**($c$) distribution specifies the allocation of contiguous segments of size $c$ cyclically to each processor.

To achieve efficient code generation and runtime support, only the regular distributions described above are included in HPF [3], though arbitrary user-defined distributions may be supported at a cost of significant runtime overhead [2]. Therefore, it is important to assess whether these two distribution patterns suffice in efficiently supporting data distribution in a wide variety of scientific applications. Figure 1 introduces a loop which is extracted from Livermore benchmark kernel eight [7].

```
DO i=1,8
   DO j =2,7
s:    B(j)=B(j)+0.5*(A(i,j-1)+A(i,j+1))
   END DO
END DO
```

Figure 1. Loop $L_1$

The loop in Figure 1 can be found in a number of applications: finding numerical solutions to heating problems using the Jacobi method and integration problems using the alternating direction implicit method [7]. In Figure 1, the number of times each array element $B(j)$ is assigned on the *left-hand-side* (lhs) of the statement is equal to the size of $i$-th dimension in the rectangular iteration space, which is the constant 8. Based on the owner-computes rule, processor workload is balanced when each processor is assigned an equally sized subset of $B$. Performance is maximized if the number of remote references for array $A$ on the *right-hand-side* (rhs) of the statement is minimized.

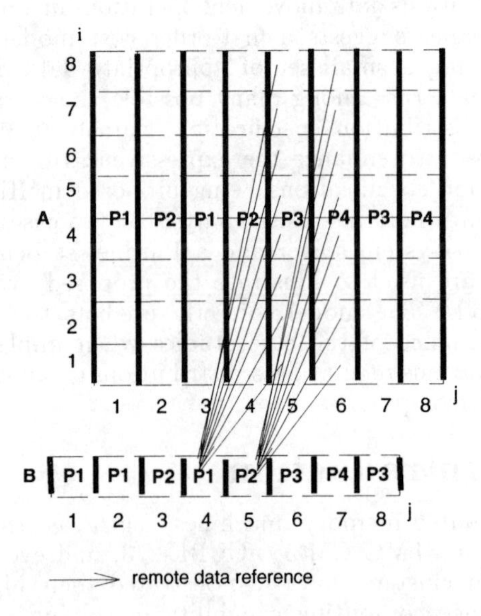


Figure 2. **CYBLOCK(2)** distribution for loop $L_1$.

The **CYBLOCK** distribution, a new regular distribution type, is introduced to minimize the number of remote references in loop $L_1$. Without loss of generality, we assume that the number of available processors is 4. As shown in Figure 2, $B$ is aligned with the $j$-th dimension of $A$ by allocating $B(j+1)$ and $A(1:8,j)$ on the same processor. The $i$-th dimension of $A$ is collapsed. The $j$-th dimension of $A$ is distributed in **CYBLOCK(2)** as follows. First, due to its access offset collection of $\{-1,1\}$, the $j$-th dimension of $A$ is partitioned in a cyclic fashion with the stride of 2. Therefore, $A(1{:}8, 1)$, $A(1{:}8, 3)$, $A(1{:}8, 5)$, and $A(1{:}8,7)$ constitute one partition, and $A(1{:}8, 2)$, $A(1{:}8,4)$, $A(1{:}8,6)$, and $A(1{:}8,8)$ constitute the other.

Consequently, four processors are partitioned into two partitions: $p_1$ and $p_3$, $p_2$ and $p_4$. Second, elements in each partition are further distributed in BLOCK with respect to the two processors in the same partition. As a result, $A(1{:}8,1)$ and $A(1{:}8,3)$ are allocated to processor $p_1$, and $A(1{:}8,5)$ and $A(1{:}8,7)$ are allocated to processor $p_3$. This two-step distribution, which follows a cyclic-first-block-second pattern, is defined as CYBLOCK for short. As shown in Figure 2, in the CYBLOCK distribution, only processors $p_1$ and $p_2$ require remote memory references, for a total of sixteen references. However, all four processors require remote references for sixteen array elements when either the BLOCK or CYCLIC distributions is used. For the loop model of $L_1$, CYBLOCK achieves better performance than either of the regular HPF distributions.

In CYBLOCK($c$) , a dimension of an array is first partitioned cyclically with stride $c$. Meanwhile, $p$ processors are also evenly partitioned into $c$ disjoint groups with $\frac{p}{c}$ processors in each group. In the second step, array elements grouped in the same partition are distributed in BLOCK with respect to $\frac{p}{c}$ processors allocated to that partition. By its definition, CYBLOCK(1) is equivalent to BLOCK , and CYBLOCK($p$) is equivalent to CYCLIC .

In the specification of CYBLOCK($c$) , $p$ may not be divisible by $c$. In fact, $p$ could even be smaller than $c$. The CYBLOCK distribution under these special conditions is resolved as follows. Regardless of the value of $p$, elements along the distributed dimension are always partitioned first cyclically with stride $c$. When $p < c$, elements within the same partition are assigned to the same processor, and distinct partitions are distributed cyclically to $p$ processors, each partition treated as a single unit for distribution. When $p > m$ but not divisible by $c$, processors are partitioned into $c$ clusters with varying size. The first $r$ clusters contain $\lceil \frac{p}{c} \rceil$ processors, while the remaining $c-r$ clusters contain $\lfloor \frac{p}{c} \rfloor$ processors, where $r$ is the remainder of $p$ divided by $c$. Elements in the same partition are further distributed in BLOCK with respect to processors available in the local cluster. Let $A(1{:}12)$ be distributed in CYBLOCK(3). Figure 3 (a) shows the data allocation pattern when $p$ is equal to 2. Figure 3 (b) shows the data allocation pattern when $p$ is equal to 4. For the formal specification of CYBLOCK($c$) and the regular HPF distribution patterns, see [8].

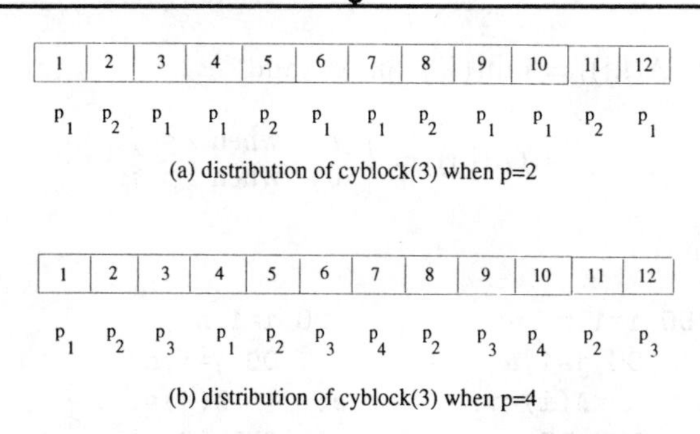


Figure 3. Special cases of CYBLOCK($c$) distribution

# 3 Computation Cost Model

An important performance factor on distributed-memory machines is achieving a balanced workload among available processors. To exploit program inherent parallelism, HPF supports processor workload distribution by data decomposition specifications. An *owner-computes rule* is employed by HPF [3], which specifies that a variable can only be written by the processor which owns it. As a result, the workload assigned to each processor is proportional to the number of times that array elements owned by that processor have been assigned. Therefore, the allocation strategy for *lhs* array elements on each assignment statement determines processor workload balance. Generally speaking, the program execution time for a given machine is unknown without actually executing it. Data decomposition, however, is determined prior to compilation. Clearly, a computation estimation function is needed to assist in data distribution selection.

Let $V(q)$ be the set of array elements owned by processor $q$ and $S$ be the set of assignment statements in the program. For simplicity, $S$ only contains those assignment statements which appear in loop bodies. $c(q)$, the computation cost estimation on the processor $q$, can be formally defined below.

$$c(q) = \sum_{s \in S, e \in V(q)} f_s(e) w_s$$

In the above definition, the *frequency*, $f_s(e)$, represents the number of times the array element $e$ is written in the assignment statement $s$, and the *weight*, $w_s$, represents the estimation for the computation on the right hand side of the statement for each assignment on $e$. $f_s(e)$ is zero if $e$ is not referenced on the *lhs* of the statement $s$. With the SPMD mode of computation, $w_s$ is likely to vary minimally for distinct *lhs* elements for a given assignment statement $s$.

The weight $w_s$ can be calculated based on the number of integer or floating-point operations involved in a particular implementation. For instance, in loop $L_1$, the weight $w_s$ is equal to two floating-point addition operations and one floating-point multiplication operation. The amount of computation can be normalized by converting non-floating point operations to the equivalent number of floating-point operations.

Suppose $e$ is an element of an array $A$. Given an assignment statement $s$ in a loop L, $f_s(e)$ is zero if $A$ does not reside on the *lhs* of statement $s$. $f_s(e)$ is a function of the dimension of $A$ and the shape of the iteration space of the loop if $A$ is the *lhs* variable of the statement $s$. Figure 4 shows a few common loop patterns extracted from scientific application programs. In Figure 4, the array element $e$ refers to either a single element $A(i)$ of a one-dimensional array or a single element $A(i,j)$ of a two-dimensional matrix. It can be proved that $f_s(\text{A(i)}) = n$ in (a), $f_s(\text{A(i)}) = i$ in (b),

$f_s$(A(i,j)) = min $(i, j)$ in (c), and

$$f_s(\text{A(i,j)}) = \begin{cases} i & \text{when } i \le j \\ 0 & \text{when } i > j \end{cases}$$

in (d).

```
 DO i=1,n                    DO i=1,n
    DO j=1,n                    DO j=i,n
s:     A(i)=...           s:       A(j)=...
    END DO                      END DO
 END DO                      END DO
         (a)                          (b)
 DO k=1,n                    DO k=1,n
    DO i=k,n                    DO i=k,n
       DO j=k,n                    DO j=i,n
s:        A(i,j)=...      s:          A(i,j)=...
       END DO                      END DO
    END DO                      END DO
 END DO                      END DO
         (c)                          (d)
```

Figure 4. Some common loop patterns

Let $V$ be the set of all *lhs* array variables in a program. Data decomposition can be formalized by partitioning $V$ into $p$ subsets such that the following conditions are satisfied:

$$\bigcup_{q=0}^{p-1} V(q) = V$$

$$V(q_i) \cap V(q_j) = \phi \text{ where } q_i \neq q_j$$

The task of optimal data decomposition is to find a partition such that the difference between $\min_{0 \le q \le p-1} c(q)$ and $\max_{0 \le q \le p-1} c(q)$ is minimal. Consider the workload on a single statement $s$. Any of the regular distributions introduced in the previous section can balance the processor workload as long as the frequency $f_s$ is uniform, such as in Figure 4 (a). Cyclic distributions usually can achieve better workload balance if the frequency $f_s$ is a linear function, such as in Figure 4 (b), (c), and (d). However, it is much more difficult to determine the optimal data decomposition pattern for an arbitrary frequency function $f_s$.

# 4 Communication Cost Model

In addition to computational load considerations discussed in Section 3, the choice of data distribution should consider data reference patterns which dictate the required data movement operations of a program. Data movement operations include scatter, gather, replication, reduction, permutation, and segmented scan [9, 10]. Minimizing remote data references at the HPF program level will reduce the communication overhead of the resulting message-passing code. If computational load is uniform over all data elements, then minimizing remote data references becomes the significant performance factor. In this section, we focus on distribution selection based solely on data movement operations. Our approach has two major steps. First, we model the data movement in the HPF code and choose the appropriate data dimensions for distribution, reducing the search space of possible distribution choices. Second, we incorporate knowledge of the target machine architecture, and model the message-passing implementations of a small set of distributions, choosing the most appropriate candidates which minimize remote data references.

Before deciding upon the type of distribution pattern, e.g. block and cyclic, the HPF programmer must determine which dimension, or dimensions, of the data template to distribute. To make an initial determination of which dimension(s) to consider for distribution, we examine the remote data references of programs by partitioning the data template into *tiles*. Our first-order cost model considers two-dimensional data. Future work will address generalizing our techniques to $n$-dimensional data.

**Definition 1** *A $n \times m$ HPF template $T$ is partitioned into a set of $t$ non-overlapping tiles, each of size $q = h \times w$, where $t \times q = n \times m$.*

Each of the four tile faces, $c_n$ (north), $c_s$ (south), $c_e$ (east), and $c_w$ (west), has an associated inter-tile data movement cost. The cost is determined by the number of data elements referenced by the tile which lie in other tiles along the given direction. Figure 5 depicts a generic tile.

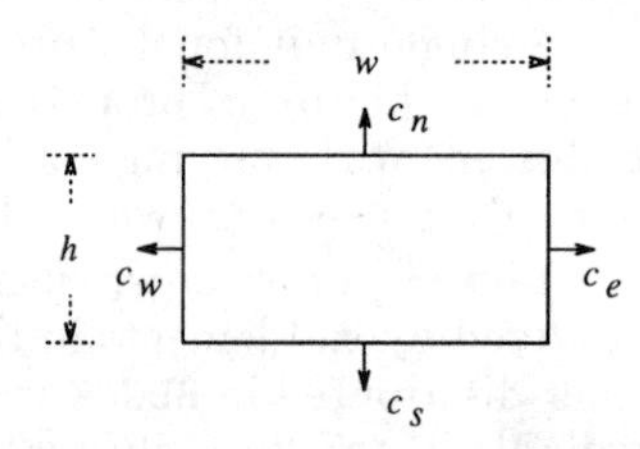


Figure 5. Generic tile with cost parameters

We look at the data movement patterns implied by specific loops in an attempt to tailor a tile's shape with the application, thus determining intertile communication cost. For the loop in Figure 6 (a), a distribution of tiles in the row dimension is optimal, since inter-tile data movement is eliminated and the data dependence between columns $j$ and $j-3$ is eliminated. The $n \times m$ matrix is partitioned into $k$ tiles, each of size $\frac{n}{k} \times m$. Distributing data by columns leads to inter-tile data movement and limits the amount of possible parallelism due to the data dependence. Therefore, we significantly reduce the search space of possible HPF distributions by only considering block and cyclic patterns for rows.

A more complicated example, common to many numerical applications, is the five-point stencil loop in

```
DO  i =  1,n
   DO  j = 4,m
      A(i,j) = A(i,j) + A(i,j-3)
   END DO
END DO
```

(a) Data movement between columns

```
DO  i =  2,n-1
   DO  j = 2,m-1
      A(i,j) = (A(i-1,j)+A(i+1,j)+
                  A(i,j-1)+A(i,j+1))/4
   END DO
END DO
```

(b) Five-point stencil problem

Figure 6. Loops to illustrate tile method

Figure 6 (b). This problem exhibits *uniform* communication since all elements require the same number of remote data references. Each data item, except for matrix boundary elements, accesses the data item immediately to its north, south, east, and west. Figure 7 illustrates the distribution choices for the five-point stencil loop. Part (a) shows a distribution of $k$ tiles in one dimension, each of size $\frac{n}{k} \times m$. Without loss of generality, assume $n \geq m$. The data movement cost is $c_{one} = 2(k-1)(m-2)$. We choose to distribute along the longer dimension since this always leads to a smaller cost than distributing along the the shorter dimension for this loop. Figure 7(b) shows a distribution of $k = i \times j$ tiles in two dimensions, each of size $\frac{m}{j} \times \frac{n}{i}$. The data movement cost in two dimensions is $c_{two} = 2(j-1)(n-2) + 2(i-1)(m-2)$. With the loop in Figure 6 (b), data movement is necessary only for those elements along a tile boundary, thus when $\frac{n}{i} \sim \frac{m}{j}$ the cost for part (b) will be minimized.

Minimizing $k$, the number of tiles, minimizes the data movement cost for both options. We assume that other factors, such as load balancing and the number of available processors will influence the choice of $k$. Assuming a fixed $k$, the tile model provides a straightforward cost analysis method based on data reference patterns which assesses the utility of distributing data in distinct dimensions. While the tile model provides an important first attempt at determining the dimensions for distribution, narrowing the search space, it has several limitations. One drawback is that the model attempts performance decisions at the HPF level based exclusively on data movement patterns, not considering the actual message passing code which must execute on the target architecture. Data movement patterns resulting from coupled subscripts for iterations present an added complication which the model does not address. Such costs would necessitate additional direction vectors besides the four already defined. Next, we examine the choice of data pattern, e.g. block, cyclic.

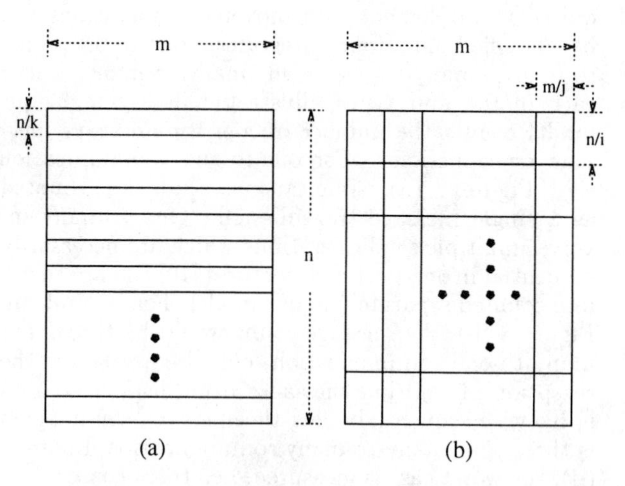


Figure 7. Two- and One-dimensional tile distributions

The possible message-passing code implementations for a given HPF program may vary greatly depending upon the target machine architecture and the optimizations exploited by the compiler. Optimization techniques such as pipelining, combining messages, and collective communication primitives can be utilized to effect performance. These factors ought to influence the choice of distribution patterns. We conclude that the tile model itself is not sufficient, it is imperative to perform data movement cost analysis at the message-passing code level to determine the most appropriate data distribution pattern.

Once the dimensions of distribution are determined, we propose a communication cost model which analyzes the amount of interprocessor communication based on the algorithm and the chosen data distribution. For a given algorithm and a set of distribution choices, the model analyzes each distribution pattern in terms of the target-architecture specific message passing operations required to implement the HPF code and selects the data pattern which minimizes the amount of data movement as with the tile method. Modification of the architecture-specific parameter values would suffice for using the model to predict performance on other machines, thus enhancing the model's portability to a wide variety of architectures. To accurately measure the unicast (one-to-one communication) message transmission time on a wormhole-routed system [11], the formula $msg(\ell) = \alpha + \beta(\ell)$ $\mu$seconds is used. $\alpha$ is subdivided into $\alpha_s$, the sending latency, and $\alpha_r$, the time required for retrieving a message from the network; $\ell$ is the message length and $\beta$ is a multiplicative factor. Network latency is the product $\beta(\ell)$. For the nCUBE-2, $\alpha_s$ is about 95 $\mu$seconds, $\alpha_r$ is approximately 75 $\mu$seconds, and $\beta$ is about 0.57. For a particular data movement operation, we determine the size of the message, apply the above formula, and then determine the number of times the message must be transmitted. Only unicast-based communication is considered here.

Fundamentally, the communication services required to implement data movement operations can be classified into four categories: one-to-one, one-to-many, many-to-one, and many-to-many. For each of the four types, illustrated in Figure 8, our model counts the number of *non-simultaneous* message transmissions. For one-to-one communication e.g., Figure 8 (a), simultaneous sends are counted as a single message transmission. One-to-many involves multiple send operations which are necessarily sequential in one-port architectures [10] and are therefore counted separately in our model. For example, in Figure 8 (b), the message count would be three. For many-to-one communication, e.g., Figure 8 (c), the reception of multiple messages from distinct sources is likewise sequential, thus the message count again is three. For many-to-many communication, Figure 8 (d), the worst case is measured, i.e., three consecutive sends by processor $p_0$.

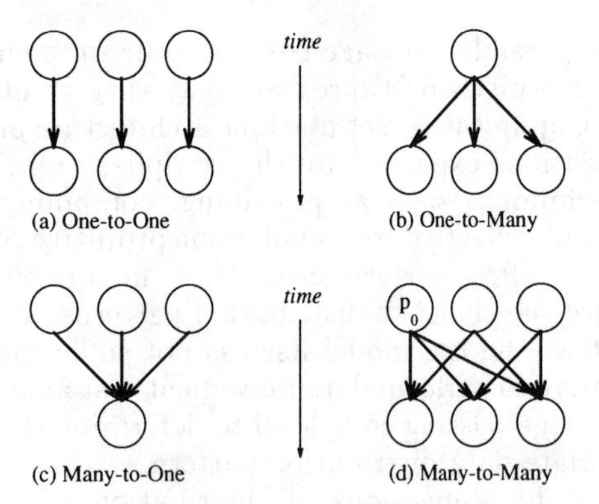


Figure 8. Interprocessor communication categories

The model has several limitations as well. First, it is a manual process since the analysis is done by the programmer without compiler assistance. Second, communication network contention among different messages which may increase the message transmission time is not considered. Thus, the communication time prediction may underestimate the actual cost of communication in the system. Third, our model considers the transmission time of $k$ $\ell$-length messages to be $k \times msg(\ell)$; however, if a non-blocking send primitive, as on the nCUBE-2, is used for consecutive message transmissions, and if there is no channel contention, then the actual transmission time could be much less than our prediction due to the overlapping of communication. Figure 9 illustrates the second and third points. All three messages represent consecutive messages from the same source. Messages $msg_1$ and $msg_2$ are routed on mutually exclusive channels and thus suffer no network contention. The source node is free to begin the sending of $msg_2$ as soon as $\alpha_s$ time has expired, producing communication overlap. $msg_3$ begins after $2\alpha_s$, but it is blocked due to channel contention and must wait, in the network, for an unknown amount of time. Effectively quantifying

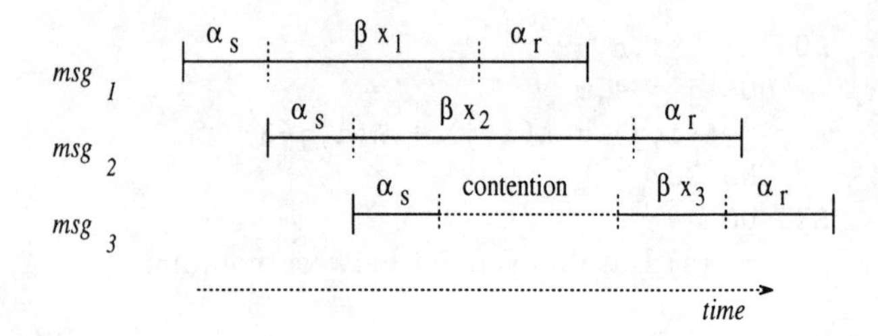


Figure 9. Communication overlap and channel contention

and incorporating dynamic network conditions is a difficult problem and merits further research. Despite these shortcomings, our approach provides a consistent prediction for the relative execution time of algorithms using different regular data distributions.

If an implementation elicits uniform computation but *non*-uniform communication, then the communication cost model is the better to compare data distributions. Conversely, *non*-uniform computation but uniform communication implies that the computation model would be preferable. Most applications, however, yield both non-uniform computation and communication. For such situations, we combine the computation and communication models by normalizing the architecture-specific parametric costs of each model and add them for an overall cost. Simply summing the normalized costs of each model, however, fails to consider the possible overlap of computation and communication in an implementation. Such a combined cost model considering the possibility of overlap must be fine-tuned for the given application. We illustrate the methodology with two real-world applications.

# 5 Case Study: Linear System Solver

The computation and communication cost models, Sections 3 and 4, are applied to solving a system of linear equations, $Ax = b$, using Gaussian elimination and backward substitution. The linear system solver is coded in HPF [8] Using our models, we estimate the computation and communication costs of the algorithm and compare our results with nCUBE-2 message-passing implementations. Our Linear System Solver was devised strictly to validate our cost model; see [12] for an extensive study of optimal matrix factorization algorithms.

Let $s$ be the elimination step (doubly-nested loop) of Gaussian elimination [8]. Based on our computational cost model, $f_s(A(i,j)) = \min(i,j) - 1$ for $1 \le i,j \le n$ where $n$ is the dimension of the matrix. Figure 10 illustrates the non-uniform computational cost per element for the algorithm. The amount of computation per element is greater for elements in the direction of the lower right corner of the matrix.

Using the tile method, we conclude that row

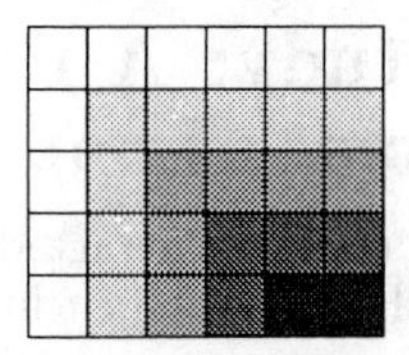

Figure 10. Non-uniform load in Gaussian elimination

distributions lead to slightly *less* data movement than column distributions. We apply our message-passing cost model to two HPF distributions for rows, (BLOCK, *) and (CYCLIC, *). For detailed analysis of column distributions, see [8]. The CYBLOCK(m) distribution would not be an appropriate choice for this application; its utility is demonstrated in Section 6.

For each pattern, the model estimates the communication cost (in $\mu$seconds) for various data sizes and processor configurations. The number of unicast messages required to implement data movement are counted and the nCUBE-2 message transmission formula $\alpha + \beta(\ell)$ is applied. Table 1 and Figure 11 show the analysis of the Linear System Solver for the two distributions. Each operation of the algorithm involv-

Table 1. Communication Cost for Linear System Solver (in $\mu$seconds)

| Operation | Data Distribution | |
|---|---|---|
| | (block, *) | (cyclic, *) |
| Repl Location | $t_1$ | $t_1$ |
| Repl Pivrow | $t_2$ | $t_3$ |
| Repl Solution | $t_4$ | $t_5$ |
| Total | $t_1 + t_2 + t_4$ | $t_1 + t_3 + t_5$ |

ing data movement is listed The first two operations are part of Gaussian elimination and the third operation is due to backward substitution. Note that the (CYCLIC, *) distribution requires slightly more data movement than (BLOCK, *) as reflected by expressions $t_3$ and $t_5$. The former distribution requires replication of data to all processors except for the last $p-1$ iterations of the algorithm whereas the latter does not.

Based on the message-passing analysis, the cost difference between the two distributions is small. An additional factor to consider is the overlapping of communication and computation of processors in the *Repl Pivrow* operation, the most significant source of communication. Figure 12 illustrates the situation. In part (a) using (BLOCK, *), processor $p_1$, replicates the first computed pivot row incurring a startup latency, $\alpha_s$. The elimination step of the algorithm follows, denoted by **comp**, requiring computation on $(\frac{n}{p} - 1) \times (n-1)$ data elements. All other processors perform the same computation on $(\frac{n}{p}) \times (n-1)$

$in$ = 4 bytes (size of integer)
$fl$ = 4 bytes (size of floating-point)
$n$ = matrix dimension
$p$ = number of processors
$\alpha$ = 170
$\beta$ = 0.57
$x$ = length of message in bytes
$m(x) = \alpha + \beta(x)$ $\mu$sec $\equiv$ msg. trans. time

$$t_1 = n(p-1)(m(in))$$
$$t_2 = \sum_{i=1}^{p-1}(p-i)(\sum_{j=1}^{\frac{n}{p}} m(fl \times (n - j - i(\tfrac{n}{p}))))$$
$$t_3 = \sum_{i=1}^{p-1}(p-1)(\sum_{j=1}^{\frac{n}{p}-1} m(fl \times (n - j - i(\tfrac{n}{p}))))$$
$$t_4 = \frac{n(p-1)}{2}(m(fl))$$
$$t_5 = [(n-p)(p-1) + \frac{p(p-1)}{2}][m(fl)]$$

Figure 11. Data Movement Formulas

data items. Due to the block distribution, $p_1$ owns rows $2..\frac{n}{p}$ of the matrix. As $p_1$ proceeds through its row block, its computation, **comp**, quickly decreases relative to other processors, e.g. for iteration $i$, $p_1$ performs the elimination step on $(\frac{n}{p} - i) \times (n - i)$ elements whereas the other processors perform computations on $(\frac{n}{p}) \times (n-i)$ elements. Thus, due to a relatively smaller data block, $p_1$ finishes its elimination step and begins replication of the next pivot row while other processors complete the elimination step for their relatively larger data blocks, resulting in a significant overlap of communication latency with computation. This is particularly significant near the beginning of the algorithm since the size of the pivot row depends on the iteration, i.e. its size is $n - i$ for iteration $i$.

Figure 12(b) illustrates the situation for (CYCLIC, *). Successive pivot rows are distributed among processors $p_1, p_2, ..., p_p$ and each processor performs the elimination step on blocks of equal size, $(\frac{n}{p} - i) \times (n-i)$ for iteration $i$, denoted by **comp**. In contrast to (BLOCK, *), $p_2$ owns row 2 and therefore must receive the first pivot row from $p_1$, incurring a latency of $\beta(\ell) + \alpha_r$ prior to pivot row replication and the elimination step. As the elimination step on all processors involves equally sized blocks, they will finish at approximately the same time. Thus, while $p_2$ incurs the stated latency and performs a pivot row replication, other processors wait idly.

To determine the most appropriate data distribution, we are concerned with assessing the *relative* performance between different distributions. The analysis from Table 1 and Figure 11 consistently predicts the relative performance of the different distributions for all values of $n$ in the 32- and 64-node configurations, accurately projecting that execution time will be greater for (CYCLIC, *). Figure 13 illustrates

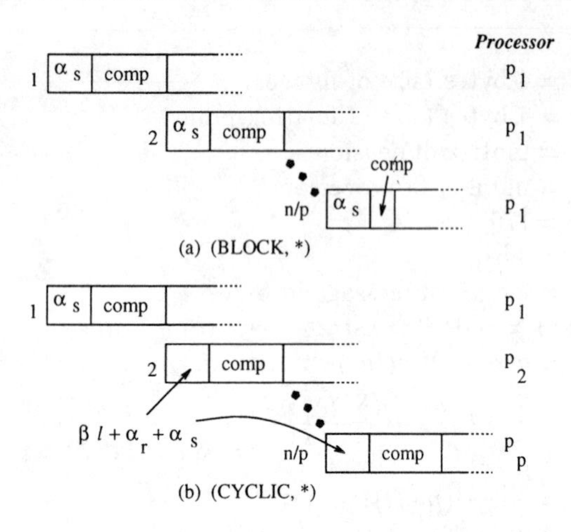


Figure 12. Computation and communication overlap

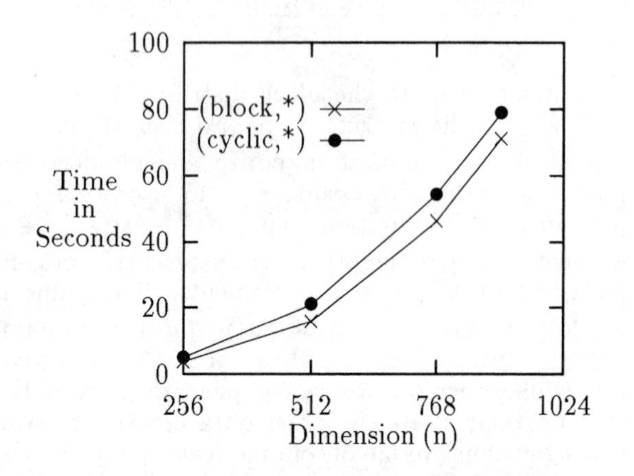


Figure 13. Execution time (32 processors)

the actual execution time for various data sizes of the row-distributed message-passing programs on an nCUBE-2 using a 32-processor configuration. For a 16-processor configuration, the model incorrectly predicts the relative performance of `(BLOCK, *)` to sustain better performance for large (i.e. $> 256$) data sizes; however, the margin of error is small. Due to the ratio of large $n$ and $p = 16$, load balancing is a more significant factor than communication, resulting in `(CYCLIC, *)` obtaining better performance. For a more detailed discussion on the significance of the ratio of $n$ and $p$, see [13]. Combining the computation and communication analysis techniques and including the overlap analysis described above would strengthen the model's predictive capability. This will be an area for future research.

# 6 Case Study: A Purdue-Set Benchmark Loop

The loop in Figure 14 is extracted from larger computations in the Purdue-set benchmark [14].

```
      REAL A(n),B(n),C(n),D(n)
      TEMPLATE T(n)
!HPF$ ALIGN A(:),B(:),C(:),D(:) WITH T(:)
      DO i =1,n-2
         D(i)=A(i)*B(i+2)+A(i+2)*B(i)-C(i)
      END DO
```

Figure 14. A Purdue-set benchmark loop

In Figure 14, the Purdue-loop for short, there exist no loop-carried dependencies, resulting in no remote data references in the loop body. Different messages from different iterations of the loop can be combined. Since each *lhs* array element $D(i)$ is only assigned once in the loop body, processor workload balance can be achieved with any one of the regular distributions. In Figure 14, $T$ is declared as the template for the virtual index space. Array $A$, $B$, $C$, and $D$ are aligned with $T$. In this section, we compare the relative performance of the Purdue-loop for different distributions of `BLOCK` , `CYCLIC` , and `CYBLOCK(`$c$`)` . Table 2 gives the communication cost for each type of the three distributions.

Table 2. Communication cost for Purdue-loop

| Distribution | Unicast Comm. Complexity |
|---|---|
| block | $4 \times msg(4)$ |
| cyclic | $2 \times \lceil \frac{n}{p} \rceil \times msg(4)$ |
| cyblock(2) | $2 \times msg(4)$ |

When the template $T$ is distributed in `CYBLOCK(2)`, two remote references are required by each processor for all iterations, one for $A$ and the other for $B$. Similarly, when $T$ is distributed in `BLOCK` , four remote references are required by each processor for all iterations, two for $A$ and two for $B$. For the `CYCLIC` distribution, two remote references are required in each iteration. The number of remote references is $2 \times \lceil \frac{n}{p} \rceil \times msg(4)$ where $\lceil \frac{n}{p} \rceil$ is the number of elements assigned to each processor. Figure 15 shows the real execution time of the Purdue-loop on the nCUBE-2.

# 7 Conclusions

The selection of a data distribution for a parallel algorithm on distributed-memory machines has a great impact on the resulting performance, particularly affecting the amount of interprocessor communication

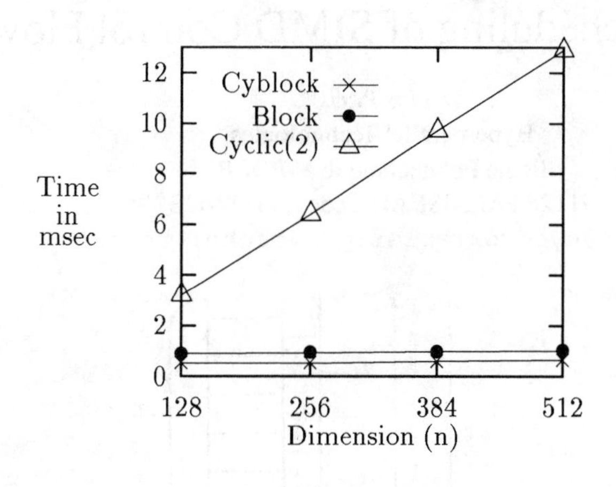


Figure 15. Execution time with 16 processors

and computational load of each processor. Currently, there are no widely-accepted techniques beyond a brute-force approach for selecting an appropriate distribution. Apart from the many possible choices, determining an appropriate data distribution using a brute-force method may be further complicated by large algorithms in which different distributions for different phases of the algorithm may result in better overall performance.

We proposed a new distribution which addresses the limitations of the current HPF data distributions. We proposed computation and communication cost models for analyzing the workload and data movement of data parallel programs implied by various distributions. The high-level models attempt to select the appropriate dimensions for distribution, while the low-level model selects a distribution pattern based on minimal communication cost. Although our techniques provide a first-order estimation, the effectiveness of the techniques are demonstrated by their accurate prediction of *relative* performance for different message-passing codes utilizing different distributions. In contrast to other techniques, we incorporate computation and communication parameters that are specific to the target architecture to better estimate a distribution's performance.

## References


1. G. Fox, S. Hiranandani, K. Kennedy, C. Koelbel, U. Kremer, C.-W. Tseng, and M.-Y. Wu, "Fortran D language specification," Tech. Rep. COMP TR90-141, Rice University, Department of Computer Science, Dec. 1990.

2. H. Zima, P. Brezany, B. Chapman, P. Mehrotra, and A. Schwald, *Vienna Fortran: A Language Specification (Version 1.1)*, 1991.

3. High Performance Fortran Forum, "High Performance Fortran Language Specification (version 1.0, draft)," Jan. 1993.

4. M. Gupta and P. Banerjee, "Demonstration of automatic data partitioning techniques for parallelizing compilers on multicomputers," *IEEE Transactions on Parallel and Distributed Systems*, vol. 3, pp. 179–193, Mar. 1992.

5. J. Li and M. Chen, "The data alignment phase in compiling programs for distributed-memory machines," *Journal of Parallel and Distributed Computing*, vol. 13, pp. 213–221, Oct. 1991.

6. S. Hiranandani, K. Kennedy, and C.-W. Tseng, "Compiling Fortran D for MIMD distributed-memory machines," *Communications of the ACM*, vol. 35, pp. 66–80, Aug. 1992.

7. Northeast Parallel Architectures Center at Syracuse University, "HPF/Fortran-D Benchmarking Suite (release 2.01)," 1992. (Available at Public Domain at Syracuse University).

8. E. T. Kalns, H. Xu, and L. M. Ni, "Evaluation of data distribution patterns in distributed-memory machines," Tech. Rep. MSU-CPS-ACS-83, Department of Computer Science, Michigan State University, East Lansing, MI., Apr. 1993.

9. J. J. Dongarra, R. Hempel, A. J. G. Hey, and D. W. Walker, "A Proposal for a User-Level, Message Passing Interface in a Distributed-Memory Environment," Tech. Rep. TM-12231, Oak Ridge National Laboratory, March 1993.

10. P. K. McKinley, H. Xu, E. T. Kalns, and L. M. Ni, "ComPaSS: Efficient communication services for scalable architectures," in *Proceedings of Supercomputing'92*, pp. 478–487, Nov. 1992.

11. L. M. Ni and P. K. McKinley, "A survey of wormhole routing techniques in direct networks," *IEEE Computer*, vol. 26, pp. 62 – 76, Feb. 1993.

12. G. A. Geist and M. T. Heath, "Matrix Factorization on a Hypercube Multiprocessor," in *Proceedings of the First Conference on Hypercube Computers and Concurrent Applications*, (Knoxville, TN), pp. 161–180, Aug. 1985.

13. L. M. Ni, H. Xu, and E. T. Kalns, "Issues in scalable library design for massively parallel computers," Tech. Rep. MSU-CPS-ACS-82, Department of Computer Science, Michigan State University, East Lansing, MI., Mar. 1993.

14. J. Rice and J. Jing, "Problems to test parallel and vector languages," Tech. Rep. CSD-TR-1016, 1990.

# Activity Counter:
# New Optimization for the dynamic scheduling of SIMD Control Flow

*Ronan Keryell* *
Centre de Recherche en Informatique
École des Mines de Paris
77305 FONTAINEBLEAU Cedex, FRANCE
keryell@cri.ensmp.fr

*Nicolas Paris* *
Hyperparallel Technologies
École Polytechnique X-POLE
91128 PALAISEAU Cedex, FRANCE
paris@hyperparallel.polytechnique.fr

## Abstract

SIMD or vector computers and collection-oriented languages, like C*, are designed to perform the same computation on each data item or on just a subset of the data. Subsets of processors or data items are implemented via an *activity* bit and a stack of activity bits when subsets of subsets are supported. This method is also used in VLIW processors through *if-conversion* to implement parallel control flow as in SIMD computers. We present a new method of dynamic sheduling of several SIMD control flow constructions which can be nested. Our implementation of activity stacks is based on *activity counters*. At a given stack depth $n$, the number of memory bits required is $\log_2 n$, whereas previous implementations require $n$ bits. The local controller is of equivalent complexity in both cases. This algorithm is useful for SIMD, vector or VLIW machines and for compilers of collection-oriented languages on MIMD computers.

## 1 Introduction

The data-parallel programming model is seen as an acceptable solution to efficiently program many parallel applications on massively parallel machines. In this model, a single program is applied on different instances of data, spread across different processors, to gain use of parallelism on SIMD or MIMD machines.

In an SIMD computer there is a unique instruction flow and thus performs the same operation on different data. But a lot of numerical problems, like solving partial differential equation problems, often need to apply different at the boundary conditions which are different from the ones used on the interior points. Such a control flow is often introduced from a sequential program through vectorization and *if-conversion* [1].

A similar problem arises in data-parallel collection-oriented languages like MPL, C* or POMPC [11] where an SIMD-like control flow must be managed even through function or procedure boundaries not known at compile time or with recursion. So a dynamic SIMD control flow is needed to follow the locked-step SIMD semantics.

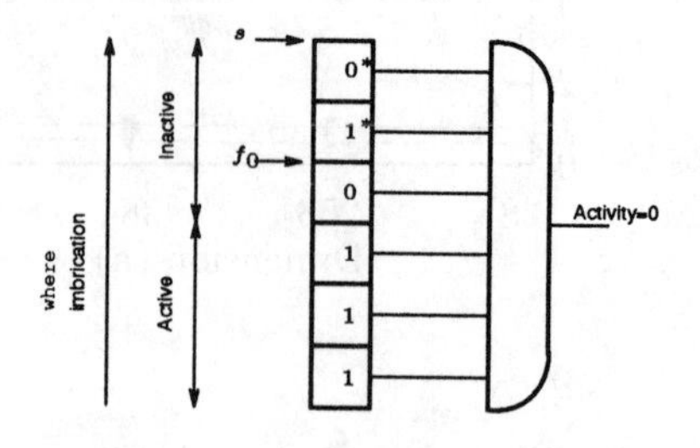


Figure 1: Example of a mask stack.

The goal of these parallel approaches is to obtain maximal performance on straight regular data parallel problems. However, it is at least as important to deal correctly with data parallel control flow and its flexibility.

There seems to be an intrinsic contradiction in the commonly used SIMD control flow model and the need for a local instruction stream, *i.e.* a bounded dissynchronization in the synchronous SIMD model, to deal with parallel control flow. This contradiction is resolved in turning off some processor elements (PEs) according to local conditions in SIMD machines, the *activity*. The nestling of several parallel `if` is usually managed with an activity stack but here we present an optimization of this method with an *activity counter* instead of a stack.

Section 2 presents our new algorithm with some examples applied to POMPC parallel control flow operators. Section 3 compares the activity stack with our method according to time and space complexity, for SIMD and MIMD, hardware and software. Section 4 presents related work.

## 2 Activity counter

If we carefully look at the activity bit stack, we see it is only used to determine the level of inactivity. Figure 1 shows an example of a nest of 6 parallel control flow statements, where the first three ones have `true` conditions (shown as "1" in the figure) and the condition is `false` after the third one (represented by "0").

Before the first `false` condition, the stack only contains 1s, indicating that the PE is executing the code. The exit of a conditional block does not change this activity:

*Major parts of this work were made when the authors were with the Laboratoire d'Informatique de l'École Normale Supérieure, 45 Rue d'ULM, 75005 PARIS, FRANCE. This research and the POMP project were partially funded by the French Research and Technology Ministry, Thomson Digital Image, the CNRS (National Center of Scientific Research), the LIENS, the École Normale Supérieure, the PRC-ANM.

Table 2: Semantic of the push and pop operations on the activity counter.

| Operation | Precondition | Action |
|---|---|---|
| push(*cond*) | $c \neq 0$ | $c \leftarrow c+1$ |
| | $(c=0) \wedge (cond=0)$ | $c \leftarrow 1$ |
| | $(c=0) \wedge (cond=1)$ | $c \leftarrow 0$ |
| pop | $c \neq 0$ | $c \leftarrow c-1$ |
| | $c=0$ | $c \leftarrow 0$ |

the PE remains active. These 1s do not have any intrinsic significance in the stack.

When a PE reaches a local *false* condition, it becomes inactive for all its included blocks. The current activity is the logical *and* of the history of activity, *i.e.* all the activity bits on the stack. Once a 0 bit is pushed on the stack, all the following bits on the stack no longer have meaning (represented with a "*" in Figure 1) since the activity is 0 (inactive).

## 2.1 Factorization

Indeed the only useful information in this stack is the nestling level of parallel conditional blocks after the first idle block, which indicates when a PE can resume execution. Therefore, it seems a waste of hardware to use a stack where a plain counter should be enough.

Let push(*cond*) and pop be the two operations controlling the stack $(a_i)_{i \in \mathbb{N}}$. We can analyze their functionality according to $f_0$, the rank of the first 0 on the stack, and $s$ the current size of the stack, according to Figure 1. The activity of a PE is defined by $\mathcal{A} = \bigwedge_{i=0}^{s-1} a_i$. The PE is active if $\mathcal{A} = 1$ and idle if $\mathcal{A} = 0$.

By definition, PEs are all active at initialization time, so $s = 1$, $a_0 = 1$ (active), $f_0 = s + 1$ when there is no 0 in any stack element. For simplicity a pop on an empty stack returns an activity *true*.

Table 1 gives an operational semantics of the activity stack. A PE is active if and only if $f_0 = s + 1$, when there is no 0 in the stack. In fact, it is more interesting to do the variable exchange $c = s+1-f_0$ because only a comparison to 0 is necessary. This form is easier to implement in hardware and often even in software [7, 8]. The basic manipulations on $c$ are the same as on $f_0$: increment or decrement, load or store, as shown on Table 2.

The push(*cond*) when $c = 0$ can be simplified to $c \leftarrow \neg cond$. A more detailed proof of the equivalence between an activity stack and an activity counter for parallel control flow can be found in [3, 10].

## 2.2 Application to a data parallel language

Now we can use this mechanism to implement classical parallel control flow operators such as those in the POMPC C-based language [11]. We present only the where and

Table 3: Implementation of the where/elsewhere with an activity counter.

| Operation | Precondition | | Action |
|---|---|---|---|
| where(*cond*) | $c \neq 0$ | *(idle)* | $c \leftarrow c+1$ |
| | $c=0$ | *(active)* | $c \leftarrow \neg cond$ |
| elsewhere | $c \leq 1$ | *(activatable)* | $c \leftarrow \neg c$ |
| | $c \nleq 1$ | | $c \leftarrow c$ |
| *End of the* where /elsewhere | $c \neq 0$ | *(idle)* | $c \leftarrow c-1$ |
| | $c=0$ | *(active)* | $c \leftarrow 0$ |

the switchwhere but the method is also used for the whilesomewhere, the return of a parallel function or procedure.

### 2.2.1 where

The basic operator is the where/elsewhere pair which is found in most data parallel languages from FORTRAN 90 to C*.

The where is equivalent to the push operator but we have to translate the elsewhere. A PE is active in an elsewhere if and only if the PE was inactive due to the *last* where, *i.e.* the inactivity level $c = 1$. The value 1 can be seen here as a special value that codes for an "activatable" state for the where or elsewhere block.

An implementation is presented in Table 3.

### 2.2.2 switchwhere

The compilation of a switchwhere, the parallel extension of the language C switch, also has several states. A PE can be:

1. inactive before the switchwhere;
2. active in a case (after matching a value) or in a default;
3. inactive in a case, waiting for a matching value;
4. inactive in the switchwhere because of a break, until the switchwhere exit.

The break is similar to the whilesomewhere one. An example of state coding we use is $c = 1$ for the state 3 and $c = 2$ for the state 4, as shown in Table 4.

# 3 Activity counters versus activity stacks

## 3.1 On an SIMD machine

The counter method needs a counter with $\log_2 c$ bits per PE if at most $c$ levels of parallel conditional blocks are nested. If each PE has an $L$-bit operator, a PE needs $\lceil \log_L c \rceil$ cycles of duration $t$ to do an activity counter operation.

The activity stack needs only 1-bit manipulation on each PE and takes a time $t$, but needs a stack pointer to manage

Table 1: Semantics of the `push` and `pop` operations on the activity stack.

| Operation | Behavior | Precondition | Action |
|---|---|---|---|
| push(*cond*) | $s \leftarrow s+1$ | $f_0 \neq s+1$ | $f_0 \leftarrow f_0$ |
| | $a_s \leftarrow cond$ | $(f_0 = s+1) \wedge (cond = 0)$ | $f_0 \leftarrow s$ |
| | | $(f_0 = s+1) \wedge (cond = 1)$ | $f_0 \leftarrow s+1$ |
| pop | *if*[a] $(s > 1), s \leftarrow s-1$ | $f_0 \neq s+1$ | $f_0 \leftarrow f_0$ |
| | return$(a_s)$ | $f_0 = s+1$ | $f_0 \leftarrow s+1$ |

[a] Note that if the program is correct, this condition is always true.

Table 4: Implementation of the `switchwhere` with an activity counter.

| Operation | Precondition | | Action |
|---|---|---|---|
| switchwhere(*value*) | $c \neq 0$ | (*idle*) | $c \leftarrow c+2$ |
| | $c = 0$ | (*active*) | $c \leftarrow 1$ |
| case *constant* : | $(c = 1) \wedge (value = constant)$ | | $c \leftarrow 0$ |
| break | $c = 0$ | (*active*) | $c \leftarrow 2^a$ |
| default : | $c = 1$ | (*activatable*) | $c \leftarrow 0$ |
| switchwhere *closing* | $c \leq 1$ | | $c \leftarrow 0$ |
| | $c \nleq 1$ | | $c \leftarrow c-2$ |

[a] Must be relative to the current `switchwhere` block, if the `break` is included in one or more `where`/`elsewhere`.

the stack. Since the execution is SIMD, all the stacks are synchronous and the stack pointer can be:

- centralized on the scalar processor which broadcasts its value to the PEs;
- distributed with local pointers which evolve synchronously.

In the first case, it takes a time $T$ on the scalar processor and the time is negligible on the PEs. In the second case, a time $t\lceil\log_L c\rceil$ is needed to control the stack pointer on each PE. The hardware complexity is $c$ for a stack of 1 bit elements in each case, plus $\lceil\log_L c\rceil$ bits for the global stack pointer in the first case and $N\lceil\log_L c\rceil$ bits for the local stack pointers in the second case, for a $N$-PE computer.

The complexity of the three previous methods are summarized up in Table 5.

If the computer has only fine grain PEs, typically $L = 1$ or 4 bits, it is more interesting to subcontract the computation to the scalar processor with a global stack pointer. Indeed, the scalar processor is often larger and more powerful, so the stack pointer computation only uses few cycles, and even the broadcast is often shorter than the $t\lceil\log_L c\rceil$ required to deal with a local stack pointer or activity counter by $L$-bit slices. Moreover, 1-bit PEs have the advantage that they easily access memory with 1-bit. This method is used on computers such as the CM-2 or the MP-1.

The activity counter algorithm is particularly interesting for coarse grain SIMD machines and could be interesting in the MP-2. This method is used in our POMP MC88100-based SIMD computer [4, 9]. These computers often have short cycle time and the local memory access is slow in comparison to the PE cycle time.

## 3.2 On an MIMD machine

The complexity of our method for an MIMD machine is the same as in table 5 except that since there is no scalar processor, it is not interesting to have a global activity stack pointer and thus only local pointers or activity counters are necessary.

As for the SIMD computers, the same conclusions arise according to the size of the PEs. Activity counters can avoid the 1-bit stack management, specially inefficient on the coarse grain PEs which are in most MIMD computers. Besides, the activity counter on each PE reduces to $\mathcal{O}(\log c)$ the hardware complexity to store the activity.

But unlike SIMD computers, it is not worth implementing the activity counter in hardware since local conditional jumps are used *in fine* to efficiently emulate the activity corresponding to the counter value.

# 4 Related work

Methods to change control dependance in data dependence statically deal more or less with activity.

In [1] a complete guard is used and in [6] a minimum number of guard is produced to control activity.

Table 5: Complexity of the activity counter and activity stack methods.

| Parallel conditioning | Computing complexity scalar | parallel | Hardware complexity | # broadcast |
|---|---|---|---|---|
| Stack (global pointer) | $T$ | $t$ | $Nc+\lceil\log_2 c\rceil$ | 1 |
| Stack (local pointers) | $\epsilon$ | $t(1+\lceil\log_L c\rceil)$ | $N(c+\lceil\log_2 c\rceil)$ | 0 |
| Activity counters | $\epsilon$ | $t\lceil\log_L c\rceil$ | $N\lceil\log_2 c\rceil$ | 0 |

In [5], all the control flow information is kept in an "Exit" variable similar to our activity counter used for complex statements like `switchwhere` or `whilesomewhere` with `break`, `case` or `return`..

But none of these methods deals with dynamic scheduling, necessary for recursion or any procedure calls.

A counter methods is also used in [2] for dynamic sheduling in a dataflow-like architecture but there is no support for recursion.

# 5 Conclusion

We have developed a new method to dynamically deal with nested parallel control flow and recursion for SIMD and MIMD computers, and compilers for languages with collection oriented data parallelism.

This technique allows a reduction to a straight logarithmic term of the size in bits of memory used to keep track of the PE history, more efficient on coarse grain parallel computers and VLIW processors.

The optimization is also interesting for compilers targeted to modern MIMD computers when the nested parallel control flow cannot be resolved at compile time. For example, if different collections are mixed, interprocedural analysis is not performed or not possible, or if complex sub-array selections cannot be determined. If the activity counter method can often be replaced by MIMD local control flow, for complex nested case it seems a better choice.

At last, it is a way to compile nested parallel flow control flow in a "flat" normal form as in F90 or HPF where such a nestling is not allowed.

The activity counters are used in the POMP computer and also in the POMPC compiler for CM-2, MP-1, iPSC/860 and ARMEN.

# 6 Acknowledgements

The authors of this paper would like to acknowledge many useful discussions with all the members of the POMP team since the beginning of the project.

Special thanks are due to Luc BOUGÉ and his team, especially Jean-Luc LEVAIRE, for their discussions on SIMD semantics in parallel control flow and for their interest for the domain and our work.

At last but not the least, the authors are indebted to Kathryn MACKINLEY, François IRIGOIN and Pierre JOUVELOT for their invaluable comments and their appropriate suggestions.

# References

[1] J. R. ALLEN, Ken KENNEDY, Carrie PORTERFIELD, and Joe WARREN. Conversion of Control Dependence to Data Dependence. In *Conference Record of the Tenth Annual ACM Symposium on Principles Of Programming Languages*. Association for Computing Machinery, January 1983.

[2] Carl J. BECKMANN and Constantine D. POLYCHRONOPOULOS. Microarchitecture Support for Dynamic Scheduling of Acyclic Task Graphs. In *The $25^{th}$ Annual International Symposium on Microarchitecture*, volume 23(1-2), pages 140–148. ACM SIG MICRO Newsletter, December 1992.

[3] Luc BOUGÉ and Jean-Luc LEVAIRE. Control structures for data-parallel SIMD languages: semantics and implementation. *Future Generation Computer Systems*, 8(3-4):363–378, 1992.

[4] Philippe HOOGVORST, Ronan KERYELL, Philippe MATHERAT, and Nicolas PARIS. POMP or How to Design a Massively Parallel Machine with Small Developments. In *PARLE '91 Parallel Architectures and Languages Europe*, volume 505(I), pages 83–100. Lecture Notes in Computer Science, Springer-Verlag, June 1991. Available by `ftp anonymous` on `spi.ens.fr` in the file `pub/reports/liens/liens-91-5.A4.ps.Z`.

[5] Bor-Ming HSIEH, Michael HIND, and Ron CYTRON. Loop Distribution with Multiple Exits. In *Proceedings of Supercomputing '90*. The Institute of Electrical and Electronics Engineers, Inc., November 1990.

[6] Ken KENNEDY and Kathryn S. MCKINLEY. Loop Distribution with Arbitrary Control Flow. In *Proceedings of Supercomputing '90*. The Institute of Electrical and Electronics Engineers, Inc., November 1990.

[7] Ronan KERYELL. POMP2 : D'un Petit Ordinateur Massivement Parallèle. Rapport de magistère, LIENS — Ecole Normale Supérieure, October 1989.

[8] Ronan KERYELL. *POMP : d'un Petit Ordinateur Massivement Parallèle SIMD à Base de Processeurs RISC — Concepts, Etude et Réalisation*. PhD thesis, Laboratoire d'Informatique de l'Ecole Normale Supérieure — Université Paris XI, October 1992.

[9] Ronan KERYELL. *POMP : d'un Petit Ordinateur Massivement Parallèle SIMD à Base de Processeurs RISC — Concepts, Etude et Réalisation*. PhD Thesis, Laboratoire d'Informatique de l'Ecole Normale Supérieure — Université Paris XI, October 1992.

[10] Jean-Luc LEVAIRE. *Deux sémantiques opérationnelles pour POMPC*. PhD Thesis, LIP — ENS Lyon, Université de Paris 7, February 1993.

[11] Nicolas PARIS. Definition of POMPC (Version 1.99). Technical Report LIENS-92-5-bis, Laboratoire d'Informatique de l'Ecole Normale Supérieure, March 1992. Available by `ftp anonymous` on `spi.ens.fr` in the file `pub/reports/liens/liens-92-5-bis.A4.ps.Z`.

# SIMD Optimizations in a Data Parallel C

Maya Gokhale
maya@super.org
Supercomputing Research Center
17100 Science Dr.
Bowie, MD 20715-4300

Phil Pfeiffer
phil@esu.edu
Dept. of Computer Science
East Stroudsburg University
East Stroudsburg, PA 18301-2999

**Abstract**

SIMD programs can devote substantial time to manipulating the underlying hardware's *context registers* – status bits that determine whether processors in the SIMD array execute or skip the current instruction. This paper describes two optimizations, implemented in a compiler for a data parallel C, that reduce the overhead of manipulating and accessing context registers. The first optimization uses observations about a program's nesting structure to eliminate context register save/restore operations performed by guarded parallel control constructs. The second uses two-version code to eliminate context register accesses performed by individual instructions. These optimizations have been implemented in the Data-parallel Bit C (DBC) compiler, which targets a variety of parallel machines, and significantly improve performance of representative benchmark programs.

## 1 Introduction

Data-parallel Bit C (DBC) is a superset of ANSI C that supports data parallel versions of the C control constructs and variable precision integer and logical operations on data-parallel operands [SG92]. DBC runs on the CM-2, Cray-2, SparcStation, Terasys, a linear SIMD "processor-in-memory array" [GHI+92], and Splash-2, an array of Field Programmable Gate Arrays [GM93]. In the SIMD computing model supported by DBC, a host controller performs operations on serial data and issues instructions to the processor array, which executes instructions in lock step on *poly* data. There are two modes of instruction execution: *unconditional* and *conditional*. the latter mode uses a *context register* associated with every processor element (PE) to determine if that PE should execute the current instruction.

This paper presents two optimizations implemented in the DBC compiler that reduce the overhead of accessing and manipulating context registers. The first uses a single recursive traversal of a program's abstract syntax tree (AST) to eliminate needless context register saves and restores. The second uses two-version code to adjust the mode in which selected instructions execute. The paper then concludes with experimental data and discussion of related work.

## 2 Efficient execution of parallel control constructs

DBC extends the standard C control constructs (`if`, `while`, `do`, and `for`) to the parallel domain. As in other data parallel C variants (such as MasPar's MPL [Mas90], Dataparallel C [HQ91], and multiC [Wav91]), the guard controlling the statement determines whether the statement is serial or parallel. If the guard is parallel, the statement is parallel.

Implementation of guarded parallel control constructs requires the compiler to emit instructions to save, modify, and restore the context register so that the "right" subset of PEs is active during execution of the statement.

We will use as an example the simple function *control* below:

```
int control(poly unsigned a:100)
{
  if (a)
    while (!a[0:]) a >> 1;
}
```

Each processor has its own instance of *a*, a 100-bit unsigned integer. All PEs active upon entry to *f* test their values of *a*. Those PEs on which *a* is 0 become inactive[1]. Then the nested **while** loop is executed: each active processor checks the 0'th bit of its

[1] If all PEs become inactive as a result of evaluating the guard, the entire if statement is skipped.

*a*. Those which have a 1 in the LSB become inactive, and the remainder execute the shift instruction. Control returns to the **while** test, with the reduced set of active processors evaluating the **while** guard. When all processors become inactive as a result of evaluating the **while** guard, the loop terminates.

The escape statements **break**, **continue**, and **return** have parallel versions. A **break** or **continue** in a parallel loop is parallel. A **return** from a function returning a parallel result (even "poly void") has parallel semantics. **Gotos** are not allowed in parallel constructs. These constructs are explained in detail in [SG92].

In addtion to the C control constructs, DBC provides an **all** block. Parallel code in an **all** block is executed by all processors regardless of previous processor activity. An **all** block may contain any serial or parallel code, including parallel **if** and parallel loops.

DBC's parallel control constructs control program evaluation by changing the values of context registers. As a rule, a context register that is modified by a parallel construct **P** must be restored to its original value when **P** finishes execution. Ordinarily, this is done by saving the registers' original values in a parallel temporary, and then restoring contexts when statements complete: *e.g.*, by compiling the above example into the following intermediate form:

```
      StoreContext(t1) - function context
      StoreContext(t2) - if-stmt context
        "and" current context with value of if-guard
      if all PEs are inactive goto L1
      StoreContext(t3) - while-stmt context
L2:     "and" current context with value of while-guard
      if all PEs are inactive goto L3
      while body
      goto L2
L3:   LoadContext(t3) - restore while-stmt context
L1:   LoadContext(t2) - restore if-stmt context
L0:   LoadContext(t1) - restore function context
```

This straightforward algorithm for generating code for parallel statements, however, gives rise to superfluous loads and stores: here, for example, the loads at L3 and L1 are killed by the load at L0, and the store into t3 is dead after the removal of the load at L3. These useless operations on context registers, and other superfluous store operations in scopes that contain multiple parallel statements, could be identified by making several careful passes over a program's control flow graph. We have, on the other hand, devised and implemented an algorithm that uses one recursive traversal of a program's AST to detect these superfluous operations on context registers. This algorithm handles programs with parallel *escapes*–parallel **break** statements, **continue** statements, and returns from the middle of functions. The algorithm first assigns to every parallel control statement and function definition a set of attributes. Five different sets of attributes are required: one for **if-then-else** statements; one for **if-then** statements with empty **else** clauses; one for **for**, **while**, and **do** statements; one for **all** statements; and one for parallel functions. A subsequent phase of compilation then uses these attributes to manage the insertion of loads and stores.

The most complex set of attributes is paired with **if-then-else** statements. Five attributes are generated for a statement S of the form "**if**(GUARD) {$SL_{THEN}$ }; **else** {$SL_{ELSE}$ };": S.store_for_then, S.store_for_else, S.restore, S.update, and S.set.

When S.store_for_then is true, the compiled code for S saves the result of evaluating GUARD in a parallel temporary variable, SV, before executing $SL_{THEN}$ . Every parallel control statement in $SL_{THEN}$ that is outside of all (*i.e.*, not in the body of any other) parallel statements in $SL_{THEN}$ uses SV to restore the context registers when it finishes execution. This attribute is set to true iff $SL_{THEN}$ contains a parallel control contruct or a (parallel) escape (that transfers control) to a statement that contains S.

S.store_for_else is similar to S.store_for_then. This attribute is set to true iff $SL_{ELSE}$ contains (1) an escape to a statement that contains S; (2) a parallel statement that is inside a sequential loop, but outside all other parallel statements in $SL_{ELSE}$ ; or (3) a parallel control construct T and a parallel statement T', such that T and T' are outside of all other parallel statements in $SL_{ELSE}$ ; T' is a function call or conditional-mode parallel statement (*cf.* the previous section); and T' is to the right of T in the program's AST.

S.restore signals that S must use a temporary variable EV created by an enclosing parallel control statement E to restore its previous context. This attribute is set to true iff E is either (1) a parallel loop; (2) the **then** branch of a parallel if-then-else statement; or (3) a guarded parallel control construct, and a parallel statement that executes in conditional mode or a function mode lies in E, lies outside of all other parallel statements in E, and follows S (in the program's AST.)

Under ordinary circumstances, a statement S simply restores the initial context when it finishes execution. PEs, however, must be deactivated by the evaluation of escapes in S that transfer control to statements containing S. S.update, when true, signals that

S must update EV to account for escapes before restoring the initial context. These updates are done by "and"ing EV with bits that flag when escapes evaluate. This attribute is set to true when S contains an escape to a statement that contains S, and a parallel statement that executes in conditional mode or a function call lies in E, lies outside of all other parallel statements in E, and follows S.

S.set, the final attribute for **if-then-else** statements, signals that S must activate all processors when it finishes executing. S.set is used in place of S.restore in **all** statements.

Four attributes manage context operations at an **if-then** statement S. One attribute, S.store_for_then, is identical in description and application to the **if-then-else** statement's S.store_for_else attribute. The remaining three attributes, S.update, S.restore, and S.set, resemble the corresponding attributes for **if-then-else** statements.

Four attributes manage context operations at a loop S. Three attributes, S.update, S.restore, and S.set, are similar to the corresponding attributes for **if-then-else** statements. The fourth, S.store_for_body, resembles the **if-then-else** statement's S.store_for_else attribute. This last attribute is set to true iff $SL_{BODY}$ contains a parallel **return**, or a parallel control statement that is outside of all other parallel statements in $SL_{BODY}$.

Two attributes manage context operations at **all** constructs. One attribute, S.restore, resembles the corresponding attribute for **if-then-else** statements. The other, S.set_context, signals that S should activate all PEs. S.set_context is set to true when S's body $SL_{BODY}$ contains statement T such that T is a function call, a guarded parallel construct, or a conditional-mode parallel statement; and T lies outside of all other parallel statements in $SL_{BODY}$. (Note: **All** statements lack the update attribute because escapes inside **all** statements may not transfer control to statements outside of **all** statements.)

Parallel functions must also save and restore the current context. The initial save and final restore of the context in a parallel function f() may be eliminated when f() contains no parallel control constructs. Otherwise, f()'s final restore context statement may be eliminated when f() contains no returns, the final **return** at the end of f() excepted.

## 3 Unconditional Mode Optimizations

As a rule, statements that are interpreted as sequences of unconditional mode instructions execute more quickly than statements that are interpreted as sequences of conditional mode instructions. The unconditional mode execution of an arbitrary statement, unfortunately, is not safe unless all PEs are active. One opportunity for translating statements into sequences of unconditional mode instructions is afforded by the **all** statement. Parallel assignment statements and guards of parallel conditional statements that are not enclosed in any parallel control constructs are also singled out for optimization by DBC. Consider, for example, the following example function f():

```
function f (poly unsigned a, b) {
  a = 1;        /** par assign **/
  if (b) {SL_THEN }; /** par if **/
}
```

The statements "a = 1" and "**if** (b)" may safely execute as unconditional mode instructions if all PEs are active when f() is called. The current implementation of the DBC compiler generates two-version code for functions that contain "exposed" parallel statements:

```
function f (poly unsigned a, b) {
  if (all processors are active) {
     a = 1;      /* unconditional mode */
/* "if (b)" in unconditional mode */
     if (b) {SL_THEN };
  } else {
     a = 1;      /* in conditional mode */
/* "if (b)" in conditional mode */
     if (b) {SL_THEN };
}}
```

Opportunities for executing doubled code in unconditional mode are detected at run-time, at the cost of a single global OR of the context register. An interprocedural flow analysis that related calling contexts to PE activity could be used to refine this technique: *e.g.*, to determine that all PEs must be active when certain functions were called. Such an analysis, however, has not been implemented.

## 4 Performance

Shown below is data to help us evaluate the effectiveness of these transformations on the CM-2 and Terasys, expressed as the ratio of (unoptimized - optimized execution time) to unoptimized execution time[2].

[2] The data presented here are actually due to three optimizations: the two context-register-related optimizations discussed

| Program | Terasys Ratio | CM-2 ratio |
|---|---|---|
| bit-inverse | 4% | 1% |
| sort | | 30% |
| DNA sequence match | 82% | 64% |

The bit-inverse program is embarrassingly parallel and contains virtually no parallel control constructs. Each processor computes the multiplicative inverse of a 97-bit integer in its memory. The small speedup for Terasys comes entirely from the "unconditional mode" optimizations. Since there is only one context operation for a 97-bit arithmetic operation, the savings is a rather small fraction of the total execution time.

The sort program contains nearest neighbor communication guarded by nested conditional statements. There is a significant speedup in the optimized code.

The DNA sequence match has conditional statements embedded in a sequential for-loop, and the optimized code for this program gives the greatest speedup of any of the benchmarks.

### 4.1 Related Work

Other work in optimization of data parallel languages include [HQ91], [ZC90], [CFR+92], [Sab92], which all target MIMD machines, and are concerned with optimizing data distribution and communication, and with relaxing the tightly synchronous semantics when safe.

SIMD optimizations are reported in [KLS90], which presents array allocation algorithms to reduce communication, and [HK92], which proposes and evaluates an algorithm to improve load balancing by applying loop transformations. The CM-2 C* compiler also does some context optimizations for parallel control constructs [Fra92].

## References


[CFR+92] A. Choudhary, G. Fox, S. Ranka, S. Hiranandi, K. Kennedy, C. Koelbel, and C.-W. Tseng. Compiling fortran 77d and 90d for mimd distributed memory machines. *Fourth Symposium on the Frontiers of Massively Parallel Computation*, pages 4–11, 1992.

[Fra92] Jamie Frankel. *Personal Communication.* Thinking Machines, Inc., 1992.


in this paper, and a simple peephole optimization that eliminates needless copy operations.


[GHI+92] Maya Gokhale, Bill Holmes, Ken Iobst, Alan Murray, and Tom Turnbull. A massively parallel processor-in-memory array and its programming environment. Technical Report TR-92-076, Supercomputing Research Center, 1992.

[GM93] Maya Gokhale and Ron Minnich. Fpga computing in a data parallel c. *Proceedings of the IEEE Workshop on FPGAs for Custom Computing Machines*, 1993.

[HK92] R. Hanxleden and K. Kennedy. Relaxing simd control flow constraints using loop transformations. Technical Report CRPC-TR92207, Center for Research on Parallel Computation, Rice University, April 1992.

[HQ91] Philip J. Hatcher and Michael. J. Quinn. *Data-Parallel Programming.* Scientific and Engineering Computation. MIT Press, 1991.

[KLS90] K. Knobe, J. Lukas, and G. Steele. Data optimization: Allocation of arrays to reduce communication on simd machines. *Journal of Parallel and Distributed Computing*, 8(2):102–118, February 1990.

[Mas90] MasPar. MasPar Application Language (MPL) Reference Manual. (9302-0000 01/90), 1990.

[Sab92] Gary Sabot. A compiler for a massively parallel distributed memory mimd machine. *Fourth Symposium on the Frontiers of Massively Parallel Computation*, pages 12–20, 1992.

[SG92] Judith Schlesinger and Maya Gokhale. DBC Reference Manual. Technical Report TR-92-068, Supercomputing Research Center, 1992.

[Wav91] Wavetracer. The multiC Programming Language. (PUB-00001-001-1.00), 1991.

[ZC90] Hans Zima and Barbara Chapman. *Supercompilers for Parallel and Vector Computers.* ACM Press Frontier Series. Addison-Wesley, 1990.

# SESSION 8B

## RESOURCE ALLOCATION/OS

# An Adaptive Submesh Allocation Strategy for Two-Dimensional Mesh Connected Systems

Jianxun Ding and Laxmi N. Bhuyan
Department of Computer Science
Texas A&M University
College Station, TX 77843-3112

## Abstract

*In this paper, we propose an adaptive scan (AS) strategy for submesh allocation. The earlier frame sliding (FS) strategy [1] allocates submeshes based on fixed orientations of incoming tasks. It also slides frames on mesh planes by fixed strides. Our AS allocation strategy differs from the FS strategy in the following two ways: (1) it does not fix the orientations of incoming tasks; (2) it scans on mesh planes adaptively. Experimental studies show that our AS strategy outperforms the FS strategy in terms of external fragmentation, completion time, and processor utilization.*

## 1 Introduction

Two-Dimensional (2D) mesh topology has become more popular because of its simplicity, regularity, and suitability for VLSI implementation [2, 3]. There are quite a few 2D mesh-based commercial or prototype supercomputer systems built or under being developed recently. Typical examples are Intel Paragon [4], Intel/DARPA Touchstone Delta [5], and Tera Computer System [6], etc. In 2D mesh connected systems, incoming jobs are allocated to submeshes. Different jobs need different sized submeshes. The sizes of submeshes range from as few as one node to as many as all the nodes in the whole mesh. As the number of processors in mesh-based systems grows, designing efficient submesh allocation strategies becomes increasingly important.

Although the results of this paper can be extended to high dimensional meshes, we limit our discussion here to two-dimensional allocations. We also assume that there is always a host processor in the mesh connected systems. The operating system in this host processor includes a task dispatcher that is responsible for allocating free submeshes to incoming tasks. An efficient task dispatcher should use a strategy which can quickly find a free submesh whose size is just sufficient to meet the requirements of any incoming tasks.

Li and Cheng proposed a Buddy based strategy [7]. This Buddy strategy is applicable only to square mesh systems whose side lengths must be power of 2. Also, it always allocates a square submesh to an incoming job, regardless the real needs of incoming tasks. The side lengths of the square submesh must again be "dilated" to power of 2. Thus, the Buddy strategy suffers from large internal fragmentations because most jobs are not necessarily in the size of $2^i \times 2^i$. To solve this problem, Chuang and Tzeng proposed a frame sliding (FS) strategy for meshes with arbitrary length and width [1]. The FS strategy always allocates a free submesh with the size exactly meeting the need of any incoming task. Hence it not only totally eliminates the internal fragmentations, but also outperforms the Buddy strategy in terms of allocation completion time and processor utilizations.

The FS strategy still suffers from some degree of external fragmentations. External fragmentation refers to the situation when task dispatcher cannot find a submesh for an incoming task even though the total number of free processors exceeds the need of the task. In this paper, we propose an adaptive scan (AS) strategy for submesh allocation. Like the FS strategy, the AS strategy has zero internal fragmentation, but it has smaller external fragmentations. Extensive simulations also show that the AS strategy has shorter allocation completion time and higher processor utilization compared to the FS strategy.

The paper is organized as follows. In section 2, we introduce notations and definitions used by submesh allocation strategies. In section 3, we propose our adaptive scan strategy and discuss the related address translation scheme. We describe our simulations and conduct performance comparisons in section 4. Finally, we conclude the paper in section 5.

## 2 Preliminaries

A two-dimensional *mesh* $M(a, b)$ is an $a \times b$ rectangular grid which consists of $a \times b$ nodes. A *node* in a mesh system refers to a processor. So the terms *node* and *processor* are used interchangeably in this paper. A node in a mesh can be represented by its coordinate $<x, y>$ where $x$ indicates its row position and $y$ its

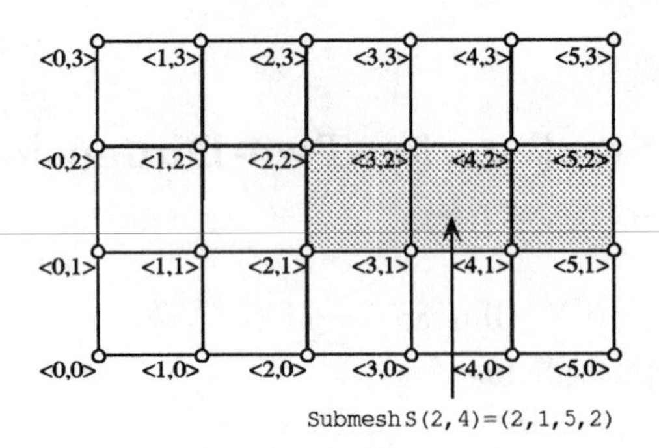


Figure 1: A Two Dimensional Mesh $M(6,4)$

column position. A non-boundary node $<x, y>$, where $0 < x < a-1$ and $0 < y < b-1$, is connected to four neighboring nodes $<x, y-1>$, $<x, y+1>$, $<x-1, y>$, and $<x+1, y>$. Boundary nodes have two or three neighbors depending on their positions. As an example, Fig. 1 shows an $M(6,4)$ mesh and the addresses of all the nodes. A submesh is shaded in the figure and is defined as follows.

A ***submesh*** in $M(a,b)$, denoted as $S(w,h)$, is a subgrid of $M(a,b)$ such that $1 \leq w \leq a$ and $1 \leq h \leq b$. If $w = a$ and $h = b$, an $M(a,b)$ mesh is its own submesh $S(a,b)$. If $w < a$, or $h < b$, or both, mesh $M(a,b)$ can have more than one $S(w,h)$ submeshes. To identify a specific submesh, we need to use the submesh address.

The ***address of a submesh*** $S(w,h)$ is a quadruple $(x, y, x', y')$, where $<x, y>$ indicates the lower left corner and $<x', y'>$ indicates the upper right corner of the submesh. Therefore, a submesh $S(w,h)$ can be identified by a 4-tuple $(x, y, x', y')$ and it consists of nodes $<i, j>$, where $x \leq i \leq x'$, $y \leq j \leq y'$, $x' = x + w - 1$, and $y' = y + h - 1$. As an example, the submesh address $(2,1,5,2)$, shown in Fig. 1, denotes an $S(4,2)$ which consists of nodes $<2,1>$, $<3,1>$, $<4,1>$, $<5,1>$, $<2,2>$, $<3,2>$, $<4,2>$, and $<5,2>$.

**Definition 1** The ***base*** of a submesh $(x, y, x', y')$ refers to node $<x, y>$.

A node can serve as bases for more than one submeshes. Node $<2,1>$ in Fig. 1, for example, is the base of submesh $(2,1,5,2)$. It is also a base for submesh $(2,1,3,3)$.

**Definition 2** A ***free submesh*** $\phi$ is a submesh in which all the processors are currently free. A ***busy submesh*** $\beta$ is a submesh in which all the processors are currently allocated to some task. The ***busy set*** of a mesh system, denoted as $B$, is the collection of all currently busy submeshes in the system.

**Definition 3** The ***coverage*** of a busy submesh $\beta$ with respect to an incoming task T, denoted as $\xi_{\beta,T}$, is a submesh. None of the processors inside this $\xi_{\beta,T}$ can serve as the base of any free submesh to accommodate task T.

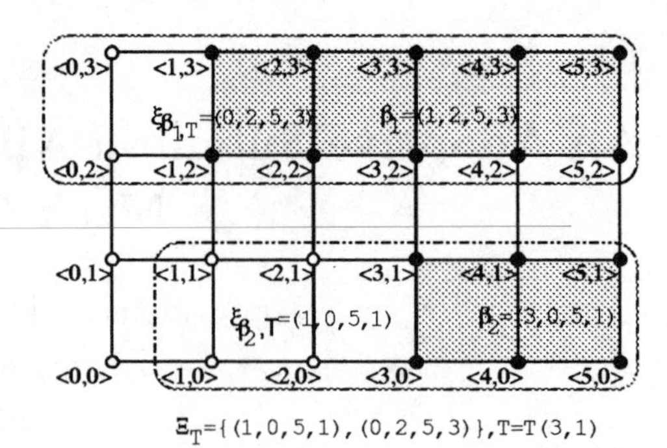


Figure 2: Coverages and Coverage Set with respect to Incoming Task $T(3,1)$

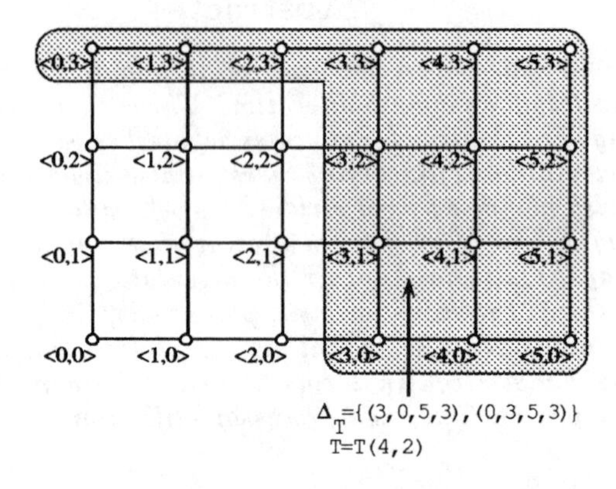


Figure 3: A Reject Set with respect to Incoming Task $T(4,2)$

An incoming task T can be represented by $T(w,h)$ where $w$ and $h$ are the width and height requirements of a task. For a busy submesh $\beta = (x, y, x', y')$ and task $T(w,h)$, the coverage $\xi_{\beta,T}$ is a submesh $(x_c, y_c, x', y')$ where $x_c = min(0, x - w + 1)$, $y_c = min(0, y - h + 1)$, and $min(i,j)$ is a function that returns the smaller value of $i$ and $j$. In Fig. 2, for example, if busy submeshes are $\beta_1 = (3,0,5,1)$ and $\beta_2 = (1,2,5,3)$, depicted as dark nodes in the figure, and the incoming task $T = T(3,1)$, then coverage $\xi_{\beta_1,T}$ is a submesh $(1,0,5,1)$ and coverage $\xi_{\beta_2,T}$ is $(0,2,5,3)$.

**Definition 4** A ***coverage set*** $\Xi_T$ with respect to task T is the union of the coverages of all busy submeshes with respect to task T, i.e., $\Xi_T = \{\xi_{\beta,T} | \beta \in B\}$.

For example, in Fig. 2, the coverage set $\Xi_T$ is the union of $\xi_{\beta_1,T}$ and $\xi_{\beta_2,T}$, i.e., $\Xi_T = \{(1,0,5,1), (0,2,5,3)\}$.

**Definition 5** A ***reject*** submesh with respect to an incoming task T, denoted as $\delta_T$, is a submesh which consists of processors that cannot serve as the base of any free submesh to accommodate task T. The ***reject set*** $\Delta_T$ is the union of all reject submeshes with respect to task T in the system.

For a mesh system $M(a,b)$, if the incoming task $T = T(w,h)$, there will be two reject submeshes $\delta'_T = (x_r, 0, a-1, b-1)$ and $\delta''_T = (0, y_r, a-1, b-1)$, where $x_r = min(0, a-w+1)$ and $y_r = min(0, b-h+1)$. In Fig. 3, for example, if $T = T(4,2)$, $\delta'_T$ will be $(3,0,5,3)$ and $\delta''_T$ will be $(0,3,5,3)$. The reject set $\Delta_T$ is the union of $\delta'_T$ and $\delta''_T$, i.e., $\Delta_T = \{(3,0,5,3),(0,3,5,3)\}$.

In order to compare the performance of different submesh allocation strategies, the following metrics will be used in this paper.

- *Allocation completion time* on an incoming task set $\{T_1, T_2, \cdots, T_n\}$ is defined as the number of time units needed to finish allocating all the tasks based on First-Come-First-Serve (FCFS) order.
- *Processor utilization* is defined as the average percentage of a processor being utilized per unit time.
- *Internal fragmentation* is defined as the ratio of overallocated processors to allocated processors. Here *overallocated* refers to those processors exceeding what are actually needed.
- *External fragmentation* is defined as the ratio of total available processors to the total number of processors in the system at each allocation failure.

The last three performance metrics are average ratios over a certain period of time $\Pi$. Usually this period of time is selected as the allocation completion time for a set of incoming tasks $\{T_1, T_2, \cdots, T_n\}$. Obviously, an efficient submesh allocation strategy must have high processor utilization, low internal and external fragmentations.

# 3 Adaptive Scan Strategy

## 3.1 Discussions on FS Strategy

For an incoming task with arbitrary size of $w \times h$, the 2D Buddy strategy [7] always allocates a $2^i \times 2^i$ submesh where $i = 2^{\lceil log_2(max(w,h)) \rceil}$. This will cause large internal fragmentations and waste processor resources. To remedy the limitations of the Buddy strategy, Chuang and Tzeng proposed a Frame Sliding (FS) strategy for submesh allocation [1]. A frame is nothing but a submesh with exactly needed size of an incoming task. The basic idea of the FS strategy is based on the fact that for any incoming task T, none of the nodes inside busy set $B$, coverage set $\Xi_T$, or reject set $\Delta_T$ can serve as the base node of free submesh to accommodate task T.

Initially, the busy set is set empty. For each incoming task $T(w,h)$ to an $M(a,b)$ system, the task dispatcher creates the coverage set $\Xi_T$ and reject set $\Delta_T$. These two sets, together with the busy set $B$ and all the nodes inside mesh $M(a,b)$, form a *mesh plane*. The task dispatcher simply slides frame over this new mesh plane. The sliding starts from the

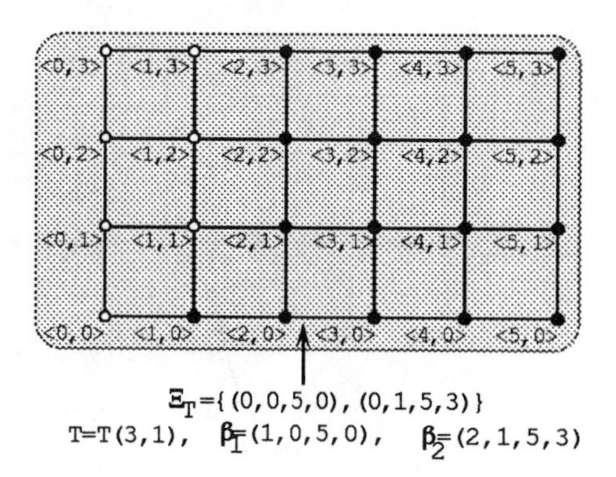


Figure 4: Situation A: Free Nodes Covered by Coverage Set

lowest leftmost node. If node $<x,y>$ belongs to none of the three sets $B$, $\Xi_T$, or $\Delta_T$, a free submesh $S(w,h) = (x, y, w-1, h-1)$ is found; the dispatcher will allocate this submesh $S(w,h)$ to task $T(w,h)$ and add it to the bust set $B$. Otherwise, the dispatcher will continue sliding the frame $S(w,h)$ to the next candidate node. The next candidate will be either $w$ hops away along the $x$-dimension or $h$ hops away along the $y$-dimension, depending on the current sliding position. The frame is first slid along the horizontal direction in stride of $w$ from left to right. If it exceeds the boundary in the right side, a vertical sliding is taken place in stride of $h$ and then the sliding will be in the direction from right to left, and so on.

Although the FS strategy is more flexible and performs better than the Buddy strategy, it still has some degree of external fragmentations. The external fragmentation happens when the total number of free processors exceeds the need of an incoming task $T(w,h)$, but an allocation procedure fails to find a base node. The external fragmentation can be classified as real and pseudo ones. Real external fragmentation refers to the situation when there is no free submesh that is big enough to accommodate the incoming task. Pseudo external fragmentation refers to the situation when there are free submeshes available to accommodate the incoming task but an allocation strategy fails to find a base node.

If we carefully examine the FS strategy, we will find that there are situations under which pseudo fragmentation can happen.

**Situation A: Free nodes covered by coverage set**

In Fig.4, if the busy set $B = \{(1,0,5,0),(2,1,5,3)\}$ and incoming task $T = T(3,1)$, the coverage set will be $\Xi_T = \{(0,0,5,0),(0,1,5,3)\}$. According to the FS strategy, the task dispatcher cannot allocate task $T(3,1)$ because all the nodes are covered by $\Xi_T$, even though there are 7 free processors left. The fragmentation in this case is 29.2%. Also, for the FCFS disci-

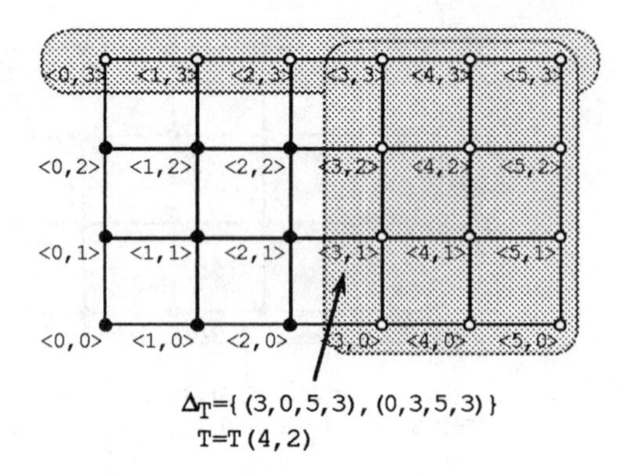


Figure 5: Situation B: Free Nodes Covered by Reject Set

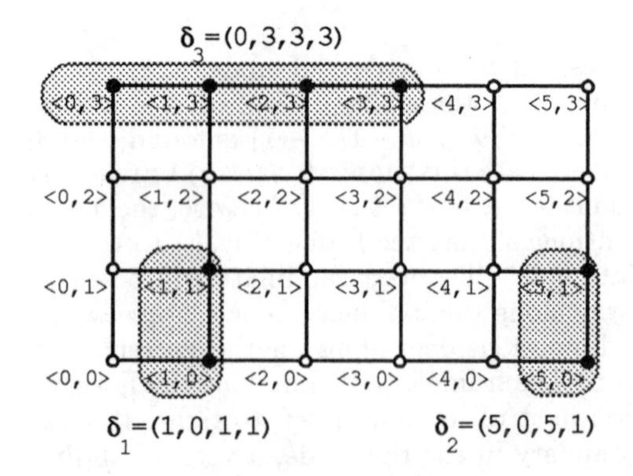


Figure 6: Situation C: A Free Node Ignored by the FS Strategy

pline, the task dispatcher must wait until a free submesh is released. This will affect allocation completion time and processor utilization as well.

**Situation B: Free nodes covered by reject set**

The similar situation happens to the reject set as well. In Fig.5, for example, if incoming task $T = T(4,2)$, the reject set $\Delta_T$ is $\{(3,0,5,3),(0,3,5,3)\}$ as we have shown in section 2. If there is a busy submesh $\beta = (0,0,2,2)$, then task $T(4,2)$ cannot be allocated based on the FS strategy even though there are 15 free processors available in the system. The fragmentation is as high as 62.5% in this case.

**Situation C: Free nodes covered by neither reject set, nor coverage set**

In Fig.6, if incoming task $T = T(3,3)$, the reject set $\Delta_T$ will be $\{(0,2,5,3),(3,0,5,3)\}$. If there are three busy submesh $\beta_1 = (1,0,1,1)$, $\beta_2 = (5,0,5,1)$, and $\beta_3 = (0,3,3,3)$ as shown in Fig. 6, the corresponding coverage set will be $\Xi_T = \{(0,0,1,1),(3,0,5,1),(0,1,3,3)\}$. Hence there is only one free node $<2,0>$ which is not covered by any set. However, based on the FS strategy, after the dispatcher checks free node $<0,0>$, it will slide over 3 nodes along the $x$-dimension. Thus the node $<2,0>$ will not be checked even though it is eligible for a free submesh $S(3,3) = (2,0,4,2)$. For the situation depicted in Fig. 6, no node will be found as the base node according to the FS strategy. The fragmentation is as high as 66.7% in this case.

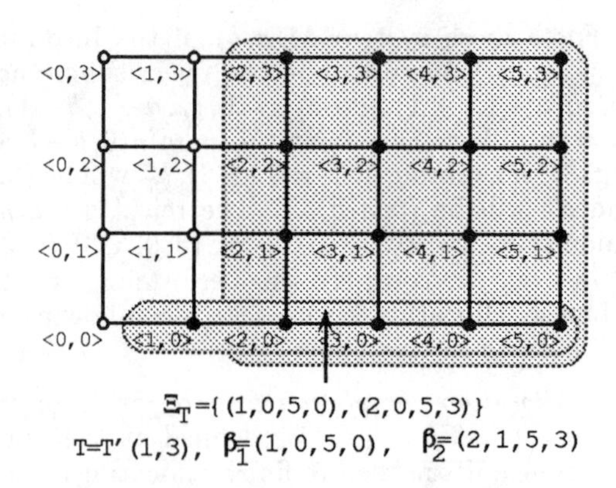


Figure 7: A Solution to Situation A: After Rotation

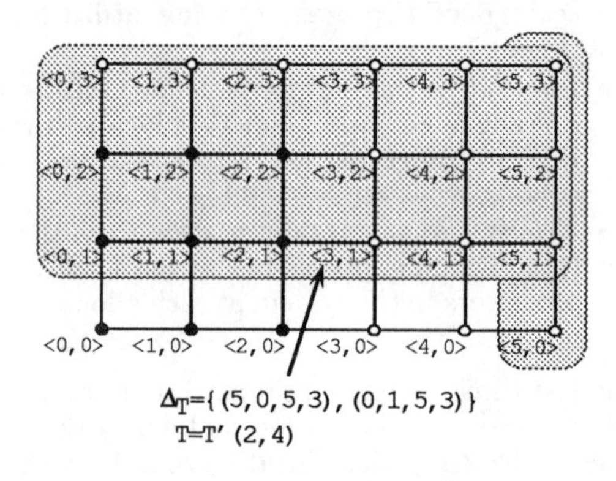


Figure 8: A Solution to Situation B: After Rotation

### 3.2 Basic Ideas of AS Strategy

Our adaptive scan strategy is aimed at reducing the allocation completion time and external fragmentation for the above three situations. In both of the cases A and B, the number of totally free processors actually exceeds the required number of processors of the incoming tasks. Furthermore, those free processors can form alternative free submeshes to host the incoming task. By "alternative", we mean that if we rotate the incoming task's orientation from $T(w,h)$ to $T'(h,w)$, the coverage set $\Xi_{T'}$ and reject set $\Delta_{T'}$ can be easily constructed accordingly. Then the mesh system in the above two cases can host the rotated incoming tasks.

Fig. 7 and Fig. 8 give the solutions to the situations A and B shown in section 3.1, respectively. For the situation A, as shown in Fig. 7, incoming task $T(3,1)$ is rotated to $T'(1,3)$. Excluding the nodes covered by busy set $B= \{(1,0,5,0)(2,1,5,3)\}$, coverage set $\Xi_{T'} = \{(1,0,5,3),(2,0,5,3)\}$, and the reject set $\Delta_{T'} = \{(4,0,5,3),(0,2,5,3)\}$, there will be three nodes $<0,0>$, $<0,1>$, and $<1,1>$ that can serve as the base nodes for free submeshes $(0,0,0,2),(0,1,0,3)$, and $(1,1,1,3)$. For the situation B, as shown in Fig. 8, the new reject set with respect to $T = T'(2,4)$ becomes $\Delta_{T'} = \{(5,0,5,3),(0,1,5,3)\}$. So two nodes $< 3,0 >$, and $< 4,0 >$ are now available to serve as the base nodes for free submeshes $(3,0,4,3)$ and $(4,0,5,3)$.

For the situation C, we may change the "sliding" to "scan". We still check the mesh plane for candidate free submeshes, but do not slide a frame. When we fail on one node and move to the next candidate node, we do not move according to the fixed strides $w$ or $h$ as it is done in the FS strategy. Rather, we always scan along the $x$-dimension from left to right. But our scan is adaptive. We do not scan every node on the mesh plane. First, we can reduce the mesh size according to two reject submeshes. Second, if the current node cannot serve as a base node, it must belong to some submeshes $\{(x'_i, y'_i, x''_i, y''_i)\}$. These submeshes are either busy submesh, or coverage submesh. Let $x_{max}$ be the maximal value of the $x_i$'s in this submeshes. The dispatcher simply jump to the $x_{max} + 1$ position along the same row. If the scanning fails to find a base node in the current row, it moves to next the row and continues doing the above procedure. The detailed allocation and deallocation procedures of our strategy will be described in the next subsection.

From the above discussion, we can see that through adaptively scanning and rotating the orientations of incoming tasks, the allocation completion time, the processor utilization, and the external fragmentations can be reduced. Since we fix neither the orientation of the incoming tasks, nor the strides during the scanning, We call our strategy the *adaptive scan* (AS) strategy.

### 3.3 Address Translation

Since our adaptive strategy may change the orientation of the incoming tasks, naturally, we need a translation mechanism to change the corresponding addresses.

Without loss of generality, suppose a node $V$ has its logical address $<x_v, y_v>$ in a task $T(w,h)$. If T is allocated to a free submesh $S(w,h) = (x_0, y_0, x_0+w-1, y_0+h-1)$ with the node $<x_0, y_0>$ as the base node, node $V$ will have a physical address $<x_0+x_v, y_0+y_v>$ according to the FS strategy. If the orientation of $T(w,h)$ has been changed during the allocation by using our adaptive strategy, node $V$ will have the physical address of $<x_0+ y_v, y_0+x_v>$.

So the address translation mechanism can use a software or hardware flag to indicate whether the incoming task's orientation has been changed or not. If the flag is unset, i.e., the incoming task's orientation is not changed, any operations regarding to logical address $<x_v, y_v>$ will be translated to physical address $<x_0+x_v, y_0+y_v>$. If the flag is set, i.e., the incoming task's orientation has been changed, any operations regarding to logical address $<x_v, y_v>$ will be translated to physical address $<x_0 + y_v, y_0 + x_v>$. Hence, this translation mechanism is very simple and can be implemented by either software or hardware. It can also be incorporated as a part of the system's logic/virtual address to physical address translation mechanism.

### 3.4 Allocation and Deallocation

For each incoming task $T(w,h)$, the allocation procedure for an $M(a,b)$ system is depicted as follows. Initially, the busy list is set empty and the rotation *flag* is set to $FALSE$.

**Allocation Procedure**

1. If *flag* $= FALSE$, then $T \leftarrow T(w,h)$, $a' \leftarrow min(0, a-w+1)$, and $b' \leftarrow min(0, b-h+1)$; otherwise $T \leftarrow T'(h,w)$, $a' \leftarrow min(0, a-h+1)$, and $b' \leftarrow min(0, b-w+1)$;
2. Based on current busy set $B$, set up coverage set $\Xi_T$ with respect to $T$.
3. Check node $<x,y>$ starting from the lowest leftmost node $<0,0>$. If node $<x,y>$ belongs to neither $B$, nor $\Xi_T$, then goto step 5; otherwise, if $<x,y>$ belongs to some submeshes inside $B$ and/or $\Xi_T$, let $d$ be the largest $x$ value of these submeshes.
   - 3.1 If $x < a'-1$, then $x \leftarrow d+1$ and go back to step 3.
   - 3.2 if $x = a'-1$ and $y < b'-1$, then $x \leftarrow 0$ and $y \leftarrow y+1$, go back to step 3.
   - 3.3 if $x = a'-1$, $y = b'-1$, and *flag* $=$ $FALSE$, then *flag* $\leftarrow$ $TRUE$ and go back to step 1; otherwise, wait until a submesh is released.
4. Set *flag* $\leftarrow FALSE$, and go back to step 1;
5. If *flag* $= FALSE$, then $S \leftarrow (x, y, w-1, h-1)$; otherwise, $S \leftarrow (x, y, h-1, w-1)$. Allocate $S$ to task $T$ and add $S$ to the busy set $B$.

In the allocation procedure, $flag = FALSE$ means task $T$'s orientation is not changed. The task dispatcher first reduces the mesh plane according to the reject set at step 1. At step 2, it sets up coverage set. At step 3, it scans the mesh plane. At step 3.1, it may jump over some nodes along the $x$-dimension if the previous node is in some busy submeshes and/or coverage submeshes. At step 3.2, it moves to next $x$ row if it finishes scanning the current row. If it fails to find a base node in the current mesh plane, the dispatcher will set flag to $TRUE$ at step 3.3, and then go back step 1. In the second round, the dispatcher first rotates task $T$'s orientation and reduces the mesh plane according to the new reject set at step 1. At step 2, it also sets up the new coverage set. Then it scans the mesh plane again at step 3. If no eligible node can be found as the base node, the dispatcher will wait at step 3.3 until a submesh is released. Then it will clear the rotation flag at step 4 and go back step 1 to start scanning on the new mesh plane again. If a base node is found at step 3, the dispatcher will go to step 5 to allocate the corresponding submesh $S$ to task $T$ and add $S$ to busy set $B$.

The procedure of deallocating a submesh $S = S(w,h)$ is very simple.

**Deallocation Procedure**

`Remove` $S$ `from the busy set` $B$

From the above procedures, we can see that the AS strategy scans on the mesh plane first. If there is no node eligible to be the base node, the task dispatcher rotates the orientation of the incoming task and scans the mesh plane again. Thus the adaptive scan strategy has the following features.

1. Like the FS strategy, the AS strategy assigns submeshes with actual sizes required by each incoming task. Thus it has zero internal fragmentation.

2. For a single incoming task, the AS strategy guarantees that its external fragmentation is smaller or equal to that of the FS strategy.

Through extensive simulations, we find that for a set of incoming tasks $\{T_1, T_2, \cdots, T_n\}$, our AS strategy has smaller external fragmentations, less allocation completion time, and higher processor utilization than the FS strategy.

# 4 Performance Comparisons

Since the FS strategy has better performance results than the Buddy strategy [1], we only compare the performance of our adaptive scan (AS) strategy with the FS strategy. Extensive simulations have been conducted for the performance comparison. The size of the simulated mesh systems ranges from $16 \times 16$ to $256 \times 256$. All the simulations use 95% confidence level with the error range of $\pm 3\%$.

First, we compare the performance for an $M(64,64)$ mesh system. In order to compare the simulation results, we use the same simulation model used by other researchers [1, 8]. Task allocation is carried out by an exclusive processor which functions as a task dispatcher. Initially the entire mesh is free, and 1000 tasks are generated and queued at the task dispatcher. Each task has a residence time requirement. The residence time is assumed to be uniformly distributed between 5 to 10 time units. Each time unit is in the magnitude such that the time needed for task dispatcher to scan the whole mesh plane is negligible. The side lengths requirement of these 1000 tasks follow a given distribution which we will discuss later. The values of these distributions are produced by different random generators.

The task dispatcher is assumed to follow the First-Come-First-Serve (FCFS) discipline, i.e., the dispatcher always tries to find a free submesh for the first task in the queue. If it fails to find a free submesh for the current first task in the queue, the dispatcher simply waits for a submesh to be released to meet the first task's requirement. After a task is assigned to a submesh, it will be removed from the queue and the next task in the queue will be served in the next time unit. No new tasks would be generated again during the course of simulation.

As we described at the end of section 2, we collect the following three performance metrics: (1) the allocation completion time $\Pi$ for the 1000 tasks; (2) the average processor utilization $U$ over the time period $\Pi$; (3) the average external fragmentation $F_{ext}$ over the time period $\Pi$. We do not care about the internal fragmentation because both FS and AS strategies have zero internal fragmentations.

Table 1 shows the results of the allocations for the $M(64,64)$ system. The side lengths of incoming tasks are assumed to follow one of the four distributions: uniform, exponential, decreasing, and increasing. For the uniform distribution, the required side lengths (width or height) of incoming tasks are unformly distributed between 1 and 64. For the exponential distribution, the mean is selected as half of the side length of $M(64,64)$, i.e., 32. Those values outside of the range $[1,64]$ were ignored. For the decreasing distribution, the probability that a side length falling into the range $[s_1,s_2]$, denoted as $P_{[s_1,s_2]}$, is distributed as follows. $P_{[1,8]} = 0.4$, $P_{[9,16]} = 0.2$, $P_{[17,32]} = 0.2$, and $P_{[33,64]} = 0.2$. For the increasing distribution, The distributions are: $P_{[1,32]} = 0.2$, $P_{[33,48]} = 0.2$, $P_{[49,56]} = 0.2$, and $P_{[57,64]} = 0.4$. The width and height based on the above distributions are generated separately.

From Table 1, we observe that the AS strategy generally performs better than the FS strategy. The completion time has improved by 8.5%, 17.3%, 18.2%, and 2.0%, respectively, for the above four distributions. The utilization has increased by 23.0%, 30.4%, 36.4%, and 8.8% comparing with the FS strategy. The external fragmentation is also reduced by 13.3%, 15.5%, 19.8%, and 2.1% comparing with the FS strategy.

Among the four distributions, the improvement of the decreasing distribution ranks the highest, the exponential distribution the second, the uniform distribution the third, and the increasing distribution the lowest. This is because for the decreasing distribution, there is totally 80% chance that the side lengths are smaller than 32. Out of this this 80% chance, there is another 50% chance that the side lengths are smaller than 8. So both side lengths of the incoming tasks have high probabilities in small sizes. Hence the rotation has higher successful probability. For the increasing distribution, there is totally 80% chance that the side lengths are bigger than 32. Out of this 80% chance, there is another 50% chance that the side lengths are bigger than 56. So the side lengths have high probabilities in large sizes. The rotation, therefore, has lower chance to be successful. The improvements for the uniform and exponential distributions are in the middle. Since the uniform and exponential distributions reflects the general situations where the nature of incoming tasks is unknown, we will further compare the performance for these two distributions under other system configurations.

Our simulation results for the FS strategy matches very well with those reported in [1]. Table 1 only compares the allocations for a fixed sized mesh system. In Fig. 9, we compare the performance for mesh systems of size $M(16,16)$, $M(32,32)$, $M(64,64)$, $M(128,128)$, and $M(256,256)$. Both width and height of the required submeshes are under the uniform distribution based on their corresponding system sizes. With the increase of the mesh sizes, the processor utilization is decreasing and the external fragmentation is increasing for both strategies. But the AS strategy always has better performance than the FS strategy. Fig. 9(a) shows that the AS strategy has about 10% higher processor utilizations than the FS strategy for all the different sized mesh systems. Fig. 9(b) shows that the AS strategy also has about 8% less external fragmentations than the FS strategy.

Fig. 10 further compares allocation performance for the above systems under the exponential distribution. The mean values of $w$ and $h$ are selected as half of mesh side lengths. For the exponential distribution, Fig. 10 shows that the processor utilization increases and the external fragmentation decreases with the increase of mesh sizes. This is in contrast to what is shown in Fig. 9. The reason is that there are more large sized submesh requirements under the uniform distribution than under the exponential distribution. The bigger the mesh system, the larger external fragmentations and lower processor utilizations will be caused. Nevertheless, the AS strategy is still consistently having higher processor utilization and lower external fragmentation than the FS strategy.

## 5 Conclusion

In this paper, we proposed an adaptive scan (AS) strategy for 2-D mesh connected systems. The previously proposed "frame sliding" strategy allocates submeshes based on fixed task orientation and slides frames by fixed strides, but performs better than the buddy strategy. Our strategy adaptively scans the mesh plane and does not allocate submeshes based on fixed orientation of the incoming tasks. The mechanism of this AS strategy is flexible, simple, and effective. Simulation results show that it delivers better performance than the existing allocation strategies for mesh connected computers.

It should be noted that the AS strategy can be easily applied to high dimensional mesh or torus based systems [9]. For the torus based systems, the reject set $\Delta_T$ can be set to empty, and the coverage set $\Xi_T$ can be obtained by modular calculations. Future research in this study involves comparing and incorporating our AS strategy to the Best-Fit strategy (BS) proposed by Zhu [10] recently. Also, scheduling schemes other than FCFS can be investigated together with the proposed task allocation schemes.

## References

[1] P. J. Chuang and N. F. Tzeng, "An efficient submesh allocation strategy for mesh computer systems," *Proc. Int'l Conf. on Distributed Computing Systems*, pp. 256–263, 1991.

[2] P. Muzumdar, "Evaluation of on-chip static interconnection networks," *IEEE Trans. Comput.*, vol. C-36, pp. 365–369, 1987.

[3] G. Randade and S. L. Johnsson, "The communication efficiency of meshes, boolean cubes and cube connected cycles for wafer scale integration," *Proc. Int'l Conf. on Parallel Processing*, pp. 497–482, 1987.

[4] "Paragon XP/S Product Overview," *Intel Corporation*, 1991.

[5] "A Touchstone DELTA System Description," *Intel Corporation*, 1991.

[6] R. Alverson et al., "The Tera computer system," *Proc. 1990 Int'l. Conf. on Supercomputing*, pp. 1–6, 1990.

[7] K. Li and K. H. Cheng, "A two dimensional buddy system for dynamic resource allocation in a partitionable mesh connected system," *Proc. ACM Computer Science Conf.*, pp. 22–28, 1990.

[8] M. S. Chen and K. G. Shin, "Processor allocation in an n-cube multiprocessor using gray codes," *IEEE Trans. Comput.*, vol. C-36, no. 12, pp. 1396–1407, Dec. 1987.

[9] "MPP Technology Preview," *Cray Research, Inc.*, 1992.

[10] Y. Zhu, "Efficient Processor Allocation Stategies for Mesh-Connected Parallel Computers," *Journal of Parallel and Distributed Computing*, vol. 16, pp. 328–337, Dec. 1992.

Table 1: Performance Comparisons for an $M(64, 64)$ System

| Performance Measure | | Uniform | Exponential | Decreasing | Increasing |
|---|---|---|---|---|---|
| Completion Time | FS | 4191.7 | 2445.5 | 1898.5 | 6450.1 |
| | AS | 3842.1 | 2021.2 | 1555.3 | 6326.6 |
| Processor Utilization | FS | 45.2 | 37.5 | 33.5 | 60.1 |
| | AS | 55.6 | 48.9 | 45.7 | 65.4 |
| External Fragmentation | FS | 34.6 | 35.4 | 34.4 | 28.9 |
| | AS | 30.0 | 29.9 | 27.6 | 28.3 |

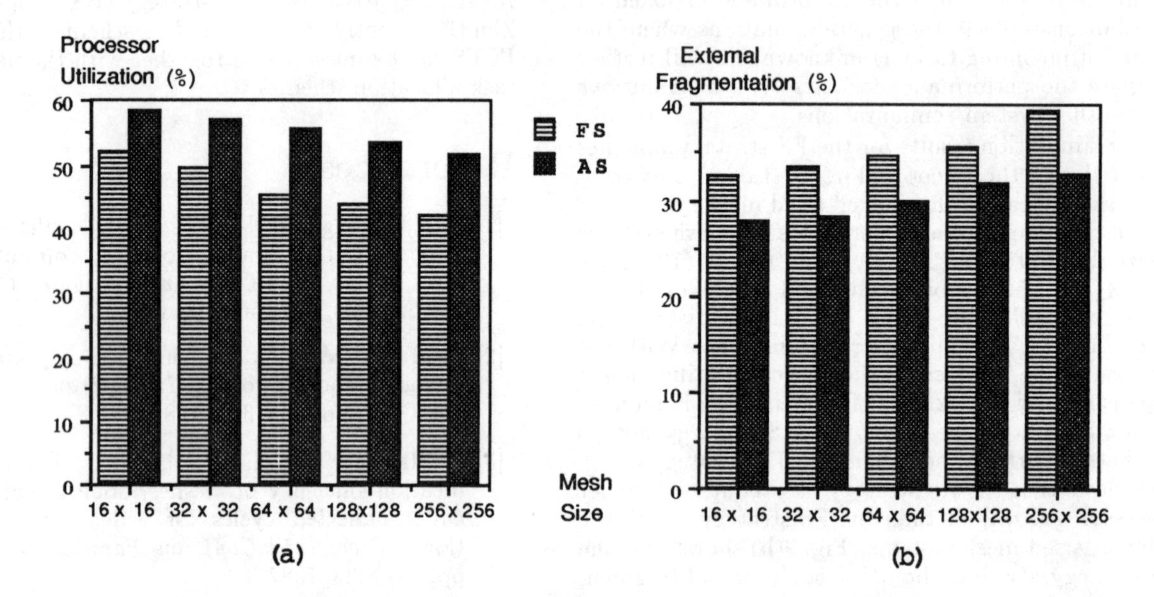


Figure 9: Performance Comparison under Uniform Distribution

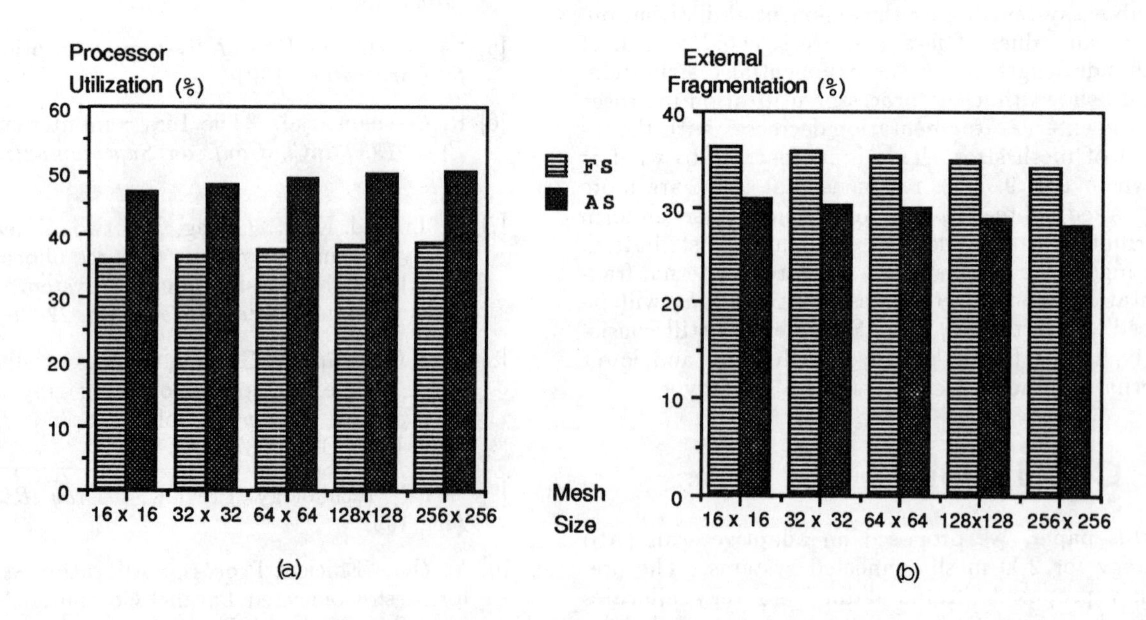


Figure 10: Performance Comparison under Exponential Distribution

# A Fair Fast Scalable Reader-Writer Lock

Orran Krieger, Michael Stumm, Ron Unrau, and Jonathan Hanna
Department of Electrical and Computer Engineering
University of Toronto, Toronto, Canada, M5S 1A4

## 1 INTRODUCTION

A reader-writer (RW) lock allows either multiple readers to inspect shared data or a single writer exclusive access for modifying that data. On shared memory multiprocessors, the cost of acquiring and releasing these locks can have a large impact on the performance of parallel applications. A major problem with naive implementations of these locks, where processors *spin* on a global lock variable waiting for the lock to become available, is that the memory containing the lock and the interconnection network to that memory will also become contended when the lock is contended.

Several researchers have shown how to implement scalable exclusive locks, that is, exclusive locks that can become contended without resulting in memory or interconnection network contention [1, 2, 5]. These algorithms depend either on cache hardware support or on the existence of *local* memory, where accesses to local memory involve lower latency than accesses to *remote* memory (and involve no network traffic).

Mellor-Crummey and Scott[6] recognized the need for scalable RW locks and developed an implementation for the BBN TC2000. Their results indicate that their algorithm performs well, however, it depends on the rich set of atomic operations provided by the BBN TC2000. In particular: 1) atomic write operations for 8, 16 and 32 bit quantities, 2) atomic **fetch_and_store** instructions, 3) atomic **compare_and_swap** instructions, and 4) **atomic_increment** and **atomic_decrement** instructions.

In this paper we describe a new fair RW locking algorithm with similar goals to that developed by Mellor-Crummey and Scott. However, our algorithm has three major advantages over their algorithm. First, in the common case of an uncontended lock, our algorithm is faster since it requires fewer atomic operations and memory references. Second, their algorithm depends on three global variables that must be accessed atomically while our algorithm depends on only one such variable. Therefore, we believe that our algorithm will scale to larger numbers of processors. Finally, the only atomic operation required by our algorithm is **fetch_and_store**. Therefore, it can be used on multiprocessors that do not support the rich set of atomic operations provided by the BBN TC2000.

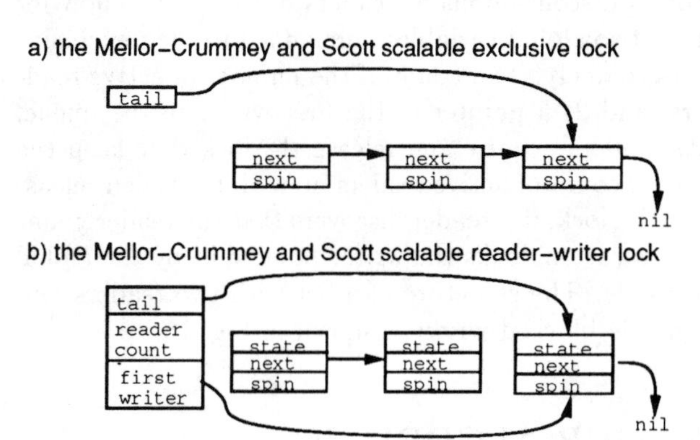


Figure 1: The Mellor-Crummey and Scott scalable locks

## 2 BACKGROUND

Mellor-Crummey and Scott's scalable RW lock is derived from their exclusive lock [6], which uses atomic operations to build a singly linked list of waiting processors (Fig. 1a). The processor at the list head has the lock and new processors add themselves to the list tail. Rather than spinning on a global lock variable, each processor spins on a variable in its local memory. A processor releases the lock by zeroing the variable on which the next processor in the queue is spinning.

For the RW variant of this algorithm, each queue element contains an additional variable to maintain the state of the request. When a new reader request arrives, the state of the previous element in the queue is examined to determine if the new request must block.

With a RW lock, readers must be able to release

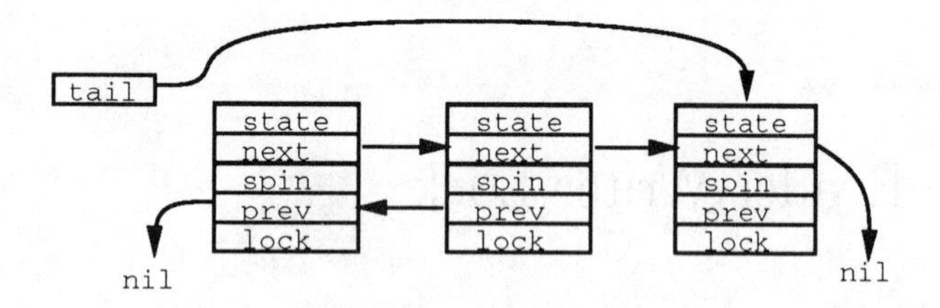


Figure 2: Our RW lock with two active readers and a single blocked writer.

the lock in any order. Hence, the singly linked list used in the Mellor-Crummey and Scott algorithm becomes discontinuous as readers dequeue. To allow for this, two global variables were added to their exclusive lock, namely: 1) a count of the number of active readers, and 2) a pointer to the first writer in the queue. As readers acquire and release the lock they keep the global count of active readers up to date. When releasing the lock, if a reader discovers that the reader count is zero, it unblocks the writer pointed to by the global variable. The structure of a list with two readers and a single blocked writer is shown in Fig. 1b.

## 3 OUR ALGORITHM

We have developed a new fair scalable RW locking algorithm which is also derived from Mellor-Crummey and Scott's exclusive locking algorithm. The key advantage of this new algorithm is that, rather than adding more global state (that can become contended), we distribute the extra state needed for a RW lock across the list associated with the lock. In particular, readers are maintained in a doubly linked list (Fig. 2).

With a doubly linked list, instead of synchronizing on a global variable, a reader that is releasing the lock can synchronize with its nearest neighbors to remove itself from the queue. This allows readers to dequeue in any order without the list becoming discontinuous. Hence, it is not necessary to keep either a global pointer to the first writer or a global count of the number of active readers.

We have developed two versions of this algorithm. The simpler (and more efficient) version requires that the hardware support atomic `compare_and_swap` operations. The more complicated version requires only `fetch_and_store` operations.

The simple version of our algorithm is shown in Figures 3 to 5. The per-processor list element structure used by our algorithm contains: 1) a state variable that indicates if the processor is an active reader, an intended reader, or a writer, 2) a local spin variable, 3) pointers to the next and previous queue elements, and 4) a spin lock used for dequeuing read requests.

```
type Lelem = record        // list element
   state : (READER, WRITER, ACTIVE_READER)
   spin  : int             // a local spin variable
   next, prev : ^Lelem     // neighbor pointers
   EL     : lock           // a spin lock

type RWlock : ^Lelem       // list tail pointer

procedure writerLock( L : ^RWlock, I : ^Lelem )
   var pred : ^Lelem
   I->state := WRITER
   I->spin := 1
   I->next := 0
   pred := fetch_and_store( I, L )
   if pred != nil
      pred->next := I
      repeat until I->spin = 0

procedure writerUnlock( L : ^RWlock, I : ^Lelem )
   var pred : ^Lelem

   if I->next = nil and
      compare_and_swap( 0, I, L ) == I
      return
   repeat until I->next != nil
   I->next->prev := 0
   I->next->spin := 0
```

Figure 3: Routines for write lock and unlock

The `writerLock` and `writerUnlock` operations (Fig. 3) are nearly identical to the `acquire_lock` and `release_lock` operations of Mellor-Crummey and Scott's exclusive lock algorithm. The only difference is that `writerUnlock` zeroes the next element's previous pointer field to signal that processor that it is now at the head of the queue. This is needed if the next processor is a reader.

On executing `readerLock`, the requesting processor constructs a doubly linked list by saving the pointer to the previous element (in the list) into its local structure and then placing a pointer to its local structure in the previous element's next pointer. After enqueueing

```
procedure readerLock( L : ^RWlock, I : ^Lelem )
   var pred : ^Lelem
   I->state := READER
   I->spin  := 1
   I->next  := I->prev = 0
   pred := fetch_and_store( I, L )
   if pred != nil
      I->prev := pred
      pred->next := I
      if pred->state != ACTIVE_READER
         repeat until I->spin = 0
   if I->next != nil and I->next->state = READER
      I->next->spin := 0
   I->state := ACTIVE_READER
```

Figure 4: The read lock routine

```
procedure readerUnlock( L : ^RWlock, I : ^Lelem )
    var prev : ^Lelem := I->prev
    if prev != nil
        exclusiveLock( &prev->EL )
        repeat until prev == I->prev
            exclusiveUnlock( &prev->EL )
            prev := I->prev
            if prev = nil break
            exclusiveLock( &prev->EL )
        if prev != nil
            exclusiveLock( &I->EL )
            prev->next := nil
            if I->next = nil and
               compare_and_swap(I->prev,I,L) != I
               repeat until I->next != nil
            if I->next != nil
               I->next->prev = I->prev
               I->prev->next = I->next
            exclusiveUnlock( &I->EL )
            exclusiveUnlock( &prev->EL )
            return
    exclusiveLock( &I->EL )
    if I->next = nil and
       compare_and_swap( 0, I, L ) != I
       repeat until I->next != nil
    if I->next != nil
       I->next->spin = 0
       I->prev->prev = 0
    exclusiveUnlock( &I->EL )
```

Figure 5: The read unlock routine

itself, the requesting processor checks if its predecessor has acquired a reader lock, in which case it can also acquire the reader lock without having to block. After acquiring the lock (and before modifies its state variable to indicate that it has done so), the requester checks to see if it has a successor that is also a reader request and if so unblocks that processor (by zeroing its spin variable).

As with `writerUnlock`, `readerUnlock` releases the lock by removing its local structure from the queue. To remove itself from the queue, the releasing processor must synchronize with both its queue neighbors. Since the order of elements in the queue is unique, it is easy to do this in a deadlock free fashion by having the releasing processor first acquire its predecessor's lock and then its own. After both these locks are acquired, the releasing processor can simply dequeue itself by modifying the previous and next fields of its neighbors in the linked list. If the releasing reader processor is at the end of the queue it swaps the pointer to its predecessor into the lock structure. (Note that `readerUnlock` differs from `writerUnlock` in that it does not unblock the next processor unless the releasing processor is at the head of the queue.)

While the algorithm described above is simple and efficient, it is not portable to all hardware bases because it depends on `compare_and_swap` operations.

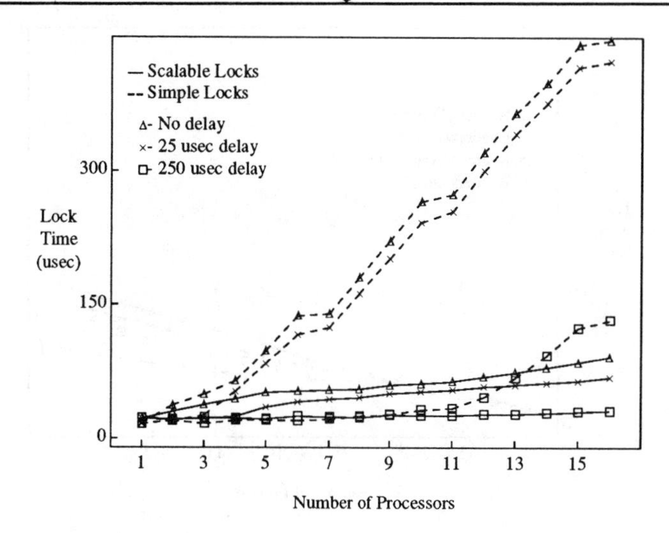


Figure 6: RW locks with only read requests.

Since most multiprocessors support `fetch_and_store` operations, we developed a more complicated version of the algorithm that depends only on these operations. The main added overhead with this version is that a global exclusive (spin) lock is required on unlock if no succeeding elements are in the linked list.

The complexity with this version arises from the fact that with only `fetch_and_store` there is no way to atomically dequeue an element at the tail of the list. That is, there is no way for an unlocking processor to atomically 1) detect that its local structure is the tail element and 2) dequeue that element. With the singly linked list used by Mellor-Crummey and Scott's exclusive lock, this problem can be solved at the cost of some requests occasionally being served out of order [5]. However, with our doubly linked list the problem becomes much more difficult to handle.

## 4 PERFORMANCE

The variant of our algorithm that uses only `fetch_and_store` was implemented in C code on the Hector multiprocessor [8]. The particular system used is a 16 processor system that runs at 16MHz and uses MC88100 processors. The experiments where performed as regular user programs running on a fully configured Hurricane operating system [7].

Figure 6 compares the performance of our scalable RW lock to a simple exponential backoff RW lock when $p$ processors continuously acquire and release locks for reading. The different curves for each lock type show the performance when the locks are held for varying amounts of time. This *lock hold* time is subtracted away in the times presented. The scalable RW lock performs much better than the simple RW lock under

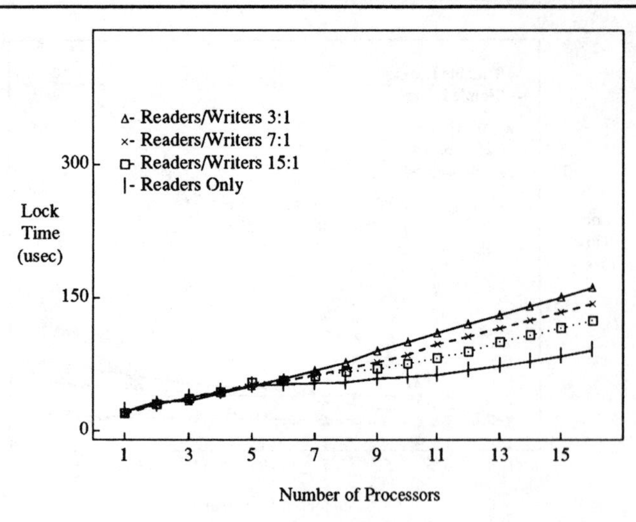


Figure 7: Scalable RW lock with various reader/writer ratios

any significant load. With a lock hold time of 0, the simple lock causes the memory containing the lock to saturate with just a small number of processors (i.e. 3 or 4) at which point the lock begins to perform very poorly. Even with a lock hold time of 250 usec, the performance of the spin lock begins to degenerate after about 11 processors.

For small numbers of processors and with a lock hold time of 0, as the number of requesting processors increases the scalable RW lock initially increases in cost quite quickly. This is because optimizations in our implementation for the uncontended lock no longer apply. However, the curve flattens as the number of processors increase further. The response time continues to increase, mainly because the memory that contains the pointer to the tail of the list becomes more contended. With a lock hold time of just 25 usec, the degradation of performance for the scalable lock is much more gradual. With a lock hold time of 250 usec, the curve for the scalable lock remains flat for all 16 processors.

The performance of the scalable RW lock with different ratios of read and write requests (and a lock hold time of 0) is shown in Figure 7.[1] Acquiring and releasing the lock for writing is (in the uncontended case) less expensive than for reading. The figure shows this, since for a small load performance is better if the ratio of readers to writers is low. However, after 5 processors contend for the lock, the greater the ratio of readers to writers the better the performance of the lock. This is expected, since the readers can acquire and release the lock concurrently. The advantage of a high ratio of readers to writers is even larger with some lock delay, hence this is a worst case experiment for readers.

[1] An off-line pseudo-random number generator was used to generate request sequences in which readers outnumbered writes according to the required ratio.

## 5 CONCLUDING REMARKS

We have developed a new scalable RW lock for shared memory multiprocessors. We believe that this lock is an improvement over Mellor-Crummey and Scott's RW lock in that 1) it involves fewer memory and atomic operations in the absence of contention; 2) it uses less global state that must be modified atomically; and 3) it requires only `fetch_and_store` operations, and hence can be used on most current multiprocessors. A full description of this algorithm is contained in [4].

To verify both versions of our algorithm, we used a state space searching tool [3] to do a full search of the state space for small numbers of requesters, and a partial search for larger numbers of requesters. Although partial searches cannot prove correctness, our tests have found all of our (previous) errors very quickly.

### Acknowledgements

We would like to thank Benjamin Gamsa for his contribution in improving the presentation of this paper.

## REFERENCES


[1] T. E. Anderson. The performance of spin lock alternatives for shared-memory multiprocessors. *IEEE Tran. on Par. and Dis. Sys.*, 1(1):6–16, 1990.

[2] G. Graunke and S. Thakkar. Synchronization Algorithms for Shared-Memory Multiprocessors. *IEEE Computer*, 23(6):60–69, June 1990. 1990.

[3] Gerard J. Holzmann. *Design and Validation of Computer Protocols.* Prentice Hall, 1991.

[4] Orran Krieger, Michael Stumm, Ron Unrau, and Jonathan Hanna. A fair fast scalable reader-writer lock. CSRI, University of Toronto, 1993.

[5] J. M. Mellor-Crummey and M. L. Scott. Algorithms for Scalable Synchronization on Shared-Memory Multiprocessors. *ACM Trans. on Comp. Sys.*, 9(1), Feb. 1991.

[6] J. M. Mellor-Crummey and M. L. Scott. Scalable Reader-Writer Synchronization for Shared-Memory Multiprocessors. In *Third ACM SIGPLAN Symp. on PPOPP*, 1991.

[7] M. Stumm, R. Unrau, and O. Krieger. Designing a Scalable Operating System for Shared Memory Multiprocessors. In *Usenix Workshop on Micro-kernels and Other Kernel Architectures*, 1992.

[8] Zvonko G. Vranesic, Michael Stumm, Ron White, and David Lewis. "The Hector Multiprocessor". *IEEE Computer*, 24(1), January 1991.

# EXPERIMENTS WITH CONFIGURABLE LOCKS FOR MULTIPROCESSORS

Bodhisattwa Mukherjee Karsten Schwan

College of Computing
Georgia Institute of Technology
Atlanta, Georgia 30332
e-mail: {bodhi, schwan}@cc.gatech.edu

Abstract – *Operating system kernels typically offer a fixed set of mechanisms and primitives. However, the attainment of high performance for a variety of parallel applications requires the availability of reconfigurable and extensible operating system kernel primitives. In this paper, we present an implementation of multiprocessor locks that can be reconfigured dynamically.*

## INTRODUCTION

Past experimentation with parallel machines has demonstrated that the attainment of high performance often requires the customization of operating system mechanisms to each class of application programs. This paper explores how an operating system kernel can support application programs in the assembly of program-specific mechanisms and policies. Specifically, we investigate how can a reconfigurable operating system kernel's abstractions be represented and what basic mechanisms are required for such reconfiguration. Furthermore, we investigate if the run-time costs incurred by such reconfiguration are justified by the possible gains due to reconfiguration[5, 2]. In this paper, we address these issues for a specific concurrency control construct in NUMA shared memory parallel programs. The configurable lock construct described in this paper permits the use of multiple strategies for lock access ranging from 'busy waiting' to 'blocking'. Tradeoffs in the use of these strategies for NUMA machines are demonstrated with measurements on a 32-node BBN Butterfly GP1000 multiprocessor.

## THREAD SYNCHRONIZATION ON NUMA MACHINES

Tradeoffs in program performance due to the use of alternative synchronization constructs have been demonstrated for most parallel architectures[1, 5, 4]. The measurements[1] in this section study the tradeoffs regarding the use of spin locks vs. blocking locks for critical sections of different lengths and accessed with different frequencies. Our experimentation with artificial work-loads imposed on a NUMA multiprocessor demonstrate that spin locks consistently outperform (Figure 1) blocking locks when the number of processors exceeds the number of threads for bursty or for uniform arrivals of lock requests. However, when multiple threads on each processor are capable of making progress, the use of blocking is indicated even for fairly small critical sections, since spinning prevents the progress of other threads not currently waiting on a critical section. The cross-over point while using blocking vs. spinning for the artificial work-loads used in our experimentation corresponds to the additional overheads of blocking on the BBN Butterfly (Figure 2).

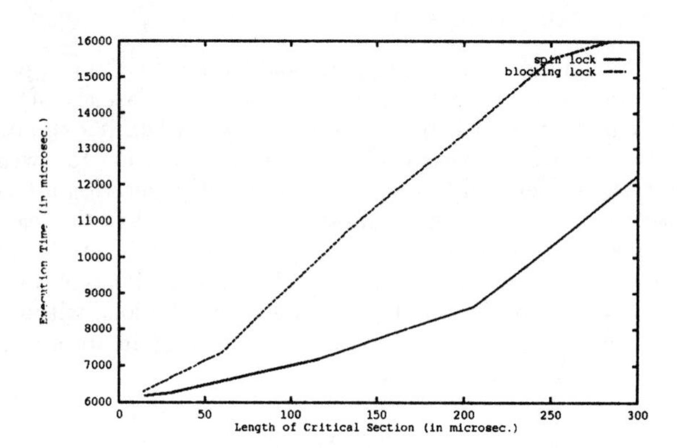


Figure 1: Critical section length Vs. Execution time

The experimental results shown above are not surprising. However, these results do imply that any locking mechanism offered by an operating system kernel should permit the mixed use of spinning, back-off spinning[1], and blocking as waiting strategies, depending on the expected or experienced lengths of critical sections protected by such locks. The design of a lock object offering such multiple waiting strategies is presented next.

## RECONFIGURABLE LOCKS

At any specific time a lock can be in one of three different states – locked, unlocked, and idle. A lock enters the "idle" state when it is free but has one or more waiting threads[5]. An idle state inhibits an application's progress

[1] All the measurements in this paper were made with a workload simulator on a 32 node BBN butterfly multiprocessor. The simulator binds one or more thread to each processor which generate locking requests following a user defined pattern.

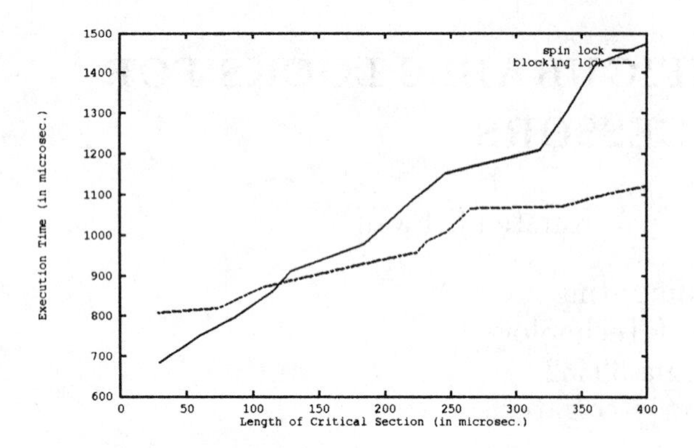


Figure 2: Critical section length Vs. Execution time when additional useful threads are present (for uniform locking requests)

by consuming processor cycles and/or blocking an application thread. A lock becomes idle when the latency of its lock and/or unlock operations are high, when it exhibits an expensive "locking cycle"(an unlock operation followed by a lock operation), and/or when the lock operation interferes with the unlock operation. A spin lock has the least idle state. However, a spin lock may result in increased bus and/or memory contention. The aim of dynamic lock reconfiguration is to reduce the idle time of a lock without significantly increasing contention resulting in improved application performance.

### Reconfigurable Locks – The Model

Any lock construct has two mechanisms and associated policies determining its behavior: (1) its *scheduling* component determines the delay in lock acquisition experienced by a thread, and (2) its *wait* component specifies the manner in which a thread is delayed while attempting to acquire the lock. Scheduling itself may be divided into three components: (a) a *registration* component logging all threads desiring lock access, (b) an *acquisition* component determining the waiting mechanism and policy to be applied to each registered thread, and (c) a *release* component that grants new threads access to the lock upon its release. Lock configuration can concern changes in each of the components mentioned above.

Reconfiguration is possible only if the components being changed offer an immutable interface to the remainder of the application program. Hence, the object model is used to describe such components. However, for reconfiguration and for attainment of high performance, application programs must be aware of additional object *attributes* that may be specified and changed orthogonally to the object's class determined by its methods. Specifically, we are concerned with object attributes that characterize an object's internal implementation and we focus on making dynamic changes to selected implementation attributes[5, 3]. Some of these attributes are discussed next for lock objects.

### Reconfigurable locks – Implementation

A reconfigurable lock's internal representation consists of the following information: *object state* (e.g., current lock state, current lock owner, registration information, etc.) and *configuration state* (e.g., timeout and spin-time parameters, list of wait methods, etc.), which is shown in detail in Table 1. These parameters implement a spectrum of locks ranging from spin to blocking locks. Configuration state not shown in the Table includes architecture-specific information like lock location, object implementation (distributed or centralized objects) etc.[5].

Table 1: Lock *Parameters*[1]

| spin time | delay time | sleep time | timeout | resulting lock |
|---|---|---|---|---|
| n | 0 | 0 | 0 | pure spin |
| n | n | 0 | 0 | spin (back-off) |
| 0 | 0 | n | 0 | pure sleep |
| x | x | x | n | cond. sleep/spin |
| n | n | n | x | mixed sleep/spin |

Given the lock implementation outlined above, each lock access request involves the following steps:

Lock. A locking operation consists of a *registration* phase followed by an *acquisition* phase. At registration time, attribute information like thread-id, priorities, ownership, *etc.* is processed by the lock's policy. The policy next performs *lock acquisition*, which implements a mapping of 'thread id' to the appropriate methods for waiting on the lock, and it also selects the appropriate lock scheduling method for delaying lock access. Both mappings may be changed by reconfiguration operations performed on the lock object described later.

Unlock. An unlock operation is primarily implemented by the *release* module, which selects the next thread that is granted access to the lock. The release module's selection policy may consist of a simple access to a thread-id noting the next thread to be executed[5] or it may execute a more complex scheduling strategy.

The reconfigurable lock object also contains a monitor module which senses or probes user-defined parameters. This monitor module can be used with an appropriate adaptation policy to implement adaptive locks[6, 2].

## PERFORMANCE EVALUATION

This section summarizes the basic costs of a non-configurable lock implementation and compares them with the costs of the operations provided by a reconfigurable lock object. A detailed performance evaluation appears elsewhere[5].

Table 2 lists the latencies of the lock and unlock operations for different lock implementations available on the BBN multiprocessor (provided by the hardware, operating system and our Cthreads library). The `atomior` function (the low-level "atomic or" operation provided by the BBN Butterfly multiprocessor hardware) is used to implement the various locks. The *spin-with-backoff* lock is a variation of the back-off spin lock suggested by Anderson et al.[1]. A thread requesting ownership of such a lock spins once, and if the lock is busy, waits (back offs) for an amount of time proportional to the number of active threads waiting

[1] n = a number, and x = "do not care"

Table 2: Basic lock operations ($\mu$secs.)

| Type | Lock | | Unlock | |
|---|---|---|---|---|
| | local | remote | local | remote |
| atomior | 30.73 | 33.86 | - | - |
| spin | 40.79 | 41.10 | 4.99 | 7.23 |
| spin-backoff | 40.79 | 41.15 | 5.01 | 7.25 |
| block | 88.59 | 91.73 | 62.32 | 73.45 |
| config. | 40.79 | 41.17 | 50.07 | 61.69 |

Table 3: Combined cost ($\mu$secs.) of successive Unlock and Lock operations lock

| Lock | local | remote |
|---|---|---|
| Spin | 45.13 | 47.89 |
| Spin-backoff | 320.36 | 356.95 |
| Block | 510.55 | 563.79 |

for the processor. The latency of the configurable lock is comparable to that of a primitive spin lock because a lock operation for configurable locks initially spins for the lock before deciding to block the requesting thread.

Table 3 lists the costs of the locking cycle (which determines the duration of the "idle state") for some static implementations of locks. As shown in Table 4, A configurable lock has the least expensive locking cycle when configured as a spin lock and has the most expensive locking cycle when configured as a blocking lock.

Table 5 lists the costs of the basic dynamic configuration operations. Scheduler reconfiguration is more expensive than waiting policy reconfiguration because simple dynamic configuration of the waiting policy requires only one memory read and one write whereas, alteration of the scheduler requires three writes for three submodules, one write to set a flag (to implement the configuration delay), and another write to reset the flag (when all the pre-registered threads are served, the old scheduler is discarded) [5]. Additional problems with dynamic configuration include configurations delay, configuration interference *etc.* which are discussed in detail in [5].

## EXPERIMENTS

In this section, we list the results of three experiments performed with reconfigurable locks. The experiments in this section demonstrate that application performance gains due to dynamic reconfiguration outweigh the performance penalties arising from the use of reconfigurable locks.

### Scheduler Configuration

This section compares the performance of the following three lock configurations using a common class of multiprocessor applications, applications structured as client-server programs.

Table 4: Combined cost ($\mu$secs.) of successive Unlock and Lock operations on a reconfigurable lock

| Configured as | local | remote |
|---|---|---|
| Spin | 90.21 | 101.38 |
| Block | 565.16 | 625.63 |

Table 5: Lock configuration operations ($\mu$secs.)

| Operation | local | remote |
|---|---|---|
| configure(wait pol.) | 9.87 | 14.45 |
| configure(scheduler) | 12.51 | 20.83 |

Table 6: Performance of Lock Schedulers

| FCFS ($\mu$sec) | Priority ($\mu$sec) | Handoff ($\mu$sec) | Perf. Gain |
|---|---|---|---|
| 463937.5 | - | 403735.69 | 13% |
| 463937.5 | 419879.49 | - | 9.5% |

FCFS and Priority lock. While FCFS lock scheduling is most common in multiprocessor lock implementations, non-preemptive priority locks can be used in applications exhibiting specific locking patterns for improved application performance. Use of priority locks has been widely discussed in the real-time domain. Such locks can also be useful in client-server models of computation. Specially, when a server is flooded with requests from many clients, the priority of its threads may be raised so that a server thread can acquire the locks for the critical sections, it shares with the clients (*e.g.*, message buffers) faster, thereby resulting in faster service.

Handoff lock. A handoff lock scheduler takes hints to select the next thread to be assigned to the critical section. The releasing thread hands off the critical section directly to the selected thread. Handoff locks, like priority locks, are useful in a client-server model of computation.

In our first experiment, one thread (executing on a dedicated processor) is designated to be a server thread serving many client threads. Communication between server and clients is performed via shared message buffers. Each of the above lock configurations are used in the implementation of the shared buffer. Table 6 compares the resulting program performance. The priority and the handoff locks perform better than the FCFS lock because they allow the server to make progress in preference to clients, therefore serving the clients at a faster rate[5].

This experiment clearly demonstrates that for improved performance an application requires lock schedulers most appropriate for its requirements.

### Waiting Policy Configuration

A few sample lock configurations obtained by varying the parameters of its internal state are spin, blocking, and combined locks. A second experiment compares the performance of these three configurations using a uniform distribution of locking request.

Spin and Blocking lock. If a lock's sleep time is zero and the spin time is nonzero, a thread spins while waiting for the lock. If the spin time is zero and the sleep time is nonzero, a waiting thread directly goes to sleep until awakened by a thread or a timeout signal.

Combined lock. A combined lock results when both spin time and sleep time are set to nonzero. A thread spins as

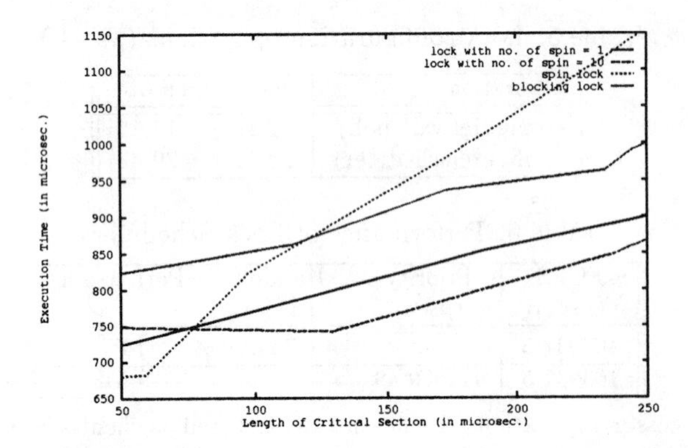


Figure 3: Critical section length Vs. Execution time

well as sleeps while waiting. The resulting waiting policy is decided by the actual values of the parameters.

The second experiment compares performance of combined locks with spin and blocking locks when multiple threads on each processor are capable of making progress (see Figure 3) using combined locks. Due to their low latency, spin locks outperform others when a critical section is small. However, for larger critical sections, the figure shows a distinct performance advantage for combined locks. The figure shows the performance of two combined locks – one that spins 10 times before blocking, another that spins once before blocking. In this experiment the former outperforms the latter for larger critical sections, However in general, the optimal number of initial spins of combined locks will depend on various application characteristics such as its locking pattern, length of critical sections, *etc.* Furthermore, the optimal waiting policy for a lock will be different during different phases of a computation from which we hypothesize that a waiting policy based on dynamic feedback is essential for better application performance[5, 6].

A third experiment studies the performance of a lock's configuration called "advisory locks". In this experiment, we study the effect of changing the waiting policy in different phases of a computation (as suggested by the result of the last experiment).

Advisory/Speculative lock. As the name suggests, the owner of such a lock advises other requesting threads on whether to spin or sleep while waiting by (dynamically) changing the parameters of the lock's internal state. In general, the length of a critical section may vary significantly in different phases of computation. The current owner of a lock advises or configures the lock for a requesting thread at different stages in the critical section depending on the remaining length of the critical section and/or waiting pattern of application's threads [5].

This experiment uses an application performing critical sections of varying lengths. Figure 4 compares the performance of such a lock with simple spin and blocking locks. As the figure shows, advisory locks outperform others when critical section is long and there are waiting threads.

We have also implemented[5] a few common lock configurations such as *read-write* locks, *recursive* locks, simple *conditional locks etc.* to demonstrate the generality of the structure of reconfigurable lock objects. The implementa-

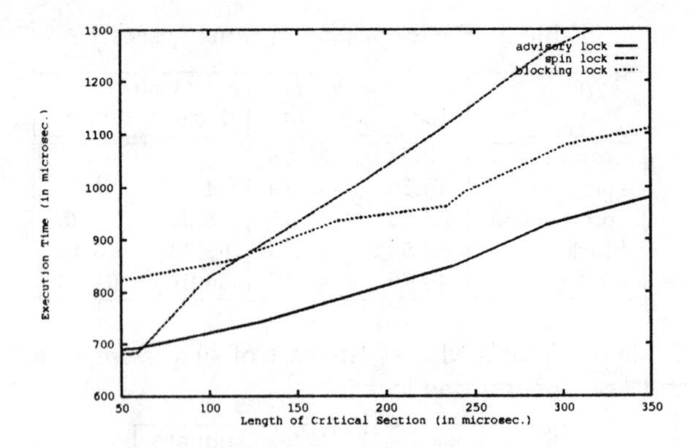


Figure 4: Critical section length Vs. Execution time

tion of a lock may also be altered using architectural hints resulting in *centralized* or *distributed semi-consistent* locks, *passive* or *active locks*, etc[5].

### CONCLUSION

This paper proposes a structure for lock objects such that their waiting and scheduling behaviors are easily changed (statically or dynamically). It then demonstrates that it is essential to have application specific lock schedulers for increased performance. Furthermore, it demonstrates that dynamic change of waiting policy often results in improved application performance. Our future research focuses in part on finding the optimal waiting policy for an application specific lock by speculative dynamic reconfiguration.

# References

[1] T. Anderson, E. Lazowska, and H. Levy. The performance implications of thread management alternatives for shared-memory multiprocessors. *IEEE Trans. on Comp.*, pages 1631–1644, December 1989.

[2] T. Bihari and K. Schwan. Dynamic adaptation of real-time software. *ACM Trans. on Comp. Sys.*, pages 143–174, May 1991.

[3] A. Gheith and K. Schwan. Chaos-arc – kernel support for multi-weight objects, invocations, and atomicity in real-time applications. *ACM Trans. on Comp. Sys.*, pages 33–72, April 1993.

[4] J. Mellor-Crummey and M. Scott. Algorithms for scalable synchronization on shared-memory multiprocessors. *ACM Trans. on Comp. Sys.*, February 1991.

[5] B. Mukherjee and K. Schwan. Experiments with a configurable lock for multiprocessors. Technical Report GIT-CC-93/05, Computing, Georgia Tech, 1993.

[6] B. Mukherjee and K. Schwan. Improving performance by use of adaptive objects: Experimentation with a configurable multiprocessor thread package. In *Proc. High Perf. Distr. Comp.*, July 1993. Also TR GIT-CC-93/17.

# ON RESOURCE ALLOCATION IN BINARY *n*-CUBE NETWORK SYSTEMS

**Nian-Feng Tzeng**† and **Gui-Liang Feng**

Center for Advanced Computer Studies
University of Southwestern Louisiana
Lafayette, LA 70504

**Abstract** —— *We introduce systematic approaches to resource allocation in a cube system so that each non-resource node is connected with one resource copy and that the allocation performance measure of interest is optimized. The methodology used is based on the covering radius results of known codes which aid in constructing desired linear codes whose codewords address nodes where resource copies are placed. Our approaches are applicable to any cube size, often arriving at more efficient allocation outcomes than prior attainable results.*

## 1. Introduction

The problem of allocating multiple copies of a resource in a binary $n$-cube network system is challenging, and a systematic allocation approach is desirable to achieve best results. Distributing resource copies in a hypercube with an attempt to optimize system performance measures of interest has been investigated. In particular, Chiu and Raghavendra have developed efficient algorithms [1] for allocating a given number of resource copies to a hypercube system in an effort to optimize a defined performance measure, called the *resource diameter*. Resource diameter is the maximum resource distance among all the cube nodes, where the resource distance of a node is the minimum number of hops from the node to a node equipped with a copy of the resource. Their algorithms partition hierarchically a system under consideration into levels, in each of which allocation is accomplished following a perfect code (like the Hamming code, the Golay code [2, Ch. 5.8]) or a basic strategy.

In this paper, we deal with allocating copies of a certain resource to cube nodes efficiently in a large system, based on the covering radius results of known codes to aid in constructing desired linear codes whose codewords specify the locations of resource copies. The resource allocation problem is translated to an integer nonlinear programming problem whose best solution can be obtained efficiently by taking advantage of basic properties derived from the known codes, giving rise to optimal or near-optimal allocation.

† This work was supported in part by the NSF under Grant MIP-9201308 and by the State of Louisiana under Contract LEQSF(1992-94)-RD-A-32.

## 2. Notations and Background

A binary $n$-cube network, denoted by $Q_n$, consists of $2^n$ nodes, each addressed by a unique $n$-bit number. The *distance* between any two cube nodes is the number of bits differing in their addresses. An allocation of multiple identical copies of a certain resource is the set of cube nodes to which these resource copies are assigned. Let $2^k$ resource copies be allocated to a binary $n$-cube network in accordance with allocation $A$, resulting in the resource diameter denoted by $d_A(n, k)$, or $d(n, k)$ for simplicity if there is no confusion. An allocation in $Q_n$ is optimized if either $d(n, k)$ is minimum for a given $k$ or $k$ is minimum for a given resource diameter.

Our resource allocation approach is based on the *binary linear block code*. A binary linear block code $\Psi(n, k)$ comprises a set of $2^k$ binary sequences of length $n$ called *codewords*. Code $\Psi(n, k)$ can be described concisely using a $k$ by $n$ matrix $\mathbf{G}$ known as the *generator matrix*. Any codeword of $\Psi(n, k)$ is a linear combination of the rows of $\mathbf{G}$. An $n$-tuple $\mathbf{c}$ is a codeword if and only if it is orthogonal to every row vector of an $(n-k)$ by $n$ matrix $\mathbf{H}$, called a *parity-check matrix* of the code, namely, $\mathbf{c} \cdot \mathbf{H}^T = \mathbf{0}$, where $\mathbf{H}^T$ is the transpose of matrix $\mathbf{H}$. A linear block code is completely defined as long as its parity check matrix $\mathbf{H}$ is known.

A basic parameter of linear codes is the covering radius [3]. Let $\Lambda(\mathbf{y})$ be the minimum distance between an $n$-tuple $\mathbf{y}$ and a codeword in $\Psi(n, k)$, then the *covering radius* of the code is the maximum $\Lambda(\mathbf{y})$ for all $\mathbf{y}$'s. As a result, the covering radius of a linear code is exactly the same as the resource diameter of an allocation obtained by following the code. Since our allocation is equivalent to finding out a desired linear code, hereinafter, the terms resource diameter and covering radius will be used interchangeably.

The allocation problem can be solved elegantly by making use of covering radius results provided in [4]. Let codes $\Psi(n_i, k_i)$, $i = 1, 2$, have covering radius $r_i$, then the direct sum of the two codes is a linear code $\Psi(n_1 + n_2, k_1 + k_2)$ with covering radius $r_1 + r_2$. Only a few codes have optimal (i.e., smallest possible) covering radii, and are referred to as perfect codes. In particular, Hamming codes $\Psi(2^\alpha - 1,\ 2^\alpha - 1 - \alpha)$ have the optimal covering radius of 1, and the Golay code (23, 12) has the optimal covering radius of 3. Hamming codes and the Golay code are the shortest codes for covering radius equal to 1 and 3, respectively. For any other covering radius, there is no method yet, by which a shortest code can be found. However, near-shortest linear codes with the covering radius of 2 have been introduced [4], where the parameters $n$ and $k$ of the near-shortest linear codes are given by

$$n = 5, \text{ if } n-k=4; \quad n=9, \text{ if } n-k=5;$$
$$n = 13, \text{ if } n-k=6; \quad n=19, \text{ if } n-k=7;$$

and for integer $p \geq 2$,

$$\begin{aligned} n &= (2^p-1)(2^{p+1}+1), && \text{if } n-k=4p; \\ n &= (2^p-1)(2^{p+1}+1)+(2^{2p}-1), && \text{if } n-k=4p+1; \\ n &= (2^p-1)(2^{p+1}+1)+(2^{2p+1}-1), && \text{if } n-k=4p+2; \\ n &= (2^p-1)(2^{p+1}+1)+(2^{2p+2}-1), && \text{if } n-k=4p+3. \end{aligned} \quad (1)$$

The parity check matrix of every near-shortest linear codes with covering radius 2 is specified in [4].

## 3. Proposed Methodology

Our basic methodology for resource allocation is to determine the parity check matrix $\mathbf{H}$ of an appropriate code whose codewords specify the node addresses at which resource copies are located, in an attempt to optimize the allocation performance measure of interest.

### 3.1. Minimizing Resource Diameter

In order to minimize resource diameter $d(n,k)$ for given $n$ and $k$, we search for an $h$ by $l$ parity check matrix, $\mathbf{H}_o$, which is composed of parity check matrices of Hamming codes, near-shortest codes with covering radius = 2 (called NS codes for simplicity), and the Golay code, such that the resultant covering radius is minimized. The constituent parity check matrices lie along the diagonal of $\mathbf{H}_o$, as depicted in Fig. 1, where the resulting linear code is constituted by the direct sum of codes with parity check matrices $H_1, H_2, \cdots, H_a$. From $\mathbf{H}_o$, we want to trivially get an $(n-k)$ by $n$ matrix, $\mathbf{H}$, which fulfills the requirement of $\mathbf{G}\cdot\mathbf{H}^T = \mathbf{0}$, so that the resource addresses can be derived (from $\mathbf{G}$). Now, if the size of $\mathbf{H}_o$ obtained satisfies $h \geq n-k$ and $l \leq n$, we may trivially get $\mathbf{H}$ from $\mathbf{H}_o$ by trimming off its lowest $h-(n-k)$ rows and adding $n-l$ columns of zero vectors to its right end, as shown in Fig. 2. This is obviously true, since $\mathbf{H}_o$ is a parity check matrix, there must be a $k$ by $l$ matrix $\mathbf{G'}$ satisfying $\mathbf{G'}\cdot\mathbf{H}^T = \mathbf{0}$, and $\mathbf{G}$ can be $\mathbf{G'}$ augmented with $n-l$ columns of arbitrary vectors to its right end.

The constraint on $h$ and $l$ values is essential, making it possible to translate the search of an appropriate code (for allocation) to an integer programming problem. Consider the case of $k = 13$ and $n = 20$ as an example ($n-k = 7$). There are many different ways to construct the parity check matrix $\mathbf{H}_o$, including the three illustrated in Fig. 3. In Fig. 3(a), the parity check matrices of four Hamming codes constitute the resulting matrix, yielding $h=7$, $l=12$, and covering radius = 4. The matrix given in Fig. 3(b) is composed of the parity check matrices of three Hamming codes, exhibiting $h = 8$, $l = 17$, and covering radius = 3. On the other hand, Fig. 3(c) consists of only the parity check matrix of an NS code. It is apparent that the matrix depicted in Fig. 3(c) is the best choice, since it leads to a minimum covering radius.

For any set of $n$ and $k$, the selection of a parity check matrix with the smallest covering radius is achieved by solving the following integer nonlinear program. Let the constituent parity check matrices involve $x_\alpha$ Hamming codes $\Psi(2^\alpha-1,\ 2^\alpha-1-\alpha)$, $y_\beta$ NS codes $\Psi(w(\beta),\ w(\beta)-\beta)$, and $z$ Golay codes, where $w(\beta)$ and $\beta$ are respectively $n$ and $n-k$ given in Eq. (1), with $\beta \geq 2$, and $x_\alpha$, $y_\beta$, and $z$ are nonnegative integers. We have

$$\begin{aligned} &\sum_{\forall\alpha} \alpha\, x_\alpha + \sum_{\forall\beta} \beta\, y_\beta + 11z \geq n-k \\ &\sum_{\forall\alpha} x_\alpha(2^\alpha-1) + \sum_{\forall\beta} y_\beta\, w(\beta) + 23z \leq n \\ &\sum_{\forall\alpha} x_\alpha + \sum_{\forall\beta} 2y_\beta + 3z \quad \textit{minimized} \end{aligned} \quad (2)$$

Solving the set of preceding equations directly is very time-consuming, as those equations involve integer variables $x_\alpha$ and $y_\beta$, for all $\alpha, \beta \geq 0$, in addition to $z$. Fortunately, we may reduce the complexity of the solution significantly by taking advantage of the basic properties associated with Hamming codes, NS codes, and the Golay code: For any given $n$, an NS code gives rise to no larger covering radius than the code resulting from a direct sum of multiple Hamming codes, and the Golay code gives rise to no larger covering radius than the code from a direct sum of multiple Hamming codes and NS codes. A solution can be obtained without searching all possibilities of $x_\alpha$ and $y_\beta$ exhaustively, by simplifying this integer program substantially, according to the next theorem, whose proof is provided in [5].

**Theorem 1:** A solution for Eq. (2) involves at most one nonzero $x_\alpha$, say $x_{\alpha^*}$, and four nonzero $y_\beta$'s. In addition, $x_{\alpha^*}$ is no larger than 1, and the nonzero $y_\beta$'s are consecutive, denoted by say $y_{\beta^*}$, $y_{\beta^*+1}$, $y_{\beta^*+2}$, and $y_{\beta^*+3}$ ($\beta^* \geq 2$).

Theorem 1 indicates that it is impossible for a solution to involve multiple Hamming code parity check matrices; a solution contains at most one Hamming code parity check matrix. The search process for the solution is thus simplified to

$$\alpha^* x_{\alpha^*} + \sum_{i=0}^{3} (\beta^*+i)\, y_{\beta^*+i} + 11z \geq n-k$$

$$x_{\alpha^*}(2^{\alpha^*}-1) + \sum_{i=0}^{3} y_{\beta^*+i}\, w(\beta^*+i) + 23z \leq n$$

such that the covering radius, $x_{\alpha^*}+2\sum_{i=0}^{3} y_{\beta^*+i}+3z$, is minimized, where $\alpha^*$ is less than $\log_2(n+2)$ (from the last inequality by letting $x_{\alpha^*}=1$, and the remaining five variables equal to 0). For a set of known $x_{\alpha^*}$ and $z$, the preceding two inequalities become respectively

$$\sum_{i=0}^{3}(\beta^*+i)\,y_{\beta^*+i}\ge m' \quad\text{and}\quad \sum_{i=0}^{3} y_{\beta^*+i}\,w(\beta^*+i)\le n' \qquad (3)$$

where $m'=n-k-\alpha^* x_{\alpha^*}-11z$ and $n'=n-x_{\alpha^*}(2^{\alpha^*}-1)-23z$, with $n'\ge m'>0$ and $\beta^*\ge 2$. The expression to be minimized becomes $\sum_{i=0}^{3} y_{\beta^*+i}$. A solution for Eq. (3) is reached efficiently by making use of the next lemma and Theorem 2, whose proofs are given in [5].

**Lemma:** Any solution for Eq. (3): $\beta^*$, $y_{\beta^*+i}$, $0\le i\le 3$, satisfies $\frac{w(\beta^*)}{\beta^*+3}<\frac{n'}{m'}$.

**Theorem 2:** Let $\beta_u$ be the largest integer satisfying the above lemma, then there is a solution set for the nonlinear program given in Eq. (3): $\beta^*$, $y_{\beta^*}$, $y_{\beta^*+1}$, $y_{\beta^*+2}$, $y_{\beta^*+3}$ such that $\beta^*$ equals $\beta_u$, $\beta_u-1$, or $\beta_u-2$.

This theorem suggests that only three possible $\beta^*$ values have to be examined in the process of search for a solution. For each $\beta^*$ value, a set of $y_{\beta^*}$, $y_{\beta^*+1}$, $y_{\beta^*+2}$, $y_{\beta^*+3}$ is identified and the covering radius calculated. The solution is thus arrived at directly without checking all possible $\beta^*$ values.

### 3.2. Minimizing the Number of Resource Copies

For a given resource diameter $d$, it is interesting to find the minimum number of resource copies required in $Q_n$. This is equivalent to deriving a linear code $\Psi(n,k)$ for a given covering radius $d(n,k)$ such that it contains as few codewords as possible (i.e., minimum $k$, as each codeword corresponds to a resource copy). In other words, a suitable parity check matrix of the code with covering radius $r$ is to be constructed using the parity check matrices of Hamming codes, NS codes, and the Golay code, in the same way as described above. However, the set of inequalities (or the nonlinear program) to be satisfied becomes

$$\sum_{\forall\alpha} x_\alpha+\sum_{\forall\beta} 2y_\beta+3z\le r$$

$$\sum_{\forall\alpha} x_\alpha(2^\alpha-1)+\sum_{\forall\beta} y_\beta\,w(\beta)+23z\le n \qquad (4)$$

$$f\triangleq\sum_{\forall\alpha}\alpha\,x_\alpha+\sum_{\forall\beta}\beta\,y_\beta+11z \quad \textit{maximized}$$

where nonnegative integers $x_\alpha$, $y_\beta$, and $z$ are as defined before, with $\beta\ge 2$. Again, this minimization problem is translated to an integer nonlinear programming problem. Like the prior one, this integer program has a very high time complexity, and significant reduction in time complexity can result from the basic properties of Hamming codes, NS codes, and the Golay code, according to the subsequent theorem, whose proof is given in [5].

**Theorem 3:** A solution for Eq. (4) involves at most one nonzero $x_\alpha$ and four nonzero $y_\beta$'s. Nonzero $x_\alpha$ is no larger than 1, and the nonzero $y_\beta$'s are consecutive, denoted by say $y_{\beta^*}$, $y_{\beta^*+1}$, $y_{\beta^*+2}$, and $y_{\beta^*+3}$ ($\beta^*\ge 2$).

Let the nonzero $x_\alpha$, if any, be denoted by $x_{\alpha^*}$. As a result of Theorem 3, the nonlinear program under consideration is simplified to

$$x_{\alpha^*}+2\sum_{i=0}^{3} y_{\beta^*+i}+3z\le r$$

$$x_{\alpha^*}(2^{\alpha^*}-1)+\sum_{i=0}^{3} y_{\beta^*+i}\,w(\beta^*+i)+23z\le n$$

such that $m\triangleq\alpha^* x_{\alpha^*}+\sum_{i=0}^{3}(\beta^*+i)\,y_{\beta^*+i}+11z$ is maximized, with $\alpha^*\le\log_2(n+2)$. For given $x_{\alpha^*}$ and $z$, the preceding two inequalities become respectively

$$\sum_{i=0}^{3} y_{\beta^*+i}\le r' \quad\text{and}\quad \sum_{i=0}^{3} y_{\beta^*+i}\,w(\beta^*+i)\le n' \qquad (5)$$

where $r'=(r-x_{\alpha^*}-3z)/2$, $n'$ is as defined in Eq. (3), $n'\ge r'>0$, and $\beta^*\ge 2$. The expression to be maximized is $m'=\sum_{i=0}^{3}(\beta^*+i)\,y_{\beta^*+i}$. We may identify the search space of the integer nonlinear program specified by Eq. (5) in two cases, depending on the value of $\sum_{i=0}^{3} y_{\beta^*+i}$.

**Case 1:** $\sum_{i=0}^{3} y_{\beta^*+i}=r'$. In this case, we have $w(\beta^*)r'=w(\beta^*)\sum_{i=0}^{3} y_{\beta^*+i}<\sum_{i=0}^{3} w(\beta^*+i)\,y_{\beta^*+i}\le n'$, leading to $w(\beta^*)<\frac{n'}{r'}$ (recall that $w(\beta)$ is a monotonically increasing function). Let $\beta_u$ be the largest integer value satisfying $w(\beta^*)<\frac{n'}{r'}$, then the $\beta^*$ value to be examined ranges from 2 to $\beta_u$ inclusively. For each $\beta^*$ (an integer) examined, the above nonlinear program becomes a linear program and a solution set can be found, with $m'$ obtained. The solution under this case is the one which gives rise to the maximum $m'$.

**Case 2:** $\sum_{i=0}^{3} y_{\beta^*+i}<r'$. In this case, let $\sum_{i=0}^{3} y_{\beta^*+i}=r'-\delta$, with $1\le\delta\le r'-1$. If $\beta^*$, $y_{\beta^*}$, $y_{\beta^*+1}$, $y_{\beta^*+2}$, $y_{\beta^*+3}$ are a set of solution for the above nonlinear program, we have the following inequality (whose proof is provided in [5]):

$\sum_{i=0}^{3} y_{\beta^*+i}\, w(\beta^* + i) > n' - \delta\, w(\beta^*)$. Rearranging the inequality yields $n' < (y_{\beta^*} + \delta)\, w(\beta^*) + \sum_{i=1}^{3} y_{\beta^*+i}\, w(\beta^* + i)$, which is $< (\delta + \sum_{i=0}^{3} y_{\beta^*+i})\, w(\beta^* + 3) = r' w(\beta^* + 3)$. This results in

$$w(\beta^* + 3) > \frac{n'}{r'}. \tag{6}$$

On the other hand, from $\sum_{i=0}^{3} y_{\beta^*+i} > 1$ (i.e., at least one of the three terms is nonzero), we have $w(\beta^*) \le \sum_{i=0}^{3} y_{\beta^*+i}\, w(\beta^* + i)$, which is $\le n'$, according to Eq. (5). Let $\beta_s$ and $\beta_u$ are respectively the minimum and the maximum values which satisfy both Eq. (6) and $w(\beta^*) \le n'$. The $\beta^*$ value to be searched ranges from $\max(\beta_s, 2)$ to $\beta_u$, where max is a maximum function. For each value searched, Eq. (5) becomes a linear program and a solution is then obtained. The solution under Case 2 is the one which yields maximum $m'$. We thus solve this problem based on the Case 1 solution and the Case 2 solution.

## 4. Conclusions

Systematic resource allocation in binary $n$-cube network systems that optimizes the measure of interest has been presented. This is made possible by constructing appropriately a parity check matrix of the desired linear code whose codewords specify the locations of resource copies, using the results of three types of known codes: Hamming codes, near-shortest codes with covering radius = 2, and the Golay code. Finding the best allocation with respect to a given measure of interest is translated to selecting a set of the codes whose covering radii satisfy two nonlinear inequalities and optimize the third one. Significant reduction in the time complexity of selecting such a set is attained by taking advantage of the basic properties of the three types of codes. We have pursued the approach to achieving minimum resource diameter in a cube system involving a given number of resource copies as well as the approach to achieving fewest resource copies for a given resource diameter. The allocation results we obtained appear interesting and are particularly useful for large-scale cube network systems.

## References


[1] G.-M. Chiu and C. S. Raghavendra, "Resource Allocation in Hypercube Systems," *Proc. 5th Distributed Memory Computing Conf.*, Apr. 1990, pp. 894-902.

[2] R. E. Blahut, *Theory and Practice of Error Control Codes*, Reading, MA: Addison-Wesley Publishing Co., 1983.

[3] G. D. Cohen *et al.*, "Covering Radius – Survey and Recent Results," *IEEE Trans. Information Theory*, vol. 31, pp. 328-343, May 1985.

[4] R. A. Brualdi, V. S. Pless, and R. M. Wilson, "Short Codes with a Given Covering Radius," *IEEE Trans. Information Theory*, vol. 35, pp. 99-109, Jan. 1989.

[5] N.-F. Tzeng and G.-L. Feng, "On Resource Allocation in Binary *n*-Cube Network Systems," Tech. Rep. *TR 93-8-3*, CACS, Univ. of Southwestern Louisiana, 1993.


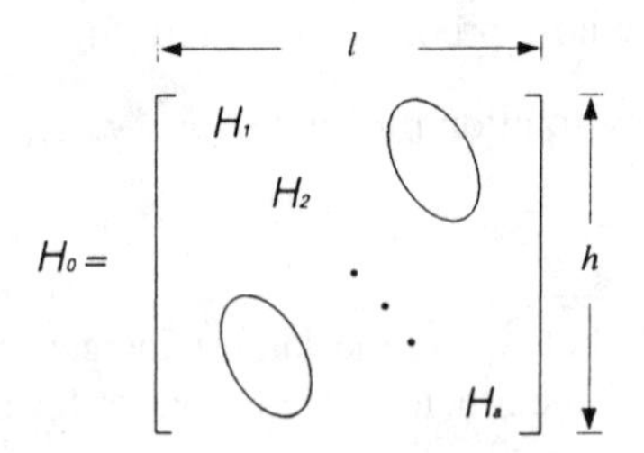


Fig. 1. $H_0$ formed by $H_1, H_2, \cdots, H_a$.

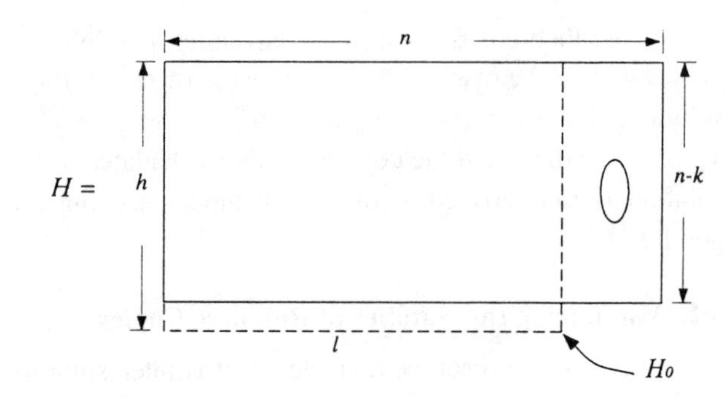


Fig. 2. $H$ constructed from $H_0$ by trimming its lowest $h - (n - k)$ rows and adding $(n - l)$ columns of zero vectors to its right end.

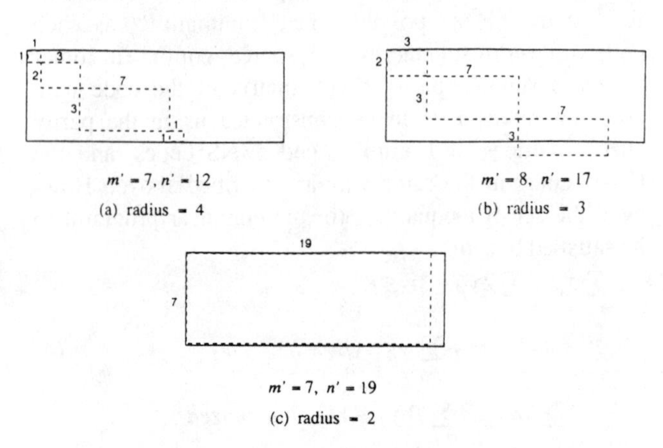


Fig. 3. Three possible ways to form a parity check matrix satisfying $n = 20$ and $k = 13$.

# A Distributed Load Balancing Scheme for Data Parallel Applications

*Walid R. Tout* and *Sakti Pramanik*

Department of Computer Science
A702 Wells Hall
Michigan State University
East Lansing, MI 48824-1027
{tout,pramanik}@cps.msu.edu

**Abstract**

*In data parallel applications, a major source of load imbalance is in the uneven distribution of data between the nodes. The major contribution of this paper is the analysis of a new distributed load balancing scheme based on random probing. Sequential probing was shown to create hot spot problems. Random probing, on the other hand, avoids these hot spots problems. We have modeled this load balancing scheme both analytically and experimentally. Distributed Hash Join was used as an example application. The experiments were done on a BBN TC2000 multiprocessor system.*

## 1 Introduction

Load imbalance is one of the major problems in data parallel applications. A major source of load imbalance in these applications is the uneven distribution of data among the various nodes in the system. The shared–memory model of computation addresses the imbalance problem by creating a large number of processes to reflect the parallelism of the application and scheduling these processes via a central ready queue. While this scheme provides dynamic load balancing, it suffers from a number of problems [2]. The centralized control of the scheme may cause bottlenecks as the number of nodes in the system increases. Another major problem is the lack of locality of references [2].

In this paper, we propose a new distributed load balancing scheme for data parallel applications where nodes that finish processing their local data, independently probe the system to help the overloaded nodes. In order to preserve the locality of references, only one process is created for each node in the system and this process will execute only on that node during its lifetime.

The rest of the paper is organized as follows. Next section introduces the load balancing scheme for data parallel applications and presents Distributed Hash Join as an example application. In section 3, we develop the analytical model for the load balancing scheme. The empirical results from the experiments are discussed and compared with those of the analytical model in section 4. Finally, the concluding remarks are presented in Section 5.

## 2 Load Balancing for Data Parallel Applications

The focus of this paper is on developing a dynamic load balancing scheme for efficiently executing data parallel applications. In this class of applications, execution usually involves processing large amounts of data in main memory where the operations are performed on the individual data items. The architecture chosen for this class of applications is the Shared–memory multiprocessors based on Non–Uniform Memory Architecture (*NUMA*). Such architectures provide a very large and scalable main memory and also scale to a large number of nodes [1].

In data parallel applications, data may be partitioned into equal or variable size blocks and distributed among the various processors in the system. In a number of applications, data may not be always evenly distributed among the processors. One such application is the relational join where the selection–projection steps may alter the initial distribution of tuples in each relation and consequently produce an unbalanced load. It is important to note that an even distribution of data items does not guarantee a balanced load. This is especially true when the input values determine the actual load a processor is assigned. Computing the transitive closure for a matrix with a clique is an example of such applications.

In a distributed load balancing scheme, each helping node, i.e. a processor that is done processing its local data, decides which other nodes to help. In order to make such a decision, a helping node checks the status of the other nodes and selects an overloaded node to help. The helping node then gets a block of data from the overloaded node and processes it locally. In a *NUMA* multiprocessor, such interactions between the nodes may cause conflicts. The next section presents the load balancing scheme in details and addresses the problem of conflicts.

### 2.1 Proposed Load Balancing Scheme

We propose a new model to better accommodate data parallel applications. In this model, we create only one thread per processor. Each thread will execute on its assigned processor during its lifetime and access its data locally, thus benefitting from the locality of references. In this scheme, load balancing is not done using a central ready queue but is performed in a distributed fashion. When load imbalance arises, each helping node will transfer an appropriately sized block of data from a selected overloaded processor and process that block locally. The selection of the overloaded node is made by each helping node independently in order to prevent any bottlenecks that may result from a centralized scheduling scheme. The distributed scheme for selecting an overloaded processor is described in the following section.

For each node (in parallel) do
**Step 1:** Process local data.
**Step 2:** Repeat
A. Select an overloaded node.
B. Transfer a block of size
Minimum(Un–processed data items, $B$)
C. Process the block locally.
Until (All nodes are done).

Figure 1. Distributed load balancing scheme

The general scheme for distributed load balancing is shown in figure 1. Here, each node maintains an index ($I_n$) to its data and a lock ($L_n$) to synchronize simultaneous access to $I_n$. A node, wishing to copy a block from node $n$, has to lock $L_n$, update $I_n$ with the size of the block (default $B$), unlock $L_n$ and then transfer the block. This means that all nodes must use the locks in order to process their own tuples. Locks can be very costly if not handled properly. Next, we discuss some strategies for selecting overloaded nodes while considering the effects of locking.

### 2.2 Sequential Probing and the Blocking Effect

Probing, to find an overloaded node, can be done either sequentially or randomly. In sequential probing, helping nodes sequentially check the status of other nodes and select the first overloaded node to help. Since the helping nodes are searching for the overloaded nodes sequentially, all the helping nodes between two successive overloaded nodes will get blocked helping one of these two overloaded nodes. This blocking phenomenon will cause an uneven distribution of helping to helped nodes. As the ratio of the number of helping to helped nodes increases, the number of helping nodes between two consecutive helped nodes also increases. This will cause a hot spot on a few overloaded nodes. To illustrate this phenomenon, let us assume that the system has 16 nodes and at time $t$, all nodes have become helping nodes except for nodes number 13 and 16 which are still overloaded. With sequential probing, node number 13 may be probed and helped by all the 12 nodes before it, while node number 16, which may have much more data than number 13, will only get help from nodes 14 and 15.

### 2.3 Scheduling by Random Probing

As the number of nodes increases, the blocking effect in sequential probing causes more performance degradation. In order to avoid this blocking problem, helping nodes probe randomly for overloaded nodes. A node is selected at random and its status is checked. If it is overloaded, the helping node will get a block of data from this node, process it locally and then randomly select another overloaded node. To determine the end of execution for all nodes, each node increments a global counter when it is done processing all its tuples. The end of execution for all nodes is signalled when this counter reaches the total number of nodes.

### 2.4 Load Balancing for Parallel Join

In this section we present the load balancing scheme for relational join on MIN architecture. The reader is referred to [4] for further details about the hash join algorithm. Let $R$ and $S$ denote the first and the second relation respectively. Each node hashes its local tuples to determine their destination nodes. Tuples that hash into the local node are further hashed to organize them into a local hash table. This table is used for probing during the second phase of the algorithm. All other tuples are marked with their destination nodes' ID ($n_i$) and later transferred in bulk to these nodes. When a node has completed distributing $R$, processing of $S$ proceeds similarly. Tuples that hash into the local node are joined with $R$ and all others are bulk transferred to their destination nodes.

Note that before a node can proceed with the join, it must wait to receive all the $R$ tuples that hashed to it. Nodes must also wait to receive all the tuples of $S$ before finishing the join. During these waiting periods, a node may randomly probe for overloaded nodes and help them.

## 3 Analytical Model

In this section we develop the analytical model for the load balancing scheme. The cost of processing all the data, without load balancing, can be computed as the total time taken by the processor that finishes last, i.e. the processor with the maximum number of data items. However, with the load balancing scheme, it is not the node with the maximum number of data items but rather the helping node, that gets the last block of data, that will finish last. In the rest of the paper, this node will be referred to as the *last helping node*.

To obtain the total time for load balancing, we will add up the processing time of each block by the last helping node. Thus, if $T_B(i)$ is the time to process the $i$th block then the total time for load balancing is

$$T = \sum_{i=1}^{L_B} (T_B(i))$$

where $L_B$ is the total number of blocks processed by the last helping node. The cost $T_B(i)$ is the sum of the following three components:

- 1. Lock $L_n$, determine the block size, update $I_n$ and then unlock $L_n$. The cost for this step is $T_{lock} + T_{unlock}$. The costs of locking and unlocking are system dependent and the costs of determining the block size and updating $I_n$ may be ignored.

- 2. Transfer a block to be processed locally, only if this is a helping node. The probability of such transfer is given by $\frac{n_0(i)}{N}$, where $n_0(i)$ is the average number of helping nodes during the time interval needed to process the $i$th block and $N$ is the total number of nodes in the system. Thus, the total cost of this step is:

$$\frac{n_0(i)}{N} \times T_{bt}(B, N),$$

where $T_{bt}(B, N)$ is the cost of transferring a block of size $B$ when $N$ nodes are active. This parameter is system dependent.

- 3. Process the data block locally. The cost for this is $T_b = B \times T_p$, where $T_p$ is the time to process a single data item. The parameter $T_p$ is application dependent.

Sequential and random probing affect the costs of items 1 and 2 differently. In sequential probing, these components contribute to the blocking effect and may result in forming *Hot-Spots*. Random probing, on the other hand, eliminates the blocking effect by randomizing the selection of overloaded nodes. However, conflicts may still occur when more than one node try to access the same location. Therefore, the time to access a memory location is $\xi_D \times T_a$, where $T_a$ is the time to execute a single access, $|D|$ is the number of data items and $\xi_D$ is the expected number of nodes involved in the access. The derivation of $\xi_D$ is based on the results of [6, 3]. The expression for $\xi_D$ is $\xi_D = \frac{N}{\phi}$ where

$$\phi = N \times \left[1 - \Pi_{i=1}^{N} \left(\frac{|D| \times N \times K - i + 1}{|D| \times N - i + 1}\right)\right]$$

$$K = 1 - \frac{1}{N}.$$

Therefore, the cost of reserving and processing a block of data is $T_B(i)$ =

$$\left[T_b + (\frac{n_0(i)}{N} \times T_{bt}(B, N) + T_{lock} + T_{unlock}) \times \xi_D\right]$$

The $L_B$ in the cost formula for $T$ is not needed. We evaluate $T$ iteratively as follows: At each step of the cost formula for $T$, the initial number of blocks in each node will be compared to $i$. If the number of blocks for a node is greater or equal to $i$ then that node becomes a helping node. Hence, $n_0(i)$ is incremented and $n_1(i)$ is decremented accordingly. When $n_0(i)$ reaches $N$, this signals the end of processing. Thus, the summation of the cost formula for $T$ is evaluated iteratively until $n_0(i)$ is equal to $N$.

Next, we present and discuss the analytical and experimental results.

# 4 Performance of the Load Balancing Scheme

In order to analyze the performance of the load balancing scheme, the data was skewed using parameterized Zipfian distribution. We introduce the formulation for Zipfian distribution and refer the interested reader to [7, 5] for more information. The probability $p_i$ that a data item belongs to a given node $n_i$ where $1 \leq i \leq N$ is $p_i = \frac{c}{i^{1-Z}}$, where

$$c = \frac{1}{\sum_{i=1}^{N} \frac{1}{i^{(1-Z)}}}$$

is a normalization constant. The value of $Z$ ranges between 0, the pure Zipfian which is highly skewed, to 1 which corresponds to the Uniform distribution.

In the rest of this section, the size of the relation is taken to be proportional to the number of nodes, i.e. $|R| = N \times F$ for some constant $F$. $F$ is set to 3000 tuples unless specified otherwise. The experiments were run on a BBN TC2000. Data items were distributed among the nodes according to Zipfian distribution as outlined above.

In order to compare the performance of the analytical model to that of the experiments, we measured the times for the system dependent parameters of Section 3 on the TC2000. The times were measured by executing the corresponding operations inside tight loops for a large number of iterations while accounting for the loop and system overheads.

Figure 2 shows the total processing time with and

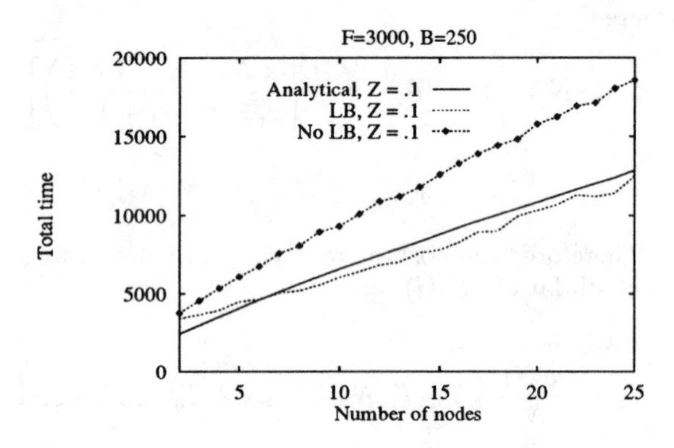


Figure 2. Total time with and without load balancing.

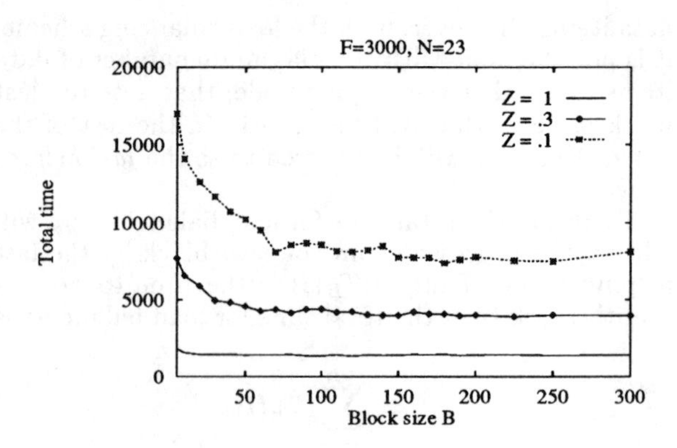


Figure 3. Effect of block size $B$ on load balancing.

without load balancing. By transferring blocks of tuples from overloaded nodes and processing them locally, the helping nodes improve the performance of the algorithm as shown in the figure. Note that the performance of the load balancing scheme keeps improving as we increase the number of nodes in the system. The reason is that, since the total number of tuples increases linearly with the number of nodes, the join algorithm without load balancing will suffer more load imbalance and its performance will deteriorate.

The performance of the analytical model is compared with the experimental results in Figure 2. Since the total number of tuples is $N \times F$, when the number of nodes increases, the number of tuples to be processed increases and the data becomes more skewed. The figure shows this trend as the number of nodes in increased up to 25. It should be noted from the figure that the analytical results closely approximate those of the experiments. The maximum difference between the model and the experimental results is 9%.

Before transferring a block of tuples for local processing, a helping node has to lock $L_n$, update $I_n$ and then unlock $L_n$. Given a fixed number of tuples (during the execution of the algorithm), a small block size means a larger number of blocks and consequently more frequent lock accesses. As mentioned earlier, lock operations are expensive on MIN systems and a large number of simultaneous lock operations severely degrade performance. The effect of the block size is shown in Figure 3 where it can be seen that for small block sizes, the increased lock conflicts degrade the performance severely. The figure also shows that the performance degradation for small block sizes is more profound with smaller values of $Z$, i.e. for higher skew rates. This is because higher skew rates mean more blocks will be transferred by the helping nodes during load balancing and thus more lock conflicts. Note that, as expected, varying the block size does not have much effect on the uniform distribution case ($Z = 1$) since the number of blocks to move in this case is insignificant.

# 5 Conclusion

In this paper, we presented a distributed load balancing scheme for data parallel applications on NUMA multiprocessors. The scheme was shown to provide a balanced load and to scale with both the number of nodes and the size of the data. We plan to investigate this scheme for a larger number of data parallel applications and compare the performance with the centralized scheme of the shared memory computation model.

# References

1. **BBN Advanced Computers, Inc., Cambridge, MA**, "Inside the TC2000 Computer", Feb. 1990.

2. **T. J. LeBlanc, and E. P. Markatos**, "Shared Memory vs. Message Passing in Shared-Memory Multiprocessors", *Technical Report, CS Dept., University of Rochester*, April 1992.

3. **H. Lu, K. Tan, and M. Shan**, "Hash-Based Join Algorithms for Multiprocessor Computers with Shared Memory," *Proceedings of the 16th VLDB Conference*, pp. 198–208, 1990.

4. **W. Tout, and S. Pramanik**, "Main Memory Hash-Based Join Algorithms for Multiprocessors with NUMA Architecture," *Tech. Report, CPS Dept., Michigan State University*, Oct., 1992.

5. **C. B. Walton**, "Four types of data skew and their effect on parallel join performance," *Technical Report TR-90-12, Dept. of Computer Science, Univ. of Texas at Austin*, 1990.

6. **S. B. Yao**, "Approximating Block Accesses in Database Organizations," *CACM* Vol. 20, No. 4, pp. 260–261, Apr. 1977.

7. **G. Zipf**, "Human Behavior and the Principle of Least Effort," *Addison-Wesley*, 1949.

# WOULD YOU RUN IT HERE... OR THERE?[†] (AHS: AUTOMATIC HETEROGENEOUS SUPERCOMPUTING)

H. G. Dietz, W. E. Cohen, and B. K. Grant
Parallel Processing Laboratory
School of Electrical Engineering
Purdue University
West Lafayette, IN 47907-1285
hankd@ecn.purdue.edu

Abstract— *Parallel programs often exhibit strong preferences for different system structures, and machines with the ideal structures may all be available within a single heterogeneous network. There is also the complication that, although a particular application might execute fastest when running by itself on one system, the best turnaround time might result from running the program on a different system that is less heavily loaded at the time the job is submitted.*

*This paper takes a very practical approach to the problem of automatically making efficient use of heterogeneous supercomputing. Rather than making a heroic effort to achieve near peak speeds on a particular machine, our system (AHS) attempts to invisibly seek out and use whatever hardware will make the user's program execute fastest. Both the theory and implementation of AHS are presented.*

## 1. Introduction

In any discussion of programming parallel machines, it is very important to distinguish between the *programming model* and the *execution model*. The programming model is simply the model of parallelism seen by the programmer and expressed in the high-level language code. In contrast, the execution model embodies machine-specific details and is generally expressed either in native assembly language or in a high-level language augmented by calls to parallelism-related functions. Parallel execution models are complicated by the combination of advances in networking and proliferation of inexpensive, high-performance, computers: the execution model is no longer one complex target machine, but a heterogeneous network of complex target machines.

In this paper, we are concerned with the mechanics of automatically mapping *any programming model* into the *best execution model available at the time the program is run*. In particular, this paper will focus on the various aspects of AHS, a prototype software system supporting automatic heterogeneous supercomputing. Section 2 describes the parallel language and programming model used to write programs for AHS. Several of the most important execution models, and the software environments needed to support them, are described in section 3. Given this language and collection of execution models, section 4 describes how programs can be automatically distributed, compiled, and run using the execution model and machine choices that result in the fastest expected execution time. Finally, section 5 summarizes the contributions of this work and suggests directions for future research.

[†] This work was supported in part by the Office of Naval Research (ONR) under grant number N00014-91-J-4013, by the National Science Foundation (NSF) under award number 9015696-CDA, and by the United States Air Force (USAF) Rome Laboratories under award number F30602-92-C-0150.

## 2. The MIMDC Language

Although any language could be used to write programs for an automatic heterogeneous supercomputing system, certain language properties can improve the performance of the system and can widen the range of execution models supported. Perhaps the best overview of what kinds of semantics allow the widest range of execution models is given in [7].

In this paper, we focus on MIMDC, a simple C-based language embodying a control-parallel programming model. However, there is no reason a data-parallel language could not be similarly supported [8]; we are currently extending AHS to support SIMDC, a data-parallel dialect of C.

Declaring a variable to have the attribute `poly` indicates that each processor has its own local value for that variable. Thus, modifying a `poly` variable's value in one process has no effect on the value of the variable in any other process. Function arguments and return values are always `poly`; the default storage class for all variables is `poly`.

When a variable is declared to have the `mono` attribute, it means that all processes see the same value for that variable. When one processor modifies a `mono` variable, all other processors that access that variable afterwards get the new value. If multiple processes store into a `mono` variable simultaneously, the race is resolved by picking a winner and storing that value. Variables defined with the `mono` attribute are never allocated on the stack; they have the same apparent address in all processes.

There are also two basic ways that synchronization can be accomplished:

1. Shared variable accesses can be used to implement semaphores, locks, or any other type of synchronization primitive.
2. All processes can be synchronized using a barrier mechanism invoked by the `wait` statement.

Effort has been made to make the barrier synchronization relatively efficient. All processes must reach a `wait` statement before any is allowed to continue executing past the `wait`. Note that all processes need not be waiting at the same `wait` statement.

Proof-of-concept compilers have been written to translate MIMDC into a variety of target languages. Most of these compilers were written using a set of locally-developed software tools called PCCTS [6]. Different code is generated for the different execution models; we do not generate the same pseudoinstructions and then interpret them.

### 3. Execution Models

In the previous section, we discussed how the MIMDC language and its compilers work. Here we discuss how the code generated by the compilers actually implements the language features in the various execution models. Since the primary goal of AHS is to be able to automatically run parallel programs in the most expedient place— be it within a single machine or across multiple networked machines — it is vital that the system be completely supported on all targets.

#### 3.1. MIMD On A MasPar MP-1

Since the MasPar MP-1 is a massively-parallel SIMD machine [1], one might assume that it would be disallowed as a MIMDC target. However, even if a SIMD machine achieves only a small fraction of its native performance while simulating MIMD execution, it may still be the fastest place to execute a MIMDC program. Several researchers [4] [5] [9] have discussed MIMD interpreters; our target environment [2] uses a more aggressive approach to optimize the performance. On the MasPar MP-1, MIMD performance is typically between 1/40$^{th}$ and 1/5$^{th}$ of peak SIMD performance [2].

The MIMDC environment for the MasPar MP-1 is organized as follows. First, the code is compiled to generate a special MIMD assembly language code. This code is then assembled by an assembler called `mimda`, producing an absolute object file. Finally, an executable shell script is produced that will invoke the MasPar's mimd interpreter (`mimd`) and feed it the object file. Thus, if the user does not look too closely, the system appears to be a native compiler for MIMDC.

Although the MasPar MP-1 does not have a shared memory, the MIMD environment has instructions that implement shared memory accesses. Surprisingly, `mono` variables are not stored in the MasPar's control unit. They are actually treated as `poly` variables that are accessed by `mono` instructions. Shared variables can be used for synchronization, but MIMDC supports a barrier synchronization mechanism that can be implemented much more efficiently as a single interpreted instruction. The `Wait` instruction disables a PE until all active PEs are similarly disabled. When there are no active PEs, the interpreter simply re-enables all PEs that were waiting at barriers.

#### 3.2. MIMD And Timeshared Machines

Perhaps the most generic target machine is a system running multiple processes within a single UNIX platform. This UNIX system may have one or more processors; for example, the same execution model is used for both a single-processor UNIX-based workstation (e.g., Sun 3/50, IBM RS6000/530, Sun 4/490, and DEC 5000/200) and a multiprocessor system (e.g., Dual Gould NP-1, 4 CPU Ardent Titan P3).

In order to make the code generated by the compiler run on nearly any UNIX platform, the compiler is designed to generate a simple stack code that is implemented using C macros. Only the most fundamental methods of process control and interprocess communication can be used, because various different versions of UNIX implement different "fancy" process communication and management. To further enhance portability, there are two implementations of this execution model: one based on UNIX pipes and another based on accessing a shared file. It is possible to create many more efficient execution models for specific UNIX boxes by using threads, shared memory primitives, etc.; the two implementations presented here are merely the most portable.

##### 3.2.1. Pipe-Based Execution Model

When asked to create *n* processes, this implementation actually creates *n*+1 processes: *n* PE processes and one control process. The control process is responsible for managing shared memory access, synchronization, and housekeeping functions; only the PE processes execute user code.

When execution begins, the control process creates the PE processes and UNIX pipes to communicate with them. All PEs send information to the manager process through a single shared pipe, but the manager has a separate pipe to respond to each PE. Each communication through a pipe is a "packet" sent using a single `write` call (to ensure atomicity). Because pipes are treated as memory buffers in UNIX, the communication overhead is generally dominated by the UNIX task switch time, and communication can be quite fast.

##### 3.2.2. File-Based Execution Model

The file-based implementation is quite different from the pipe-based version, and is nearly always more efficient — if it works. In the pipe-based model, all PEs write into the same pipe (to the control process), but each PE reads from a different pipe; in the file-based model, all PE processes read and write the same file. Although this should function correctly on any UNIX system, and even across UNIX systems, we have found it to be erratic on some multiprocessors (e.g., Dual Vax 11/780 and the Sun 4/600) and on computers using files mounted via NFS (the Network File System).

When execution begins, the first step is the creation of a file to hold the combined state of all PEs and shared (`mono`) memory. Once this file has been initialized, *n*-1 additional processes are created for PEs 1 through *n*-1; the original process becomes PE 0. There is no control process mediating between the processes during execution, but only the contents of the shared file from which each PE can determine the state of all other PEs. Thus, this file acts as a shared memory image for the entire parallel machine.

#### 3.3. Distributed Computing On A Network

Because networked UNIX workstations are so plentiful, there has been a great deal of interest in using their idle cycles as an economical alternative to parallel supercomputers. The execution model discussed in this section can distribute the PE processes of a program over a group of networked UNIX machines. Each of these UNIX machines may be a uniprocessor or a multiprocessor system. Further, each machine may execute any number of PE processes. For example, a 10 PE process program might be run with 7 PE processes within a 4 CPU multiprocessor machine, 2 more on a uniprocessor workstation, and the final PE process on a heavily loaded timeshared UNIX mainframe.

Unlike the other execution models discussed, AHS's distributed execution model is not responsible for allocating *n* PE processes. Rather, the allocation is managed by the techniques described in section 4.2. The execution model is responsible for creating PE programs that can act together even if some PE processes share the same node and others are on different nodes. Interactions use UDP packets sent via a BSD Socket (henceforth referred to as the "UDP Socket" model). PE processes communicate directly, rather than using persistent "daemon processes" as in PVM [3]. Because there are no daemons, the PE code must manage asynchronous communication through the UDP Socket, and this requires fairly complex signal-driven event handling code in the PE program. However, the MIMDC compiler generates this code for the user. Likewise, the compiler takes into account the fact that UDP Socket communications might not be received in the order in which they were sent.

## 4. Target Selection And Program Execution

Most research in parallel processing focuses on achieving a large fraction of the target machine's rated peak performance; however, we want the system to minimize the expected execution time. For example, if all we have is a Sun Sparc workstation and a 16,384-PE MasPar MP-1, most MIMDC programs with parallelism width 128 should probably be run on the MasPar because it will be faster — although the utilization of the MasPar will be less than 1%. However, if the MasPar has a multitude of jobs waiting and the Sun is idle, running this code on the Sun may result in the smallest expected execution time, so the system should run the program on the Sun in this case.

There are a multitude of problems that need to be addressed in order to minimize execution time using heterogeneous supercomputing. The goal of AHS is to take all the key factors into account, yet be as unobtrusive as possible. We want the user to be able to be blissfully ignorant of how the system works; this approach may sacrifice a little performance, but it should make the user more productive.

The following section discusses how the AHS prototype records the vital information about each combination of execution model and machine. Section 4.2 explains how each program is analyzed and, using the execution model and machine information, how the target is selected. Finally, section 4.3 discusses how the program is actually made to execute on the automatically selected target.

### 4.1. Execution Model And Machine Database

In the AHS prototype, there is a file that contains all vital information about each combination of execution model and machine known to the system. At the time the system is configured, an entry is made for each combination. Much like PVM's machine database file [3], each entry contains information about how each machine can be used:

- **The name of the machine.** This is typically the internet address of the machine.
- **The width of the machine.** This number is the maximum number of PEs that can execute on the machine. For a parallel computer with a fixed set of PEs and no support for virtual PEs, the width is recorded as the number of actual PEs. For a uniprocessor or multiprocessor UNIX system, we use a width of 0 to indicate that an essentially unlimited number of processes can be executed. Further, because the distributed execution model requires UNIX support, only machines with a width of 0 are able to host PEs for distributed execution.
- **The compile and run script.** This information describes how to initiate a user program using this host and execution model.

However, in AHS, each entry also contains detailed information that can be used to accurately predict relative performance of any given program. This performance-prediction information is composed of two types of timing information: load-independent properties and load-dependent properties.

#### 4.1.1. Load-Independent Timing Properties

The load-independent timing properties of a combination of execution model and machine are represented as a list giving the **execution time for each basic operation**. Although the various targets might use different instruction sets, etc., AHS defines a basic set of operations that are used for predicting execution time. The approximate execution time, in seconds, is recorded for each of these basic operations. Normally, the times are determined experimentally, and are entered at the time the system is configured.

A support program (called `timer`) has been created to use UNIX process timing facilities to measure the execution time for each basic operation. UNIX timing is only accurate to $1/60^{th}$ second, so accurate estimates are obtained by timing a set of long-running sections of code and then solving the resulting set of equations to determine the time for each operation. Accuracy is limited by the averaging effect of the long runs and by UNIX scheduling anomalies (e.g., being charged for time spent processing another processes' interrupt). We attempt to minimize these imperfections by using 5-point median filtering on the computed times, but even this gives a typical accuracy of only about +/-10%. The good news is that even a 50% error in one of these estimates is unlikely to have a significant adverse affect on the performance of AHS.

Table 1 samples the basic operation times for a variety of machines available within Purdue University's School of Electrical Engineering. The first four machines are UNIX-based uniprocessors. The second four are UNIX-based multiprocessors with two or four processors each. The next is a massively-parallel supercomputer with 16,384 processing elements. Finally, the last line is for a typical network of UNIX-based systems (e.g., Sun 4) connected by a single Ethernet. In each case, the times quoted are single-process times for unoptimized operations.

Several interesting observations can be made from this table. Perhaps the most obvious is that, with the exception of the MasPar, communication time (LDS Time) is always much greater than compute time (ADD Time). It is somewhat surprising that the UDP Socket model can perform communication over an Ethernet nearly as fast most UNIX machines could communicate between processes within a single machine.

| Machine | Exec. Model | ADD | LDS |
|---|---|---|---|
| Sun 3/50 | Pipes | $5\times10^{-6}$s | $1\times10^{-3}$s |
| | Shared File | $5\times10^{-6}$s | $7\times10^{-4}$s |
| IBM RS6000/530 | Pipes | $2\times10^{-6}$s | $5\times10^{-4}$s |
| | Shared File | $2\times10^{-6}$s | $3\times10^{-4}$s |
| Sun 4/490 | Pipes | $1\times10^{-6}$s | $3\times10^{-4}$s |
| | Shared File | $1\times10^{-6}$s | $1\times10^{-4}$s |
| DEC 5000/200 | Pipes | $8\times10^{-7}$s | $2\times10^{-4}$s |
| | Shared File | $8\times10^{-7}$s | $1\times10^{-4}$s |
| Dual Vax 11/780 | Pipes | $1\times10^{-5}$s | $5\times10^{-3}$s |
| Dual Gould NP-1 | Pipes | $2\times10^{-6}$s | $1\times10^{-3}$s |
| | Shared File | $3\times10^{-6}$s | $1\times10^{-3}$s |
| Sun 4/600 | Pipes | $1\times10^{-6}$s | $8\times10^{-4}$s |
| 4 CPU Titan P3 | Pipes | $5\times10^{-7}$s | $1\times10^{-3}$s |
| | Shared File | $2\times10^{-7}$s | $8\times10^{-4}$s |
| 16K MasPar MP-1 | MIMD Int. | $5\times10^{-5}$s | $5\times10^{-5}$s |
| *UNIX network using Ethernet* | UDP Socket | *Same time as Pipes* | $3\times10^{-3}$s |

**Table 1:** Sample Operation Times For Some Targets

The speed of UDP Sockets is still more surprising in that if we had used PVM [3] to construct the AHS distributed model, the `LDS` Time would have been about $1.6\times10^{-1}$s (as we measured it on the same systems used for the UDP Socket timing). However, AHS's UDP Socket model avoids most of PVM's system overhead, and that is actually the dominant portion of the PVM communication time. This is demonstrated by the fact that using PVM for an `LDS` of a variable that *resides on the requesting machine* also yields a time of about $1.6\times10^{-1}$s.

#### 4.1.2. Load-Dependent Timing Properties

While the above numbers approximate the *best* execution time for each basic operation (i.e., they correspond to UNIX "user time" plus "system time"), they do not provide an accurate prediction of what performance will be when a program is run. The reason is simply that other programs may be running on these machines, hence, execution may be slower by a factor proportional to the number of processes currently sharing each machine. This information is recorded for each combination of execution model and machine as:

- **The last known load average..** This number is a multiplicative factor that indicates how much slower the machine was last known to be executing due to other programs being run on the system. Because not all programs are compute bound, the load average is rarely an integer.
- **The load average increment.** Whenever this system assigns a job to a machine, the load average for that machine may change to reflect that another process must be scheduled. This value is the increment by which the load average changes for each additional process scheduled on the machine. Under UNIX on a uniprocessor, it is assumed to be 1.0; for an *n*-processor UNIX-based multiprocessor, it is assumed to be 1.0/*n*.

Notice that only the load average is likely to change after the system has been configured.

Ideally, one might like to automatically update the load average information just before deciding where to run each user program. However, this is usually impractical because one needs to obtain the load average for *every* machine, and there may be many machines. For example, on Purdue University's Engineering Computer Network, there are over 500 machines that could be available to the system. Thus, AHS allows the user to explicitly issue a command to update the load average database information.

### 4.2. Minimizing Expected Program Execution Time

As discussed above, the machine load and execution time of each pseudocode operation is available for each of the potential targets. When a MIMDC program is compiled, a cost formula for that program is also computed. The cost formula is simply a weighted sum; there is a constant factor for each instruction type. The weighting is determined by a version of the compiler that does not generate code, but simply records expected execution counts for each type of operation. Currently, the rules by which execution counts are computed for MIMDC are very simple:

- The expected execution count at the beginning of `main()` (or any other function) is assumed to be 1.0.
- In an `if` construct, the then clause frequency is assumed to be 0.51 (and 0.49 for the `else` clause, if one is present).
- The body of any looping construct is assumed to be executed 100.0 times. Note that the conditional test in some loops is executed one more time than the body code; in such a case, the condition test code's execution count is multiplied by 101.0 instead of 100.0.

Thus, a program is reduced into a table giving estimated expected execution counts for each of the different operations.

When the program is to be run, the user specifies the number of PEs desired. Given that, the expected execution counts, and the execution model and machine database, the fastest target (or set of distributed targets) is selected by:

**Target Selection Algorithm**

1. **Pick the best single machine target.** For each target that has a specified maximum execution width ≥ the number of PEs requested or that uses the pipe or shared file execution model:

1.1 Temporarily adjust the load average by adding the product of the number of PEs requested and the load increment.

1.2 Compute the sum, over all operations, of the product of the operation time (from the database) and expected execution count (for the given program). Multiply this result by the adjusted load average.

1.3 If the resulting time estimate is less than the best so far, make this target the new best target.

2. **Pick the best set of distributed targets.** For *i*=1 to the number of PEs requested:

2.1 Set the time of bestPE to infinity.

2.2 For each target that has a width of 0 and uses the UDP Socket execution model:

2.2.1 Temporarily adjust the load average by adding the load increment, since we are about to add one PE process to this target.

2.2.2 Compute the sum, over all operations, of the product of the operation time (from the database) and expected execution count (for the given program). Multiply this result by the adjusted load average.

2.2.3 If the resulting time estimate is less than bestPE, make this target the new bestPE.

2.3 Record that this target was selected as the $i^{th}$ bestPE and make the adjustment to the load average from step 2.2.1 permanent.

3. Since the time for the program is the maximum time for any PE, if the time for last value of bestPE > the value of best, then best is the target and we are done.
4. The sequence of distributed targets selected for bestPE are selected. However, rather than representing this as a list that says where each PE's target is, we convert this into a list of targets that specifies which PEs are assigned to each target.

### 4.3. Controlling Program Execution

Having selected a target, or set of distributed targets, the final step is to actually cause the program to execute there. This is done by:

1. When the user "compiles" a MIMDC program, it is not actually compiled, but is analyzed and packaged into a "master shell script." This shell script contains the expected execution counts, as well as the full source of the MIMDC program.
2. The user views the master shell script as an ordinary executable object file, and hence runs it by simply typing its filename with at least one command line argument — the desired number of PEs.
3. In execution, the first thing done by this master shell script is to apply the above algorithm to select the fastest target(s). Once target(s) are selected, the program will run to completion on those target(s); running processes are never migrated.
4. Finally, for each of the selected target(s), the master shell script uses `rsh` to send and execute a second shell script that contains both the MIMDC program and the sequence of commands needed to compile and execute it for all the PEs assigned to that target. Hence, there is no need to keep track of paths to user object files, or to use NFS mounting across machines. The method used in AHS has the overhead of sending and recompiling the source program every time the program is run, but the MIMDC compilers run very fast, and compile time is nearly always small compared to the runtime of typical supercomputing applications.

Notice that, unlike PVM, there are no daemon processes running on the target(s) when the system is not running user code — in fact, there are no daemons running even when user programs are using the target(s). The process overhead is just the initial shell script that the user executed and the shared memory manager in some execution models (e.g., the UNIX pipes model).

## 5. Conclusions

This paper has taken a very practical approach to the problem of automatically making efficient use of heterogeneous supercomputing. Rather than making a heroic effort to achieve near peak speeds on a particular machine, our system attempts to invisibly seek out and use whatever hardware will make the user's program execute fastest.

Toward this goal, it is necessary that the programming model, and language, be defined to facilitate automatic porting of programs to a wide range of targets. For each potential type of target, there must be an execution model that can support the programming model. Finally, there must be a procedure by which the performance of a program on each potential target can be predicted, and a method by which the system can automatically select the fastest target and cause the program to be executed there. Each of these aspects is covered both as an abstract problem and as a summary of how these issues have been managed by a prototype system.

The prototype system, AHS, accepts a control-parallel dialect of C called MIMDC. It can automatically select the execution model and machine(s) with the fastest expected execution time for any given program, and invisibly causes the program to execute using that target. Execution models support a massively-parallel supercomputer, individual uniprocessor or multiprocessor UNIX machines, and even groups of networked UNIX systems running as a distributed computer. Little benchmarking of the system has been done, but the execution models for the SIMD MasPar MP-1 (as a MIMD) and for groups of UNIX systems are surprisingly efficient.

As presented, AHS is functional, but not complete. It is intended to be the basis for a more sophisticated system. That system will analyze and schedule individual functions within a program and will support additional programming models and languages. When AHS is mature enough, we intend to distribute it as a full public domain software release.

## References

[1] T. Blank, "The MasPar MP-1 Architecture," 35th IEEE Computer Society International Conference (COMPCON), February 1990, pp. 20-24.

[2] H.G. Dietz and W.E. Cohen, "A Control-Parallel Programming Model Implemented On SIMD Hardware," in Proceedings of the *Fifth Workshop on Programming Languages and Compilers for Parallel Computing*, August 1992.

[3] B.K. Grant and A. Skjellum, "The PVM Systems: An In-Depth Analysis And Documenting Study," in the 1992 MPCI Yearly Report: Harnessing the Killer Micros, Lawrence Livermore National Laboratory, August 1992, pp. 247-266.

[4] M. S. Littman and C. D. Metcalf, *An Exploration of Asynchronous Data-Parallelism*, Technical Report, Yale University, July 1990.

[5] M. Nilsson and H. Tanaka, "MIMD Execution by SIMD Computers," Journal of Information Processing, Information Processing Society of Japan, vol. 13, no. 1, 1990, pp. 58-61.

[6] T.J. Parr, H.G. Dietz, and W.E. Cohen, "PCCTS Reference Manual (version 1.00)," *ACM SIGPLAN Notices*, Feb. 1992, pp. 88-165.

[7] M.J. Phillip, "Unification of Synchronous and Asynchronous Models for Parallel Programming Languages" Master's Thesis, School of Electrical Engineering, Purdue University, West Lafayette, Indiana, June 1989.

[8] M. Quinn, P. Hatcher, and B. Seevers, "Implementing a Data Parallel Language on a Tightly Coupled Multiprocessor," Advances in Languages and Compilers for Parallel Processing, edited by A. Nicolau, D. Gelernter, T. Gross, and D. Padua, The MIT Press, Cambridge, Massachusetts, 1991, pp. 385-401.

[9] P.A. Wilsey, D.A. Hensgen, C.E. Slusher, N.B. Abu-Ghazaleh, and D.Y. Hollinden, "Exploiting SIMD Computers for Mutant Program Execution," Technical Report No. TR 133-11-91, Department of Electrical and Computer Engineering, University of Cincinnati, Cincinnati, Ohio, November 1991.

# SESSION 9B

## TASK GRAPH/DATA FLOW

# A Concurrent Dynamic Task Graph

Theodore Johnson
Dept. of Computer and Inf. Science
University of Florida
Gainesville, Fl 32611-2024
ted@cis.ufl.edu

**Abstract**

*Task graphs are used for scheduling tasks on parallel processors when the tasks have dependencies. If the execution of the program is known ahead of time, then the tasks can be statically and optimally allocated to the processors. If the tasks and task dependencies aren't known ahead of time (the case in some analysis-factor sparse matrix algorithms), then task scheduling must be performed on the fly. We present simple algorithms for a concurrent dynamic-task graph. A processor that needs to execute a new task can query the task graph for a new task, and new tasks can be added to the task graph on the fly. We present several alternatives for allocating tasks for processors and compare their performance.*

## 1 Introduction

A common method for expressing parallelism is through a task graph. Each node in the task graph represents a unit of work that needs to be performed, and edges represent dependencies between tasks. If there is an edge from task $T_1$ to task $T_2$ in the task graph, then $T_1$ must complete before $T_2$ can begin. Previous work [3, 17, 11, 10, 20, 9, 12] has assumed that the task graph is specified ahead of time. This is often a reasonable assumption, since the task graph can be generated by a parallelizing compiler, or by a symbolic analysis of the problem to be solved (i.e, analysis-only LU decomposition algorithms).

Since the task graph is specified ahead of time, it can be analyzed for static scheduling purposes. The scheduling can be static or dynamic. In *static* scheduling, the tasks are allocated to the processors before the computation starts [3, 17, 10, 20]. In *dynamic* scheduling, the tasks are allocated to processors on the fly [9, 12]. If good task execution time estimates can be made in advance, static scheduling will outperform dynamic scheduling, but dynamic scheduling will adjust to the actual execution conditions.

In this paper, we propose a scheduling structure, the *dynamic-task graph*, or *DTG*, that allows the task graph to be specified during the program execution. A DTG is useful when the structure of the problem instance is determined at execution time. This work was motivated by the problem of parallelizing analysis-factor LU decomposition algorithms for asymmetric sparse matrices. A common approach is the multifrontal method, in which portions of the matrix are gathered into fronts for factoring, and these fronts make contributions to other fronts. The tasks in the DTG represent the fronts, and the links represent the contributions passed between fronts. In some analysis-factor multifrontal algorithms [4, 5], the tasks and their dependencies are determined during execution.

In section 2, we present the basic concurrent DTG algorithm, and some extensions. In section 3, we explore algorithms for scheduling eligible tasks, and in section 4 we examine some performance issues. Finally, in section 5 we draw our conclusions.

## 2 Concurrent Dynamic-Task Graph

A *dynamic-task graph*, or DTG, consists of a set of labeled vertices $V$ and a set of arcs on the vertices $A$. The arcs in $A$ are the dependencies among the tasks. If the arc $(t1, t2)$ is in $A$, then task $t1$ must complete execution before task $t2$ can start execution. We call $t1$ the *prerequisite* task, and $t2$ the *dependent* task. Obviously, the DTG must be an acyclic digraph. The nodes correspond to tasks, and are labeled:

**U** : if the task is unexecuted,
**E** : if the task is executing,
**F** : if the task has finished execution, or
**N** : if the task is not ready.

A task $t_0$ is *eligible* for execution only if all tasks $t_i$ such that $(t_i, t_0) \in A$ are labeled $F$ (similar to a *mature* node in [17]).

There are three operations on a DTG:

1 **add_task(T,D)**: The add_task operation adds task T to the DTG, and specifies that the set of tasks $D = \{t_1, \ldots, t_n\}$ must finish execution before T can start execution. The set D is task T's *dependency set*.

2 **t=get_task()**: The get_task operation returns a task that is eligible for execution. If there is no eligible task, the processor blocks until a task becomes eligible.

3 **finished_task(T)**: The finished_task operation declares that task T has completed its execution.

When a task is added to the DTG, it must be uniquely named. This requirement is not a problem, since a processor can name the task it adds to the DTG with a sample from a local counter appended to its processor id, or with a pointer to a description of the task. The application might naturally provide a unique task name. For example, in a sparse matrix solver the name of the task can be the row of initial pivot of the frontal matrix.

When a task is added to the DTG (via the `add_task` operation), it is labeled **U**. When a task is selected for execution, its label is changed to **E**. When a task completes its execution, it performs the `finished_task` operation, which changes the task state to **F**. If in the `add_task` operation, task $t$ specifies that it is dependent on task $t'$ but $t'$ has not yet been added to the DTG, then task $t$ must create an entry for $t'$ and specify that $t'$ is *not ready* by setting the state of $t'$ to N. When `add_task(t',D)` is executed, the state of $t'$ changes to **U**.

We initially assume that all tasks in a task's dependency set have already been added to the DTG, and later extend our algorithms to handle not-ready tasks. Since all tasks in the DTG have been determined, this assumption is reasonable. Furthermore, it lets us reclaim tasks from the DTG. Whenever a task finishes execution, it is dropped from the DTG. If an `add_task` operation can't find a task $t_d \in D$ in the DTG, then $t_d$ has finished.

Figure 1 illustrates a sample execution sequence. Tasks $T1$, $T2$, $T3$, and $T4$ are added to the DTG, and task $T4$ depends on tasks $T1$ and $T2$. Next, tasks $T1$ and $T2$ are selected for execution in response to `get_task` requests. Task $T1$ finishes, and its state changes to **F**. Task $T3$ is selected for execution in response to a `get_task` request. Task $T2$ finishes, and task $T4$ becomes eligible.

A task is represented by a *task record* in the DTG. The task record contains a field for the *name* of the task, information necessary for executing the task, the number of unfinished prerequisite tasks `ND`, and a list of the dependent tasks `dependent`.

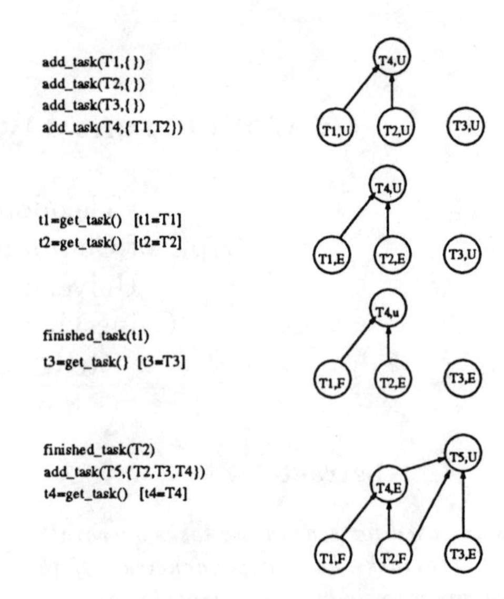


Figure 1: Example execution sequence

The first issue to address is the problem of finding the tasks in the DTG without explicitly searching the DTG. Since the queries that need to be made are simple look-up queries, a hash table is the best data structure to maintain the translation table. We use a static-sized open hash table. The primary purpose of the hash table is to permit parallel access to the tasks in the DTG, so the number of hash table buckets only needs to be proportional to the number of processors (as opposed to the number of tasks). If the DTG has few buckets, the buckets of the DTG might be required to store many task records, so the records should be stored in some fast access structure, such as a binary tree. The bucket data structure doesn't need to be a concurrent data structure, instead the entire bucket can be locked. The hash table operations are:

1 **enter_task(T)**: put a new task in the hash table.

```
a lock hash table bucket
b insert T into bucket.
c unlock hash table bucket.
```

2 **p=translate_task(T)**: search the hash table for the task, and return a pointer to the task.

```
a lock hash table bucket.
b find and lock T.
  i If T is not found, release all locks and
    return NIL.
c unlock hash table bucket.
```

3 **delete_task(T)**: remove T from the hash table.

```
a lock hash table bucket.
b find and lock T.
c remove T from bucket.
d unlock T and hash table bucket.
```

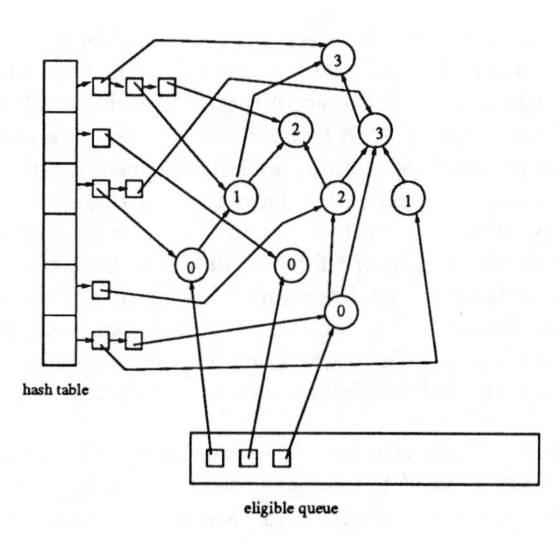


Figure 2: Concurrent task graph data structures

The next issue is where to store the dependency pointers: in the prerequisite task record or in the dependent task record. Storing the dependency pointer in the dependent task record simplifies the operation of adding a task, but greatly complicates the operation of getting an eligible task. We choose the option of storing the pointers in the prerequisite task records. This choice requires that when a task is added to the DTG, all tasks in the dependency set must be modified. Fortunately, the hash table permits a fast lookup.

The last issue is finding tasks that are eligible for execution. We assume that pointers to these tasks are stored in a separate data structure, the *eligible queue*. A task can be inserted into the eligible queue, and the eligible queue can be queried for an operation to execute. We leave the implementation and the semantics of the eligible queue unspecified for now, since there are many possible allocation heuristics.

The psuedo-code for the add_task, get_task, and delete_task operations follows. We assume that the tasks and the hash table buckets can be locked (we later discuss a non-locking implementation). The lock can be a simple busy-wait lock, or the contention-free MCS lock [15]. Each task graph entry $t$ has three fields: a field for the lock, a count of the number of unfinished prerequisite tasks (`ND`), and a list of tasks that depend on $t$ (`dependent`). When a `translate_task` operation is performed, the lock on the hash table entry is retained. This ensures that the task remains in the task graph until it is modified (in spite of the concurrent execution of `finished_task` operation). The data structures for the concurrent dynamic-task graph are shown in Figure 2.

```
add_task(T,D)
  T.ND=0
  enter_task(T)
  for i=1 to |D| do
    t=translate_task(i'th task in D)
    if t is null  // i.e., finished
      number_finished++
    else
      add T to t->pointers
      unlock(t)
  if number_finished > 0
    lock(T)
    T.ND -= number_finished
    if T.ND == 0
      add T to the eligible queue
get_task()
  get a task t from the eligible queue
  return(t)
finished_task(T)
  delete_task(T) // from hash table only
  for i = 1 to |T.dependent|
    t = i'th task in T.dependent
    lock(t)
    t->ND- -
    if t->ND == 0
      add t to the eligible queue
  reclaim the space used by T
```

## 2.1 Correctness

In this section, we make some correctness arguments. In particular we show that a task is added to the eligible queue before all of its dependent tasks complete the `finished_task` operation, but after all of its dependent tasks start the `finished_task` operation.

We will say that task T *exists in* DTG $\mathcal{D}$ between the times that `add_task(T)` and `delete_task(T)` are executed. This period of time is well defined because of the locks which make the hash table operations atomic. When $T$ executes the `translate_task` operation for $t$ in the `for` loop of the `add_task` operation, we say that $T$ *accesses* $t$.

**Lemma 1** *The* `add_task(T,D)` *inserts a dependency link from* $t$ *to* $T$ *if and only if* $t \in$ D *and* $t$ *exists in* $\mathcal{D}$ *when* $T$ *accesses* $t$.

*Proof:* The `for` loop in the `add_task` operation scans through the tasks $t \in$ D and adds a dependency pointer to $t$ if and only if $t$ is in the hash table when $T$ accesses $t$. Since $t$ is in the hash table if and only if $t$ exists in $\mathcal{D}$, the lemma follows.□

**Lemma 2** *When a task completes, it decrements the dependency count of all tasks that add a dependency pointer to it.*

*Proof:* When a task completes, it executes the `finished_task` procedure. The `finished_task` operation removes the completed task's record from the hash table, so no further pointers are added to it. The finished_task operation is atomic due to the locks it sets, so after the `finish_task` operation, a tasks record will contain all dependency pointers that are inserted. The remainder of the `finish_task` operation decrements the dependent tasks. □

**Theorem 1** *Every task is added to the eligible queue exactly once, and only after all prerequisite tasks are marked $F$.*

*Proof:* Consider a new task $T$ executing the `add_task` operation. Task $T$ accesses every $t \in$ D(T). By lemma 1, $T$ adds a pointer to $t$ if $t \in \mathcal{D}$. If $t \notin \mathcal{D}$, then $t$ must be finished, since we have assumed that all tasks in D(T) have been added to $\mathcal{D}$ already. Therefore, $T$ correctly counts the number dependent tasks that are finished when accessed, and the remaining tasks receive a pointer to $T$.

The `add_task` operation decrements $T$'s dependency count by the number of finished dependent tasks. By Lemma 2, the unfinished dependent tasks decrement $t$'s dependency count by one. The use of locks makes the decrement atomic. Since the task graph is a DAG, all prerequisite tasks of $T$ will finish. The last task to perform the decrement will find that the dependency count is zero, and will add $T$ to the eligible queue. □

## 2.2 Extensions

In this section, we discuss some possible extensions and optimizations of the concurrent DTG.

**Not-Ready Tasks** The algorithms that we presented depend on the assumption that all tasks in a new task's dependency set exist in the DTG. One can imagine that a new task might depend on tasks that have not yet been added to the DTG. For example, a parallel algorithm for the LU factorization of sparse asymmetric matrices might assign the task of adding pivots to frontal matrices to one set of processors, and assign the task of composing and factoring the frontal matrices to a different set [4]. Processors $p$ and $q$ might build frontal matrices $A$ and $B$ concurrently, where elements of $B$ depend on the factorization of $A$. Since $A$ and $B$ are built concurrently, $B$ might be added to the DTG before $A$.

To distinguish between not-ready and finished tasks the state of a task is explicitly stored in the task. A finished task is retained in the DTG and is marked **F**. When a task $t \in D(T)$ is accessed in the `for` loop of the `add_task` operation, the following protocol is used: If $t$ doesn't exist in the hash table, a task record for $t$ is created, its state is set to $N$, and a pointer to $T$ is added. If a record for $t$ exists in the hash table, its state is tested to determine whether or not the task has finished. The `enter_task` hash table operation must be modified to account for the possibility that the task $T$ already exists as a not-ready task.

**Dense Task Names** The concurrent DTG requires a hash table if the range of task names is large and the names of the actual tasks is sparse. In some applications, the tasks in the DTG are relatively dense in their name space. An example are frontal matrices in an asymmetric sparse matrix algorithm. The task can be named by the row of the upper left hand pivot, so there are $O(n)$ possible task names. Sparse matrix algorithms contain several $O(n)$ supplementary data structures, so allocating a bucket for each possible task name does not create an excessive space overhead. Allocating a bucket for each task greatly simplifies the implementation of the DTG, since the hash table operations become simple $O(1)$ procedures. In addition, the bucket lock serves as the task record lock, so only half the number of locks need to be set as would otherwise be needed.

**Non-locking Algorithms** A non-locking algorithm uses atomic read-modify-write instead of locking to ensure correctness in spite of concurrent accesses. Non-locking algorithms have the attractive property that they avoid busy-waiting, which can degrade performance [1, 6]. These algorithms typically use the *compare_and_swap* or the *compare_and_swap double* instruction to commit modifications [8, 16, 18, 19], although some algorithms use the fetch-and-add instruction [7]. The correctness of the DTG algorithms depends on the atomicity of the hash table operations. Fortunately, many practical non-locking list and search structure algorithms exist in the literature [16, 19].

One place where care must be taken involves access to a tasks list of dependent tasks. The `finished_task` operation should declare that a task is finished with a decisive operation, so that the correctness of Lemma 2 is maintained. We can modify the technique of Prakash, Lee and Johnson by maintaining the list of dependent tasks in a task record as a non-locking stack, and use one bit of the pointer to the head of the stack as a *deleted* bit. The `finished_task` operation sets the deleted bit as its decisive operation, and a `add_task` operation that reads a set deleted bit is a task's dependent list considers the task to be finished.

## 3 Eligible Queue

The eligible queue is responsible for scheduling tasks for execution. The goal in designing the eligible queue is to maximize the speedup of the parallel computation. Maximizing the speedup requires that we minimize the blocking that occurs at the `get_task` procedure, and that we minimize the overhead of the scheduler. However, minimizing blocking and minimizing overhead are conflicting goals. As previous works have shown, careful scheduler design can reduce response time by reducing the amount of time that a processor is blocked waiting for a task to become eligible for execution. However, optimal scheduling is NP-complete, and the best heuristics require that the entire task graph be known in advance and require considerable overhead. Scheduling overhead also reduces performance, and should be kept to a minimum.

In order to provide guidance on implementing an eligible queue, we ran simulation experiments to compare scheduling algorithm performance. We wrote a simulation of a parallel dynamic-task graph execution. The simulation is initialized with 80 initial tasks. An additional 80 tasks are created, each with dependencies on the preceding tasks. These tasks have a random number of dependencies (a truncated normal distribution with a mean of 4.0 and standard deviation of 5.2), and a prerequisite task is chosen by subtracting a randomly chosen *backwards distance* from the new task's number. The backwards distance has a truncated Erlang distribution with a mean of 80. Given this initial task graph, $p$ processors execute by repeatedly getting an eligible task, finishing the task, and adding a random number of new tasks. The number of new tasks added has a binomial distribution with mean 2 for the first 2000 tasks to execute, and mean .5 for the remainder. We set the task execution time so that the scheduling overhead has little effect.

We tested six eligible queue algorithms:

**FIFO:** The eligible queue is managed as a FIFO queue.

**Max weight:** The eligible queue always returns the task with the greatest execution time in response to a `get_task` operation.

**Min weight:** The eligible queue returns the task with the minimum execution time.

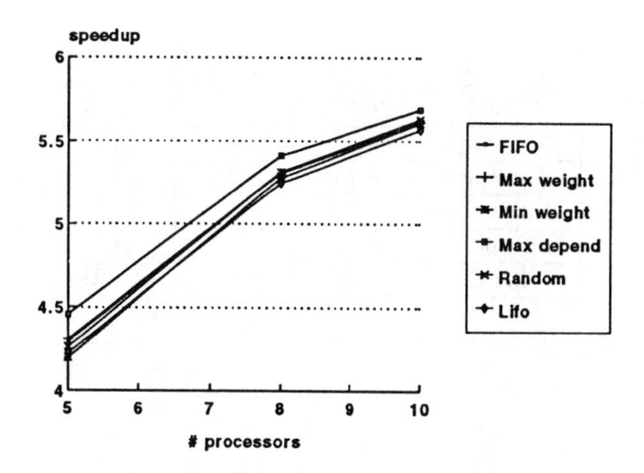


Figure 3: Comparison of DTG scheduler performance.

**Max depend:** The eligible queue returns the task with the maximum number of dependent tasks.

**Random:** The eligible queue returns a uniformly randomly chosen task.

**LIFO:** The eligible queue is maintained as a LIFO queue.

Figure 3 shows a comparison of the speedup produced by each algorithm given 5, 8, and 10 processors. To collect a data point, we used the average of ten runs. Figure 3 shows that Max depend is consistently the best algorithm, while LIFO is consistently the worst. However, no algorithm is significantly better than another. The summary in in Table 1 shows that there is only about a 5% difference in the speedup offered by the best and the worst algorithm.

In Table 2, we list the speedup from static-task graph algorithms. We collected the dynamic-task graph by recording the dependency edges, and added a dependency link from task $T$ to all tasks that $T$ creates. We simulated the following algorithms on the equivalent static-task graphs using eight processors:

**CP:** Critical path method [17]. An eligible node's priority is the weight of the heaviest weighted path to any exit node.

**FIFO:** Schedule tasks in the order that they become available.

**Largecalc:** Schedule the heaviest task first [2].

**Heavy:** An eligible node's priority is the sum of its weight and the weight of its immediate successors [2].

**Levelfifo:** Assign BFS levels to tasks, and schedule the tasks within a level by FIFO.

**Levellarge:** Assign BFS levels to tasks, and schedule the heaviest task in a level first [17].

| Scheduler | 5 Proc. | 8 Proc. | 10 Proc. |
|---|---|---|---|
| FIFO | 4.297 | 5.311 | 5.632 |
| Maxweight | 4.309 | 5.309 | 5.616 |
| Minweight | 4.198 | 5.280 | 5.610 |
| Maxdepend | 4.456 | 5.414 | 5.690 |
| Random | 4.263 | 5.315 | 5.631 |
| LIFO | 4.227 | 5.246 | 5.631 |

Table 1: Comparison of DTG scheduler performance.

| Scheduler | speedup |
|---|---|
| Max | 7.072 |
| CP | 7.036 |
| FIFO | 5.470 |
| Largecalc | 5.470 |
| Heavy | 5.456 |
| Levelfifo | 5.474 |
| Levellarge | 5.492 |
| Maxdep | 5.440 |

Table 2: Comparison of static-task graph scheduler performance (eight processors).

**Maxdep:** Schedule the task with the greatest number of dependent tasks first [2].

Comparing Tables 1 and 2, we note that there is little difference among most of the dynamic and static scheduling policies. While the best speedup from the static-task graphs heuristics is considerably greater than that provided by the DTG scheduling algorithms, the only static-task graph scheduler that provides this performance is CP. The remaining static-task graph scheduling policies provide a speedup similar to that of the DTG scheduling policies. Thus, there is little performance difference between local scheduling on a static-task graph and scheduling on a dynamic-task graph. We note that the CP method requires $O(n^2)$ time for processing, and may not always be reasonable.

Since there is little difference in DTG scheduler performance, we recommend a simple or low-overhead method. For example, [16] presents a simple lock-free FIFO queue which can serve as the eligible queue. Manber [14] has proposed *concurrent pools* as a low overhead method for implementing shared queues that doesn't require a FIFO ordering.

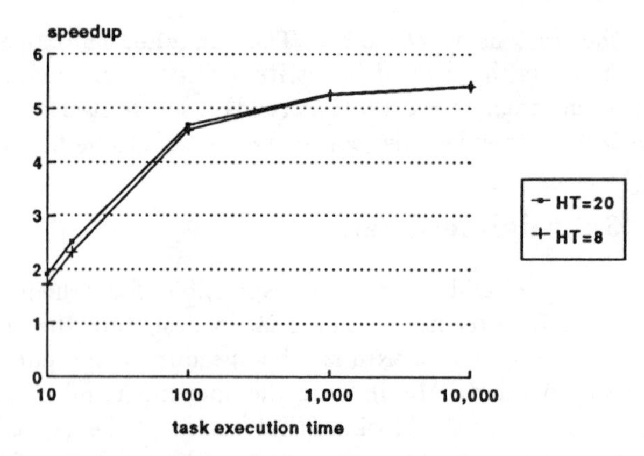


Figure 4: Speedup vs. task execution time.

## 4 Performance

In this section, we investigate the effect of DTG overhead on speedup. Figure 4 shows a plot of speedup versus task execution time. The simulator and parameters are the same as those used in the previous section, but 8 processors are used and the task execution time varied. In addition, we ran one set of experiments with 8 buckets in the hash table, and another set with 20 buckets. The leftmost point on the chart is a close approximation to the speedup obtainable without scheduler overhead. If $S_{max}$ is the speedup without scheduler overhead, each task executes for $E$ seconds, and the total time to process a task in the DTG is $O$, then the speedup is approximately

$$S_{actual} = S_{max}\frac{E}{E+O}$$

In the simulations, each task was dependent on about 4.0 prerequisite tasks, and adding each task pointer required 2 time units. Each finished task was required to access about 4.0 tasks, each access requiring 2 time units. The `add_task` procedure required 3 time units to modify the tasks's own record, and the `finished_task` procedure required 1. Finally, each task was added to and later removed from the eligible queue, and both of these actions required 1 time unit. In total, processing each task required about 22 time units of DTG overhead. In addition, there was a certain mbount of overhead due to lock contention.

When the task execution time is 10 units, the formula for $S_{actual}$ predicts a speedup of about 1.70. The observed speedup is somewhat greater, because the increased scheduler overhead replaces some of the task blocking. In general the formula holds, and we see that the DTG is appropriate for medium to coarse grain parallelism, but not for fine grain parallelism.

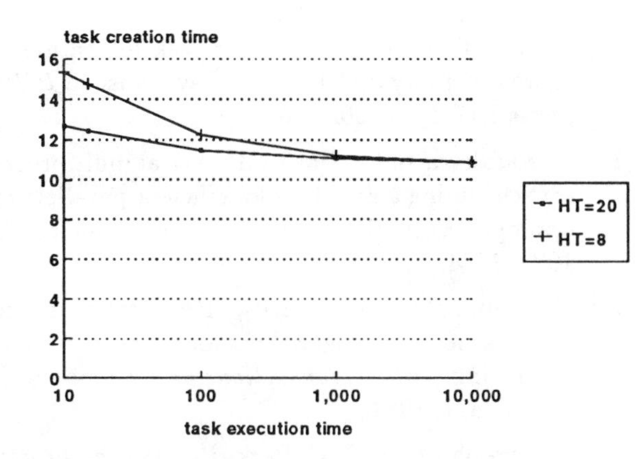


Figure 5: Add_task execution time vs. task execution time.

In Figure 4, we show the speedup for two cases: when the hash table contains 8 entries and when the hash table contains 20 entries. When the task execution time is small, the smaller hash table shows a slightly smaller speedup. The reduction in speedup is due to the increased lock contention, As a result, the overhead for processing a task in the DTG is greater. Figure 5 shows a plot of the time to execute the `add_task` operation against the task execution time. When the task size becomes moderately large, Figure 4 shows that the difference in speedup becomes negligible.

We develop a performance model to determine an appropriate size for the hash table. Let

$D$: be the average number of task dependencies,
$E$: be the average task execution time,
$T_q$: be the average time to access the eligible queue,
$T_a$: be the average time to access a task,
$S$: be the speedup, and
$H$: be the number of hash table entries.

Suppose that we model each hash table bucket as an M/M/1 queue (a crude but workable approximation). We first calculate the arrival rate. Each dependency link causes two task accesses: one in the `add_task` operation and one in the `finished_task` operation. Since every task is added once and finishes once, we assume that a processor cycles between executing and accessing the DTG. After executing a task, a processor declares that the task is finished (requiring $D+1$ task accesses) and adds one task to the eligible queue, adds a new task to the DTG ($D+1$ task accesses), and gets a new task from the eligible queue. On average, $S$ processors are executing, and the task graph accesses are hashed among $H$ buckets. Therefore, the arrival

| | analytical | simulation |
|---|---|---|
| 8 buckets, E=10 | .374 | .755 |
| 20 buckets, E=10 | .103 | .267 |
| 8 buckets, E=100 | .065 | .197 |
| 20 buckets, E=100 | .024 | .0723 |

Table 3: comparison of analytical and simulation predictions of bucket waiting times.

rate at a bucket is:

$$\lambda_b = \frac{2S(D+1)}{E+2((D+1)T_a+T_q)} * \frac{1}{H}$$

The time to execute a task access is $T_a$, so

$$\mu_b = 1/T_a$$

The waiting time at an M/M/1 queue is [13]

$$W = \lambda/(1-\rho)$$

where $\rho = \lambda/\mu$. Therefore

$$W_b = \frac{2(D+1)S}{H(E+2(D+1)T_a+2t_q)-2(D+1)ST_a}$$

To find the minimum number of hash table buckets to support a speedup of $S$, we solve $\rho = 1$ for $H$ and find

$$H_{min} = \frac{2(D+1)ST_a}{H(E+2(D+1)T_a+2T_q)}$$

and a rule of thumb is to use at least enough buckets so that $\rho < .5$, or

$$H_{half} = \frac{4(D+1)ST_a}{H(E+2(D+1)T_a+2T_q)}$$

We used the parameters from the example in Figure 5 to generate Table 3. The analytical model is uniformly optimistic. The DTG accesses in the simulation are very non-uniform, being much heavier in the initial part of the simulation. For $E = 10$, the analytical model recommends at least 6.85 buckets, and for $E = 100$, the model recommends at least 1.83 buckets.

## 5 Conclusion

In this paper, we present algorithms for performing *dynamic-task graph scheduling*. We present a concurrent data structure, the *concurrent dynamic-task graph* that allows the task graph of the computation to be specified while the parallel computation proceeds. Such a capability is useful for certain classes of parallel computations, such as analysis-factor LU factorizations of asymmetric sparse matrices. We study some aspects of the performance of the DTG algorithms. We conclude that simple scheduling strategies work well, that the DTG is appropriate for medium to coarse grained computations, and provide a rule of thumb for determining the number of hash table buckets for the DTG.

**Acknowledgements** We'd like to than Steve Hadfield for his help and for the use of his static-task graph scheduler.

## References

[1] R. Anani. LR-algorithm: Concurrent operations on priority queues. In *Proceedings of the Second IEEE Symposium on Parallel and Distributed Processing*, pages 22–25, 1990.

[2] C.D. Polychronopolous U. Bannerjee. Processor allocation for horizontal and vertical parallelism and related speedup bounds. *IEEE Transactions on Computers*, C-36:410–420, 1987.

[3] E.G. Coffman. *Computer and Job-Shop Scheduling*. John Wiley and Sons, 1976.

[4] T. A. Davis and P. C. Yew. A nondeterministic parallel algorithm for general unsymmetric sparse LU factorization. *SIAM J. Matrix Anal. Appl.*, 11(3):383–402, 1990.

[5] T.A. Davis and I.S. Duff. Unsymmetric-pattern multifrontal methods for parallel sparse LU factorization. Technical report, University of Florida, Dept. of CIS TR-91-23, 1991. Available at anonymous ftp site cis.ufl.edu:cis/tech-reports.

[6] R.R. Glenn, D.V. Pryor, J.M. Conroy, and T. Johnson. Characterizing memory hotspots in a shared memory mimd machine. In *Supercomputing '91*. IEEE and ACM SIGARCH, 1991.

[7] A. Gottlieb, B. D. Lubachevsky, and L. Rudolph. Basic techniques for the efficient coordiantion of very large numbers of cooperating sequential processors. *ACM Trans. on Programming Languages and Systems*, 5(2):164–189, 1983.

[8] M. Herlihy. A methodology for implementing highly concurrent data structures. In *Proceeding of the Second ACM SIGPLAN Symposium on Principles and Practice of Parallel Programming*, pages 197–206. ACM, 1989.

[9] J. Ji and M. Jeng. Dynamic task allocation on shared memory multiprocessor systems. In *ICPP*, pages I:17–21, 1990.

[10] H. Kasahara and S. Narita. Practical multiprocessor scheduling algorithms for efficient parallel processing. *IEEE Trans. on Computers*, C-33:1023–1029, 1984.

[11] D. Klappholz and S. Narita. Practical multiprocessor scheduling algorithms for efficient parallel processing. *IEEE Transactions on Computers*, C-33:315–321, 1984.

[12] D. Klappholz and H.C. Park. Parallelized process scheduling for a tightly-coupled mimd machine. In *Int'l Conf. on Parallel Processing*, pages 315–321, 1984.

[13] L. Kleinrock. *Queueing Systems*, volume 1. John Wiley, New York, 1975.

[14] W. Massey. A probabilistic analysis of a database system. In *ACM SIGMETRICS Conference on Measuring and Modeling of Computer Systems*, pages 141–146, Aug. 1986.

[15] J.M. Mellor-Crummey and M.L. Scott. Algorithms for scalable synchronization on shared-memory multiprocessors. *ACM Trans. Computer Systems*, 9(1):21–65, 1991.

[16] S. Prakash, Y.H. Lee, and T. Johnson. A non-blocking algorithm for shared queues using compare-and-swap. In *Proc. Int'l Conf. on Parallel Processing*, pages II68–II75, 1991.

[17] Shirazi, Wang, and Pathak. Analysis and evaluation of heuristic methods of static task scheduling. *Journal of Parallel and Distributed Computing*, 10:222–232, 1990.

[18] J. Turek, D. Shasha, and S. Prakash. Locking without blocking: Making lock based concurrent data structure algorithms nonblocking. In *ACM Symp. on Principles of Database Systems*, pages 212–222, 1992.

[19] J.D. Valois. Towards Practical Lock-free Data Structures. Submitted for publication, 1993.

[20] Z. Yin, C. Chui, R. Shu, and K. Huang. Two precedence-related task-scheduling algorithms. *Int'l Journal of High Speed Computing*, 3(3):223–240, 1991.

# UNIFIED STATIC SCHEDULING ON VARIOUS MODELS †

Liang-Fang Chao
Department of Computer Science
Princeton University
Princeton, NJ 08544

Edwin Hsing-Mean Sha
Dept. of Computer Science & Engineering
University of Notre Dame
Notre Dame, IN 46556

## Abstract

*Given a behavioral description of an algorithm represented by a data-flow graph, we show how to obtain a rate-optimal static schedule with the minimum unfolding factor under two timing models, integral grid model and fractional grid model, and two design styles for each model, pipelined design and non-pipelined design. We present a simple and unified approach to deal with the four possible combinations. A unified polynomial-time scheduling algorithm is presented, which works on the original data-flow graphs without really unfolding. The values of the minimum rate-optimal unfolding factors for all the four combinations are also derived.*

## 1 Introduction

We consider the problem of scheduling a recursive or iterative algorithm as being studied in [1, 4], which is particularly useful in DSP applications. *Scheduling* is important to compilers for parallel machines, and an important task during the architecture synthesis of VLSI design. The scheduling of iterative algorithms or loops in a program is usually critical to the system performance. We design algorithms to obtain rate-optimal static schedules and prove the minimum rate-optimal unfolding factors under various models and design styles.

A data-flow graph (DFG) is represented by a directed weighted graph $G = (V, E, d, t)$ where $V$ is the set of computation nodes, and $E$ is the edge set which defines the precedence relations. The graph in Figure 1 is an example of DFG. The integer number attaches to a node $v$ is its computation time, denoted by $t(v)$, and dark bars on an edge $e$ is called *delays*, the number of which is denoted by $d(e)$.

The execution of all computation nodes once is called one *iteration*. An edge $e$ from $u$ to $v$ with delay count $d(e)$ means that the computation of node $v$ at iteration $j$ depends on the computation of node $u$ at iteration $j-d(e)$. A *static schedule* is a schedule to be executed repeatedly. Each instance of the static schedule starts at a fixed time interval, called *cycle period* (or initiation interval).

In order to improve execution rate, we can schedule $f$ iterations in an instance of a static schedule. This is called *unfolding* (or unrolling) by an *unfolding factor* $f$. Each cycle now contains $f$ iterations. Therefore, the *iteration period*, which is the average computation time per iteration (cycle period / $f$), can be reduced. For example, the schedule in Figure 2 is a static schedule with unfolding factor 2 and cycle period 3.

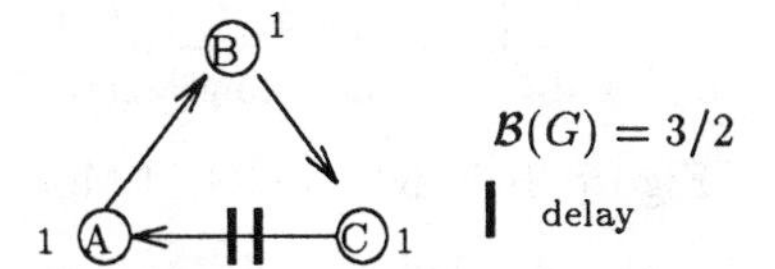


Figure 1: A simple data-flow graph (DFG)

It is well known that any DFG which involves loops, feedbacks or recursions has a lower bound on the iteration period [5]. This *iteration bound* $\mathcal{B}(G)$ for a DFG $G$ is given by $\mathcal{B}(G) = \max_{\forall\, loop\, l} T(l)/D(l)$, where $T(l)$ is the sum of computation times in loop $l$, and $D(l)$ is the sum of delay counts in loop $l$. For example, the $\mathcal{B}(G)$ in Figure 1-(a) is 3/2. A schedule is *rate-optimal* if the iteration period of this schedule equals the iteration bound.

There have been many efforts on finding the rate-optimal static schedule with the minimum unfolding factor [6, 1, 2]. This paper tries to clarify this problem by defining two timing models (*integral/fractional*) and two design styles (*pipelined/nonpipelined*). We first present an efficient unified scheduling algorithm for both timing models and design styles with time complexity $O((|V| + f)|E|)$. This algorithm works on the original DFG instead of the larger unfolded DFG. For each of the four combinations, we derive the minimum rate-optimal unfolding factor, and the relationship between any unfolding factor and its minimum iteration period. The corresponding schedules can be obtained by the above scheduling algorithm. Because of the space limit, we skip all the proofs; the details can be found in [3].

Given a DFG with the iteration bound $\mathcal{B}(G) = \sigma/\rho$ in its irreducible form, we summarize the minimum rate-optimal unfolding factors in the following table.

| Min Rate-optimal | Timing Models | |
|---|---|---|
| Unfolding Factor | Fractional | Integral |
| Pipelined Design | 1 | $\rho$ |
| Nonpipelined Design | $\left\lceil \frac{\max_v t(v)}{\mathcal{B}(G)} \right\rceil$ | $\left\lceil \frac{\max_v t(v)}{\sigma} \right\rceil \cdot \rho$ |

†The work of Chao was supported in part by DARPA/ONR contract N00014-88-K-0459 and NSF award MIP90-23542.

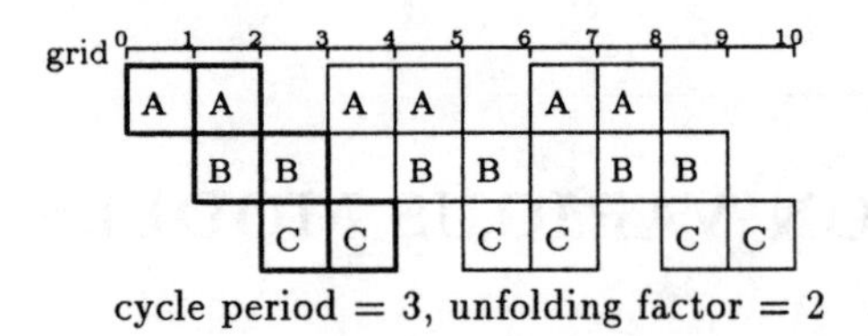


cycle period = 3, unfolding factor = 2

Figure 2: Integral-grid model

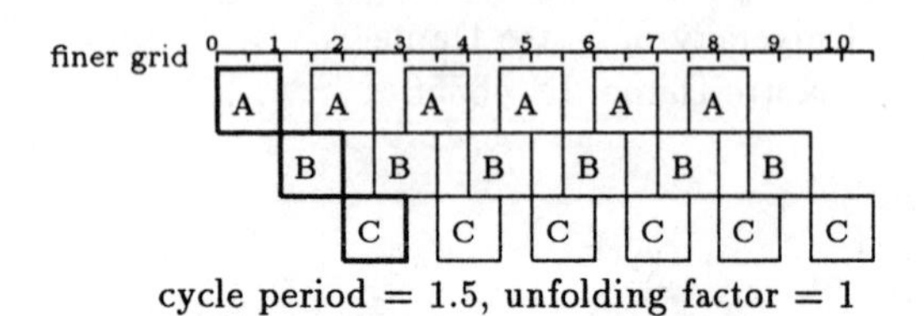


cycle period = 1.5, unfolding factor = 1

Figure 3: Fractional Grid Model

## 2 Various Models and Design Styles

A *schedule* is represented by a function $S : V \times N \longrightarrow R$. The starting time of node $v$ in the $i$-th iteration is $S(v,i)$. A schedule is *legal* if for every edge $u \overset{e}{\longrightarrow} v$ and iteration $i$, we have $S(u,i) + t(u) \leq S(v, i + d(e))$. A *repeating* schedule for *unfolding factor* $f$ and *cycle period* $c$ is a legal schedule such that $S(v, i + f) = S(v,i) + c$ for every $v$ in $V$ and iteration $i$. Thus, a repeating schedule can be represented by the partial schedule of the first $f$ iterations. A new instance of this partial schedule of $f$ iterations can be initiated for every interval of length $c$ to form a legal repeating schedule. Moreover, if in every instantiation, an operation of the partial schedule is assigned to the same processor, this type of schedule is called a *static schedule* or a *processor-static schedule.*

The *iteration period* of a repeating schedule is the average computation time per iteration. A repeating schedule with cycle period $c$ and unfolding factor $f$ has iteration period $c/f$. If the value of $c/f$ equals the iteration bound $\mathcal{B}(G)$, we call such an $f$ to be a *rate-optimal unfolding factor.*

The measurement of computation time and cycle period is in terms of a pre-defined *time unit,* which may be a machine cycle or a clock cycle under different models. Without loss of generality, we assume that every node has an integral computation time. There are two models for the timing style of scheduling.

**Integral-grid model** We imagine there is an integral grid (time unit) for the schedule. An operation can only be issued in the beginning of a time unit. Under this model, a schedule has an integral cycle period.

**Fractional Model (Gridless Model)** The starting time of a node can be any fractional number, i.e. the schedule can have infinitely fine grid, and an operation can be issued at any time. Under this model, a schedule may have a fractional cycle period. Actually, we will show that a grid of size $1/\rho$ gives schedules as good as gridless schedules, where $\rho$ is the denominator of $\mathcal{B}(G)$ in its irreducible form.

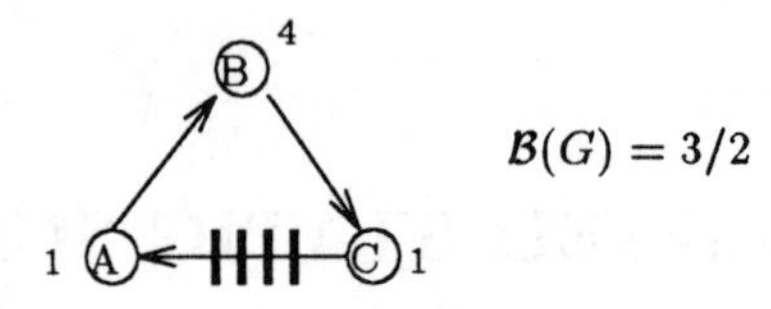


Figure 4: An exemplary DFG $G$

Consider the DFG in Figure 1 with iteration bound $\mathcal{B}(G) = 3/2$. Figure 2 and Figure 3 show a schedule under the integral grid model and a schedule under the fractional model, respectively. These two repeating schedules have the same iteration period, 3/2.

In order to implement processor-static schedules, there are two types of design styles under every timing model.

**Pipelined design** There is no restriction on the scheduling of copies of the same node besides precedence relations. If the second copy of a node starts its execution on the same processor as the first copy before the first copy is finished. we can use pipelined hardware to implement this processor. This model is the same as the concept of software pipelining for parallel compilers.

**Nonpipelined design** Since the hardware or processors are non-pipelined, the next copy of a node cannot start execution before the previous copy has finished execution. Thus, this constraint causes an implicit precedence relations between these two copies. For a static schedule with the unfolding factor $f$, the $i$-th copy of a node has an implicit precedence relation with the $i + f$-th copy because they are assigned to the same processor.

The exemplary DFG which is used in the rest of the paper is shown in Figure 4, and its iteration bound $\mathcal{B}(G)$ is 3/2. The minimum rate-optimal unfolding factors derived from theorems in this paper for these four combinations are as follows.

| Min Rate-optimal Unfolding Factor | Timing Models | |
|---|---|---|
| | Fractional | Integral |
| Pipelined Design | 1 | 2 |
| Nonpipelined Design | 3 | 4 |

## 3 Unified Scheduling Algorithm

Here, we present an algorithm to find a schedule for given cycle period $c$ and unfolding factor $f$ under fractional and integral models. For a DFG $G = (V, E, d, t)$ and given $c$ and $f$, we define a modified graph with different weights on edges: *scheduling graph* $G^s = (V, E, w)$ where $w(e) = d(e) - t(u) \cdot f/c$ for every edge $u \overset{e}{\longrightarrow} v$ in $E$. This modified graph was used in paper [1] to compute legal retimings for a unit-time DFG. Here, we use the modified graph to compute legal schedules for different models.

We add a node $v_0$ and directed edges from $v_0$ to every other node with zero weight in the scheduling graph. Let $sh(v)$ be the length of the shortest path from $v_0$ to $v$ in the scheduling graph $G^s$. The values of $sh(v)$ can be computed by any single-source shortest path algorithm. Notice that

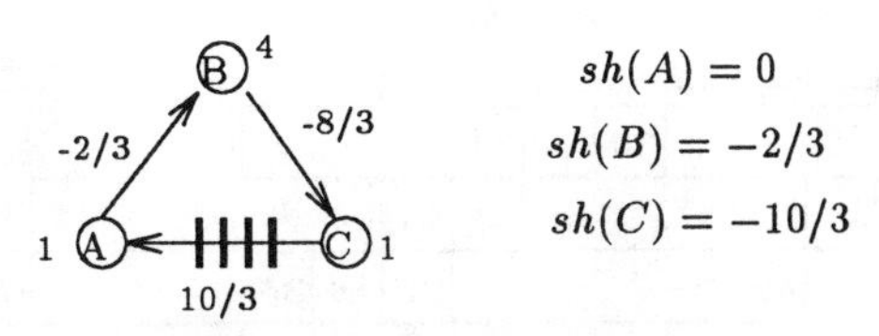


$sh(A) = 0$

$sh(B) = -2/3$

$sh(C) = -10/3$

Figure 5: The scheduling graph $G^s$ with $c/f = 3/2$

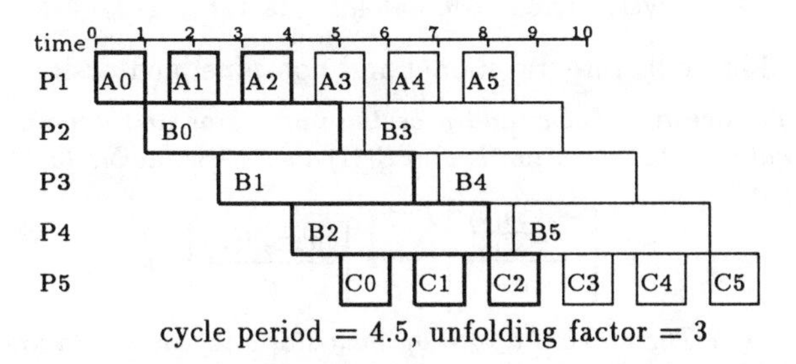


Figure 6: Fractional model and nonpipelined design

for every node $v$ we have $sh(v) \leq 0$, and there exists a node $u$ in $V$ such that $sh(u) = 0$.

The following theorem obtains legal fractional and integral schedules from the scheduling graph.

**Theorem 1** *Let $G$ be a DFG, $c$ a cycle period, $f$ an unfolding factor and $sh(v)$ the length of the shortest path of node $v$ found in $G^s$.*

**(a)** $S^f(v,i) = -sh(v) \cdot c/f + i \cdot c/f$ *for every $v$ and $i$ is a legal schedule under fractional model.*

**(b)** $S^i(v,i) = \lceil -sh(v) \cdot c/f + i \cdot c/f \rceil$ *for every $v$ and $i$ is a legal schedule under integral grid model.*

Note that another integral schedule can be obtained by substituting the *floor* operation for the *ceiling* operation. Since there exists a node with $sh(v) = 0$, the schedule $S^f$ and $S^i$ both start from time 0. Consider the DFG in Figure 4 as an example. The iteration bound $\mathcal{B}(G)$ of this DFG is 3/2. The scheduling graph $G^s$ is depicted in Figure 5 [1], where the number beside an edge $e$ is weight $w(e)$. We first compute all the $sh(v)$ for every $v$ in $V$, as shown in Figure 5. For $c = 4.5$ and $f = 3$, we obtain the fractional schedule in Figure 6, where $S^f(A,0) = 0$, $S^f(B,0) = 1$, and $S^f(C,0) = 5$. And, $S^f(A,1) = 0 + c/f = 1.5$, $S^f(B,1) = 2.5$, $S^f(C,1) = 6.5$.

The following lemma shows that the schedules $S^f$ and $S^i$ are repeating schedules with unfolding factor $f$ and cycle period $c$.

**Lemma 2** *The schedules $S^f$ and $S^i$ both are repeating schedules, i.e. they have the following properties: For every node $v$ and every positive integer $i$,*

**(a)** $S^f(v, i+f) = S^f(v,i) + c$.

**(b)** $S^i(v, i+f) = S^i(v,i) + c$.

It takes time $O(|V|\,|E|)$ to compute $sh(v)$ or detect negative-weight loops, and time $O(f\,|V|)$ to generate a schedule with unfolding factor $f$. The time complexity of the scheduling algorithm is $O((|E| + f)\,|V|)$. This algorithm is particularly efficient because our scheduling algorithm for finding repeating schedules with unfolding factor $f$ works on the scheduling graph $G^s$, the size of which is the same as the original DFG. We do not unfold the DFGs while computing schedules.

[1] Notice that the scheduling graph $G^s$ is the same for finding schedules of the same rate, i.e. the same $c/f$ ratio. All rate-optimal schedules found in this paper for the exemplary DFG will use the same scheduling graph in Figure 5.

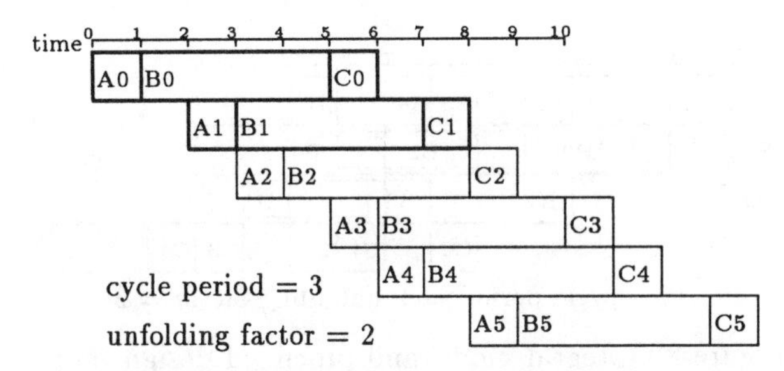


Figure 7: Integral model and pipelined design: repeating schedule

In our scheduling algorithm, a schedule exists if there is no loop with negative weight. The existence of negative-weight loops depends on the formula $c/f \geq \mathcal{B}(G)$ [1]. The following theorem provides the necessary and sufficient condition for the existence of a repeating schedule with cycle period $c$ and unfolding factor $f$.

**Theorem 3** *Let $c$ be a fractional/integral cycle period, and $f$ an unfolding factor. There exists a legal fractional/integral repeating schedule with unfolding factor $f$ and cycle period $c$ if and only if $c/f \geq \mathcal{B}(G)$.*

## 4 Static Scheduling for Pipelined Design on Integral Model

All legal repeating schedules can be implemented as static schedules by using pipelined processors. Therefore, the minimum unfolding factors for the pipelined design can be derived from Theorem 3.

**Theorem 4** *Let $G$ be a general-time DFG and $\rho$ the denominator of $\mathcal{B}(G)$ in its irreducible form. The minimum rate-optimal unfolding factor under integral model and pipelined design is $\rho$.*

Consider the example in Figure 4. Since $\mathcal{B}(G)$ is 3/2, we know $\rho$ is 2. Figure 7 shows a rate-optimal schedule with unfolding factor 2. From Theorem 4, we know that 2 is the minimum rate-optimal unfolding factor under integral model. Figure 8 shows the corresponding static schedule.

Assume that we would like to find a rate-optimal schedule. We first find the iteration bound $\mathcal{B}(G)$ in its irreducible form. The iteration bound for a DFG can be found in time $O(|V||E|\log|V|)$. Then, we use the scheduling algorithm to find a schedule with unfolding factor $f$ and cycle period $c$ in time $O((|E| + f)|V|)$.

In practice, the minimum unfolding factor to achieve rate-optimal non-pipelined design may be too large to be realistic. It is necessary to find a schedule of the highest rate achievable under practical design requirements. For a

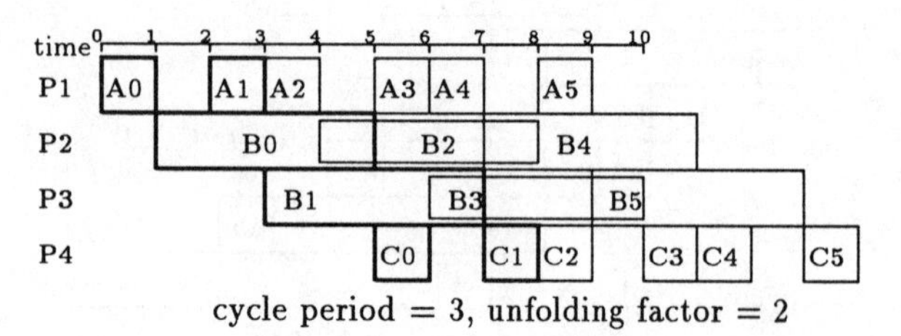


cycle period = 3, unfolding factor = 2

Figure 8: Integral model and pipelined design: static schedule

given unfolding factor, we can derive its minimum iteration period from Theorem 3.

**Lemma 5** *Let $f$ be a given unfolding factor. The pipelined schedule using unfolding factor $f$ has the minimum cycle period $MCP_p(f) = \lceil f \cdot \mathcal{B}(G) \rceil$.*

The larger the unfolding factor is, the larger the memory used to store the repeating schedule is. When there is a bound for the amount of available memory, we have the following.

**Theorem 6** *Let $F$ be the maximum unfolding factor feasible under resource requirements. The pipelined schedule with minimum iteration period using an unfolding factor not exceeding $F$ has unfolding factor $\hat{f}$ and cycle period $\hat{c}$, where $\hat{c}/\hat{f}$ is the irreducible form of the value*

$$\min_{1 \le f \le F} \lceil f \cdot \mathcal{B}(G) \rceil / f.$$

For example, for a DFG with $\mathcal{B}(G) = 15/11$, the minimum rate-optimal unfolding factor is 11. But, the maximum feasible unfolding factor under design requirements is only 6. From the above theorem, we know the minimum iteration period is obtained from $\hat{f} = 5$ and $\hat{c} = 7$.

## 5 Static Scheduling for Non-Pipelined Design on Integral Model

Since the hardware or processors are non-pipelined, for a static schedule with the unfolding factor $f$, the $(i+f)$-th copy of a node cannot start execution before the $i$-th copy has finished execution. Thus, a schedule with unfolding factor $f$ can be implemented by using non-pipelined processors if for every $v$ in $V$ and positive integer $i$, we have $S(v, i+f) \ge S(v, i) + t(v)$. A new copy of a node assigned to the same processor starts execution every $c$ time units in a static schedule of cycle period $c$. Thus, the cycle period must be larger than the computation time of a node.

**Theorem 7** *Let $c$ be a cycle period and $f$ an unfolding factor. There exists a legal static schedule under non-pipelined design with cycle period $c$ and $f$ if and only if $c/f \ge \mathcal{B}(G)$ and $c \ge \max_v t(v)$.*

The following theorem combines the requirements for rate-optimality and non-pipelined design to derive the rate-optimal minimum unfolding factor.

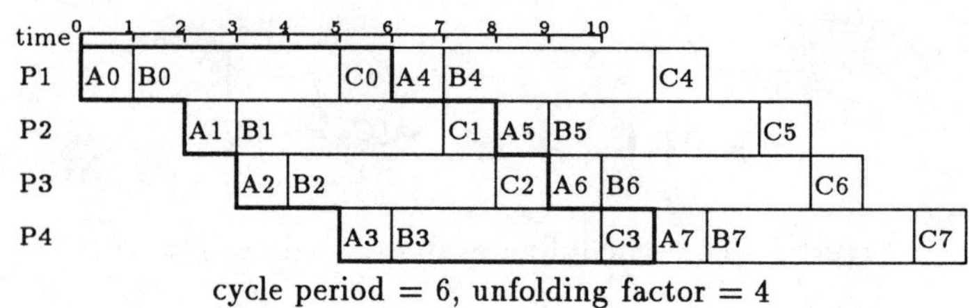


cycle period = 6, unfolding factor = 4

Figure 9: Integral model and non-pipelined design

**Theorem 8** *Let $\sigma$ and $\rho$ be the numerator and denominator of the iteration bound $\mathcal{B}(G)$ in its irreducible form. Set*

$$\hat{f} = \left\lceil \frac{\frac{\max_v t(v)}{\mathcal{B}(G)}}{\rho} \right\rceil \cdot \rho = \left\lceil \frac{\max_v t(v)}{\sigma} \right\rceil \cdot \rho.$$

(a) *There exists a rate-optimal static schedule with unfolding factor $\hat{f}$ under non-pipelined design.*

(b) *$\hat{f}$ is the minimum rate-optimal unfolding factor to deliver a static schedule under non-pipelined design.*

Our scheduling algorithm can find a repeating schedule $S^i$. This schedule is a static schedule under non-pipelined design. Consider the example in Figure 4. The iteration bound $B(G)$ is 3/2, and the $\max_v t(v)$ is 4. From Theorem 8, we derive the minimum rate-optimal unfolding factor is $\hat{f} = \lceil \max_v t(v)/\sigma \rceil \cdot \rho = \lceil 4/3 \rceil \cdot 2 = 4$. From the scheduling algorithm in the previous section, we obtain $S^i(A, 0) = 0$, $S^i(B, 0) = 1$, $S^i(C, 0) = 5$, $S^i(A, 1) = \lceil 3/2 \rceil = 2$, $S^i(B, 1) = \lceil 1 + 3/2 \rceil = 3$, and $S^i(C, 1) = \lceil 5 + 3/2 \rceil = 7$. Figure 9 shows the rate-optimal schedule with unfolding factor 4. This schedule is static under non-pipelined design, where the operation B5 starts after B1 is finished.

When the minimum rate-optimal unfolding factor is too large, we can find the minimum cycle period for any given unfolding factor.

**Lemma 9** *Let $f$ be a given unfolding factor. The static schedule under non-pipelined design using unfolding factor $f$ has the minimum cycle period $MCP_n(f)$, where*

$$MCP_n(f) = \begin{cases} \max_v t(v) & \text{if } \max_v t(v) \ge f \cdot \mathcal{B}(G), \\ \lceil f \cdot \mathcal{B}(G) \rceil & \text{otherwise.} \end{cases}$$

For example, given a DFG with $\mathcal{B}(G) = 15/11$, $f = 5$, if $\max_v t(v) = 8$, we have $MCP_n(5)$ is 8; if $\max_v t(v) = 6$, we have $MCP_n(5)$ is 7.

**Theorem 10** *Let $F$ be the maximum unfolding factor feasible under resource requirements. The non-pipelined schedule with the minimum iteration period using an unfolding factor not exceeding $F$ has minimum unfolding factor $\hat{f}$ and cycle period $\hat{c}$. The values of $\hat{f}$ and $\hat{c}$ can be found as follows.*

(a) *If $\max_v t(v) \ge F \cdot \mathcal{B}(G)$, we have $\hat{f} = F$ and $\hat{c} = \max_v t(v)$.*

(b) *If $\max_v t(v) < F \cdot \mathcal{B}(G)$, the minimum iteration period $\hat{c}/\hat{f}$ is the irreducible form of the value*

$$\min\left\{ \frac{\max_v t(v)}{\left\lfloor \frac{\max_v t(v)}{\mathcal{B}(G)} \right\rfloor}, \min_{\frac{\max_v t(v)}{\mathcal{B}(G)} < f \le F} \frac{\lceil f \cdot \mathcal{B}(G) \rceil}{f} \right\}.$$

For example, given a DFG with $\mathcal{B}(G) = 15/11$, $F = 6$, and $\max_v t(v) = 4$, we have $\max_v t(v) = 4 < 6 \cdot 15/11$. Thus, we use the formula in (b). The first term is $4/\left\lfloor \frac{4}{15/11} \right\rfloor = 4/2 = 2$; the second term is $\min_{2<f\leq 6} \lceil f \cdot 15/11 \rceil /f = 7/5$. Therefore, $\hat{c} = 7$ and $\hat{f} = 5$.

# 6 Static Scheduling on Fractional Model

Under some implementation model, the starting time of a node can be any fractional number, i.e. the schedule can have finer grid. We can reduce the unfolding factor by using a finer time slice.

We use our algorithm to find a schedule with a fractional cycle period $c$ and an (integral) unfolding factor $f$. First, we find the smallest integer $\rho$ such that $\rho c$ is integral, i.e. $\rho$ is the denominator of $c$ in its irreducible form. Then, we use a finer time slice: $1/\rho$ time unit is a time slice. All the computation times in the DFG are modified to refer to time slices. Thus, we have a new DFG $G' = (V, E, d, t')$, where $t'(v) = \rho \cdot t(v)$. The lower bound of the new DFG becomes $\mathcal{B}(G') = \rho \cdot \mathcal{B}(G)$. Thus, we can use the theorems for integral models to compute rate-optimal unfolding factors.

By using the fractional model, we can reduce the unfolding factor caused by the iteration bound $\mathcal{B}(G)$. The new iteration bound is $\rho \cdot \mathcal{B}(G)$. If we choose a right value for $\rho$, we can reduce the denominator of the iteration bound by a factor of $\rho$. Thus, under pipelined design, the rate-optimal unfolding factor is reduced by a factor of $\rho$. Theoretically, we can always choose $\rho$ to be the denominator of $\mathcal{B}(G)$ in its irreducible form so that a static schedule without unfolding can be found. However, the fractional model does not relieve the inequality for non-pipelined design $c \geq \max_v t(v)$.

The minimum rate-optimal unfolding factor for pipelined and non-pipelined design under fractional model are given in Theorem 11 and Theorem 12.

**Theorem 11** *Under the fractional model and pipelined design, there always exists a rate-optimal pipelined static schedule without unfolding.*

**Theorem 12** *The minimum unfolding factor to deliver a rate-optimal static schedule under non-pipelined design on the fractional model is* $\hat{f} = \left\lceil \max_v t(v)/\mathcal{B}(G) \right\rceil$.

Consider the example in Figure 4. Under fractional model and pipelined design, Theorem 11 shows that there always exists a rate-optimal schedule without unfolding. The rate-optimal schedule obtained by our scheduling algorithm is shown in Figure 10. Under fractional model and non-pipelined design, we derive the rate-optimal minimum unfolding factor from Theorem 12 is $\hat{f} = \lceil \max_v t(v)/\mathcal{B}(G) \rceil = \left\lceil \frac{4}{3/2} \right\rceil = 3$. Figure 6 shows the rate-optimal schedule with unfolding factor 3 obtained by our scheduling algorithm.

The minimum cycle period for a given unfolding factor or a maximum unfolding factor can be similarly derived for both design styles as the results for integral-grid model. The interested reader is referred to [3].

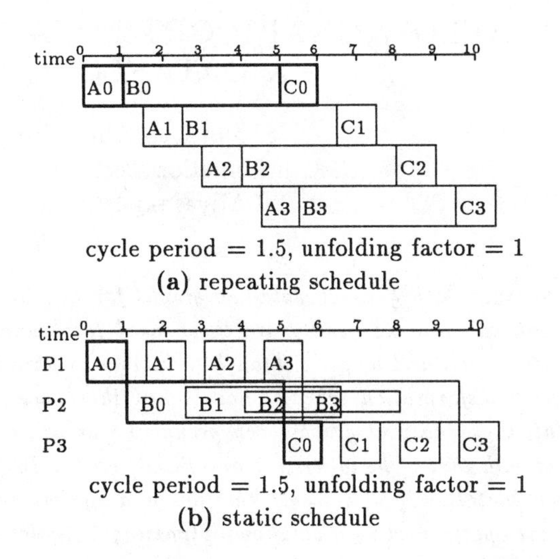


Figure 10: Fractional model and pipelined design

# 7 Conclusion

We have proposed a scheduling algorithm to generate static schedules with implicit unfolding. Our scheduling algorithm has also incorporated the effect of retiming implicitly; so, explicit retiming is not necessary. Currently we are exploring the interplay of scheduling and allocation under a limited number of processors.

# References

[1] L.-F. Chao and E. H.-M. Sha, "Unfolding and retiming data-Flow DSP programs for RISC multiprocessor scheduling," *Proc. 1992 IEEE Int'l Conf. on Acoustic, Speech, and Signal Processing,* San Francisco, March 1992, pp. V565-V568.

[2] L.-F. Chao and E. H.-M. Sha, "Retiming and unfolding data-flow graphs," *Proc. 1992 International Conference on Parallel Processing,* St. Charles, Illinois, August 1992, pp. II 33-40.

[3] L.-F. Chao and E. H.-M. Sha, "Static scheduling for synthesis of DSP algorithms on various models," Technical Report, Department of Computer Science, Princeton University, 1993.

[4] K. K. Parhi and D. G. Messerschmitt, "Static rate-optimal scheduling of iterative data-flow programs via optimum unfolding", *IEEE Trans. on Computers,* Vol. 40, No. 2, Feb. 1991, pp. 178-195.

[5] M. Renfors and Y. Neuvo, "The maximum sampling rate of digital filters under hardware speed constraints," *IEEE Trans. on Circuits and Systems,* Vol. CAS-28, No. 3, March 1981, pp.196-202.

[6] D. J. Wang and Y. H. Hu, "Fully static multiprocessor realization for real-time recursive DSP algorithms," *International Conf. Application-Specific Array Processors,* Aug. 1992, pp. 664–678.

# DATAFLOW GRAPH OPTIMIZATION FOR DATAFLOW ARCHITECTURES — A DATAFLOW OPTIMIZING COMPILER —

Sholin Kyo, Shin'ichiro Okazaki and Masanori Mizoguchi
C&C Information Technology Research Laboratories, NEC Corportaion
1-1 Miyazaki 4-Chome, Miyamae-ku, Kawasaki, 216, Japan
*e-mail: sholin@pat.cl.nec.co.jp*

Abstract— *In general, dataflow graphs for which it is difficult to optimize quality are those used for branches (conditionals) and loops. Although a number of dataflow graph transformation schemes have taken these into account, the quality of graphs been produced was not compared with that of hand-written optimized graphs. In this paper, dataflow graph transformation and optimization schemes performed by a dataflow optimizing compiler are described, and the results of an evaluation of the degree of optimization been achieved are presented.*

## I. INTRODUCTION

Although any dataflow architecture can of itself extract parallelism of any grain size, the actual computation power of a dataflow computer greatly depends on the ability of the compiler, which generates dataflow graphs. The authors' research group had originally developed ImPP($\mu$PD7281) [1], a one-chip static dataflow processor, which was first put into commercial use in 1984, with a compiler subsequently developed. But as the compiler employed conventional methods of dataflow graph generation, the time required for execution of those graphs was, depending on the specific application, 2 to 13 times that of hand-written dataflow graphs. In response to this disadvantage, the group began to work on a new, C-like language called DPC (Dataflow Pipelining C), whose primary goal was the achievement of a DPC compiler which could generate dataflow graphs of a quality nearly equivalent to that of hand-written graphs.

In general, dataflow graphs for which it is difficult to optimize quality are those used for the structure statements of branches (conditionals) and loops. With branches, the main difficulty is the distribution of switch(-node)s within the graph for data stream controls. With loops, the main difficulty is trying to achieve maximum loop unfolding while at the same time avoiding the exhaustion of hardware resources. A number of dataflow graph transformation schemes [2]-[5] which take these problems into consideration have previously been proposed, but they do not compare the quality of graphs produced with that of hand-written graphs.

In the section below, after a brief overview of DPC and the DPC language processor, the schemes used in the 2nd and 3rd pass of DPC compiler for transforming and optimizing dataflow graphs (hereafter, referred to as *PG*(Program Graphs)) are described. With regard to the 2nd pass, the scheme for transforming conditionals and loops into *PG*s, and a brief overview of the whole pro-

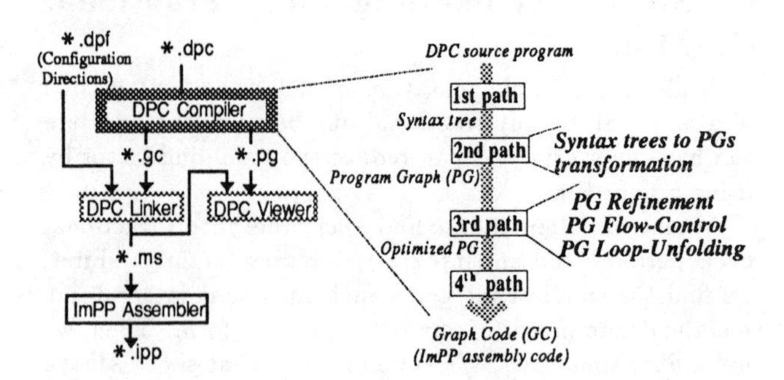


Figure 1: The DPC Language Processor

cess of the pass are given. With regard to the 3rd pass, each of the *PG* optimization processes: *PG* Refinement, *PG* Flow-Control, and *PG* Loop-Unfolding are described. Finally, the degree of *PG* optimization achieved by the DPC compiler and the feasibility of its practical use in real applications are evaluated.

## II. DATAFLOW GRAPH OPTIMIZATION

### A. Overview of DPC and the DPC Language Processor

DPC is a langauge which aims at achieving a brevity of description for low to medium level image processing. Its syntax is based on C, but contains additional extentions and restrictions for stream-like descriptions and implementations. The DPC language processor is shown in Fig.1. The DPC compiler is comprised of four processing passes. The 1st pass translates DPC descriptions into syntax trees; the 2nd pass tranforms syntax trees into *PG*s, which are particularly well-suited to the optimizations in the continuing 3rd pass; the 3rd pass performs *PG* Refinement, *PG* Flow-Control, and *PG* Loop-Unfolding as optimization processes; and finally, the 4th pass translates each node of the *PG*s into *GC*s(the ImPP assembly code). The DPC linker accomplishes function-level resource allocations of *GC*s, and the DPC viewer is available for the visualization of *PG*s and *GC*s used in debugging.

### B. *PG* Transformation for conditionals and loops

Generally, transformation of syntax trees into *PG*s is easy, except those for such structure statements as branches and loops, for which data stream control must be accomplished. Such data stream control is usually achieved in dataflow architectures by matching each data token with a boolean token at a switch, at which point the destination of the data token may or may not be changed, depending upon information contained in the boolean to-

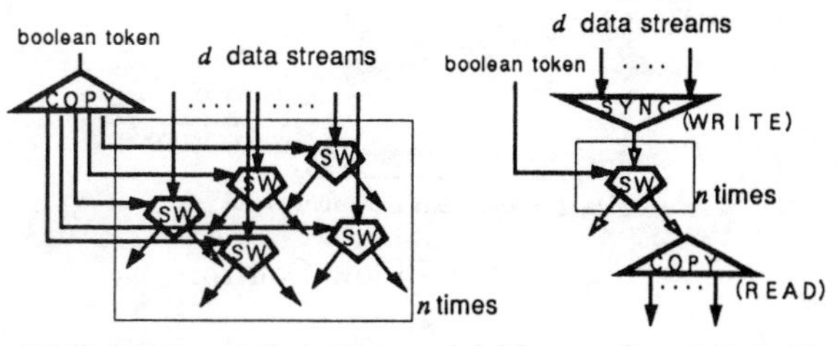


*(b) Token-Based Switching* *(c) Memory-Based Switching*

Figure 2: Two ways for controlling data streams in dataflow architectures.

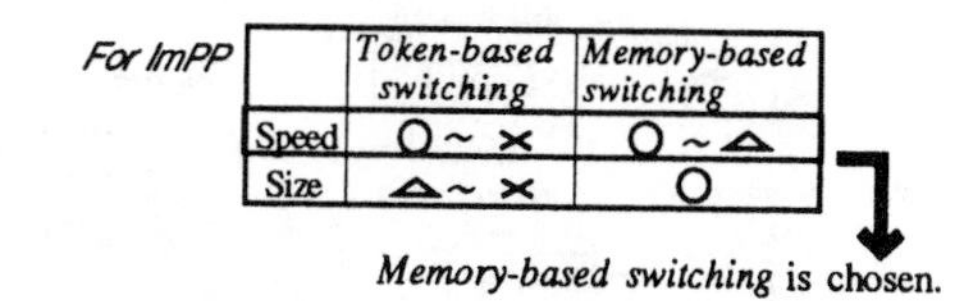

For ImPP
Token-based switching
Memory-based switching
Speed
Size
Memory-based switching is chosen.


Figure 3: Decision on the stretegies for *PG* transformation.

ken. Note that this process will be repeated for each data token as many times as there are nest levels ($n$) for each branch. This scheme for controlling data streams is called *token-based switching* (Fig.2(a))[6].

On the other hand, data stream control can also be achieved, employing the same operations as are found in Von Neumann architectures, by using memory spaces. Here, it is first necessary to *write* data tokens for corresponding memory spaces. Second, **switch**s are activated as many times as there are nest levels ($n$) for each branch, and the same number of boolean tokens used. After that, data streams are reformed by *reading* from the corresponding memory spaces, and the implementation continues. This scheme for data stream control is called *memory-based switching* (Fig.2(b))[6].

Since individual data streams are controlled entirely independently in *token-based switching* scheme, a large amount of parallelism can be achieved. However, assuming that the number of data streams to be controlled (branched) is $d$ and that the nest level for the branch is $n$, then **switch**s must be activated a total of $n \times d$ times, which may create a large overhead (**sw**-overhead). On the other hand, in *memory-based switching* scheme, **sw**-overhead is small since the number of times **switch**s must be activated is $n$, but execution from *write* to *read* is sequentialized, which creates further overhead.

While *token-based switching* has been the traditional default scheme for producing branches in dataflow architectures, *token-based switching* will not necessarily produce the least overhead, since the amount of overhead depends upon the specific architecture and application to be used. In many cases, a *memory-based switching* may produce less overhead, and in fact, our investigation of results for ImPP have shown that dataflow graphs produced with a *memory-based switching* scheme are superior, in execution time and as well in size, to those produced with a *token-based switching* scheme (Fig.3)[6]. For this reason, we chose *memory-based switching* as the basic scheme for *PG* transformation in the 2nd pass of the DPC compiler. This choice eliminates the necessity of distributing a series of **switch**s in the dataflow graph, and as well allows the use of conventional register-based compile techniques for Von Neumann computers, which greatly facilitates *PG* transformation.

The 2nd pass transformation of syntax trees into *PG*s may be described in brief overview as consisting of the following three stages:

**stage 1.** PDG (Program Dependence Graph) [7] methodology is used to extract data dependencies and control dependencies of the program and to construct conventional block structures.

**stage 2.** The content of each block is transformed into a *PG*, which is a dataflow graph constrained purely by data dependencies.

**stage 3.** Previously extracted data and control dependencies are used in a *memory-based switching* scheme to connect individual *PG*s into a structure which may then be considered the program's overall destination *PG*.

As we are using *memory-based switching* scheme, data transfer within the structure is accomplished by *read* from and *write* to memories. Although this will initially cause multiple **read**(-node) and **write**(-node) to exist, redundant **read**s will be detected and deleted in the *PG* Refinement process of the 3rd pass of the DPC compiler.

C. <u>*PG* Refinement</u>

*PG* Refinement is the first process in the optimizing pass (the 3rd pass) of the DPC compiler. The objective of *PG* Refinement is to detect and delete redundant nodes in *PG* produced by the previous pass. *PG* Refinement can be classified into three sub-transformations: *Basic-Refinement(BR)*, *Identical-Refinement(IR)*, and *Synchronization-Deletion-Refinement(SDR)*, which are conducted as follows:

**Step 1.** Traverse the *PG* from the **start**(-node) and apply to each node, in order, the rules of *BR* and *IR*. Repeat this until no furter refinement is possible.

**Step 2.** Traverse the *PG* from the **start** and apply to each node, in order, the rules of *SDR*. Repeat this until no further refinement is possible. Go back to **Step 1.** if any refinement has taken place.

The main function of *BR* is either to combine some number of individual nodes into a single compound node, or to parallelize nodes for reduction of the critical path (see Fig.4), whereas the main function of *IR* is either to delete **read**s or to move them outside loops. Before starting **Step.1**, value-attributes($=va$) of **read**s and **write**s, which stand for the value of their input data, are

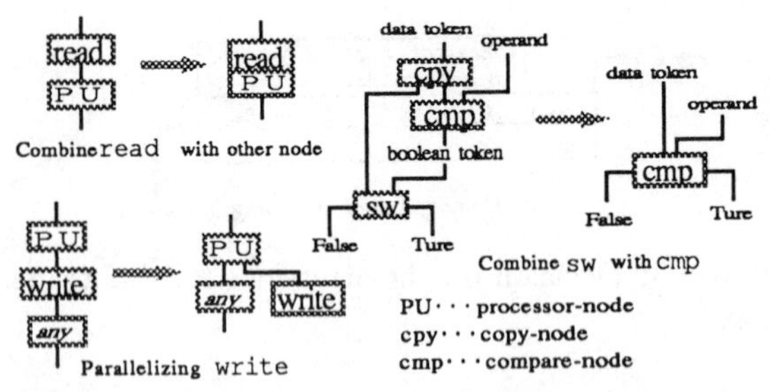


Figure 4: *Basic Refinement rule(BR)* samples.

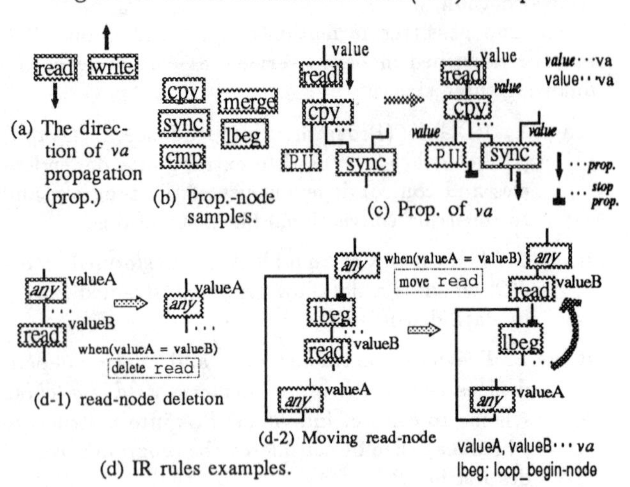


Figure 5: An Overview of *Identical Refinement(IR)*.

propagated downward (for **reads**) or up-ward (for **writes**) (Fig.5(a)). Nodes other than **reads** or **writes** either propagate *va* or stop propagation(Fig.5(b),(c)). After the completion of *va*-propagations, the rules of *IR* delete **reads** or move out side of loop **reads** in the condition shown in Fig.5(d-1) and Fig.5(d-2) respectively in **Step 1.**.

The main function of *SDR* is the deletion of redundant **syncs**(synchronization nodes). The rank-values of each node in the *PG* are first found, and the presumed static execution order for each node connected to a **sync** is then determined on the basis of these rank-values. This information then indicates which connection to the **sync** is redundant. Redundant connections are then deleted, and after these deletions, those **syncs** with less than two connections are deleted.

Fig.6 shows an example of *PG* Refinement in a simple DPC program for finding Fibonacci sequences. In this example, after applying *IR* and *BR*, the rank-values of each node are found and the deletion of a **sync** by the *SDR* is performed.

## D. *PG* Flow-Control

In order to prevent exhaustion of hardware resources, timing control for data token production and consumption must be done. As nodes in *PG* can be classified into producer-nodes(**pr**), operation-nodes(**op**), and consumer-nodes(**cn**), where data produced by **prs** are bounded so that they are absorbed by **cns**, *PG* Flow-Control is

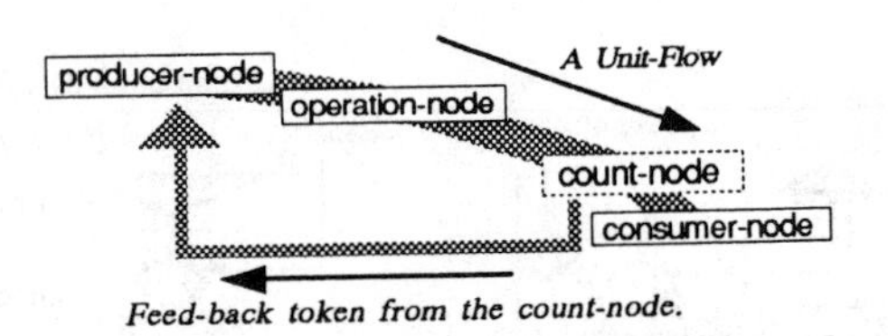


Figure 7: *PG* Flow-Control

achieved in the following manner: first, the value for the Unit-Flow (the upper-boundary on the amount of data for one production action) is determined on the basis of the amount of hardware resource available. Next, for each **pr**, the **cn** having the largest rank-value among all corresponding **cns**, i.e. the last node to be executed among corresponding **cns**, is chosen, and a **count** (a token counter) is inserted before it. The role of the **count** is to send a re-activation (feed-back) token back to the corresponding **pr** after a Unit-Flow of tokens has passed (Fig.7). Since detailed flow analyses are required for optimizing the value of the Unit-Flow, this value has been left unoptimized in the current version of DPC compiler, and it is assumed that individual users will specify an optimum value.

## E. *PG* Loop-Unfolding

While loop unfolding is the conventional way of speeding up loop executions on dataflow architectures, it can also lead to the exhaustion of hardware resources due to its overwhelming parallelism. For the control of parallelism in loop unfolding, the control of the two parameters, i.e. the unfolding degree $k$, and the initiation interval for each iteration $d$, is important. The objective of *PG* Loop-Unfolding is to perform graph reconstructions for each loop structure in *PG* in order to achieve proper control on these two parameters.

For the control of $k$, a technique derived from early ImPP assembly programs is employed, where a **sync** is used to synchronize the delivery of boolean tokens to the **switchs** with the termination of every iteration, while $k$ tokens are preloaded on the input link to the **sync** from the **sync**-tree, which produces the termination signal token from every iteration (Fig.8). The actual value for $k$ is adjusted to $L/d$ provided that $2 \leq k \leq 5$ under the consideration of matching memory capacity of ImPP, where $L$ is the average execution time for each iteration. The effect is that, the number of iterations which proceed in parallel is limited to a constant value $k$, i.e. a so called $k$-bounded loop execution[8] is achieved.

The control of $d$ is accomplished by adjusting the value of $d$, which initially depends on individual iteration content(e.g. the update latency of an inductive variable), to an adequate value in order to reduce both execution time for the loop and consumption of hardware resources (the matching memory). Simulation studies previously conducted by the authors[9] indicates that for *do*across-loops, i.e. loops in which there exists at least one anti-dependency-path (a path from a **read** to a **write** for the same variable), let $r_{min}$ be the minimum length for all

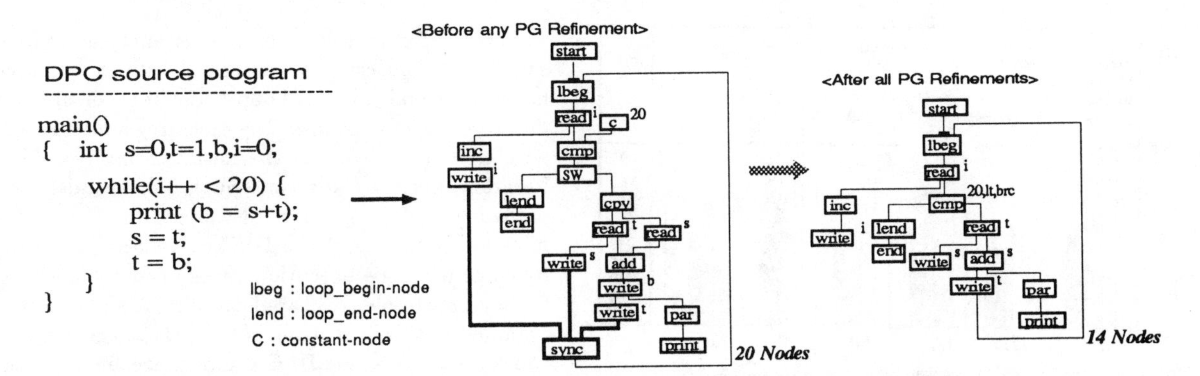


Figure 6: Example of *PG* refinement in a simple DPC program for finding 20 Fibonacci-Sequences.

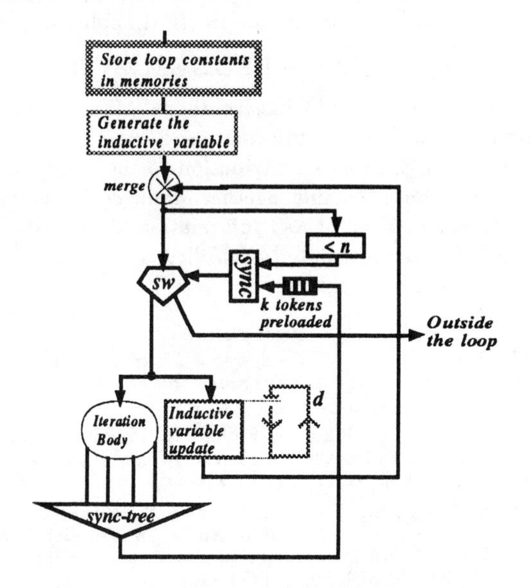


Figure 8: Parameter $d$ and $k$ in Loop-Unfolding.

anti-dependency-paths (lengths are found by taking into account conditionals) in the iteration, then for dataflow architectures, if initially $d < r_{min}$, enlarge $d$ to $r_{min}$ will make both the execution time and the consumption of matching memory be optimumly reduced.

Note that for *memory-based switching PGs*, since data are delivered via memories, there exist precedence constraints which give $d$ a lower bound, which is the maximum length $r_{max}$ of all anti-dependency-paths in *do*across-loops[10][11]. Since $r_{max} \geq r_{min}$, the optimal loop unfolding described above may not necessarily be obtained in all cases for *memory-based switching PGs*. In order to overcome this drawback, for those anti-dependency-paths whose path-lengths exceed $r_{min}$, we reconstruct *PG* so as to modify the delivery of data from that using normal memory to using matching memory to prevent the occurence of precedence constraints. This is accomplished as described below and illustrated in Fig.9. First, for small $d$ values, **nop**s with a total latency of $a$ are inserted in order to modify $d$ to $r_{min}$ where $(d+a) = r_{min}$. Next, for those anti-dependency-paths whose path-lengths exceed $r_{min}$,

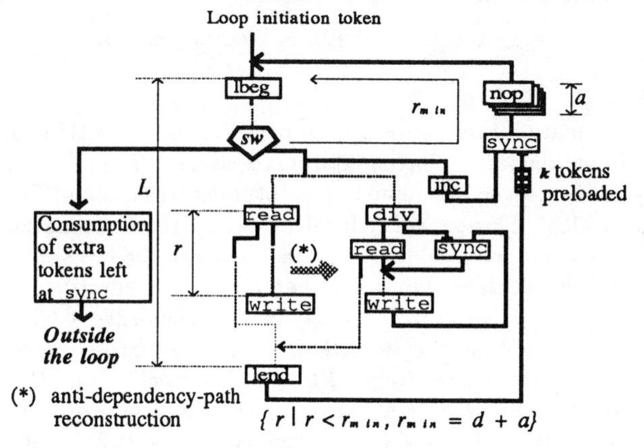


Figure 9: Graph reconstruction in *PG* Loop-Unfolding for *memory-based switching PGs*.

data delivery between iterations, i.e. delivery from the **write** for the current iteration to the **read** for the next iteration, is modified to using matching memories. Note that as seen in Fig.9, **div** works as a distribution node which passes its first input to the output on its left, and passes other inputs to the output on its right.

## III. EVALUATION OF DPC COMPILER

We have evaluated *PG* transformation and optimization processes of the DPC compiler by measuring the execution time for each of the four ImPP code categories ((*Unoptimized*)~(*Hand-optimized*) below) for the seven applications shown in Fig.10.

[*unoptimized* ] The code acquired from *PG*s after the accomplishment of a single graph traversal for applying *BR*.

[*optimized(memory-based)* ] The code acquired from *PG*s after the accomplishment of the full pass of optimization by the DPC compiler.

[*optimized(token-based)* ] The code acquired from *PG*s after the full pass optimization of the DPC compiler, and modification of *memory-based switching PG*s to *token-based switching*.

[*hand-optimized* ] The hand-written optimized code.

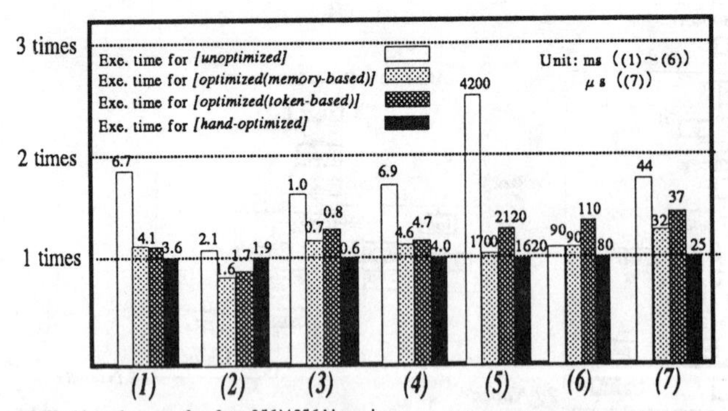


Figure 10: DPC compiler evaluation results.

Evaluation results are as shown in Fig.10. Note that for [*unoptimized*], a single graph traversal for *BR* application has been executed to make possible the translation of *PG*s to *GC*s. This generally involves merely the combination of an **read** with another node (as shown in the left side of Fig.4), which may decrease the difference in effectiveness between [*unoptimized*] and [*optimized(token-based)*]. The execution time for [*unoptimized*] and [*optimized(token-based)*] are, respectively, 52% and 11% greater than that for [*optimized(memory-based)*], whereas the execution time for [*optimized(memory-based)*] is only 13% greater than that for [*hand-optimized*], which indicates the feasibility of the compiler's use in real applications. With regard to the comparison between execution times for [*optimized(memory-based)*] and [*optimized(token-based)*], in applications as (1)–(5), in which a small amount of data and small nest-level for branches exist, the time for [*optimized(token-based)*] is 8% greater than that for [*optimized(memory-based)*], whereas in applications (6) and (7), in which a greater amount of data and more nest-levels for branches exist, the time for [*optimized(token-based)*] is 19% greater than that for [*optimized(memory-based)*]. This indicates that the more complicated an application is, the greater the difference in effectiveness between codes [*optimized(memory-based)*] and [*optimized(token-based)*].

## IV. CONCLUSION

In this paper, the dataflow graph transformation and optimization schemes which comprised of *PG* transformation, *PG* Refinement, *PG* Flow-Control, and *PG* Loop-Unfolding for a dataflow optimizing compiler, the DPC compiler, are mainly described. In the *PG* transformation process, *a memory-based switching scheme* is adopted, and the transformation of such structures as branches and loops are accomplished for minimizing **sw**-overheads. In the *PG* Refinement process, which can be further divided into three sub-refinement processes, we accomplished detection and deletion of redundant nodes in *PG*s. In *PG* Flow-Control, the relation between production and consumption of data tokens in *PG*s is analyzed, and token counters are inserted to prevent hardware resource exhaustion. Finally, in *PG* Loop-Unfolding, the unfolding of *do*across-loops is optimized by executing a $k$-bounded loop with the modification of the initiation interval $d$ of each iteration, and by the reconstruction of anti-dependency-paths inside *do*across-loops. Evaluation in these *PG* optimization processes of the DPC compiler is accomplished by comparing the execution times of four kinds of codes for several simple, real applications. A significant result obtained in this evaluation is that, the execution times of codes provided by the DPC compiler are, on average, only 13% greater than that of hand-written optimized codes, which indicates the feasibility of the DPC compiler with its optimization processes in practical applications.


## ACKNOWLEDGEMENTS

The authors would like to thank Masaaki Takizawa from NEC Scientific Information System Development, Ltd. for his immense contribution to the construction of the DPC compiler, and together Manager T. Temma and many others at the C&C Information Tech. Res. Labs. at NEC who contributed both directly and indirectly to the completion of this work.


## References


[1] Temma et al : Data Flow Processor Chip for Image Processing, *IEEE Trans. Electron Devices*, Vol.ED-32,1985,pp.1784-1791.

[2] M. Beck, R. Johnson, R. and K. Pingali : From Control Flow to Dataflow, *Journal of Parallel and Distributed Computing*, No.12,1991,pp.118-129.

[3] S. Sekiguchi, T. Shimada, and K. Hiraki : A Decision Principle of Switch Nodes in Parallel Language DFC, *Trans. of Information Processing Society of Japan*, Vol.31, No.10,1990,pp.1454-1462.

[4] R. A. Ballance, A. B. Maccabe, and K. J. Ottenstein : The Program Dependence Web: A Representation Supporting Control-,Data-, and Demand-Driven Interpretation of Imperative Languages, *Proc. of ACM SIGPLAN'90 Conf. on PLDI*, 1990,pp.257-271.

[5] A. H. Veen and R. van den Born : The RC Compiler for the DTN Dataflow Computer,*Journal of Parallel and Distributed Computing*, Vol.10, No.4, 1990, pp.319-332.

[6] S. Kyo, S. Sekiguchi and M. Sato : Data Stream Control Optimization in Dataflow Architecutures,*Proc. of the 7th ACM Int. Conf. on Supercomputing*, July, 1993.

[7] J. Ferrante, K. J. Ottenstein and J. D. Warren: The Program Dependence Graph and Its Use in Optimization, *ACM Trans. on Programming Language and Systems*, Vol.9, No.3,July 1987, pp.319-349.

[8] Arvind, R. Nikhil : Executing a Program on the MIT Tagged-Token Dataflow Architecture, *IEEE Trans. on Computers*, Vol.39, No.3, 1990,pp.300-318.

[9] S. Kyo, M. Mizoguchi : The Optimization of Flow Graphs for Dataflow Computers — Dataflow pipelining —, *Proc. of the 45th Annual Meeting of IPSJ*, Vol.6, 2L-7, Oct 1992,pp.123-124.

[10] M. Lam : Software Pipelining: An Effective scheduling Technique for VLIW Machines, *Proc. of ACM SIGPLAN'88 Conf. on PLDI*,1988, pp.318-328.

[11] A. Aiken, A. Nicolau : Optimal Loop Parallelization, *Proc. of ACM SIGPLAN'88 Conf. on PLDI*, 1988, pp.308-317.

# Increasing Instruction-level Parallelism through Multi-way Branching

Soo-Mook Moon (moon@cs.umd.edu)
IBM Thomas J. Watson Research Center
P.O.BOX 218, Yorktown Heights, NY 10598

## Abstract

*Sequential execution of conditional branches in non-numerical code limits the exploitation of instruction-level parallelism (ILP). In order to cope with this limitation, exploitation of parallelism must be extended to concurrent execution of data and branch instructions, as well as parallel execution of multiple branches in a single cycle. This paper introduces a new representation of concurrency, and describes an empirical study aimed to evaluate the performance improvement that can be expected from it. The results indicate that, as more data instructions are executed in parallel, the performance benefits arising from the proposed representation become quite notable.*

## 1 Introduction

Very long instruction word (VLIW) and superscalar machines derive their performance advantages mainly from parallel execution of *data* instructions (e.g. ALU and memory instructions) within a single instruction cycle. After data instructions are executed in parallel, sequential execution of *branch* instructions takes place. Such a sequential execution might become a bottleneck for increasing performance, due to the high frequency of conditional branches in non-numerical code. Therefore, in order to achieve better performance, *branch parallelism* must be exploited by concurrent execution of data and branch instructions, and by parallel execution of multiple branches in a single cycle.

Concurrent execution of data and branch instructions has already been allowed in some superscalar machines. For example, the IBM Risc System/6000 can overlap the execution of a conditional branch with a few data instructions [1]. In addition, *conditional execution* is widely used to hide the execution of a conditional branch, so that an instruction is executed based on a value stored in a condition register [2]. Parallel execution of multiple branch instructions means a mechanism to specify a number of "conditions" and "targets", so that the program counter value is selected from among the targets depending on a combination of the conditions. This mechanism has been known as *multi-way branching* in the VLIW context [3, 4, 5]. Most VLIW machines, thus, allow the concurrent execution of multi-way branching and data instructions [4, 6].

Even though concurrent execution and multi-way branching have already been used as described above, there have been fewer results reported on the performance improvements to be expected from employing these techniques. In this paper, we introduce a generalized representation of concurrency called a *tree* representation, and perform a comprehensive empirical study to evalute the tree representation. This study is based on the *selective scheduling* compiler in [7], and thus, the results described here characterize performance improvements for VLIW and statically scheduled superscalar machines. The rest of the paper is organized as follows. Section 2 describes our approach to concurrent execution and multi-way branching and makes comparison with other approaches. Section 3 describes our experiments and the results are described in Section 4. Conclusion follows in Section 5.

## 2 Our Approach

### 2.1 Exploiting Branch Parallelism: Static vs. Dynamic Scheduling

Branch parallelism has been exploited in both dynamic scheduling (run-time) and static scheduling (compile-time). Dynamic scheduling usually resorts to branch prediction and allows speculative execution of data instructions at one target of predicted branches. However, as more data instructions are issued in parallel, with increased degree of speculation, the penalty for mispredicted branches becomes too large. Therefore, the limiting factor in increasing performance is the complexity of the dynamic scheduling hardware to issue/execute/commit independent instructions.

On the other hand, static scheduling is advantageous in exploiting branch parallelism, since the compiler can rearrange code so that independent instructions including branches are grouped together. If the target machine can execute all instructions in a group during a single cycle, the execution of conditional branches will no longer be a bottleneck to increasing performance. The resulting performance of static scheduling, therefore, depends on how the compiler group independent branch and data instructions together and on how to execute them concurrently.

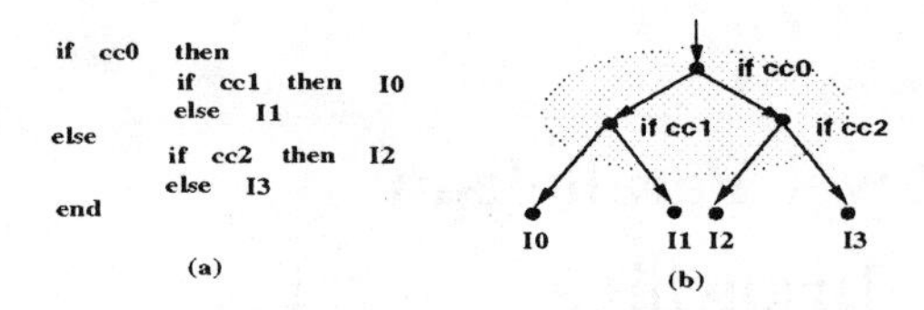


Figure 1: An example code segment scheduled in a group and its corresponding condition tree

## 2.2 Multi-way Branching: Generalized vs. Linear

Let us first consider multi-way branching, which is statically grouping independent branches and executing them in a single cycle. Static scheduling techniques fall into two categories: those that allow code motions across only one path of a branch based on branch probability [6, 8, 9] and those that allow code motions across both paths of a branch [7, 10, 11].

Static branch prediction is useful in numerical code since most branch outcomes are predictable. The usefulness of branch prediction in non-numerical code, however, is quite controversial. There have been results that static branch prediction on non-numerical benchmarks is shown to be accurate over 85 % [8, 9]. However, these figures include the loop control branch outcomes, and most of other branches are still unpredictable, such as `compare-and-branch` in sorting. In general, it is advantageous to allow code motions across all execution paths, and branch probability can be used as an option to distinguish priorities of speculative code motions. The selective scheduling takes this approach.

These optimization techniques may render the control structure of the executable code quite different from that programmed. If code motions are allowed across only one target of a branch, those conditional branches grouped together form a *linear condition tree* [3] (a tree which is skewed to the left (right), where only the left (right) offspring of a node can have further offsprings). In contrast, if code motions are allowed across all execution paths when a group of independent instructions is created, those conditional branches that are scheduled in the group form an arbitrary condition tree. The evaluation of this control structure of tree in a single cycle forms our *generalized multi-way branch* problem, which is decide the next target in a single cycle based on the truth values of the tree nodes.

For example, consider a code segment scheduled in a group of instructions and its corresponding condition tree of four-way branching as shown in Figure 1. Before executing this group, condition registes cc0, cc1 and cc2 have been set as a result of previously executed Boolean or comparative operations. A particular set of values for the condition registers determines which target instruction is executed next. This entire tree is represented as a multi-way branch instruction, and it can be either included in a long word (VLIW), or grouped together with independent data instructions (superscalar).

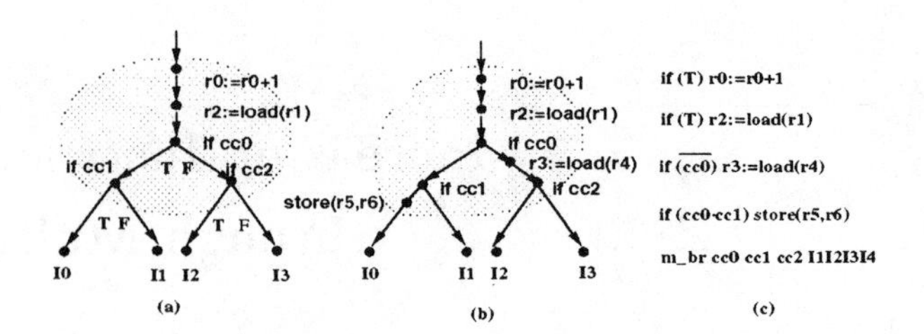


Figure 2: Concurrent execution

## 2.3 Concurrent Execution: Conditional vs. Independent

Let us now address the issue of executing data instructions concurrently with a multi-way branch instruction, in a single cycle. The solution depends on the manner how the group of independent instructions is created. In Figure 2 (a), for example, a group of independent instructions is created where independent data instructions are at the root of the condition tree. The execution of the group in a single cycle is straightforward in both VLIW and superscalar machines; data and multi-way branch instructions are executed *independently*. The compiler can generate this group on the control flow graph by scheduling data instructions first and then scheduling branch instructions.

A more advanced form of concurrent execution consists of allowing data instructions to be scheduled at any edges of the condition tree, as shown in Figure 2 (b). This representation can minimize the number of cycles required to execute a code, due to its high degree of parallelism. Namely, a machine that incorporates this tree representation can *conditionally* execute operations from a branch target before even branching to this target, and thus gaining extra cycles. However, the execution of the group requires architectural support of conditional execution. A VLIW machine can easily incorporate conditional execution due to its tightly coupled execution mode [4]. In contrast, conditional execution in superscalar is not straightforward. One solution is to extend the support of *predicated execution*.

Predicated execution or guarded execution refers to including a condition register (predicate) as a source operand of a data instruction. When the predicate is TRUE, the data instruction is executed normally; when the predicate is FALSE, the instruction is treated as a *no_op*. Predicated execution allows the compiler to remove many conditional branches in the code by converting control dependence to data dependence. This technique is called *if-conversion* and has already been used for vectorization of loops with conditional branches, or to facilitate code scheduling in both numerical and non-numerical code [2].

Our approach consists of extending this technique by including more than one predicate in a data instruction, so that the whole group of independent instructions is converted into a sequence of predicated data instructions, along with a multi-way branch instruction. For example, the group in Figure 2 (b) can be converted into the straight line code in Figure 2 (c).

# 3 Experiments

We experimented in the VLIW environment that is shown in a companion paper [12]. The tree representation of conditional execution and generalized multi-way branching is experimented through the *tree* VLIW instruction [7]. Our machine model is based on parametric resource constraints of $n$ ALU and $m$-way branching. We have evaluated four machine models, with $n$ = 2, 4, and 8,16, with different branching degree values of $m$. For each machine model, we also evaluated the case of sequential branch execution, that executes each branch operation alone.

# 4 Results

## 4.1 Sequential Branch Execution

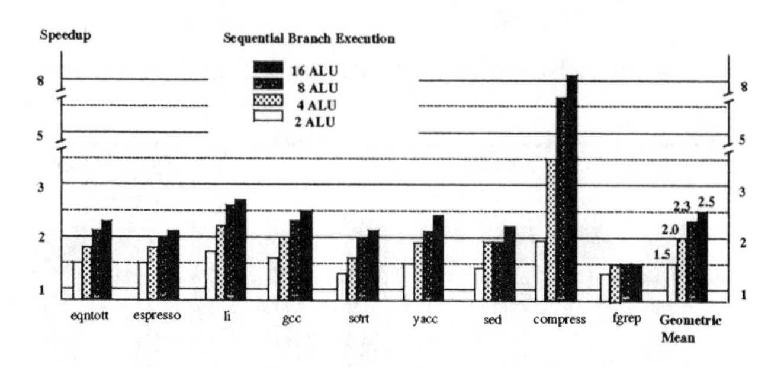


Figure 3: Speedup with sequential branch execution

The first experiment examines the performance of the four machine models when each branch operation is executed sequentially; branch parallelism is not exploited at all and performance relies only on data operation parallelism. The code for this experiment is generated by the selective scheduling compiler as follows: if the starting operation scheduled in a group is a data operation, the group is scheduled only by data operations; otherwise, it is scheduled by a single conditional branch. Therefore, a tree instruction is composed of either data operations or a single branch.

Figure 3 shows the speedup of the four machine models (four vertical bars for each benchmark) and the corresponding geometric mean for each one. The graph shows that in most benchmarks the speedup does not increase significantly, even though the number of ALUs is doubled in each step. The only exception is `compress`, whose performance is not limited by branch parallelism because its dynamic frequency of conditional branches is smaller (3 %) than the other benchmarks. This experiment indicates that sequential execution of conditional branches becomes a limiting bottleneck in increasing performance.

## 4.2 Parallel Branch Execution

Next, we examine the effect of concurrent execution of branch and data operations with varying maximum branching degree. For large-resource machines such as

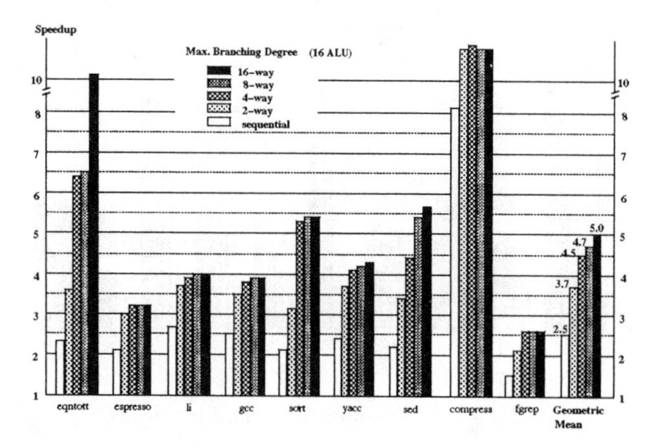


Figure 4: Speedup of `16-ALU` machines

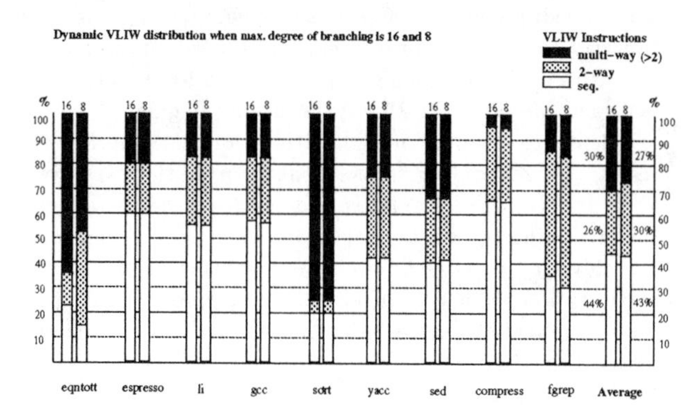


Figure 5: Dynamic branch frequency of `16-ALU` code

`16-ALU` and `8-ALU`, we evaluate four branching degrees ($m$ = 2, 4, and 8, 16). For small-resource machines such as `4-ALU` and `2-ALU`, we evaluate two branching degrees ($m$ = 2, 4), because higher branching degrees might be an overkill for small-resource machines. We also measure the frequency of multi-way branch instructions in some of the experiments, to understand how many times multi-way branching really occurs.

### 4.2.1 Large-Resource Machines

Figure 4 shows the speedup of the `16-ALU` machine with different branching degrees. The speedup of sequential branch execution is also included for comparison. There are five bars in each benchmark that correspond to the four branching degrees, and the case of sequential branch execution. Again, the geometric mean of speedup is included. For example, the conditional execution with two-way branching increases the speedup of sequential branch execution from 2.5 to 3.7. This means that the `16-ALU` machine increase its performance by 50 %, when a single conditional branch can be scheduled in a group. The only architectural support needed in this case is the con-

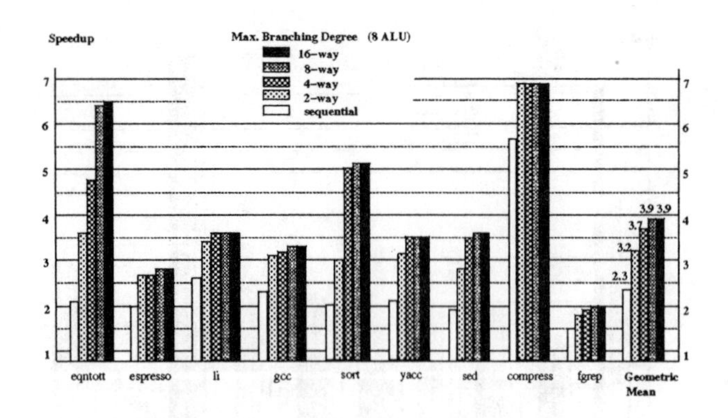


Figure 6: Speedup of 8-ALU machines

ventional predicated execution[1] on the two-way branching. This speedup can also be increased to 4.5 when going from two-way to four-way branching, which is scheduling maximum three branches in a group. Finally, a five fold speedup is achievable with the 16-way branching degree.

There are a couple of explanations for this experimental result. By scheduling branches as soon as they are available, unnecessary speculative code motions can be reduced, thus reducing the wastage of ALUs. Also, some inner loops are software pipelined into a tree instruction of one cycle/iteration with a high branching degree. If the maximum branching degree is smaller than this one, the tree instruction should be split into two or three cycle/iteration tree instructions. This actually happens in `eqntott`[2]. Figure 5 shows the frequency of multi-way branch instructions in this experiment, when the maximum branching degree is constrained to 16 and 8.

We evaluated the same branching degrees in the 8-ALU machine, and the corresponding speedup is shown in Figure 6. We have observed similar speedup increase with conditional execution and multi-way branching with higher branching degrees. However, the 8-ALU machine does not show any speedup increase from 8-way branching to 16-way branching. This indicates that the branching degree should be balanced with the number of ALUs.

### 4.2.2 Small-Resource Machines

Figure 7 shows the speedups of 4-ALU and 2-ALU machines with four-way and two-way branching degrees. Again, the speedup of sequential branch execution is included for comparison. The speedup of the 4-ALU machine improves by 20 % when data instructions are executed concurrently with a single conditional branch (two-way branching). Moreover, the four-way branching capability improves this performance by another 20 %. The 2-ALU machine can also

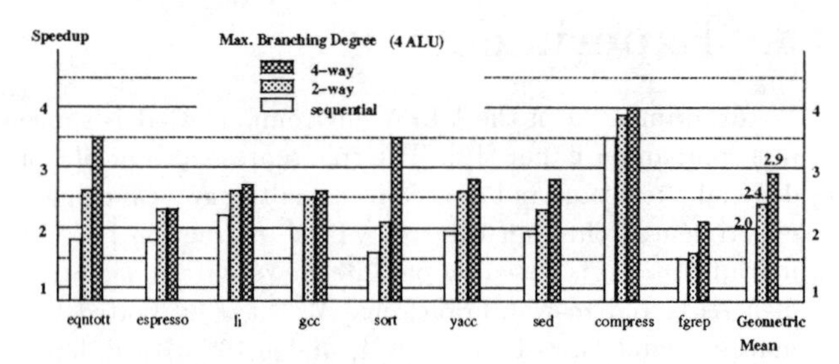


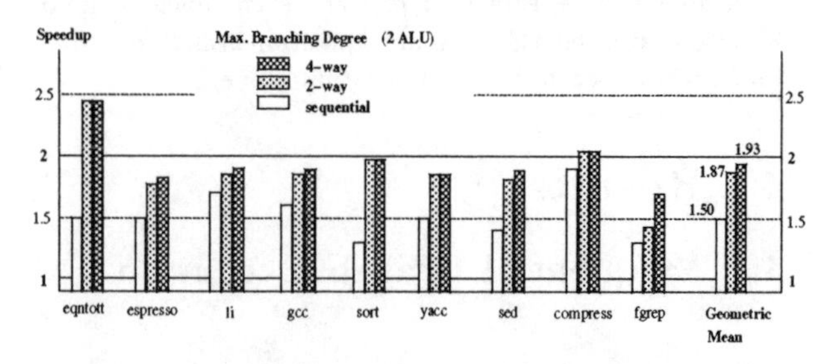


Figure 7: Speedup of 4-ALU and 2-ALU machines

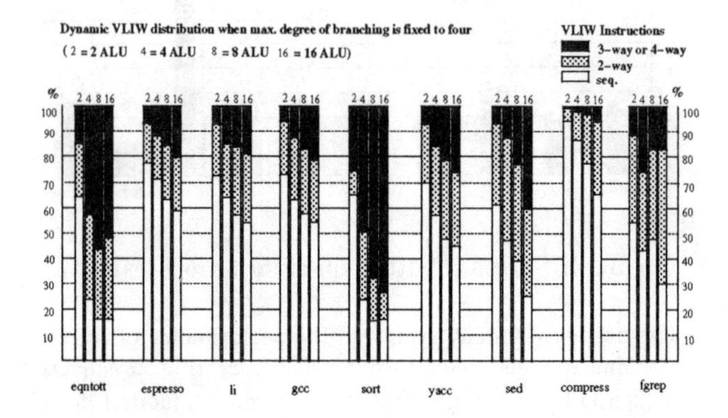


Figure 8: Dynamic branch frequency with the fixed branching degree of four

benefit from conditional execution with two-way branching, as shown in Figure 7. However, there is no significant speedup increase when going from two-way to four-way branching, compared to other machine models. This is mainly due to the limited number of code motions possible, which is not enough to cluster independent conditional branches together in a tree instruction. This is also confirmed by the graph shown in Figure 8, which compares the dynamic branching frequencies of the four machine models when the maximum degree of branching is fixed to four. In many of the benchmarks, the difference of multi-way branching frequencies when going from 2-ALU to 4-ALU is more than that of other cases.

## 4.3 Discussion

The graph in Figure 9 summarizes the experimental results by comparing the geometric mean of speedup based on different branching degrees. We can observe a few prop-

[1] It should be noted that our if-conversion is performed **after** code scheduling, while conventional if-conversion is performed **before** code scheduling to remove conditional branches [2]. Therefore, all speculative code motions can be performed on unpredicated RISC operations.

[2] The frequent inner loop in `cmppt` is parallelized into a highly compacted tree instruction with 11-way branching.

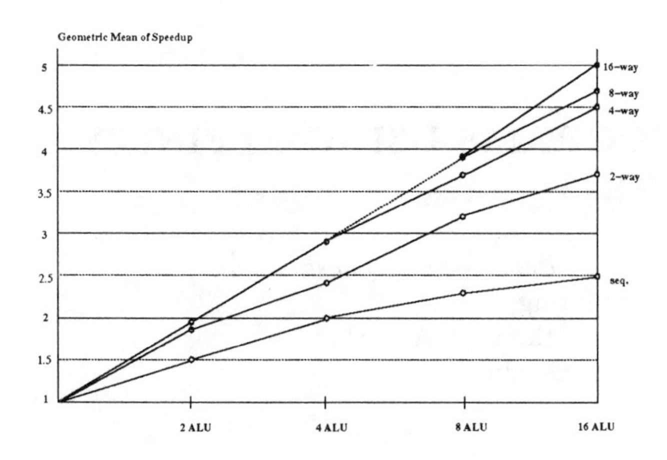


Figure 9: Speedup of different branch degrees with conditional execution and multi-way branching

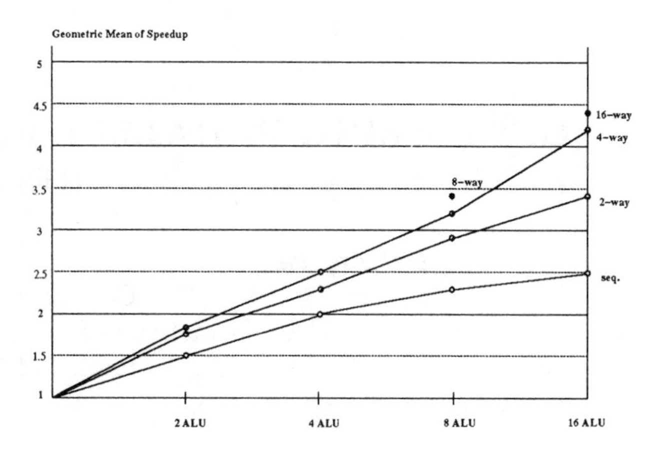


Figure 10: Speedup of different branch degrees only with multi-way branching (Independent Execution)

erties in all machine models. First, there are two large speedup increases: when going from sequential branch execution to conditional execution with two-way branching, and when going from two-way to four-way branching capabilities. Second, the speedup increase itself gets wider in larger resource machines compared to smaller resource machines. The two observations indicate that, as more data operations are executed in parallel, the performance gain of conditional execution and multi-way branching becomes quite notable. Moreover, most of the performance improvement even in large resource machines is achievable only through four-way branching capabilities, which is relatively simple to implement in the CPU.

We performed another set of preliminary[3] experiments to examine the effect of conditional execution. Figure 10 shows the speedup of independent execution depicted in Section 2.3, which employs only the multi-way branching capabilities. We experienced around 10 % performance degradation.

Although the results presented above are not completely general to determine the optimal degree of branching in VLIW or superscalar machines, they do demonstrate the value of multi-way branch capability in exploiting irregular ILP. Multi-way branching has already been shown to be implemented efficiently in [4, 5].

# 5 Conclusion

In this paper, we have addressed the issue of increasing instruction-level parallelism through conditional execution and multi-way branching. Our experimental results show that sequential execution of conditional branches becomes a worst limiting constraint in increasing speedup. In contrast, the proposed tree representation can relax this constraint significantly. Consequently, superscalar and VLIW are required to employ the representation.

[3] Geometric means are obtained without the result of gcc. Also, the machines of `8 ALU 16-way` and `16 ALU 8-way` were not evaulated due to time constraints.

# References

[1] IBM. A Special Issue on IBM RISC System/6000. *IBM Journal of Research and Development*, 34(1), Jan 1990.

[2] S. Mahlke et al. Effective Compiler Support for Predicated Execution Using the Hyperblock. In *Proceedings of the Micro-25*, pages 45–54, Dec 1992.

[3] J. Fisher. VLIW Architecture and the ELI-512. In *Proceedings of the 10th ISCA*, pages 140–150, 1983.

[4] K. Ebcioğlu. Some Design Ideas for a VLIW Architecture for Sequential Natured Software. In M. Cosnard et al., editor, *Parallel Processing*, pages 3–21. North Holland, April 1988.

[5] S.-M. Moon, S. Carson, and A. Agrawala. Hardware Implementation of a General Multi-way Jump Mechanism. In *Proceedings of the Micro-23*, pages 38–45, 1990.

[6] J. Ellis. *Bulldog: A Compiler for VLIW Architecture.* PhD thesis, Yale University, Feb 1985.

[7] S.-M. Moon and K. Ebcioğlu. An Efficient Resource-Constrained Global Scheduling Technique for Superscalar and VLIW processors. In *Proceedings of the Micro-25*, pages 55–71, Dec 1992.

[8] P. Chang et al. IMPACT: An Architectural Framework for Multiple-Instruction-Issue Processors. In *Proceedings of the 18th ISCA*, pages 266–275, 1991.

[9] M. Smith, M. Horowitz, and M. Lam. Efficient Superscalar Performance Through Boosting. In *Proceedings of the ASPLOS-5*, pages 248–259, 1992.

[10] A. Aiken and A. Nicolau. A Development Environment for Horizontal Microcode. *IEEE Transactions on Software Engineering*, 14(5):584–594, May 1988.

[11] D. Bernstein and M. Rodeh. Global Instruction Scheduling for Superscalar Machines. In *Proceedings of the SIGPLAN 1991 Conference*, pages 241–255, 1991.

[12] S.-M. Moon. On Performance and Efficiency of VLIW and Superscalar. In *Proceedings of 1993 ICPP*, Aug 1993.

# OPTIMIZING PARALLEL PROGRAMS USING AFFINITY REGIONS

*Bill Appelbe* *Balakrishnan Lakshmanan*
College of Computing,
Georgia Institue of Technology, Atlanta, GA 30332
e-mail: bill@cc.gatech.edu

Abstract – *Affinity regions ensure that a shared processor scheduling, mapping iterations to processors, is used in consecutive parallel loop nests. Using affinity regions can significantly improve locality of reference and hence the performance. This paper discusses the use and effectiveness of affinity regions. We introduce a general algorithm, based upon dependence analysis, to determine the feasibility and profitability of affinity regions. The algorithm is being implemented in PAT[4, 6], an interactive parallelization tool.*

## Introduction

The most common parallel construct is the PARALLEL LOOP. On shared memory multiprocessors, the access times for data of any category, PRIVATE, SHARED or possibly REDUCTION, is essentially the same (ignoring the cache effects[5]). However, in some multiprocessors like KSR-1, the memory is collocated with the processors so that access to local shared data is faster than access to non-local shared data(on another processor), although there is no explicit communication[1, 2].

To maximize performance on KSR-1 it is necessary to improve locality of reference. On the KSR-1, access to memory can be optimized across loop nests.

Many parallel programs consist of sequences of PARALLEL LOOPS accessing common data arrays with regular access patterns and overlapping references. The scheduling of loops can be either pre-scheduling or self-scheduling. Although self-scheduling results in proper load balancing, pre-scheduling, with proper knowledge of the program semantics can result in better locality of reference[8]. An affinity region is a sequence of loop nests which share a schedule[3] (either pre-scheduled or self-scheduled). The KSR-1 provides support for affinity regions by using compiler directives with some restrictions (described below).

## Related Primitives

In practice, aside from affinity regions, there are two other ways of achieving data locality between loop nests - *LOOP FUSION* and PARALLEL REGIONS.

Loop fusion is a transformation that combines two adjacent loops into one loop, provided the two loops have no serial fusion preventing dependences and parallel fusion preventing dependences in the case of DO loops and DOALL loops respectively[8]. The obvious result from fusing the two loops is a reduction in the overhead caused by loop control. Also it has been shown that loop fusion can be used to reduce traffic in a memory hierarchy[9]. Since fusion preventing dependences are present in manyloops, loop fusion frequently cannot be used in practice.

Parallel regions are SPMD style parallelization[8] in which all code within the region is executed in parallel by all processors except when processors explicitly synchronize on loop iterations. The parallel region construct necessitates the programmer using explicit low level primitives and non-portable calculations to obtain the proper mapping of the iterations to the processors or threads. In certain cases, even if a thread is mapped to the correct iteration, the thread may not execute in the correct processor due to automatic loadbalancing[2, 1]. Merging loops into parallel regions also means that any intervening serial code between the two loops will be executed redundantly by all processors (or a SERIAL SECTION construct used). Explicit barriers are necessary if there are loop carried dependences between the two loops.

## Affinity Regions

Affinity regions are an alternative approach that can be used to specify that two loop nests should use the same processor scheduling. The compiler and operating system are free to choose any scheduling strategy for the first loop nest in the region. This implicitly enforces a schedule for the subsequent loop nests constrained by the mapping between the indices in the loop nests as specified in the directive.

Program-1, that is shown below, uses the KSR Fortran syntax for affinity regions and tiling. KSR Parallel Fortran specifies Parallel Loop Nests using the PTILE directive whose arguments include:

- The parallel loop indices (I and K in both loop nests above).
- The size and shape of the tiles allocated to threads.
- If the loop nest is a member of an enclosing affinity region.

```
         DO 10 J = 1, N
            DO 20 I = 1, N
               A(I,J) = 0
   20       CONTINUE
   10       CONTINUE
C*KSR* AFFINITY REGION(I:1,N, K:1,N)
         DO 30 J = 1, N
C*KSR* PTILE(I,K,TILESIZE=(I:1,K:N),
     &                       AFFMEMBER=1)
           DO 40 I = 1, N
              DO 50 K = 1, N
                A(I,J) = A(I,J) + B(K,I)*C(K,J)
   50         CONTINUE
   40      CONTINUE
   30    CONTINUE
C*KSR* PTILE(I,K,TILESIZE=(I:1,K:N),
     &                       AFFMEMBER=1)
         DO 60 I = 1, N
            DO 70 K = 1, N
               SUM(I) = SUM(I) + B(K,I)
   70       CONTINUE
   60    CONTINUE
C*KSR* END AFFINITY REGION
```

Program 1: Affinity Regions

C*KSR*AFFINITY REGION is a directive which defines the beginning of an affinity region and the properties of the affinity region. Its arguments include:

- The index variables (I and K) along which the tile families in the affinity region are to be tiled, and the maximum bounds along each index (1:N for both I and K).

KSR Parallel Fortran requires that the loops in the affinity region use the same loop indices, but does not require that the loop bounds are the same (though they must overlap). Either the affinity region or the first loop in the affinity region (as in the example above), can specify the scheduling strategy and tile sizes.

## Feasibility of Affinity Regions

The goal of inserting affinity regions is to ensure that the same processors access the same array elements in different loop nests. To use affinity regions effectively we need to develop algorithms to determine both the feasibility and profitability of introducing affinity regions. We first establish a set of criteria that has to be satisfied by the loop nests to be eligible candidates for inclusion in affinity region. Our algorithm assumes that the following information is available:

*Loop nest information* We assume that all loops are `DO` loops, or `PARALLEL DO` loops, with a loop L at a nesting level of $k$ having an index variable $I_k$, and lower bound, upper bound and stride of $LB(I_k)$, $UB(I_k)$ and $STRIDE(I_k)$ respectively. In a loop nest $\mathcal{L}$, of depth $k$, the set of index variables is denoted by $\mathcal{L}_{index} = (I_1, I_2, I_3, ..., I_k)$.

*Loop array access summary* For each loop level a list of each array variable accessed, whose subscripts are a function of loop indices.

*Dependences* We assume that for each variable access (definition or use), in a loop a list of the uses and the definitions reaching this access is available.

We assume initially that all parallel loops have been identified. Placing two loop nests in the same affinity region may require only choosing to parallelize loop nests in a subset of the dimensions in which they can be parallel.

There is no gain in placing two loops in an affinity region unless they share access to atleast one subscripted variable. Also, it is necessary that these shared variables are not defined in the intervening code. Therefore we define the criteria for *affinity adjacent loops.*

Two loops nests $\mathcal{L}^1$ and $\mathcal{L}^2$ are affinity adjacent if:

**A** There is a dependence from a shared array variable $A(\tilde{S}^1)$ in $\mathcal{L}^1$ to an access $A(\tilde{S}^2)$ in $\mathcal{L}^2$, such that all their subscripts $\tilde{S}^1$ and $\tilde{S}^2$ are not loop invariants. $\tilde{S}^1$ and $\tilde{S}^2$ are a list of subscript expressions. If $A$ is a m-dimensional array, then the $j$th subscript expression in $\tilde{S}^1$ and $\tilde{S}^2$ is denoted by $s_j^1$ and $s_j^2$ respectively, and therefore $\tilde{S}^1 = s_1^1, s_2^1, ...s_m^1$ and $\tilde{S}^2 = s_1^2, s_2^2, ...s_m^2$

Even if two loops are affinity adjacent, it may not be feasible to place them in an affinity region. It must be possible to align the schedules of the parallel loops so that corresponding schedules in the loop nests access the same partition of the shared array $A$.

Let the set of parallel loops in $\mathcal{L}^1$ and $\mathcal{L}^2$ be $\mathcal{L}^1_{Par-index}$ and $\mathcal{L}^2_{Par-index}$ respectively. The loops must satisfy two more criteria:

**B1** The loop nests must be parallelizable at one or more loop levels, that is $\mathcal{L}^1_{Par-index}$ and $\mathcal{L}^2_{Par-index}$ must be non-empty.

**B2** There must be a mapping between a subset of the parallel index sets, $\mathcal{L}_{MAP} : \mathcal{L}^1_{Par-index} \mapsto \mathcal{L}^2_{Par-index}$ such that

**B2.1** The mapping is one-to-one, that is $\mathcal{L}_{MAP}(I_{k1}) \equiv \mathcal{L}_{MAP}(I_{k2})$ implies $k1 \equiv k2$.

**B2.2** The mapping is between indices with overlapping bounds but with the same stride, that is,

1. $[LB(I_k)..UB(I_k)]$ and $[LB(\mathcal{L}_{MAP}(I_k))..UB(\mathcal{L}_{MAP}(I_k))]$ must intersect.
2. $STRIDE(I_k) = STRIDE(\mathcal{L}_{MAP}(I_k))$.

Although criteria B2.2 is mandatory for the correct execution on the KSR-1, in theory, the range of the loop indices that have a mapping in the region need not overlap. The mapping $\mathcal{L}_{MAP}$ may not use some of the levels of parallelism in either the first or the second loop nests or both. However, a shared array $A$

(from criteria **A**) must be within the parallel loops of $\mathcal{L}_{MAP}$. Finally for each dimension of the shared array $A$, the subscript expressions under the loop index mapping must either be the same or be independent of the parallel loop indices. These two criteria can be summarized as:

**C1** The subscript $\tilde{S}^1$ must not be invariant, or independent, of the parallel loop indices in $(Domain(\mathcal{L}_{MAP}))$.

**C2** For each subscript dimension $j$ of $A$:
Either $s_j^2 \equiv \mathcal{L}_{MAP}(s_j^1)$, or $s_j^1$ does not contain any subscripts in $Domain(\mathcal{L}_{MAP})$ and $s_j^2$ does not contain any subscripts in $Range(\mathcal{L}_{MAP})$.

The above criteria is being incorporated into an algorithm which decides the feasibility of placing two loops in an affinity region [3].

Affinity regions can be nested. Affinity regions can also overlap. In principle, it should be possible and profitable in a sequence of loop nests $\mathcal{L}^1, \mathcal{L}^2, \mathcal{L}^3, \mathcal{L}^4$ to place the first and third loop nests in one affinity region and the second and fourth loop nests in another, the only requirement being that the intervening loop nest should not to the shared subscripted variable of the other region. But, such cases are not supported in the KSR (although the nests might be exchanged if there were no dependences). Thus, once two loop nests are placed in an affinity region, there are two cases to be considered:

**Intervening loop nests** These loop nests could be placed in the same affinity region if they share the same mapping indices or could be placed in a separate affinity region if they do not write to the common subscripted variable.

**Non-intervening or adjacent loop nests** The affinity region can be extended to include these loop nests provided they also have similar index set mappings or they could be put in another affinity region. The extension of the above algorithm to incorporate such loop nests is fairly simple.

## Profitability

The algorithm described above can be used to determine if two loop nests can be placed in an affinity region. Estimating the profitability of affinity regions is relatively straight forward. A simple estimator is the number of local references generated by introducing the affinity region. If this is modeled as the profitability measure, then this gives us the upper bound on the profitability since under the worst case assumptions, without an affinity region, none of the accesses for shared variables in the second loop nest are local. The profitability estimate is also constrained by the available number of processors and the number of threads of control, or tiles in the case of KSR, that are going to be involved. Also, there is potential gain due to the avoidance of synchronization overhead in scheduling the loop nests (except the first one) which outweighs the cost of saving the schedule from the first loop nest for use in subsequent loop nests in the affinity region. We intend to incorporate these in a sophisticated performance model for the KSR later.

## Transformations for Affinity

There may be more than one possible affinity region to encompass two loop nests. This in turn implies that the affinity regions differ in the common subscripted variable being used. It would be profitable if the affinity regions can be merged in some way. The merging of affinity regions can be made possible by the use of transformations. Consider, for example, program-2.

```
PARALLEL LOOP I = 1,N
  A(I) = FUNC1(I)
  B(I) = FUNC2(I)
ENDLOOP
(SERIAL CODE)
PARALLEL LOOP J = 1,N
  A(J) = FUNC3(J)
  B(N-J) = FUNC4(J)
ENDLOOP
```

Program-2:Need for Transformation

In program-2, if the two loops are placed in an affinity region considering A as the common variable, then the locality of references to B is lost. On the other hand, if B is considered, then the locality of A is lost. However, if the J loop is transformed as shown in program-3

```
PARALLEL LOOP J = 1,N
  A(J) = FUNC3(J)
  B(J) = FUNC4(N-J)
ENDLOOP
```

Program-3:Transformed J Loop

the data locality of accesses to both A and B is preserved resulting in the merge of the two possible affinity regions for program-2

So far we have assumed that the subscripts are linear functions of loop indices. Affinity regions could however be inserted even if the subscripts are non-linear. The simple criteria for considering such cases is that the subscript expressions in the corresponding dimensions to be identical after the application of the index mapping to the indices.

In general, inclusion of affinity regions is better done after all other transformations to maximize parallelization and data localization. The two main reasons for this are:

- The application of transformations after the inclusion of affinity regions may prove the affinity regions not useful.
- Also, for the profitability analysis to precisely measure the trade-off in obtaining better locality of reference at the cost of fewer levels of parallelism, the transformations should have already been applied to the program to maximize parallelism.

| Dimension (N) | Without Affinity | With Affinity |
|---|---|---|
| 15 | 78.8 | 21.4 |
| 20 | 100.1 | 24.15 |
| 30 | 132.75 | 38.40 |
| 90 | 364.7 | 167.2 |

Table 1: Results

## Results

The code in program-1 was run on a 32 node KSR. The program was run varying the dimension N of the arrays used. For each N, time spent on the loop nests were collected both with and without affinity regions. The results of the experiments are shown in table 1.

## Related Work

In the case of pure shared memory multiprocessors, the problem of data locality occurs due to the difference in the overhead associated with accessing the cache and main memory. Locality of reference, and hence performance, can be improved by minimizing the number of cache lines loaded from memory[5]. However, because of the relatively small size of the cache, this optimization is done on a per-loop nest basis. It has also been shown that a loop scheduling algorithm that attempts to simultaneously balance the workload, minimize synchronization and co- locate loop iterations with the necessary data does significantly better on modern shared memory multiprocessors than those that ignore the location of data when assigning iterations to processors[10].

## Conclusion

Affinity regions are more general and effective than either loop fusion or parallel regions for parallelizing loop nests on machines with non-uniform memory access. Using affinity region directives on the KSR-1 obtained a speedup of a factor between 2 and 3 for the example program mentioned. We have come up with a general algorithm and the criteria necessary to profitably insert affinity regions encompassing one or more loop nests to increase the locality of reference of parallel programs. Further work is needed in improving the accuracy of the estimates for profitability of affinity regions and reducing the complexity of the algorithm. The algorithm is being implemented in PAT, and tested upon suites of Benchmarks including the Perfect Benchmarks. We also intend to generalize the results to distributed memory architectures.

## References


[1] *KSR Fortran Programming.* Kendall Square Research, Waltham, Massachusetts, 1991.

[2] *KSR Parallel Programming.* Kendall Square Research, Waltham, Massachusetts, 1991.

[3] Appelbe, B., and Lakshmanan, B. Optimizing parallel programs using affinity regions. Tech. Rep. GIT-ICS-92/59, Georgia Institute of Technology, Nov. 1992.

[4] Appelbe, B., McDowell, C., and Smith, K. Start/Pat: A parallel-programming toolkit. *IEEE Software 6*, 4 (July 1989), 29–38.

[5] Kennedy, K., and McKinley, K. Optimizing for parallelism and data locality. In *International Conference on Supercomputing* (July 1992), pp. 323–334.

[6] Smith, K., Appelbe, B., and Stirewalt, K. Incremental dependence analysis for interactive parallelization. In *International Conference on Supercomputing* (June 1990), pp. 330–341.

[7] Wolfe, M. *Optimizing Supercompilers for Supercomputers.* The MIT Press, Cambridge, Massachusetts, 1989.

[8] Zima, H., and Chapman, B. *Supercompilers for Parallel and Vector Computers.* ACM Press, New York, New York, 1990.

[9] Kuck et al *The structure of an advanced retargetable vectorizer.* IEEE Society Press, SilverSpring, Maryland, 1984.

[10] Markatos, E.P., and LeBlanc, T.J. *Using Processor Affinity in Loop Scheduling on Shared Memory Multiprocessors* In *International Conference on Supercomputing* (Nov 1990), pp. 104–113.

# SESSION 10B

## DATA DISTRIBUTION/PARTITIONING

# A Model for Automatic Data Partitioning *

Paul D. Hovland
Department of Computer Science
University of Illinois at Urbana-Champaign
Urbana, IL 61801
hovland@cs.uiuc.edu

Lionel M. Ni
Department of Computer Science
Michigan State University
East Lansing, MI 48824-1027
ni@cps.msu.edu

**Abstract**

In order to efficiently exploit global parallelism, it is essential to find a good way to distribute data among the processors in distributed-memory parallel computer systems. A formal technique utilizing *augmented data access descriptors* (ADADs) to determine this distribution is presented. This technique differs from previous approaches in that it views the problem of finding a good distribution as an extension of data dependence analysis. The importance of this difference is demonstrated through an explanation of how ADADs facilitate interprocedural analysis, directed loop transformations, and incremental analysis, which may lead to improvements in the efficiency of both program development and the program itself.

## 1 Introduction

The dominant variety of supercomputer today is the distributed-memory parallel computer. This dominance arises in great part from the scalability of these machines. But, programming these computers is an extremely difficult task, with the adaptation of existing programs to run on them an even more haunting challenge. The data to be used for the parallel operations must be distributed to the processor where the computation will take place and the result of the computation must be stored in its destination. Accessing variables stored in local memory requires significantly less time than accessing variables located in the memory associated with some other processor, as the data must be communicated between the processors via some sort of network.

If we wish to achieve good performance on a distributed-memory computer, we must try to minimize the communication of data from one processor to another while preserving parallelism within our computations. This problem reduces to deciding where data should be stored (that is, where it is computed) among the various processors.

In order to aid parallelizing compilers in the task of code generation for distributed-memory computers (or any computer with non-uniform memory access time), several languages have been developed that enable the programmer to include information concerning the way in which data should be distributed among the various processors [1, 2]. If a proper alignment of data is chosen, then communication costs may be minimized. In this paper, we shall use Fortran D language extension proposed by Rice University for the purpose of illustration. In Fortran D, the `DECOMPOSITION` statement indicates how an array is to be used as a frame of reference, or *template*. The `ALIGN` statement indicates how an array should be aligned with the template. The `DISTRIBUTE` statement indicates how the template should be distributed to different processors. For a more complete explanation of the syntax of this and other Fortran D instructions, see [1].

Li and Chen have proven that the general problem of data distribution on distributed-memory machines is NP-complete [3]. However, even to find a good, not necessary the best, distribution for the arrays of a complex program on a distributed-memory parallel computer may be a formidable task if the good distribution must be determined by the programmer. Instead, we would like to determine a good distribution of data in an automatic fashion. In order to assist in the automatic generation of data alignment information, several heuristic techniques have been developed to determine a good data alignment in a mechanical fashion. Li and Chen model the data alignment problem as a graph problem, and present a heuristic algorithm for solving the problem [3]. Gupta and Banerjee use a different approach, attacking the problem from the perspective of the whole program, and solving for an optimal combination of parallelism and communication costs, subject to certain constraints [4]. Both approaches have their merit and may be better suited for many applications than our technique to be described later. However, the approach we examine has the advantage that it is very simple (and thus suitable for use in a compiler, which must process a program in a reasonable amount of time) and can treat statements, regions of programs, and entire programs in an identical manner. This flexibility is an important feature, because it enables the separate examination of sections of a program. In many cases, one alignment is appropriate for one stage of a computation,

*This work was supported in part by the NSF grants CDA-9121641 and MIP-9204066.

while a different alignment is appropriate for a later stage. Using both alignments can result in better performance than just using one alignment or the other for the entire program.

We address this problem, proposing the augmented data access descriptor as a means to determine a good distribution for arrays on a distributed memory parallel computer. We propose a formal technique utilizing *augmented data access descriptors* to compute a close approximation to this distribution. The method presented provides a good heuristic, in that it is often able to identify a communication free distribution, if one exists. Our technique differs from previous approaches in that it views the problem of finding a good distribution as an extension of data dependence analysis.

The organization of this paper is as follows. Section 2 provides a brief explanation of two subjects that are the underpinnings of augmented data access descriptors–data dependence analysis and Data Access Descriptors. The following sections present the augmented data access descriptor, specifying its format, computation, meaning, and application to modular programs. Section 5 describes the effects of various loop transformations. Section 6 concludes this paper.

# 2 Background

This section serves as a brief introduction to data dependence analysis and Data Access Descriptors. This background information is necessary for a complete understanding of augmented data access descriptors, and why they are useful.

## 2.1 Data Dependence Analysis

Instead of treating automatic data alignment as a completely new problem, we can view data alignment as an extension of data dependence analysis. Data dependence analysis is the investigation of the manner in which the execution of one statement influences the results of another. In order for statements to be executed in parallel, it is necessary that the result of each statement be independent of the results of all other statements. Data dependence may take one of three forms: flow dependence, antidependence, or output dependence.

Traditionally, parallelizing compilers have relied upon data dependence analysis to determine whether particular sections of code or iterations of a loop may be executed in parallel. If a dependence exists, then the loop is not parallelized. However, in the global parallelization scheme, it is assumed that occasionally a variable will be defined by one processor and used by another. What is most important is that the number of times a dependence occurs (thereby creating a need for communication) is minimized. Thus, data alignment may be viewed as an advanced type of data dependence analysis, through which we attempt to minimize some function, which serves as an approximation to communication costs.

## 2.2 Data Access Descriptors

In developing the Data Access Descriptor, Balasundaram takes a slightly different approach to data dependence analysis [5]. Rather than focusing upon the dependence of one statement on another, Balasundaram examines the manner in which regions of a program influence one another. This approach allows traditional statement-to-statement data dependence analysis to be unified with interprocedural data dependence analysis, which is extremely important in modular programs.

The Data Access Descriptor contains a variety of information regarding the way in which a particular variable is accessed. An extremely important aspect of the Data Access Descriptor is the simple section, a set of hyperplanes that form convex hulls that completely enclose the regions of arrays accessed by a section of a program (which may be a single statement or the entire program). By applying intersection and union operations to these descriptors, it is possible to determine whether a data dependence exists between two regions of a program.

# 3 Augmented Data Access Descriptors

We now explore a mechanism for finding a good data alignment. We define a good alignment as one that leads to a distribution that is free from communication. In cases where we cannot eliminate communication, we attempt to minimize it.

One approach to quantifying the amount of communication overhead associated with various data distributions is to extend the data access descriptor so that it describes both the sections of an array that are accessed by a particular statement and also the manner in which they are accessed. As explained in Section 2.2, the original data access descriptor consists of a set of hyperplanes forming a convex hull within which all array accesses are guaranteed to occur. This information can be very useful, but may be expensive to compute. However, because of the restrictive nature of data alignment languages like Fortran D and DINO [1, 2], it is sufficient to restrict our convex hulls to hypercubes.

We are most interested in accesses that proceed along one dimension of the array (as opposed to cases where the subscripts are coupled [6]). Such accesses are easy to identify, because they occur when one of the subscripts of the array is a function of only one of the iteration variables; e.g., `A(3*I+7)=1.0`. These accesses are also important, because they indicate that the array may be distributed across that dimension (and parallelized along that dimension) without any need for communication.

The *augmented data access descriptor* (ADAD) for an array referenced in a particular statement will consist of a $k \times n$ array of tuples, where $k$ is the level of nesting for the statement and $n$ is the number of dimensions in the array, plus an integer element, whose value is set equal to $k$ (Strictly speaking, this integer element is not necessary, as we may deduce $k$ from the number of rows in the ADAD. However, it is included to facilitate the use of a statement's degree of nesting as a tie-breaker in determining alignments. The use of this heuristic for situations in which a communication-free alignment is not possible will be discussed later). For example, in statement S1 in the following code:

```
        DO I=1, 100
          DO J=2, 50
S1:         A(I,J,I+J) = B(I,J)
          ENDDO
        ENDDO
```

the ADAD describing A would contain a $2 \times 3$ array of tuples, plus the integer 2, and the ADAD describing B would contain a $2 \times 2$ array of tuples, plus the value 2. For a $k \times n$ tuple array, each tuple is denoted by $\mathcal{D}(A, S, I, D)$, where $A$ is the array being described, $S$ is the statement with which we are concerned, $I$ is an iteration variable, and $D$ is a dimension of array $A$.

The first two members of each tuple indicate the lower and upper bounds, respectively, on the value subscript $D$ may assume (note that this value is independent of $I$). The third element gives some indication of whether parallelizing the iteration variable to which that row of the ADAD corresponds will create a need for communication if the array in question is distributed across that dimension. If this element is an asterisk (*) then communication is necessary, and the fourth element in the tuple is the empty set. If the element is a positive integer, then communication is not necessary, if the array is distributed in bands of width equal to this positive integer. In this case, the fourth element of the tuple is a set of offsets indicating which rows (or columns, depending on perspective) are referenced by that statement. This information is important for combining ADADs, a procedure to be described later.

If the third element is zero, no communication is necessary, so long as this is the only reference to the array. The value zero is used as an indicator that the subscript is a nonlinear function of the particular iteration variable. Distribution is complicated, but possible, as discussed in Section 4.4. As in the case of an asterisk, the tuple's fourth element is the empty set.

The ADADs for the example above are:

$\mathcal{D}$(A,S1)

| 2 | dimension 1 | dimension 2 | dimension 3 |
|---|---|---|---|
| I | (1,100,1,{0}) | (2,50,*,∅) | (3,150,*,∅) |
| J | (1,100,*,∅) | (2,50,1,{0}) | (3,150,*,∅) |

$\mathcal{D}$(B,S1)

| 2 | dimension 1 | dimension 2 |
|---|---|---|
| I | (1,100,1,{0}) | (2,50,*,∅) |
| J | (1,100,*,∅) | (2,50,1,{0}) |

The details of how these ADADs are computed will be discussed in the next section.

# 4 Finding a Good Alignment Using ADADs

## 4.1 Program Model

In describing ADADs and how they should be used, several assumptions are made about the programs being evaluated. One assumption is that loops have a step size of one. Such a loop is said to be normalized. Since it is possible to normalize any loop [7], this restriction is not significant. We also assume that arrays are primarily indexed by iteration variables. Constants and induction variables are also often used to index arrays. Since constants do not provide a means for distribution and induction variables can always be replaced by expressions involving iteration variables, this assumption does not impose unnecessary restrictions. The third assumption is that those statements located in the innermost loops are executed most frequently. In general, this is a reasonable assumption, especially if loops involve more than a few iterations. However, if a statement is guarded by a conditional, and the guard is usually false, then it may be executed less frequently than a statement with a smaller degree of nesting. However, since this assumption only governs our decision as to how data should be aligned when a communication-free distribution does not exist, correctness is not affected, and it is hoped that the effect on performance is minimal.

## 4.2 Computing Augmented Data Access Descriptors

The computation of ADADs is very simple. Given a particular reference to a particular array in a particular statement, for each loop within which that statement is nested, we compute a tuple, $\mathcal{D}(A, S, I, D)$, representing the dependences of a particular subscript with respect to the iteration variable corresponding to that loop, as well as the upper and lower bounds of the subscript. The upper and lower bounds, which as mentioned earlier are the first and second elements of a tuple and which we denote $\mathcal{D}(A, S, I, D)[1]$ and $\mathcal{D}(A, S, I, D)[2]$, can often be determined in a straightforward fashion using the upper and lower bounds of the iteration variables. For example, if a subscript is a monotonic function, $G$, of iteration variables $I$ and $J$, then the lower bound of the subscript, $\mathcal{D}(A, S, I, D)[1]$ is equal to the lower bound of the function:

$$\mathcal{D}(A, S, I, D)[1] = \min(G(L(I), L(J)), G(L(I), U(J)), G(U(I), L(J)), G(U(I), U(J)))$$

Similarly, the upper bound of the subscript, $\mathcal{D}(A,S,I,D)[2]$, is:

$$\mathcal{D}(A,S,I,D)[2] = \max(G(L(I),L(J)), G(L(I),U(J)), G(U(I),L(J)), G(U(I),U(J)))$$

In cases where $G$ is not monotonic, or where the bounds of the iterations are not known, a conservative assumption regarding these bounds (0 for the lower bound and positive infinity ($\infty$) for the upper bound) may be made.

The computation of the third element of the tuple may be expressed as: $\mathcal{D}(A,S,I,D)[3] =$

$$\begin{cases} m & \text{if D is a linear function of } I \text{ (i.e., } mI+b) \\ 0 & \text{if D is a nonlinear function of } I \\ * & \text{otherwise} \end{cases}$$

Similarly, we have $\mathcal{D}(A,S,I,D)[4] =$

$$\begin{cases} \{b\} & \text{if D is a linear function of } I \text{ (i.e., } mI+b) \\ \emptyset & \text{otherwise} \end{cases}$$

As an example, the augmented data access descriptor for statement S1 in the following code segment.

```
      DO 100 I=1,100
        DO 110 J=1,100
          A(I,J) = SQRT(I*J)        (S1)
110     CONTINUE
100   CONTINUE
```

$D$(A,S1)

| 2 | dimension 1 | dimension 2 |
|---|---|---|
| I | (1,100,1,{0}) | (1,100,*,∅) |
| J | (1,100,*,∅) | (1,100,1,{0}) |

This is equivalent to saying: "Statement `S1` has a nesting level of 2. With respect to iteration variable `I`, array `A` may be distributed across its first dimension, since the first subscript is a linear function ($1I+0$) of `I`, but not across its second dimension. With respect to iteration variable `J`, array `A` may be distributed across its second dimension, since the second subscript is a linear function ($1J+0$) of `J`, but not across the first dimension." This is in complete agreement with the actual syntax of S1.

### 4.3 Combining Augmented Data Access Descriptors

Combining two ADADs that describe the same array is also a rather straightforward task, as long as the convex hulls described by the the first two elements of the tuples intersect. If they do not intersect, it may be possible to avoid communication by distributing those two regions of the array in different ways. However, since the ability to do this also implies that the program could be re-written in terms of two different arrays, each of which would have its own augmented data access descriptor, we can ignore this case and assume that the hulls intersect.

**Definition 1** *Merged ADADs consist of an $l \times n$ array of tuples, where $l$ is the number of iteration variables in common to the ADADs being combined, and $n$ is the number of columns in the original ADADs (also equal to the number of dimensions in the variable being described). As in regular ADADs, each tuple is a 4-tuple. For each iteration variable $I$ and dimension $D$, we derive a new tuple according to the following procedure.*

1. $\mathcal{D}(A,\{S1,S2\},I,D)[1] = \min\{\mathcal{D}(A,S1,I,D)[1], \mathcal{D}(A,S2,I,D)[1]\}$
2. $\mathcal{D}(A,\{S1,S2\},I,D)[2] = \max\{\mathcal{D}(A,S1,I,D)[2], \mathcal{D}(A,S2,I,D)[2]\}$

The reason for choosing these elements in this manner is that accesses to the array by these statements must lie within the hypercube whose bounds correspond to the maximum and minimum, respectively, of the upper and lower bounds of the hypercubes containing the accesses of the statements being merged.

3\.

**if** $(\mathcal{D}(A,S1,I,D)[3] \neq \mathcal{D}(A,S2,I,D)[3])$
  **or** $(\mathcal{D}(A,S1,I,D)[3] = 0)$
  **or** $(\mathcal{D}(A,S2,I,D)[3] = *)$
**then**
  $\mathcal{D}(A,\{S1,S2\},I,D)[3] = *;$
  $\mathcal{D}(A,\{S1,S2\},I,D)[4] = \emptyset;$
**else**
  **let** $B = \mathcal{D}(A,S1,I,D)[4] \cup \mathcal{D}(A,S2,I,D)[4];$
  **if** $\max(B) - \min(B) < \mathcal{D}(A,S1,I,D)[3]$
  **then**
    $\mathcal{D}(A,\{S1,S2\},I,D)[3] = \mathcal{D}(A,S1,I,D)[3];$
    $\mathcal{D}(A,\{S1,S2\},I,D)[4] = B;$
  **else**
    $\mathcal{D}(A,\{S1,S2\},I,D)[3] = *;$
    $\mathcal{D}(A,\{S1,S2\},I,D)[4] = \emptyset;$
  **endif**
**endif**

As an example, consider the following ADADs:

$D$(A,S1)

| 3 | dimension 1 | dimension 2 | dimension 3 |
|---|---|---|---|
| I | (1,49,2,{-1}) | (1,100,*,∅) | (1,100,*,∅) |
| J | (1,49,*,∅) | (1,100,1,{-1}) | (1,100,*,∅) |
| K | (1,49,*,∅) | (1,100,*,∅) | (1,100,2,{0}) |

$D$(A,S2)

| 3 | dimension 1 | dimension 2 | dimension 3 |
|---|---|---|---|
| I | (2,100,2,{0}) | (1,200,*,∅) | (1,100,*,∅) |
| J | (2,100,*,∅) | (1,200,1,{0}) | (1,100,*,∅) |
| K | (2,100,*,∅) | (1,200,*,∅) | (1,100,3,{0}) |

Then the merged augmented data access descriptor is:

$D$(A,{S1,S2})

| 3 | dimension 1 | dimension 2 | dimension 3 |
|---|---|---|---|
| I | (1,100,2,{-1,0}) | (1,200,*,∅) | (1,100,*,∅) |
| J | (1,100,*,∅) | (1,200,*,∅) | (1,100,*,∅) |
| K | (1,100,*,∅) | (1,200,*,∅) | (1,100,*,∅) |

**Definition 2** *A* **global augmented data access descriptor** *is a merged augmented data access descriptor for all statements in the section of a program being analyzed.*

### 4.4 Using ADADs to Develop Alignment Statements

After all of the ADADs describing a particular array have been combined to form one global ADAD, we can make some decision as to what the best distribution for that array might be.

**Lemma 1** *If the third element of the tuple corresponding to the iteration variable whose loop is to be parallelized and the dimension of the array across which we wish to distribute is a positive integer, m, then the array should be distributed in bands of width equal to m. In addition, they should be offset by an amount corresponding to the smallest member of the fourth element. If this procedure is followed, then communication will not be necessary. This may be formalized as follows:*

**Given**:

$S$ – a section of code.

$I$ – an iteration variable in section $S$.

$n$ – the number of arrays accessed in section $S$.

$A_k(k = 1 \cdots n)$ – the arrays accessed in section $S$.

$D_k$ – an index corresponding to a dimension of array $A_k$.

$\mathcal{D}(A, S, I, D)[j]$ – the value of the $j$th element of the tuple corresponding to iteration variable $I$ and dimension $D$ in the ADAD describing the accesses to array $A$ in section $S$.

**if**:

$$(\exists\, I : \mathit{Loop}(I) \subseteq S : (\forall A_k : 1 \leq k \leq n : (\exists D_k : 1 \leq D_k \leq \mathit{Numdims}(A_k) : \mathcal{D}(A_k, S, I, D_k)[3] \geq 0)))$$

**then**:

The loop may be parallelized in a communication free manner.

**Algorithm**:

Given any code segment $S$ for which the constraint holds, we can construct a data parallel Fortran D version that is free from communication in the following manner:

1. Change the DO loop associated with $I$ to a parallel loop.
2. For each array $A_k$ choose some dimension $D_k$ satisfying the above constraint, and let $T_k = \mathcal{D}(A_k, S, I, D_k)[3]$. Also, let $T_{lcm}$ equal the least common multiple of the nonzero $T_k$s.
3. Add a decomposition statement of the form

   DECOMPOSITION X($U$)

   to the code, where X is an unused variable name, and
   $U = \max\{\mathcal{D}(A_k, S, I, D_k)[2] \times (T_{lcm}/T_k)\}$.
4. For each array $A_k$, if $T_k \neq 0$ add an alignment statement of the form

   ALIGN $A_k$(I,J,...) with X(*st**M−(*st*× *offset*))

   to the program, where *st* (stride) $= T_{lcm}/T_k$, *offset* $= \max(\mathcal{D}(A_k, S, I, D_k)[4])$, and M corresponds to the placeholding index variable (I, J, etc.) representing dimension $D_k$. If $T_k = 0$, array $A_k$ is referenced only once and its distribution should be handled independently and in a manner consistent with that reference. If the reference pattern is complicated enough, this may require specifying the distribution of each row individually.
5. Add a distribution statement, such as

   DISTRIBUTE X(BLOCK_CYCLIC($T_{lcm}$))

   to the program.

**Example**:

As an example of the application of this algorithm, suppose we have the following ADADs.

- $D$(A,*global*,I,1) = (1,100,*,∅)
- $D$(A,*global*,I,2) = (2,100,1,{0})
- $D$(B,*global*,I,1) = (2,199,2,{-2,-1})
- $D$(B,*global*,I,2) = (1,100,*,∅)

Then, we should create a Fortran D version of the program by adding the following lines to the program.

```
DECOMPOSITION X(200)
ALIGN A(I,J) with X(2*I)
ALIGN B(I,J) with X(I+1)
DISTRIBUTE X(BLOCK_CYCLIC(2))
```

A value of 0 for the third element indicates a special case for which the requisite distribution is difficult–the rows being accessed by successive iterations must be distributed to successive processors. Describing such distributions is possible in languages like Fortran D, but very complicated. Because of the large amount of work involved, as well as the potential for error, one would not typically attempt to add the necessary code by hand. However, since the alignment statements describing this distribution can be generated automatically, the presence of such references is a strong argument for the use of a tool that can generate alignment information automatically.

If the third element of a tuple is an asterisk, then communication will be necessary if the array is distributed across that dimension and the loop designated by that iteration variable is to be parallelized. In order to determine which distribution will result in the least communication, we re-evaluate the global ADAD, considering only those ADADs whose level of nesting is maximal. This approach is approximate, based on the assumption that the most deeply nested statements will be executed most frequently. If the new global ADAD still has an asterisk for the third element, much communication will be necessary if we insist on using that iteration variable and distributing across that dimension. Any of the distributions favored by an individual ADAD of maximal nesting is acceptable (if no such distribution exists, then any will suffice).

### 4.5 Interprocedural Analysis

It is often the case that a section of code that we wish to analyze contains one or more calls to subroutines. Therefore, we must establish a framework for computing ADADs when such a call occurs. It is a fairly simple task to include procedure calls in the scheme for combining ADADs. First, we compute the ADAD(s) for the subroutine, as usual. Then, the ADAD(s) for the call statement can be computed by replacing the formal parameters in the ADAD(s) computed for the subroutine with the actual parameters of the call. If the arrays undergo reshaping due to the nature of the call, then this reshaping and the indexing used need to be taken into account. Any rows corresponding to iteration variables local to the subroutine can be ignored, since we are restricting ourselves to an analysis at a higher level. If we wish to consider these variables in our alignment analysis, we should determine the best alignment for the section of code before the call, the best alignment for the subroutine, and the best alignment for the section of code after the call. Finally, the ADAD(s) for the call statement can be merged with the ADADs for the rest of the section.

## 5 The Effects of Loop Transformations on ADADs

Most parallelizing compilers attempt to perform a number of different loop transformations, in the hope of enabling the parallelization of loops that previously could not be parallelized. Loop transformations are also used to increase the granularity of a program. When we use a data parallel programming paradigm, our concerns are different. In particular, our primary goal is to reduce the amount of communication required. Loop transformations may help us to achieve that goal. Therefore, the effects of some of the standard loop transformations on augmented data access descriptors are presented, not because it is easier to modify ADADs than it is to recalculate them, but so that our transformations may be guided by the effect they will have on the ADADs.

### 5.1 Promotion

Promotion entails the addition of a dimension to a variable, which is indexed by the iteration variable of a loop within which the variable is defined. This technique is particularly appropriate for intermediate variables. For example, consider the following section of code:

```
        DO I=1, 100
          DO J=1, 10
S1:          TEMP(J) = COS(A(I,I+J))
S2:          B(I,J) = 2.0*TEMP(J)
          ENDDO
        ENDDO
```

The variable **TEMP** is not indexed by **I** and therefore can not be distributed among the processors using that variable. However, if we perform a promotion, we get the following code:

```
        DO I=1, 100
          DO J=1, 10
S1':         TEMP(I,J) = COS(A(I,I+J))
S2':         B(I,J) = 2.0*TEMP(I,J)
          ENDDO
        ENDDO
```

which enables **TEMP**, **A**, and **B** to be distributed among up to 100 different processors.

After a promotion, the ADADs associated with the variable being promoted should have a dimension (column) added. The tuple for the iteration variable with which we are indexing the new dimension will be $(L, U, 1, \{0\})$, where $L$ and $U$ are the lower and upper bounds, respectively, on the iteration variable. The tuple for all other rows is $(L, U, *, \emptyset)$. So, for **TEMP** in the example above, the original ADADs were:

- $D$(TEMP,S1,I,1) = (1,10,*,∅)
- $D$(TEMP,S1,J,1) = (1,10,1,{0})
- $D$(TEMP,S2,I,1) = (1,10,*,∅)
- $D$(TEMP,S2,J,1) = (1,10,1,{0})

After promotion, the ADADs are:

$D$(TEMP,S1')

| 2 | dimension 1 | dimension 2 |
|---|---|---|
| I | (1,100,1,{0}) | (1,10,*,∅) |
| J | (1,100,*,∅) | (1,10,1,{0}) |

$D$(TEMP,S2')

| 2 | dimension 1 | dimension 2 |
|---|---|---|
| I | (1,100,1,{0}) | (1,10,*,∅) |
| J | (1,100,*,∅) | (1,10,1,{0}) |

## 5.2 Loop Unrolling

In loop unrolling, the number of iterations of the loop is reduced by changing the step size, and adding statements to accommodate the values between steps. Since we have restricted actual step sizes to a length of 1, the effective step size may be changed by dividing the loop's upper bound by some factor, which we will refer to as the unrolling factor, and multiplying the iteration variable by that factor. For example,

```
        DO I=1, 100
S1:       A(2*I) = B(I+2)
        ENDDO
```

might become (with an unrolling factor of 4)

```
        DO I=1, 25
S1a:      A(8*I-6) = B(4*I-1)
S1b:      A(8*I-4) = B(4*I)
S1c:      A(8*I-2) = B(4*I+1)
S1d:      A(8*I) = B(4*I+2)
        ENDDO
```

If a loop is unrolled by a factor $M$, the ADADs for statements within the loop could be expanded to $M$ ADADs, one for each statement introduced by the unrolling. However, it is simpler to modify the original ADAD to one which describes the $M$ derived statements. Only tuples in the row corresponding to the loop being unrolled are affected. The first and second elements in each tuple remain unchanged. If the third element, $\mathcal{D}(A,S,I,D)[3]$, is * or 0, it should remain unchanged. Otherwise, $\mathcal{D}(A,S,I,D)[3]$ should be multiplied by $M$ and the fourth element modified according to the following procedure. For each element $E$ in the original set, $\mathcal{D}(A,S,I,D)[4]$, introduce $M$ elements to the new set, $\mathcal{D}(A,\{S_a,S_b,...\},I,D)[4]$, equal to $E + k \times \mathcal{D}(A,S,I,D)[3], (k = [1-M, 0])$. The original ADADs for the example above are:

- $D$(A,S1,I,1) = (2,200,2,{0})
- $D$(B,S1,I,1) = (3,102,1,{2})

The new augmented data access descriptors would be:

- $D$(A,{S1a,S1b,S1c,S1d},I,1) = (2,200,8,{-6,-4,-2,0})
- $D$(B,{S1a,S1b,S1c,S1d},I,1) = (3,102,4,{-1,0,1,2})

So, the Fortran D alignment statements might look like:

```
      DECOMPOSITION X(200)
      ALIGN A(I) with X(I)
C The expression 2*I-4 in the next statement
C derives from the fact that 8/4=2 and B has
C offset 2, so we need to
C align B(I) with X(2*(I-2)).
      ALIGN B(I) with X(2*I-4) overflow (WRAP)
      DISTRIBUTE X(BLOCK_CYCLE(8))
```

## 5.3 Loop Fusion

In loop fusion, two adjacent loops are merged to form one loop. The iteration variables of the original loops are replaced by the new iteration variable. For example,

```
        DO I=1, 100
S1:       B(I) = 2*I*A(2*I)
        ENDDO
        DO J=1, 100
S2:       C(J) = A(2*J-1) - 1
        ENDDO
```

could become

```
        DO K=1, 100
S1':      B(K) = 2*K*A(2*K)
S2':      C(K) = A(2*K-1) - 1
        ENDDO
```

In order to perform loop fusion, while preserving program correctness, there are certain criteria which must be satisfied. The formal verification that transformations are correctness-preserving is left to books on parallel compiler construction [7]. Instead, we restrict ourselves to examples that are simple enough that correctness is readily apparent.

Modifying the ADADs affected by a loop fusion, once it has been determined that one may occur, is simple. The rows of the ADADs that corresponded to the original iteration variables now correspond to the new iteration variable. Also, the ADADs describing a region of code may change, and should be computed. Thus, for the example above, the original ADADs of

- $D$(A,S1,I,1) = (2,200,2,{0})
- $D$(B,S1,I,1) = (1,100,1,{0})
- $D$(A,S2,J,1) = (1,199,2,{-1})
- $D$(C,S2,J,1) = (1,100,1,{0})
- $D$(A,{S1,S2},,) = NULL

become:

- $D$(A,S1',K,1) = (2,200,2,{0})
- $D$(B,S1',K,1) = (1,100,1,{0})
- $D$(A,S2',K,1) = (1,199,2,{-1})
- $D$(C,S2',K,1) = (1,100,1,{0})
- $D$(A,{S1',S2'},K,1) = (1,200,2,{-1,0})

## 5.4 Loop Interchange

As the name implies, loop interchange entails the interchange of two loops. The outer loop becomes the inner loop, and the inner loop becomes the outer loop. Thus,

```
DO I=1, 100
  DO J=1, 100
     A(I,J) = 0.0
  ENDDO
ENDDO
```

would become

```
DO J=1, 100
  DO I=1, 100
     A(I,J) = 0.0
  ENDDO
ENDDO
```

As with loop fusion, the conditions under which loop interchange may safely be performed depend on sophisticated data dependence analysis [7]. However, assuming loop interchange can be performed, the ADADs associated with individual statements and with regions of the program will remain unchanged.

## 5.5 Strip Mining

Strip mining is often used in parallelizing compilers to increase the amount of work performed inside of a loop. It is analogous to loop unrolling, except that rather than increasing the number of statements by some factor $M$, a new loop is introduced to cover the range of iterations between the outer steps. So,

```
DO I=1, 1000
   A(I) = 3.14*(R(I)**2)
ENDDO
```

becomes

```
DO I=1, 10
  DO J=1, 100
   A(100*(I-1)+J) = 3.14*(R(100*(I-1)+J)**2)
  ENDDO
ENDDO
```

This is an extremely useful technique when we want to parallelize a particular loop for execution on a shared memory computer with a relatively small number of processors. However, the process of strip mining results in all tuples associated with the original iteration variable, as well as those associated with the new iteration variable, having a third element of *. Thus, when we are attempting to exploit global parallelism on a large distributed memory computer, strip mining can have an adverse effect. Instead of increasing the work load per processor through strip mining, we should utilize the distribution facilities of the data parallel language being used.

## 5.6 Invariant Code Movement

If a statement within a loop uses variables that do not change within the loop, defines a variable that is not defined elsewhere in the loop nor used earlier in the loop, and the statement does not occur within a conditional, then that statement may be moved to a position immediately before the loop. This is useful for eliminating extra calculations. For example,

```
       DO J=1, 5
S1:      V(J) = 0.0
         DO I=1, 100
S2:         A(J) = 3.14*R(J)*R(J)
S3:         V(J) = V(J) + A(J)*F(I)
         ENDDO
       ENDDO
```

can be changed to

```
       DO J=1, 5
S1:      V(J) = 0.0
S2':     A(J) = 3.14*R(J)*R(J)
         DO I=1, 100
S3:         V(J) = V(J) + A(J)*F(I)
         ENDDO
       ENDDO
```

When a statement like this is moved outside of a loop, the row in the ADAD for the statement corresponding to that loop's iteration variable should be eliminated, and the integer representing the nesting level of the statement should be decremented. So, in the previous example,

- $D$(A,S2,I,1) = (1,5,*,$\emptyset$)
- $D$(A,S2,J,1) = (1,5,1,{0})

becomes

- $D$(A,S2',J,1) = (1,5,1,{0})

Although this technique does not assist in the parallelization of the program, it can result in significant performance improvements, by eliminating unnecessary work.

## 5.7 An Example

To understand the usefulness of directed loop transformations, consider the following section of a program:

```
       DO I=1, 100
         DO J=1, 100
S1:         A(J,I) = A(J,I) + SQRT(J*I)
         ENDDO
         DO K=1, 50
S2:         A(2*K,I) = A(2*K-1,I)
         ENDDO
         DO L=1, 100
S3:        A(L,I+1) = A(L,I)
         ENDDO
       ENDDO
```

If we attempt to determine the best alignment for this code, we have to obtain all ADADs (due to space limitation, please refer to [8] for detailed ADADs). However, there is no way in which to distribute this code without the need for communication being introduced. However, if we unroll the first and third loops by a factor of two, then fuse the three loops, we get:

```
        DO I=1, 100
          DO J=1, 50
S1a':           A(2*J-1,I) = A(2*J-1,I)
                           + SQRT((2*J-1)*I)
S1b':           A(2*J,I) = A(2*J,I)+SQRT(2*J*I)
S2':            A(2*J,I) = A(2*J-1,I)
S3a':           A(2*J-1,I+1) = A(2*J-1,I)
S3b':           A(2*J,I+1) = A(2*J,I)
          ENDDO
        ENDDO
```

Now we can obtain a different set of ADADs [8]. After the transformations, we may distribute A along the first dimension, using a stride size of 2, and we will have no need for communication. This could be achieved using:

## 6 Conclusions

We presented a formal technique for determining a good distribution of data using the augmented data access descriptor (ADAD). This technique differs from previous approaches in that it views the problem of data alignment as an extension of data dependence analysis, rather than a completely new problem. Thus, the Data Access Descriptor [5], used for data dependence analysis, may be augmented to facilitate data alignment analysis, yielding the augmented data access descriptor. This approach is simple, efficient, and quite accurate.

Ways to handle interprocedural analysis and the effects of loop transformations were presented. The former was discussed so that programs with good modularity are not penalized by incomplete analysis. The effects of loop transformations were examined so that they might guide decisions as to which transformations should be performed. An example showing how this knowledge could be utilized was provided. We also explained how ADADs can be used to discover a good data alignment for a program and presented an algorithm for generating Fortran D alignment and distribution statements automatically.

Possible subjects for future research include the development of a tool to automatically compute ADADs, plus the development of a tool that can use ADADs and the original source code to automatically generate source code in some language with data alignment constructs, such as Fortran D or DINO [1, 2]. Also, analysis of real programs would facilitate an assessment of the viability of ADADs in the automatic data partitioning of existing programs. Should ADADs prove inadequate, we may employ a more sophisticated analysis (at the cost of performance) which preserves the notion of data alignment as data dependence analysis and recognizes the importance of interprocedural analysis and directed loop transformations. An important feature of ADAD analysis is that it can be done incrementally. Analysis is performed on small regions of a program, and the ADADs are then merged so that they describe larger regions of the program. This feature makes ADAD analysis particularly well suited for a programming environment, because small changes to the program do not mandate another complete analysis of the program, but instead incur a small additional cost. In addition, incremental merging of ADADs may allow a programmer to detect what statements are blocking a communication free distribution. This knowledge may enable the programmer to rewrite the program in a manner better suited for global parallelization. Thus, future studies should examine what hierarchy of ADAD analysis best facilitates these two goals.

## References


1. G. Fox, S. Hiranandani, K. Kennedy, C. Koelbel, U. Kremer, C.-W. Tseng, and M.-Y. Wu, "Fortran D language specification," Tech. Rep. COMP TR90-141, Rice University, Department of Computer Science, Dec. 1990.

2. M. Rosing, R. B. Schnabel, and R. P. Weaver, "The DINO parallel programming language," *Journal of Parallel and Distributed Computing*, vol. 13, pp. 30–42, Sept. 1991.

3. J. Li and M. Chen, "The data alignment phase in compiling programs for distributed-memory machines," *Journal of Parallel and Distributed Computing*, vol. 13, pp. 213–221, Oct. 1991.

4. M. Gupta and P. Banerjee, "Demonstration of automatic data partitioning techniques for parallelizing compilers on multicomputers," *IEEE Transactions on Parallel and Distributed Systems*, vol. 3, pp. 179–193, Mar. 1992.

5. V. Balasundaram, "A mechanism for keeping useful internal information in parallel programming tools: The data access descriptor," *Journal of Parallel and Distributed Computing*, vol. 9, pp. 154–170, 1990.

6. Z. Li, P.-C. Yew, and C.-Q. Zhu, "An efficient data dependence analysis for parallelizing compilers," *IEEE Transactions on Parallel and Distributed Systems*, vol. 1, pp. 26–34, Jan. 1990.

7. M. Wolfe, *Optimizing Supercompilers for Supercomputers*. Cambridge, Massachusetts: The MIT Press, 1989.

8. P. D. Hovland and L. M. Ni, "A model for automatic data partitioning," Tech. Rep. MSU-CPS-ACS-73, Dept. of Computer Science, Michigan State University, Oct. 1992.

# MULTI-LEVEL COMMUNICATION STRUCTURE FOR HIERARCHICAL GRAIN AGGREGATION

D. H. Gill*, T. J. Smith*, T. E. Gerasch*, C. L. McCreary**

*The MITRE Corporation
McLean, Virginia

**Auburn University
Auburn, Alabama

Abstract -- *Algorithms for the partitioning and scheduling of parallel programs have been the topic of intense research for numerous years. It is well understood that the organization and size of the parallel tasks are critical to the efficiency of a parallel program and to the effective use of parallel resources. Various algorithms have been proposed for partitioning and scheduling that are optimal under severely constrained assumptions. Under realistic assumptions, the general partitioning and scheduling problems are intractable. In this paper we present an approach that uses hierarchical analysis of communication locality in the computation graph and grain selection based on architectural metrics. Our approach analyzes communication structure in making decisions at different levels of the graph hierarchy to decide whether to use or to ignore potential parallelism, taking into account communication overhead due to parallel execution.*

## 1. Introduction

In ongoing research, we are studying the problem of structuring computations for parallel execution. The primary goal of this research is to improve the reconfigurability of parallel algorithms. We seek to support portability and architecture-independence of software. Our approach is to use structural information about the locality of communication in a graph, together with metrics that model overhead characteristics of target architectures, in order to tailor the granularity of parallelism in a computation. Previously, we developed a local heuristic partitioning algorithm [10] that improves parallel execution time of an algorithms by eliminating local communication overhead when sequential execution is predicted to yield a shorter execution time. In the earlier work we presented a simple approach for making aggregation decisions. In this paper, we present an improved analysis of communication overhead that examines more exact aggregation decisions for hierarchical grain decomposition.

## 2. Background

Algorithms for the partitioning and scheduling of parallel programs have been the topic of intense research for numerous years. It is well understood that the organization and size of the parallel tasks are critical to the efficiency of a parallel program and to the effective use of parallel resources. Various algorithms have been proposed for partitioning and scheduling [3]. Some are optimal subject to constrained assumptions [11], [15] such as constant task delay due to synchronization and communication requirements of the algorithm, or absence of contention. However, with realistic assumptions, such as varying computation and communication cost or resource contention, both optimal partitioning and optimal scheduling are intractable, even for restricted types of computations. The inadequacy of optimal methods under realistic overhead assumptions is demonstrated in [15], where transformation techniques that improve data locality are shown to improve execution. Heuristic attempts to manage communication and other overhead have been an important component of parallelizing compiler research [2], [1], [12]. Our work has sought practical heuristics that tailor the amount of parallelism used to obtain good performance under diverse architectural and operating system assumptions. We are developing partitioning techniques for static analysis of parallel task structure, but we do not require the computation to be fully statically mapped. Like most others, our basic analysis technique applies to the directed acyclic graph (DAG) representation of a computation. It must be supplemented by graph manipulation techniques such as unrolling.

The key elements of our approach are hierarchical analysis of locality in the computation graph and architecture-based metrics to estimate parallel performance of an algorithm. The decomposition techniques expose subgraphs with common communication and scheduling requirements. Although the analysis presented in this paper can apply to any hierarchy of subgraphs exhibiting communication locality, we use the hierarchy obtained by graph parsing that identifies subgraphs with common data dependence. The first step of our analysis uses a *clan parsing* technique that identifies a hierarchy of subgraphs (*clans*) having the "hereditary" property [4], [7]. That is, the nodes of the subgraph are related to each node in the rest of the graph in the same way: as ancestor, descendant, or unrelated. Nodes in the interior of the clan subgraph communicate with the rest of the graph only through the source and sink nodes of the subgraph. This property captures the essential data locality in the computation. The result is a bipartite ordered parse tree. Independent clans have, as children, clans that can be executed in parallel. Linear clans order the execution of the clans they contain.

The clan-based graph parsing technique has been extended [10] to allow edge augmentation, permitting irregular program graphs to be fully decomposed. The advantage of the extended parsing technique is that it identifies the independent and linearly related components of naturally irregular computation graphs, such as the data flow or program dependence graph of a computation. Thus, regularly structured control flow graphs need not be assumed as in other hierarchical techniques [14], [12], and dependence graphs that represent only essential sequencing constraints can be analyzed. However, the analysis we present is useful for any hierarchical technique that must resolve communication structure at subgraph boundaries. Furthermore, the assumption of alternating layers of independent and linearly ordered subgraphs is characteristic of graph hierarchies for analysis of parallelism [14], [6].

*Aggregation* is a decision to ignore opportunities for parallelism and combine potentially parallel computations into a sequential grain, so that communication and other sources of overhead due to parallel execution can be reduced or removed. In [9] and [10] we presented simple local metrics for determining the granularity of parallel computations. The simple metric improved on [9] and gave good results for a variety of computations [8]. At linear clans, this metric considers aggregation of the linearly ordered children. The children represent adjacent sets of independent computations that are immediately related (linearly ordered) by the program dependence relation. (That is, data and control flow from one set to another.) The decomposition algorithm makes a *parallelize* or *aggregate* decision by evaluating a cost function that considers potential communication latency between the linearly related sets of subgraphs. The decomposition algorithm allows parallelize vs. aggregate decisions to be revisited when one of the sets is subsequently paired for evaluation with another adjacent set that yields a different aggregation decision. The original metric computation considers aggregation of all nodes contained in the linear children of an independent clan even if some of them contain independent nodes for which a parallelize decision has been previously made [9], [10], [13]. An improved decision structure which preserves lower-level parallelize decisions is presented below. For the remainder of the paper, we examine aggregation decisions for independent and linear levels of bipartite graph hierarchies, regardless of whether they were generated by clan parsing or other techniques.

## 3. Improved Metric

In the improved metric presented in this paper, a more precise aggregation decision is examined, in which the entire subgraph is not aggregated, but only the specific operations for which communication between subgraphs can be eliminated. Let $C_1$ and $C_2$ be linearly related independent subgraphs for which an aggregation decision is to be evaluated. Both C1 and C2 may have lower level subgraphs for which aggregation decisions have been made. Figure 1 illustrates this situation.

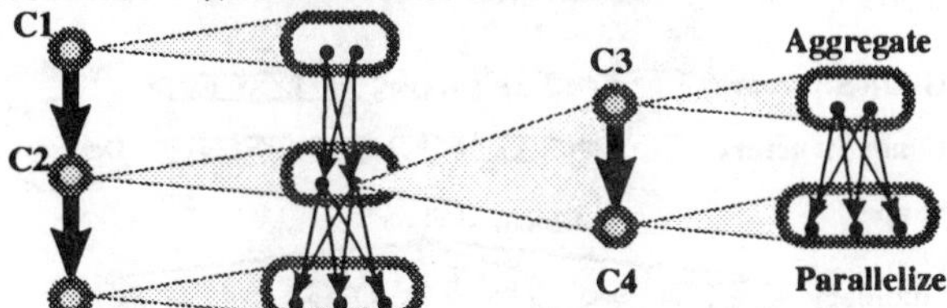


**Figure 1. Linearly related independent subgraphs**

Adjacent sets of independent subgraphs are considered during aggregation. Only the potentially communicating nodes (or aggregates) at the beginning or ending boundary of a subgraph need be considered for further aggregation. Interior parallel groups are left as they are, and communication to them is left unaltered. After a parallelize decision, each linear child is transformed into a normalized *grain structure* with at most three segments: a fully aggregated prefix, a parallel segment, and fully aggregated suffix. The parallel segment begins and ends with parallelized grains, possibly with interior aggregation. Some segments may be missing. If so, the normal form is fully aggregated segment. We call this a *simple* grain structure. Figure 2 illustrates a normalized grain structure with three segments. Figure 3 illustrates the various forms the normalized grain structure can take. The labels are two characters, P (parallelized) or A (aggregated), that indicate the aggregation conditions at the communicating "ends" of the grain structure.

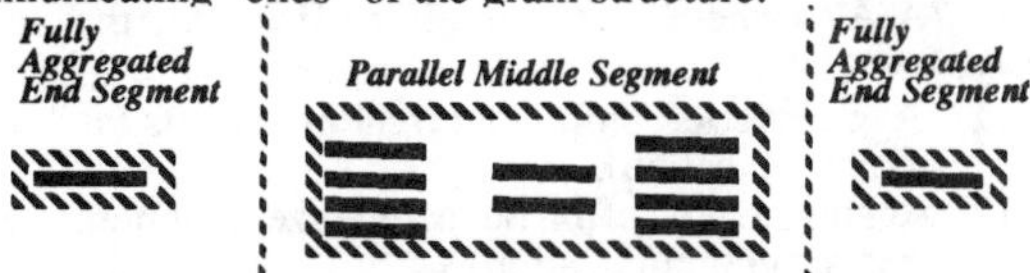


**Figure 2. A normalized linear grain structure with three segments**

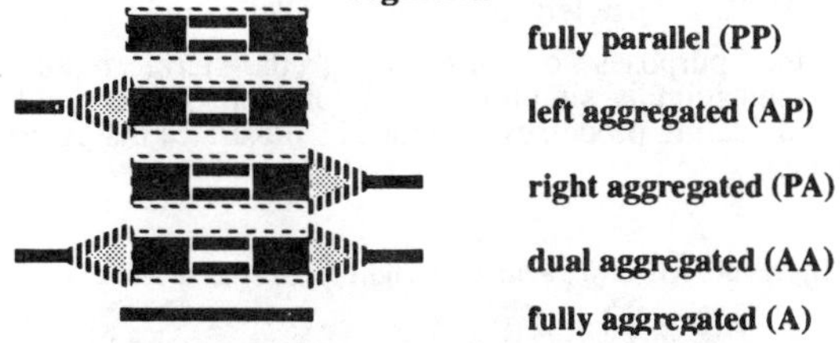


**Figure 3. Forms of normalized grain structures**

Figure 4 shows the scope of an aggregation decision for linearly related independent subgraphs whose children in the hierarchy are normalized grain structures. Only the aggregated suffix elements of the first set of grain structures and the aggregated prefix components of the second set of grain structures are considered for further aggregation. The decision is whether to execute these components sequentially as a single grain or not. This would eliminate their internal communication.

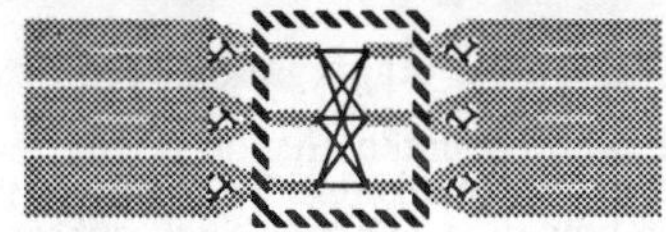

**Figure 4. Elements considered for aggregation**

The improved metric limits the decision space. In the improved metric, the decision space uses alternative decision rules for making aggregation decisions in the presence of lower-level *parallelize* decisions. The computation considers only (further) aggregation of sets of simple segments. We make aggregate vs. parallelize decisions involving independent sets of end segments, but do not re-visit previous decisions to parallelize independent sets at a lower-level grain structure. That is, if the grain structure has a missing simple suffix (prefix) for any child of the first (respectively, second) independent graph, that graph is always left parallel at the corresponding end.

## 4. Decision Labelling

Metric analysis determines the aggregation decisions for subgraphs, and yields a grain structure for each decision. We introduce two labellings: a shape labelling and a summary labelling. The *shape* labelling describes the normalized grain structure that results from the aggregation decision. The *summary* labelling summarizes the shapes of a set of independent grain structures prior to the aggregation decision. The parse tree used by the improved metric is bipartite, alternating linearly and independently related subgraphs. Thus, each linear level consists of the sequence of its independent child subgraphs. The leaf nodes of the tree are the computation nodes of the graph. The root of the parse tree is classified as linear. The labels are computed recursively (leaves-up), as follows.

A leaf node has a simple normalized grain structure, and shape labelling A. At a linear level of the graph, we examine adjacent sets of independent grain structures. A summary labelling is assigned to each set before the aggregation decision is made. After the aggregation decisions for the linear level are made, the grain structure for the linear level is computed, and the final shape labelling assigned.

### 4.1 Independent subgraphs: summary and shape labels

Before the aggregation decision, the labelling of an independent level of the hierarchy summarizes the aggregation status of its constituent child subgraphs. The summary labelling of the independent level depends upon the shape labellings of the (linear) grain structures beneath it. This labelling is the least upper bound of the labellings of the grain structures for the children, with respect to the complete partial order of Figure 5.

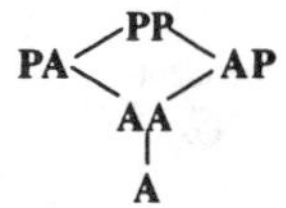


**Figure 5. Complete partial order for summary label**

If any linear child's grain structure begins (ends) with a parallelize decision, the first (second) component of the summary labelling is P. Figure 6 gives an example that illustrates the summary label (PP) for an independent level containing grain structures of all types. After an aggregation decision, the beginning (or ending) set of fully aggregated prefix (suffix) components may either become aggregated or parallelized. If the decision is to aggregate, the corresponding end is marked A, otherwise marked parallel, P. Figure 6 illustrates the summary labellings for $C_{l1}$ and $C_{l2}$. Figure 7 illustrates the post-decision shape labels that result from the aggregation decisions shown.

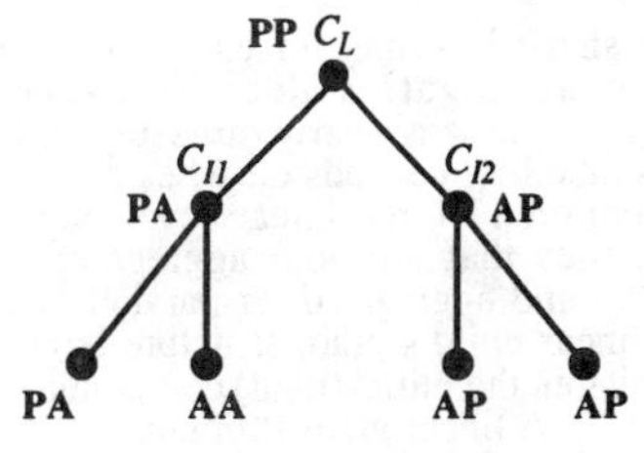


**Figure 6. Example summary label of independent level derived from children**

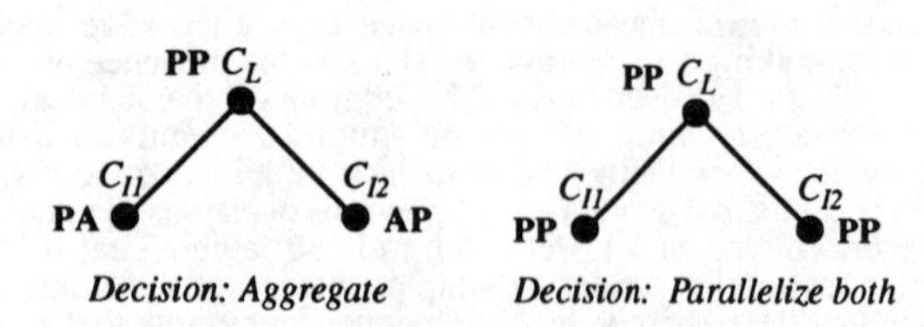


**Figure 7. Example shape labels resulting from aggregation decisions**

We now consider an example using a bipartite parse tree. In Figure 8, $C_{l1}$ and $C_{l2}$ have the same structure and because their children are leaves (hence, fully aggregated simple form), both have type A before the first-level decision. The lowest level independent nodes have summary labels before the aggregation decision and shape labels afterward. When the aggregation decision is made for an independent level, its labelling is changed to reflect its structure after the decision and any normalization.

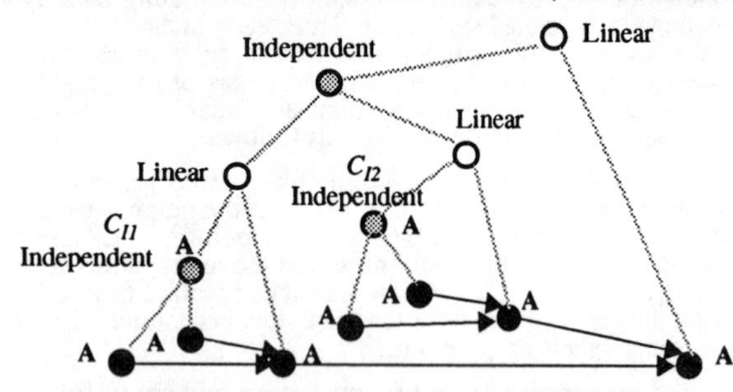


**Figure 8. Bipartite parse tree with summary labels**

Figure 9 illustrates the situation after a typical decision. Suppose the decisions are to aggregate $C_{l1}$ and to parallelize $C_{l2}$, giving them shape labels of A and P, respectively. After the decision $C_{L1}$ has type AA, which is subsequently normalized to type A, and $C_{L2}$ has type PA.

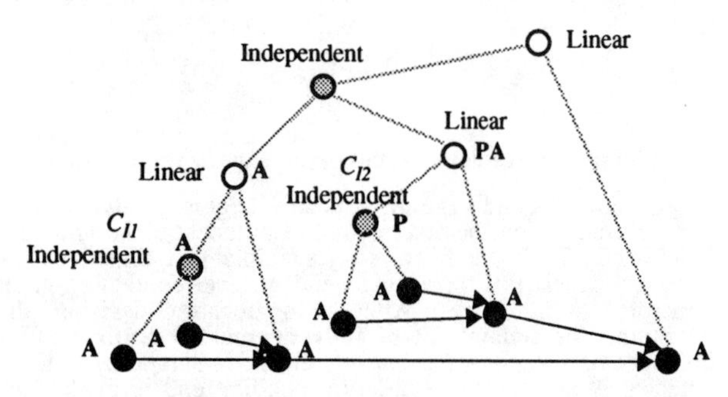


**Figure 9. Shape labelling resulting from possible decisions**

### 4.2 Linear subgraphs: shape labels

The shape labelling for each linear level is obtained from the two aggregation decisions (the shape label) corresponding to the grain structures of its initial and final children. This labelling depends on whether the initial and final independent children of the linear level begin and end with independent nodes that are both aggregated, aggregated and parallel, parallel and aggregated, or parallel and parallel. If the initial (final) linear child's grain structure begins (ends) with a parallelize decision, the initial (final) component of the label is P; otherwise, it is A. A linear grain structure is labelled as AA, AP, PA, PP, or A. (The labelling does not describe aggregation decisions interior to the linear level because their communication is not relevant to the current aggregation decision.) Figure 3 illustrated the linear grain structure of each type, together with an example thread set for the type.

## 5. Aggregation Decisions

### 5.1 Decision alternatives

We now turn to the nature of the aggregation decisions. Consider a linear level $C_L$ of the hierarchy. Let $C_{li}$ and $C_{li+1}$ be adjacent independent subgraphs for which the aggregation decision must be made. Figure 10 illustrates the linear grain structure and typical shape labellings of its subordinate independent subgraphs after aggregation decisions have been made.

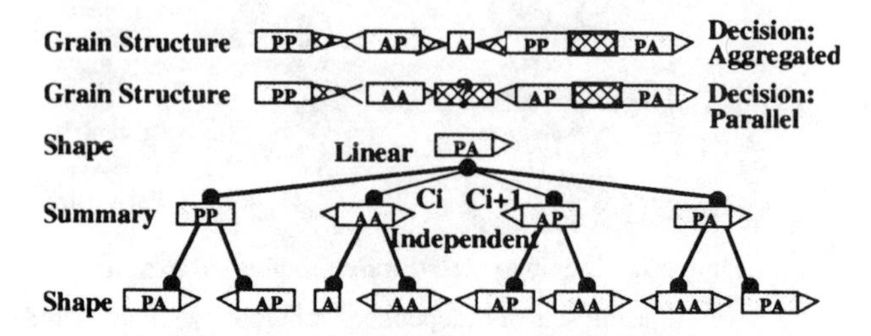


**Figure 10. Labelled linear grain structure and decision alternatives**

The metric determines the best aggregation decisions for $C_{li}$ and $C_{li+1}$, given the overhead of communicating between them. Each child is normalized to the form (*prefix, mid, suffix*). Given a linear grain structure N, the selectors *pref(N)*, *mid(N)*, and *suff(N)* designate these components. The selector *thread(N)* designates the entire sequence of components.

The aggregation decision determines whether to combine the adjacent aggregated prefixes and (or) aggregated suffixes into a single aggregated grain, in order to remove (reduce) their interior communication overhead. For such a decision involving aggregated prefix and suffix sets, which we denote ?A-A? (where ? means either P or A), the possible decisions are:

1. Aggregate left suffix and right prefix sets (and remove communication).
2. Aggregate left suffix, but parallelize right prefix set (and reduce communication).
3. Parallelize left suffix, but aggregate right prefix set (and reduce communication).
4. Parallelize left suffix, and parallelize right prefix set.

For the purposes of this paper, the term representing communication is simplified to a function of the number of communicating processes. Figure 11 illustrates the cases to be consider.

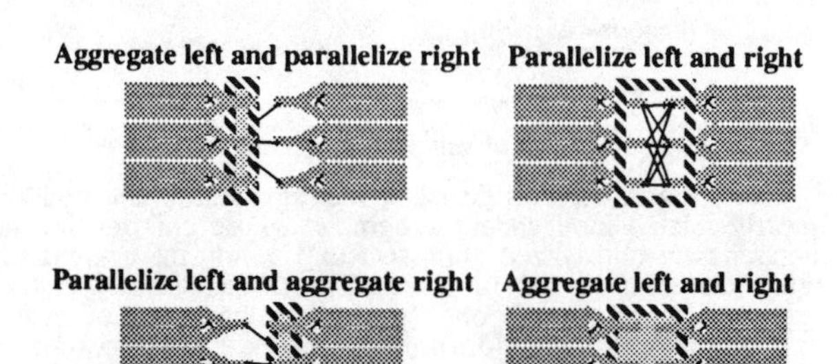


**Figure 11. Aggregation decisions**

It should be noted that a generalization of these aggregation decisions would consider additional partitions of the sets into communicating subsets. Such subsets would range between the aggregation decision extremes illustrated here, singletons (fully aggregated) vs. non-proper subsets (fully parallel), and could examine other subsets of the independent node sets. A more detailed technique containing such a subset decision procedure is described in [5].

### 5.2 Execution cost functions

The cost functions used by the metric correspond to the decision space alternatives. A simplified cost function considers only the number of communicating processors and the amount of data communicated. Figure 12 gives the estimated execution time associated with each decision. The decision that minimizes execution time is chosen.

**Decisions for ?A-A?:**

$$T_{PP} = max_i\{t_{thread}(C_i(N_1))\} + max_j\{t_{thread}(C_j(N_2))\}$$
$$+ Comm(|lastnodes(N_1)|, |firstnodes(N_2)|)$$

$$T_{AA} = max_i\{t_{pref}(C_i(N_1)) + t_{mid}(C_i(N_1))\}$$
$$+ Comm(|lastnodes(pref(N_1))|, |firstnodes(mid(N_1))|)$$
$$+ \sum_i t_{suff}(C_i(N_1)) + \sum_j t_{pref}(C_j(N_2))$$
$$+ max_j\{t_{mid}(C_j(N_2)) + t_{suff}(C_j(N_2))\}$$
$$+ Comm(|lastnodes(mid(N_2))|, |firstnodes(suff(N_2))|)$$
$$+ Comm(|lastnodes(mid(N_1))|, 1)$$
$$+ Comm(1, |firstnodes(mid(N_2))|)$$

$$T_{AP} = max_i\{t_{pref}(C_i(N_1)) + t_{mid}(C_i(N_1))\}$$
$$+ Comm(|lastnodes(pref(N_1))|, |firstnodes(mid(N_1))|)$$
$$+ \sum_i t_{suff}(C_i(N_1))$$
$$+ max_j\{t_{thread}(C_j(N_2))\}$$
$$+ Comm(|lastnodes(mid(N_1))|, 1)$$
$$+ Comm(1, |firstnodes(thread(N_2))|)$$

$$T_{PA} = max_i\{t_{thread}(C_i(N_1))\}$$
$$+ \sum_j t_{pref}(C_j(N_2))$$
$$+ max_j\{t_{mid}(C_j(N_2)) + t_{suff}(C_j(N_2))\}$$
$$+ Comm(|lastnodes(mid(N_2))|, |firstnodes(suff(N_2))|)$$
$$+ Comm(|lastnodes(N_1)|, 1)$$
$$+ Comm(1, |firstnodes(mid(N_2))|)$$

**Figure 12. Estimated execution time for each decision**

The communication term is determined by the number of processors involved in communicating from the final parallel components of $C_{li}$ to the aggregation form determined for the set of trailing aggregates at $C_{li}$, and between the initial aggregates of $C_{li+1}$ to the processors executing the initial parallel components of $C_{li+1}$, as well as possible communication resulting from the aggregation decision between trailing fully aggregated segments of $C_{li}$ to the initial aggregated segments of $C_{li+1}$. For purposes of this explanation, the accumulation of data size is omitted.

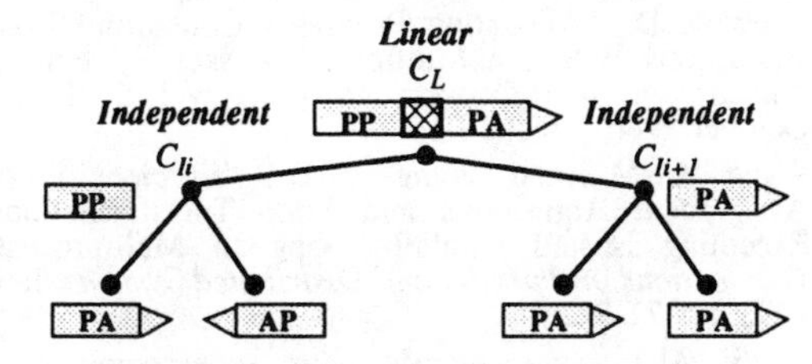


**Figure 13. Parallel-Parallel decisions**

Now, let $C_{li}$ and $C_{li+1}$ be as illustrated in Figure 13. In this situation, the decision case at $C_L$ is ?P-P?. That is, some child node of $C_{li}$ ends with a parallel component, rather than a trailing aggregate, and some child of $C_{li+1}$ begins with a parallel component rather than a leading aggregate. For simplicity in this paper, parallelize decisions are not revisited. The only decision in this case is to leave the end parallel, and communication is assumed to take place between the unaggregated threads of $C_{li}$ and $C_{li+1}$.

The set of linear children with A labels may cause conflicting decisions, called unstable adjacencies [10], in two ways. First, suppose the set consists only of linear threads labelled A. The clan represents a set of fully aggregated independent elements. This set cannot be at the same time parallel and aggregated. Thus, if different aggregation decisions (P, A) are made, a conflict exists and must be resolved As in [10], we evaluate the decision for both adjacent subgraphs, one on either side. When any of the set of linear children has a parallel mid-segment, a communication will occur (between the parallel mid segment and the aggregated leading prefix or trailing suffix segments) regardless of the aggregation decisions involving the ends. If either (or both) of the decisions is P (parallelize), there is no conflict. An A decision at either end aggregates all the adjacent maximal aggregated segments at that end. Simple grain structures (those labelled A), become aggregated; however, if both decisions are A, a conflict again exists. This second type of conflict is easier to resolve. Since some of the linear clans have parallel mid segments, communication must occur. Thus, the simple grain structure may be executed alone as a parallel *mid* segment, retaining both communications, or may be associated with one aggregated end segment or the other. Figure 14 illustrates the resolution of the possible types of conflict.

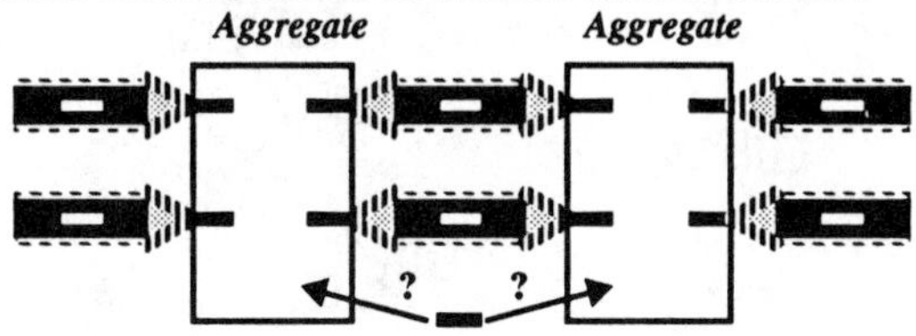


**Figure 14. Resolution of conflicting decisions**

## 6. Examples

In [13], a comparison is made of methods for partitioning and scheduling flow graphs that are trees. Tree methods permit advantage to be taken of special structure. The following binary expression tree in Figure 15 was decomposed using several competing methods, including our earlier DAG decomposition technique (a grain size determination method). Figure 15 gives the decomposition and finish time (59 time units) for this method. Zhu's workload scheduling and the Shirazi-Chen Libra scheduling achieved a finish time of 45 time units.

**McCreary-Gill (Grain Size Determination Method)**
**Time = 59**

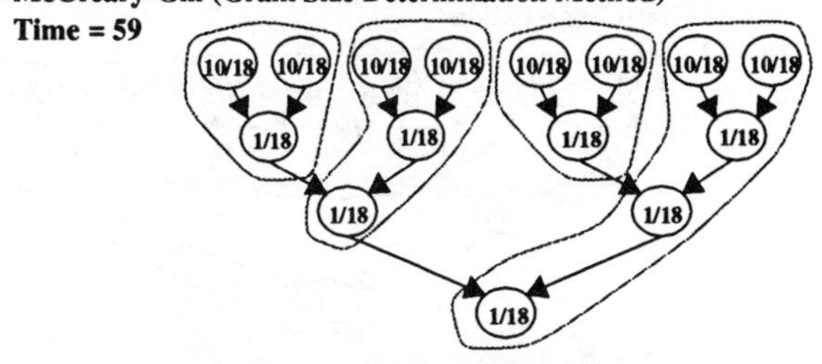


**Figure 15. Example binary tree**

This example illustrated the need for a more detailed analysis of the communication structure of a grain. The use and result of the metric with the improved analysis of communication structure are illustrated in Figure 16 and Figure 17. Computation nodes are labelled *a,...,z*, and are the leaves of the graph hierarchy. Interior nodes are labelled *1,...,99*, and represent subgraphs at higher levels of the hierarchy. Following our new decision procedure, the graph hierarchy is traversed, and the following decisions are made. First, at the linear level represented by node *5*, the parallel vs. sequential scheduling of the independent subgraph at node *1* is examined. This node consists of computation nodes *a* and *b*, which are considered for aggregation with the singleton node *i*. The best choice is to aggregate, giving a cost of 21 at node *5*. Linear nodes *6*, *7*, and *8* yield identical results. At linear node *11*, the decision is whether to further aggregate the (fully aggregated) children of independent node *9* with the singleton. Since parallel execution of the aggregates with communication overhead is less costly than sequential execution (21+18+1 vs. 21+21+1), that is chosen, yielding a parallelize decision at *9*. The maximal aggregated suffix of node *11* is the singleton computation node *m*. The computation for node *12* is identical, and has singleton suffix n. At linear node *14*, the children *11* and *12* of independent node *13* are considered for aggregation with singleton node *o*. In this case, the normal forms for both *11* and *12* have a non-null parallel component and a non-null aggregated suffix. The independent nodes have pre-decision labels of type PA, and the singleton is of type A. The aggregation decision considers aggregation of only nodes *m* and *n* (the maximal aggregated suffix strings), with node *o*. Thus, the decision has the form ?A-A?. Because this cost (1+1+1) is less than that of parallelizing (1+18+1), aggregation is chosen. The labellings in Figure 16 illustrate the final decisions and resulting labellings at each node of the hierarchy and the cost labellings used to determine these decisions. Figure 17 gives the final decomposition and finish time for the improved metric.

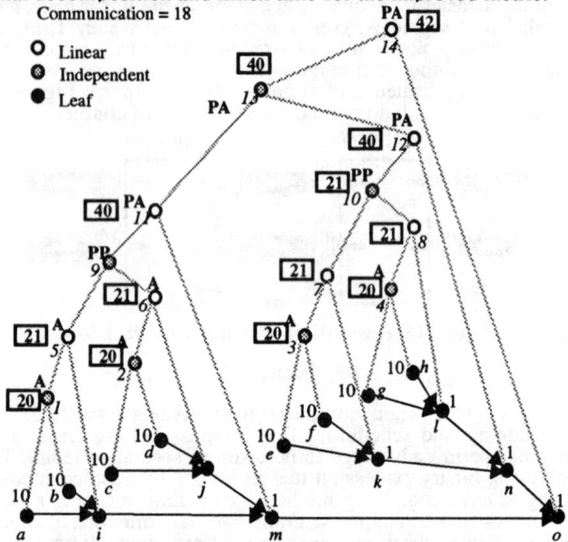


**Figure 16. Metric computation on labelled graph hierarchy**

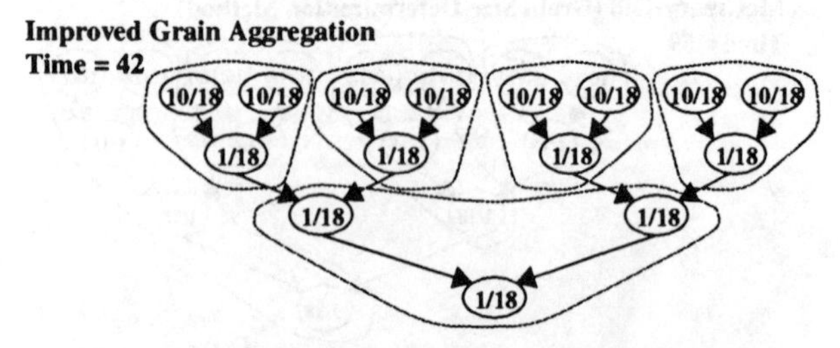


**Figure 17. Decomposition result and finish time for the improved metric**

## 7. Conclusions and Future Directions

The primary strategy of this partitioning method is to identify and exploit structural data locality in the computation. This aids in identifying overhead that can profitably be eliminated and avoids useless (excess) parallelism. The result illustrated in the example given shows how different regions of the computation graph may be chosen to be scheduled together. In [5] we give new methods for combining metric-based grain aggregation with linear thread extraction and static scheduling to identify further opportunities for removing overhead. In future work, we expect to generalize the treatment of aggregation decisions given here to include more general subset decisions.

## 8. Acknowledgements

This research is sponsored by The MITRE Corporation and the National Science Foundation grant no. NSF-CCR-9203319.

## References


[1] Allen, F., Burke, M., Cytron, R., Ferrante, J., Hseih, W., Sarkar, V., "A Framework for Determining Useful Parallelism," *ACM International Conference on Supercomputing '88,1988*, pp. 207-215.

[2] Allen, J. R. and Kennedy, K., "Automatic Translation of Fortran Programs to Vector Form," *ACM Transactions on Programming Languages and Systems*, Vol. 9, No. 4, October 1987,pp. 491-542.

[3] Bokhari, S., H., *Assignment Problems in Parallel and Distributed Computing*, Kluwer Academic Publishers, Norwell, Massachusetts, 1987.

[4] Ehrenfeucht, A. and Rozenberg, G., "Theory of 2-Structures, Part I: Clans, Basic Subclasses, and Morphisms," *Theoretical Computer Science* (1990), 277--303.

[5] Gill, D. H., Smith, T. J., McCreary, C. L., Stirewalt, R.E.K., Warren, J.V., "Spatial-Temporal Analysis of Program Dependence Graphs," (to appear), *Journal of Parallel and Distributed Processing*, April, 1993.

[6] Girkar, M. and Polychronopoulos, C., "Automatic Extraction of Functional Parallelism from Ordinary Programs," *IEEE Parallel and Distributed Systems*, March 1992, pp. 166-178.

[7] McCreary, C. L., "An Algorithm for Parsing a Graph-Grammar," Ph.D. dissertation, University of Colorado, Boulder, 1987.

[8] McCreary, C. L., and McArdle, M. E., "A Comparison of Task Partitioning Techniques," Technical Report CSE-91-14, Auburn University, 1991.

[9] McCreary, C. L. and Gill, D. H., "Automatic Determination of Grain Size for Efficient Parallel Processing," *Communications of the ACM* (1989), 1073--1078.

[10] McCreary, C. L. and Gill, D. H., "Efficient Exploitation of Concurrency Using Graph Decomposition," *Proceedings of the 1990 International Conference on Parallel Processing*.

[11] Polychronopoulos, C.D., *Parallel Programming and Compilers*, Kluwer, Boston, 1988.

[12] Sarkar, V., *Partitioning and Scheduling Parallel Programs for Multiprocessors*, Cambridge: MIT Press, 1989.

[13] Shirazi, B., and Kavi, K., "Parallelism Management: Synchronization, Scheduling, and Load Balancing," *(Tutorial Notes) 1991 International Conference on Supercomputing*, December 1991.

[14] Towsley, D., "Allocating Programs Containing Branches and Loops Within a Multiple Processor System," *IEEE Transactions on Software Engineering*, Vol. SE-12, No.10, October 1986, pp. 1018-1986.

[15] Wang, C.-M., and Wang, S.-D., "Efficient Processor Assignment Algorithms and Loop Transformations for Executing Nested Parallel Loops on Multiprocessors," *Transactions on Parallel and Distributed Systems*, January, 1992, pp. 71-82.

[16] Wolfe, M.J., *Supercompilers for Supercomputers*, MIT press, Cambridge, 1989.

# DEPENDENCE-BASED COMPLEXITY METRICS FOR DISTRIBUTED PROGRAMS

**Jingde Cheng**

Dept. of Computer Science and Communication Engineering, Kyushu University
6-10-1 Hakozaki, Fukuoka 812, Japan
cheng@csce.kyushu-u.ac.jp

**Abstract** - - *This paper proposes some new metrics, which are defined based on various primary program dependences, for measuring complexity of distributed programs. Since different program dependences reflect different attributes of programs, the metrics can be used to measure various complexities of a distributed program from various viewpoints.*

## 1. INTRODUCTION

Metrics for measuring software complexity have many applications in software engineering activities including analysis, testing, debugging, and maintenance of programs, and management of project. However, although a number of complexity metrics have been proposed and studied for sequential and centralized software [3,4,7], until recently little attention was focused on complexity metrics for concurrent and distributed software [6].

"Measurement is the process by which numbers or symbols are assigned to attributes of entities in the real world in such a way as to describe them according to clearly defined rules" [4]. When we intend to measure some attributes of some entities, we must be able to capture information about the attributes, and therefore, we must have some representation and/or model of the entities such that the attributes can be explicitly described in the representation and/or model. To measure the complexity of a concurrent program, we must capture information about not only control structure and data flow in every process but also synchronization structure and interprocess communication among processes. It is obvious that those program representations proposed and studied for sequential programs are inadequate to our purpose.

This paper presents some graph-theoretical representations for distributed programs and defines some new metrics based on the representations for measuring complexity of distributed programs. A major feature of our approach is to measure complexity of distributed programs by capturing various program dependences in the programs. Since different program dependences reflect different attributes of programs, the proposed metrics can be used to measure various complexities of a distributed program from various viewpoints.

## 2. CONTROL-FLOW AND DEFINITION-USE NETS

A *nondeterministic parallel control-flow net* (CFN for short) is a 10-tuple $(\mathbf{V}, \mathbf{N}, \mathbf{P_F}, \mathbf{P_J}, \mathbf{A_C}, \mathbf{A_N}, \mathbf{A_{P_F}}, \mathbf{A_{P_J}}, \mathbf{s}, \mathbf{t})$, where $(\mathbf{V}, \mathbf{A_C}, \mathbf{A_N}, \mathbf{A_{P_F}}, \mathbf{A_{P_J}})$ is a simple arc-classified digraph such that $\mathbf{A_C} \subseteq \mathbf{V} \times \mathbf{V}$, $\mathbf{A_N} \subseteq \mathbf{N} \times \mathbf{V}$, $\mathbf{A_{P_F}} \subseteq \mathbf{P_F} \times \mathbf{V}$, $\mathbf{A_{P_J}} \subseteq \mathbf{V} \times \mathbf{P_J}$, $\mathbf{N} \subset \mathbf{V}$ is a set of elements, called *nondeterministic selection vertices*, $\mathbf{P_F} \subset \mathbf{V}$ $(\mathbf{N} \cap \mathbf{P_F} = \Phi)$ is a set of elements, called *parallel execution fork vertices*, $\mathbf{P_J} \subset \mathbf{V}$ $(\mathbf{N} \cap \mathbf{P_J} = \Phi$, and $\mathbf{P_F} \cap \mathbf{P_J} = \Phi)$ is a set of elements, called *parallel execution join vertices*, $\mathbf{s} \in \mathbf{V}$ is a unique vertex, called *start vertex*, such that in-degree$(\mathbf{s}) = 0$, $\mathbf{t} \in \mathbf{V}$ is a unique vertex, called *termination vertex*, such that out-degree$(\mathbf{t}) = 0$ and $\mathbf{t} \neq \mathbf{s}$, and for any $v \in \mathbf{V}$ $(v \neq \mathbf{s}, v \neq \mathbf{t})$, there exists at least one path from $\mathbf{s}$ to $v$ and at least one path from $v$ to $\mathbf{t}$. Any arc $(v_1,v_2) \in \mathbf{A_C}$ is called a *sequential control arc*, any arc $(v_1,v_2) \in \mathbf{A_N}$ is called a *nondeterministic selection arc*, and any arc $(v_1,v_2) \in \mathbf{A_{P_F}} \cup \mathbf{A_{P_J}}$ is called a *parallel execution arc*.

A usual (deterministic and sequential) control-flow graph can be regarded as a special case of nondeterministic parallel control-flow nets such that $\mathbf{N}$, $\mathbf{P_F}$, $\mathbf{P_J}$, $\mathbf{A_N}$, $\mathbf{A_{P_F}}$, and $\mathbf{A_{P_J}}$ are the empty set.

A *nondeterministic parallel definition-use net* (DUN for short) is a 7-tuple $(\mathbf{N_C}, \Sigma_\mathbf{V}, \mathbf{D}, \mathbf{U}, \Sigma_\mathbf{C}, \mathbf{S}, \mathbf{R})$, where $\mathbf{N_C} = (\mathbf{V}, \mathbf{N}, \mathbf{P_F}, \mathbf{P_J}, \mathbf{A_C}, \mathbf{A_N}, \mathbf{A_{P_F}}, \mathbf{A_{P_J}}, \mathbf{s}, \mathbf{t})$ is a CFN, $\Sigma_\mathbf{V}$ is a finite set of symbols, called *variables*, **D**: $\mathbf{V} \rightarrow P(\Sigma_V)$ and **U**: $\mathbf{V} \rightarrow P(\Sigma_V)$ are two partial functions from **V** to the power set of $\Sigma_V$, $\Sigma_C$ is a finite set of symbols, called *channels*, and **S**: $\mathbf{V} \rightarrow \Sigma_C$ and **R**: $\mathbf{V} \rightarrow \Sigma_C$ are two partial functions from **V** to $\Sigma_C$.

A DUN can be regarded as a CFN with the information concerning definitions and uses of variables and communication channels. A usual (deterministic and sequential) definition-use graph can be regarded as a special case of nondeterministic parallel definition-use nets such that $\mathbf{N}$, $\mathbf{P_F}$, $\mathbf{P_J}$, $\mathbf{A_N}$, $\mathbf{A_{P_F}}$, $\mathbf{A_{P_J}}$, $\Sigma_C$, $\mathbf{S}$, and $\mathbf{R}$ are the empty set.

Note that the above definitions of CFN and DUN are graph-theoretical, and therefore, they are independent of any programming language.

As an example, Fig. 1 shows a simple program fragment written in Occam 2 and Fig. 2 shows a textual representation of the DUN of the Occam 2 program. In Fig. 2, each line represents how a vertex, denoted by the first number in the line, is adjacent to other vertices in the DUN and which variables and communication channels are defined and/or used at the vertex. It is not so difficult to draw a graph representation of the DUN based on the information given in Fig.2.

Having CFN and DUN as representations for distributed programs, those well-known complexity metrics for sequential programs

```
1   PAR
2     SEQ
3       input1 ? x; y
4       IF
5         (x<0) OR (y<0)
6           error1 ! 14 :: "minus operator"
7           ce ! 0.0
8         y=0
9           error1 ! 11 :: "zero divide"
10          ce ! 0.0
11        TRUE
12          c ! x/y
13    SEQ
14      input2 ? n
15      sum := 0
16      WHILE n<>0
17        input2 ? data
18        sum, n := sum+data, n-1
19      ALT
20        c ? factor
21          result ! sum*factor
22        ce ? factor
23          error2 ! 14 :: "invalid factor"
24  STOP
```

Fig. 1 A fragment of Occam 2 program

```
1 p-adjacent-to 3, 14;
3 adjacent-to 5 def x, y receive-from input1;
5 adjacent-to 6, 8 use x, y;
6 adjacent-to 7 send-to error1;
7 p-adjacent-to 24 send-to ce;
8 adjacent-to 9, 12 use y;
9 adjacent-to 10 send-to error1;
10 p-adjacent-to 24 send-to ce;
12 p-adjacent-to 24 use x, y send-to c;
14 adjacent-to 15 def n receive-from input2;
15 adjacent-to 16 def sum;
16 adjacent-to 17, 19 use n;
17 adjacent-to 18 def data receive-from input2;
18 adjacent-to 16 def sum, n use sum, data, n;
19 s-adjacent-to 20, 22;
20 adjacent-to 21 def factor receive-from c;
21 p-adjacent-to 24 use sum, factor send-to result;
22 adjacent-to 23 def factor receive-from ce;
23 p-adjacent-to 24 send-to error2;
24 adjacent-to t;
```

Fig. 2 A textual representation of the DUN of the Occam 2 program of Fig. 1

defined based on usual (deterministic and sequential) control-flow and definition-use graphs can be easily redefined for distributed programs based on the representations.

## 3. PROCESS DEPENDENCE NET

Program dependences are dependence relationships holding between statements in a program [5]. Based on the DUN of a distributed program, we can formally define five kinds of primary program dependences, i.e., *control dependence*, *data dependence*, *selection dependence*, *synchronization dependence*, and *communication dependence* [1,2]. Because of the limitation of space, here we only give some informal descriptions for these primary program dependences.

Informally, a statement u is directly control-dependent on the control predicate v of a conditional branch statement (e.g., an if statement or while statement) if whether u is executed or not is directly determined by the evaluation result of v; a statement u is directly data-dependent on a statement v if the value of a variable computed at v has a direct influence on the value of a variable computed at u; a statement u is directly selection-dependent on a nondeterministic selection statement v if whether u is executed or not is directly determined by the selection result of v; a statement u is directly synchronization-dependent on another statement v if the start and/or termination of execution of v directly determines whether or not the execution of u starts and/or terminates; a statement u in a process is directly communication-dependent on another statement v in another process if the value of a variable computed at v has a direct influence on the value of a variable computed at u by an interprocess communication.

If we represent all five kinds of primary program dependences in a distributed program within an arc-classified digraph such that each type of arcs represents a kind of primary program dependences, then we can obtain an explicit dependence-based representation of the program. The author named such a representation the "Process Dependence Net" [1,2].

The *Process Dependence Net* (PDN for short) of a program is an arc-classified digraph (**V**, **Con**, **Sel**, **Dat**, **Syn**, **Com**), where **V** is the vertex set of the CFN of the program, **Con** is the set of control dependence arcs such that any (u,v)∈**Con** iff u is directly weakly control-dependent on v, **Sel** is the set of selection dependence arcs such that any (u,v)∈**Sel** iff u is directly selection-dependent on v, **Dat** is the set of data dependence arcs such that any (u,v)∈**Dat** iff u is directly data-dependent on v, **Syn** is the set of synchronization-dependent arcs such that any (u,v)∈**Syn** iff u is directly synchronization-dependent on v, and **Com** is the set of communication dependence arcs such that any (u,v)∈**Com** iff u is directly communication-dependent on v.

As an example, Fig. 3 shows a textual representation of the PDN of the Occam 2 program

```
24 syn-dep-on 23,21,12,10,7;
23 sel-dep-on 19;
22 sel-dep-on 19 syn-dep-on 10,7;
21 dat-dep-on 20,18 sel-dep-on 19 com-dep-on 3;
20 sel-dep-on 19 syn-dep-on 12;
19 con-dep-on 16;
18 con-dep-on 16 dat-dep-on 18,17,15,14;
17 con-dep-on 16;
16 con-dep-on 16 dat-dep-on 18,14 syn-dep-on 1;
15 syn-dep-on 1;
14 syn-dep-on 1;
12 con-dep-on 8 dat-dep-on 3;
10 con-dep-on 8;
 9 con-dep-on 8;
 8 con-dep-on 5 dat-dep-on 3;
 7 con-dep-on 5;
 6 con-dep-on 5;
 5 dat-dep-on 3 syn-dep-on 1;
 3 syn-dep-on 1;
```

Fig. 3 A textual representation of the PDN of the Occam 2 program of Fig. 1

of Fig. 1. In Fig. 3, each line represents how a vertex, denoted by the first number in the line, is dependent on other vertices in the PDN. It is not so difficult to draw a graph representation of the PDN based on the information given in Fig.3.

## 4. DEPENDENCE-BASED COMPLEXITY METRICS

Below, we will use the following notations of relational algebra:

$\mathbf{R}^+$ : the transitive closure of binary relation **R**.

$\sigma_{[1]=v}(\mathbf{R})$ : the selection of binary relation **R** such that $\sigma_{[1]=v}(\mathbf{R})=\{(v_1,v_2)\,|\,(v_1,v_2)\in\mathbf{R} \text{ and } v_1=v\}$.

$\sigma_{[1]\in S}(\mathbf{R})$ : the selection of binary relation **R** such that $\sigma_{[1]\in S}(\mathbf{R})=\{(v_1,v_2)\,|\,(v_1,v_2)\in\mathbf{R} \text{ and } v_1\in S\}$.

$\sigma_{[2]\in S}(\mathbf{R})$ : the selection of binary relation **R** such that $\sigma_{[2]\in S}(\mathbf{R})=\{(v_1,v_2)\,|\,(v_1,v_2)\in\mathbf{R} \text{ and } v_2\in S\}$.

Based on the PDN=(**V**, **Con**, **Sel**, **Dat**, **Syn**, **Com**) of a distributed program, we propose the following dependence-based metrics for measuring complexity of the program, where $|A|$ is the cardinality of set A, $\mathbf{D_p}\in\{\mathbf{Con}, \mathbf{Sel}, \mathbf{Dat}, \mathbf{Syn}, \mathbf{Com}\}$, $\mathbf{D_u}=\mathbf{Con}\cup\mathbf{Sel}\cup\mathbf{Dat}\cup\mathbf{Syn}\cup\mathbf{Com}$, and P is the set of all statements of a process named P:

$|\mathbf{D_p}|/|\mathbf{D_u}|$ : This is the proportion of a special primary program dependence to all primary program dependences in a program, and therefore, it can be used to measure the degree of concurrency of the program from a special viewpoint. For example, the larger are $|\mathbf{Con}|/|\mathbf{D_u}|$ and $|\mathbf{Dat}|/|\mathbf{D_u}|$ of a program, the less concurrent is the program; the larger is $|\mathbf{Sel}|/|\mathbf{D_u}|$ of a program, the more nondeterministic is the program; the larger are $|\mathbf{Syn}|/|\mathbf{D_u}|$ and $|\mathbf{Com}|/|\mathbf{D_u}|$ of a program, the more concurrent is the program.

$|\mathbf{Sel}\cup\mathbf{Syn}\cup\mathbf{Com}|/|\mathbf{D_u}|$ : This is the proportion of those primary program dependences concerning concurrency to all primary program dependences in a program, and therefore, it can be used to measure the degree of concurrency of the program from a general viewpoint.

$|\mathbf{D_p}^+|/|\mathbf{D_u}^+|$ : This metric is different from $|\mathbf{D_p}|/|\mathbf{D_u}|$ in that $|\mathbf{D_p}|/|\mathbf{D_u}|$ only concerns direct dependences while $|\mathbf{D_p}^+|/|\mathbf{D_u}^+|$ concerns not only direct dependences but also indirect dependences.

$|(\mathbf{Sel}\cup\mathbf{Syn}\cup\mathbf{Com})^+|/|\mathbf{D_u}^+|$ : The difference between this metric and $|(\mathbf{Sel}\cup\mathbf{Syn}\cup\mathbf{Com})|/|\mathbf{D_u}|$ is similar to the difference between $|\mathbf{D_p}^+|/|\mathbf{D_u}^+|$ and $|\mathbf{D_p}|/|\mathbf{D_u}|$.

$\max/\min\{|\sigma_{[1]=v}(\mathbf{D_p})| \mid v \in \mathbf{V}\}/|\mathbf{V}|$ : These are the proportions of the maximal/minimal number of statements, which a statement is directly control, data, nondeterministic selection, synchronization, or communication dependent on, respectively, to the total number of statements in a program. Obviously, the larger is this metric of a program, the more complex is the program. In particular, the larger is $\max/\min\{|\sigma_{[1]=v}(\mathbf{D_{pc}})| \mid v \in \mathbf{V}\}/|\mathbf{V}|$, where $\mathbf{D_{pc}} \in \{\mathbf{Sel}, \mathbf{Syn}, \mathbf{Com}\}$, of a program, the more complex is the concurrency in the program.

$\max/\min\{|\sigma_{[1]=v}(\mathbf{D_u})| \mid v \in \mathbf{V}\}/|\mathbf{V}|$ : These are the proportions of the maximal/minimal number of statements, which a statement is somehow directly dependent on, to the total number of statements in a program.

$\max/\min\{|\sigma_{[1]=v}(\mathbf{D_p}^+)| \mid v \in \mathbf{V}\}/|\mathbf{V}|$ : These are the proportions of the maximal/minimal number of statements, which a statement is directly and indirectly control, data, nondeterministic selection, synchronization, or communication dependent on, respectively, to the total number of statements in a program.

$\max/\min\{|\sigma_{[1]=v}(\mathbf{D_u}^+)| \mid v \in \mathbf{V}\}/|\mathbf{V}|$ : These are the proportions of the maximal/minimal number of statements, which a statement is somehow directly and indirectly dependent on, to the total number of statements in a program.

The following metrics can be used to measure the degree of interaction among processes in a program. The larger is the maximal/minimal value of such a metric of a program, the more complex is the concurrency in the program.

$|\sigma_{[1]\notin P}(\mathbf{Syn} \cup \mathbf{Com})|/|\mathbf{V}|$ : This is the proportion of the number of statements of other processes, which a process is directly dependent on, to the total number of statements in a program.

$|\sigma_{[1]\notin P}(\mathbf{Syn} \cup \mathbf{Com})^+|/|\mathbf{V}|$ : This is the proportion of the number of statements of other processes, which a process is directly and indirectly dependent on, to the total number of statements in a program.

$|\sigma_{[1]\in P1}(\mathbf{Syn} \cup \mathbf{Com}) \cap \sigma_{[2]\in P2}(\mathbf{Syn} \cup \mathbf{Com})|/|\mathbf{V}|$ : This metric is the proportion of the number of statements such that two processes directly depends on each other, to the total number of statements in a program.

$|\sigma_{[1]\in P1}(\mathbf{Syn} \cup \mathbf{Com})^+ \cap \sigma_{[2]\in P2}(\mathbf{Syn} \cup \mathbf{Com})^+|/|\mathbf{V}|$ : This metric is the proportion of the number of statements such that two processes directly and indirectly depends on each other, to the total number of statements in a program.

Some of the above dependence-based complexity metrics can be used to measure the concurrency complexity of a distributed program in various aspects, some of them can be used to measure the overall complexity of the program. Some other complexity metrics similar to the above metrics can also be considered.

## 5. CONCLUDING REMARKS

Compared with those complexity metrics defined on control-flow and definition-use graphs, the dependence-based complexity metrics proposed in this paper can be used to measure "deeper" or "more intrinsic" attributes of programs.

Since the proposed complexity metrics are defined based on graph-theoretical representations, they can also be applied to software design and specification if a design and specification can be represented by these graph-theoretical representations.

## REFERENCES


[1] J. Cheng, "Task Dependence Net as a Representation for Concurrent Ada Programs", in J. van Katwijk (ed.) "*Ada: Moving towards 2000*", Lecture Notes in Computer Science, Vol.603, Springer-Verlag, (June, 1992), pp.150-164.

[2] J. Cheng, "Slicing Concurrent Programs", *Proc. 1st International Workshop on Automated and Algorithmic Debugging* (May, 1993), to appear.

[3] S. D. Conte, H. E. Dunsmore, and V. Y. Shen, "*Software Engineering Metrics and Models*", Benjamin/Cummings, (1986), 396 pp.

[4] N. E. Fenton, "*Software Metrics: A Rigorous Approach*", Chapman & Hall, (1991), 337 pp.

[5] J. Ferrante, K. J. Ottenstein, and J. D. Warren, "The Program Dependence Graph and Its Use in Optimization", *ACM TOPLAS*, Vol.9, No.3, (July, 1987), pp.319-349.

[6] S. M. Shatz, "Towards Complexity Metric for Ada Tasking", *IEEE-CS TSE*, Vol.14, No.8, (August, 1988), pp.1122-1127.

[7] H. Zuse, "*Software Complexity: Measures and Methods*", Walter de Gruyter, (1990), 605 pp.

# Automatically Mapping Sequential Objects to Concurrent Objects: The Mutual Exclusion Problem

David L. Sims Debra A. Hensgen
Department of Electrical and Computer Engineering
University of Cincinnati
Cincinnati, Ohio 45221-0030
USA
david.sims@uc.edu debra.hensgen@uc.edu

**Abstract**

*We show how to automatically map an object, given its sequential implementation, to a concurrent object with no points of interference. Automation is needed because without it programmers must determine where mutual exclusion is needed, where synchronization is needed, and how to ensure liveness, fairness, and absence of deadlock. A high degree of confidence in the concurrent implementation can be obtained only through a formal proof of correctness or thorough testing, and it is extremely computationally complex to thoroughly test a concurrent implementation due to the inherent non-determinism. This paper focuses on the mutual exclusion problem. Our system locates all potential interference points in a sequential object and inserts an appropriate mutual exclusion algorithm. Two mutual exclusion algorithms are considered. We show how to automatically map each one onto a sequential object. The resulting concurrent object has no points of interference.*

## 1 Introduction

The major contribution presented here is a technique for automatically transforming arbitrary sequential objects into concurrent objects that contain no points of interference. Our technique requires no intervention from the programmer. We present two different lock-based mutual exclusion algorithms that transform sequential objects to concurrent objects. The transformations prevent interference and enable several object methods to execute simultaneously to increase throughput. Our model of a sequential object is not restrictive. It allows such constructions as integers, records, arrays, and pointers. Besides assignment statements, the model permits any control structure that can be constructed from branching statements such as the **WHILE**, **EXIT-LOOP**, and **RETURN** statements.

Object-oriented systems that are most efficiently implemented using non-determinism, such as servers in a distributed system, must use concurrent objects. Object-oriented programs are also used to take advantage of multiprocessor architectures. These systems require objects to asynchronously communicate with one another. The asynchrony in these environments allows several object methods to be active simultaneously, and precautions must be taken to ensure that the concurrently executing object methods do not interfere with one another.

Correct, efficient, concurrent objects are inherently more complex to develop than sequential objects [6, 3]. Several problems that do not exist in sequential programming must be solved in concurrent programming. These problems include mutual exclusion, fairness, liveness, deadlock, and synchronization.

Instead of developing concurrent software objects directly, it is easier to first develop a sequential object and then automatically map that object to a concurrent object. Using our system, the sequential object must be developed and then tested or verified. Then the sequential object can be automatically mapped to a correct, concurrent implementation.

Concurrent objects derived from sequential ones have a useful application in servers in a distributed

```
PROCEDURE Server() =
VAR message: Message;
BEGIN
  Initialize();
  LOOP
    ReceiveMessage(message);
    CASE message.type OF
    | RequestType1 => Thread.Fork(Service1);
    . . .
    | RequestTypeN => Thread.Fork(ServiceN);
    END CASE;
  END LOOP;
END Server;
```

Figure 1: Concurrent Server

system. The most efficient way for these servers to respond to requests is to fork a lightweight process or thread each time a request arrives [8]. Such servers essentially consist of a main loop that waits for an incoming message, and reacts to it by asynchronously invoking a method of a sequential object to respond to the service request. The server then waits for the next message. For example, the server program in Figure 1 forks a thread to service each request. Because there may be several requests being serviced simultaneously, the programmer of the server must worry about synchronization, mutual exclusion, and deadlock. However, by automatically deriving concurrent objects from sequential objects, we relieve the programmer of these additional concerns so he can concentrate on the simpler task of writing a correct sequential server object.

Our system addresses the primary problems involved in mapping a sequential object to a concurrent object: mutual exclusion without deadlock and synchronization.

- **Mutual exclusion.** If the methods of an object are allowed to execute concurrently, with no restrictions, they may interfere with one another and cause incorrect results. Our system inserts mutual exclusion primitives to guarantee the deadlock-free serialization of concurrently executing object methods.

- **Synchronization.** Methods that require an object to be in a certain state before commencing execution must wait for the object to enter the required state. We derive synchronization conditions from the exception handling statements in the sequential object. Then the synchronization conditions are inserted into the concurrent object.

Herlihy has developed a methodology to automatically convert sequential objects to concurrent objects. His methodology uses non-blocking algorithms to generate concurrent objects without points of interference [5]. Unlike our lock-based mutual exclusion algorithms, non-blocking mutual exclusion provides a higher degree of fault tolerance. For small objects, this technique must copy the entire state of an object to another region of memory. For large objects, only certain regions of the object's state are copied, and this copying must be performed under the control of the programmer. Therefore, for arbitrary objects, this method requires a large amount of copying and is not intended to parallelize objects that have already been written since programmer intervention is required. In addition, our system generates concurrent objects that achieve higher levels of concurrency than Herlihy's concurrent objects because his system permits at most one simultaneous writer while our system can permit multiple writers to different fields of a single object.

Barnes improves on Herlihy's work [2]. He is able to reduce the amount of copying required in the concurrent object.

In this paper we consider only the problem of automatically mapping mutual exclusion algorithms onto sequential objects. The task of automatically inserting synchronization conditions into concurrent objects is considered elsewhere [7].

## 2 Mutual Exclusion

In this section we consider two mutual exclusion algorithms and show how to automatically map each one onto a sequential object. The resulting concurrent object has no points of interference. We take a conservative approach when inserting mutual exclusion primitives. If two methods can potentially interfere with each other, then they are prevented from doing so under all circumstances.

Each of the following algorithms improves upon the naïve technique of making entire methods mutually exclusive for every method in an object. While the naïve method prevents interference in concurrent objects, it prevents methods from *overlapping* with one another. If two methods do not both need to atomically read or update a variable, they can overlap their executions.

The next two algorithms show how to apply two phase locking without deadlock to objects. Both algorithms avoid deadlock. In this section we do not address how to map methods containing array or pointer variables. They are considered in Sections 3 and 4.

### 2.1 Conservative Two Phase Locking without Deadlock

Two phase locking [4] can be used to implement mutual exclusion for concurrent object methods. Two phase locking does not prevent deadlock; however, if all locks are acquired according to a predefined ordering, then deadlock will be avoided. The *conservative two phase locking* technique places its growing phase, the period when locks are acquired, before the first statement of the sequential object. The shrinking phase, during which locks are released, occurs after the last statement in the object.

Automatic insertion of conservative two phase locking is simple and inexpensive to implement. Unfortunately, it can lock variables for a long time before they are actually used. This conservative approach reduces concurrency because fewer overlapped executions are permitted.

### 2.2 Liberal Two Phase Locking without Deadlock

In *liberal two phase locking* the growing phase overlaps with the program statements in each object method, i.e., the growing phase is not completed before the first statement in each method. To avoid deadlock, locks are acquired in the usual predefined order.

Automatic insertion of liberal two phase locking is more difficult to implement than the conservative technique, but it can increase the level of concurrency in an object method because locks are acquired later and released earlier. Unfortunately, the amount of run time overhead can be greater than conservative two phase locking because extra program statements may be needed to downgrade write locks to read locks.

## 3 Array Variables

Arrays are similar to ordinary scalar program variables. They are of fixed length and can be indexed by a constant expression such as `array[3]`. Arrays are different from other program variables because array elements can also be indexed by ordinal type expressions of arbitrary complexity such as `array[i + j]`.

In our system we treat each case differently. If an array is indexed either by an expression involving constants only or by an expression that can resolved to a constant by constant folding and constant propagation [1], then the array reference can be resolved to a unique array element. However, if an array index cannot be resolved to a constant statically, then it could potentially refer to any array index.

Array indices that can be resolved to constants (*constant indices*) require no special attention. They are no different from regular program variables because they have unique names.

On the other hand, array references that are not indexed by constant indices (we call them *variable indices*) can be resolved at run time to any array element.

For the mutual exclusion algorithms considered in Section 2 to work correctly with arrays, we must acquire locks on individual array elements. If an object method accesses an array with a constant index, a single lock must be acquired. If a variable index is used, locks on all array elements must be acquired.

## 4 Pointer Variables

In this section we examine how to map sequential objects that use pointer variables to concurrent objects. First, pointers to statically allocated variables are examined. Then we address pointers to dynamically allocated variables.

### 4.1 Static Allocation

Pointers have essentially two functions. First, addresses can be written to and read from pointer variables. Second, pointer variables can be dereferenced multiple times before reading or writing.

The first case is handled like any other global variable. A pointer variable's value is an address, which can be read and assigned to other pointer variables. For example, if `p` and `q` are pointer variables, then the statement `p := q` requires no special mutual exclusion algorithms. The mutual exclusion algorithms introduced in Section 2 suffice.

The second case requires further consideration. Whenever a pointer variable is dereferenced and a value is assigned to, or read from, the dereferenced address, we must determine the set of addresses that the pointer variable could possibly point to so that appropriate locks can be obtained in the mutual exclusion algorithms.

Before continuing, we must define a few terms and conditions. An *address* is the location in memory where some variable resides. The address of a variable is computed using the `ADR` function. Our system allows neither pointer arithmetic nor type casting with pointers.

To build this set of addresses, we use the *reaching algorithm*. First, we collect statements of the form `p := q` and `q := ADR(x)`. Next, we take the transitive closure to determine which pointer variables have access to which program variables. The set of variables that a pointer `p` can potentially reach is denoted *Reach(p)*.

Once the set *Reach* has been computed for each pointer variable, the mutual exclusion algorithms discussed previously in Section 2 may be used to prevent interference among methods using pointers. Whenever a pointer variable `p` is dereferenced, we must assume that any variable in *Reach(p)* is dereferenced. Appropriate locks must be obtained for all variables in *Reach(p)*. As usual, the locks must be obtained in a predefined order to avoid deadlock.

### 4.2 Dynamic Allocation

Since dynamically allocated variables are created at run-time, it is difficult to acquire locks on them in a predefined order. Because the risk of deadlock is high, we take a conservative approach and lock all dynamically allocated memory locations whenever one of them is referenced.

## 5 Conclusion

Concurrent objects are more complex than sequential objects. It is fundamentally simpler to write and verify a sequential object, and then map it onto a concurrent object. Towards this end, we have shown how to impose several mutual exclusion algorithms on sequential objects in order to derive concurrent objects with no points of interference. Our system automatically maps sequential objects to concurrent objects in this manner, guaranteeing that interference cannot occur. Such a guarantee relieves the software developer of extensive testing or formal proof to verify that all interference points have been eliminated from a concurrent object.

## References

[1] Alfred V. Aho, Ravi Sethi, and Jeffrey D. Ullman. *Compilers: Principles, Techniques, and Tools.* Addison Wesley, 1986.

[2] Greg Barnes. A method for implementing lock-free shared data structures. In *Proceedings of the Symposium on Parallel Algorithms and Architectures*, 1993.

[3] Carla S. Ellis. Concurrent search and insertion in 2-3 trees. *Acta Informatica*, 14:63–86, 1980.

[4] K.P. Eswaran, J.N. Gray, R.A. Lorie, and I.L. Traiger. The notions of consistency and predicate locks in a database system. *Communications of the ACM*, 19(11):624–633, November 1976.

[5] Maurice Herlihy. A methodology for implementing highly concurrent data objects. Technical Report 91-10, Cambridge Research Laboratory, Digital Equipment Corporation, Massachusetts, October 1991.

[6] Maurice Herlihy. Wait-free synchronization. *ACM Transactions on Programming Languages and Systems*, 11(1):124–149, January 1991.

[7] David L. Sims and Debra A. Hensgen. Automatically mapping sequential objects to concurrent objects: The synchronization problem. Technical Report TR 143/2/93/ECE, Department of Electrical and Computer Engineering, University of Cincinnati, February 1993.

[8] C.Q. Yang, T. Thomas, D. Hensgen, and R. Finkel. Supporting utility services in a distributed environment. In T.L. Casavant and M. Singhal, editors, *Readings in Distributed Computing.* IEEE Computer Society Press, 1992.

# Communication-Free Data Allocation Techniques for Parallelizing Compilers on Multicomputers*

Tzung-Shi Chen and Jang-Ping Sheu

Department of Electrical Engineering, National Central University, Chung-Li 32054, Taiwan, R.O.C.
sheujp@ncuee.ncu.edu.tw

**Abstract** — In this paper, we devote our efforts to the techniques of allocating array elements of nested loops onto multicomputers in a communication-free fashion for parallelizing compilers. The arrays can be partitioned under the communication-free criteria with non-duplicate or duplicate data. In addition, the performance of the strategies with non-duplicate and duplicate array data is compared.

## 1. Introduction

For distributed memory multicomputers, the memory access time from a processor to its own local memory is much faster than the time to local memory of the other processors. An efficient parallel executing programs thus requires the goal of low communication overhead. To achieve this goal, various compiler techniques have, therefore, been developed to reduce communication traffic on multicomputers. The purpose of exploiting a large amount of parallelism in sequential programs has been the previous focus of a number of researchers [13] [14]. However, exploiting a large amount of parallelism in sequential programs may not promise that the parallelized programs for parallel execution can obtain more efficiency on multicomputers. The main reason is that those extracted parallelism may possibly cause more communication overhead during parallel execution. Under the above considerations, several researchers developed parallelizing compilers in which programmers must explicitly specify data allocation and the codes could then be generated with appropriate communication constructs [1] [6].

Achieving automatic data management in designing parallelizing compilers is, nevertheless, difficult since the data must be attentively distributed so that communication traffic is minimized in parallel execution of programs. Therefore, several researchers [3] [4] [7] [12] focus the data allocation problem on automatically allocating the data or restructuring the programs in order to improve the efficiency of usage of memory hierarchy or reduce the interprocessor communication overhead in parallel machines. For distributed memory multicomputers, large amounts of communication overhead may cause the poor performance during parallel execution of programs. Some researchers, such as King, Chou and Ni [5], Ramanaujam and Sadayappen [9], and Sheu and Tai [11], studied the problems of transforming programs into the parallel form and reducing the interprocessor communication overhead. Furthermore, Ramanaujam and Sadayappen [10] focused on analyzing the For-all loops and partitioning these loops and the corresponding data such that the partitioned programs are executed without communication overhead in the distributed memory multicomputers.

In this paper, we concentrate on automatically allocating the array elements of nested loops with uniformly generated references [3] on distributed memory multicomputers. First, we analyze the pattern of references among all arrays referenced by a nested loop, and derive the sufficient conditions for communication-free partitioning of arrays. Two communication-free partitioning strategies, non-duplicate data and duplicate data, will be proposed. Our method can obtain more parallelism than the method proposed by Ramanaujam and Sadayappen [10] in For-all loops with uniformly generated references. Finally, the performance of the data allocation with non-duplicate and duplicate data strategies is discussed.

* This work was supported by the National Science Council of the Republic of China under grant NSC 82-0408-E-008-010.

## 2. Basic Concepts and Assumptions

A normalized $n$-nested loop [14] is considered in this paper. Let $\mathbf{Z}$ and $\mathbf{R}$ denote the set of integers and the set of real numbers, respectively. The symbols $\mathbf{Z}^n$ and $\mathbf{R}^n$ represent the set of $n$-tuple of integers and the set of $n$-tuple of real numbers, respectively. The *iteration space* [14] of an $n$-nested loop is a subset of $\mathbf{Z}^n$ and is defined as $I^n = \{(I_1, I_2, \ldots, I_n) \mid l_j \leq I_j \leq u_j, \text{ for } 1 \leq j \leq n\}$. The vector $\vec{i} = (i_1, i_2, \ldots, i_n)$ in $I^n$ is represented as an iteration of the nested loop. In the nested loop, there may exist *input, output, flow dependences* or *antidependence* [8] which are referred to as *data dependence* in the following discussions. Let the linear function $h : \mathbf{Z}^n \longrightarrow \mathbf{Z}^d$ be defined as a *reference function* $h(I_1, \ldots, I_n) = (a_{1,1}I_1 + \cdots + a_{1,n}I_n, \ldots, a_{d,1}I_1 + \cdots + a_{d,n}I_n)$ and be represented by the matrix

$$H = \begin{bmatrix} a_{1,1} & \cdots & a_{1,n} \\ \vdots & \vdots & \vdots \\ a_{d,1} & \cdots & a_{d,n} \end{bmatrix}_{d \times n}$$

where $a_{i,j} \in \mathbf{Z}$, for $1 \leq i \leq d$ and $1 \leq j \leq n$. In the loop body, a $d$-dimensional array element $A[h(\vec{i_1}, i_2, \ldots, i_n) + \vec{c}]$ may be referenced by the reference function $h$ at iteration $(i_1, i_2, \ldots, i_n)$ in $I^n$, where $\vec{c}$ is known as the constant offset vector in $\mathbf{Z}^d$ [12]. The *data space* of array $A$ is a subset of $\mathbf{Z}^d$ and is defined over the user-defined array subscript index set. For array $A$, all $s$ referenced array variables $A[H_p\vec{i} + \vec{c}_p]$, for $1 \leq p \leq s$, are called *uniformly generated references* [3] [12] if $H_1 = H_2 = \cdots = H_s$ where $H_p$ is the linear transformation function from $\mathbf{Z}^n$ to $\mathbf{Z}^d$, $\vec{i} \in I^n$, and $\vec{c}_p$ is the constant offset vector in $\mathbf{Z}^d$. Since little exploitable data dependence exists between nonuniformly generated references, we focus the data allocation to each array on the same reference function in a nested loop. The different arrays may have different reference functions.

**Example 1:** Consider a 2-nested loop $L1$.

```
for i = 1 to 4
    for j = 1 to 4
        S1 : A[2i, j] := C[i, j] * 7 ;
        S2 : B[j, i + 1] := A[2i - 2, j - 1] + C[i - 1, j - 1] ;
    end
end          (L1)
```

In this example, the iteration space is $I^2 = \{ (i,j) \mid 1 \leq i,j \leq 4 \}$. In loop $L1$ with three arrays $A$, $B$, and $C$, the respective reference functions are

$$H_A = \begin{bmatrix} 2 & 0 \\ 0 & 1 \end{bmatrix}, \; H_B = \begin{bmatrix} 0 & 1 \\ 1 & 0 \end{bmatrix}, \text{and } H_C = \begin{bmatrix} 1 & 0 \\ 0 & 1 \end{bmatrix}.$$

A flow dependence exists between the variables $A[2i, j]$ at statement $S_1$ and $A[2i-2, j-1]$ at statement $S_2$ with the different offset vectors $(0,0)$ and $(-2,-1)$, respectively. For array $C$ only read by loop $L1$, an input dependence exists between the variables $C[i,j]$ at statement $S_1$ and $C[i-1, j-1]$ at statement $S_2$ with the different offset vectors $(0,0)$ and $(-1,-1)$, respectively. The array variable $B[j, i+1]$ is only generated at statement $S_2$ and its offset vector is $(0,1)$. Loop $L1$ thus has the uniformly

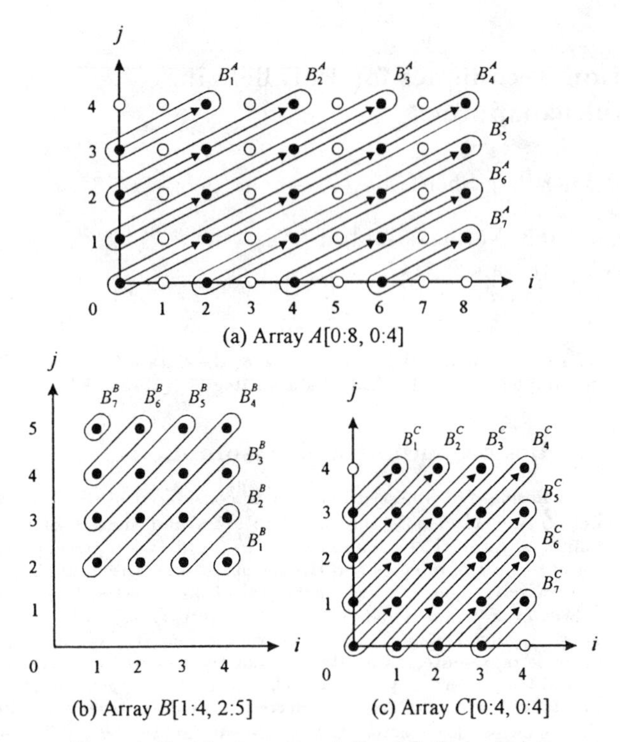

(a) Array $A[0{:}8, 0{:}4]$
(b) Array $B[1{:}4, 2{:}5]$
(c) Array $C[0{:}4, 0{:}4]$


Figure 1. Partitioning arrays $A$, $B$ and $C$ of loop $L1$ into their corresponding data blocks.

generated references on arrays $A$, $B$, and $C$. □

**Definition 1: [Data referenced vector]**
In an $n$-nested loop $L$ with uniformly generated references, if there exist two referenced array variables $A[H\bar{i} + \bar{c}_1]$ and $A[H\bar{i} + \bar{c}_2]$ for array $A$, then the vector $\bar{r} = \bar{c}_1 - \bar{c}_2$ is called *data referenced vector* of array $A$. □

The data referenced vector $\bar{r}$ represents the difference between two array elements $A[H\bar{i} + \bar{c}_1]$ and $A[H\bar{i} + \bar{c}_2]$ which are referenced by an iteration $\bar{i}$. Note that any data dependence in loop $L$ exists between two distinct referenced array variables $A[H\bar{i}+\bar{c}_1]$ and $A[H\bar{i}+\bar{c}_2]$, i.e., two iterations $\bar{i}_1$ and $\bar{i}_2$ can reference the same array element, if and only if $H\bar{i}_1 + \bar{c}_1 = H\bar{i}_2 + \bar{c}_2$, i.e., $H(\bar{i}_2 - \bar{i}_1) = \bar{r}$. Communication overhead is therefore not to be incurred if the iteration space is partitioned along the direction $\bar{i}_2 - \bar{i}_1$ into iteration blocks and the data space of array $A$ is partitioned along the direction $\bar{r}$ into data blocks.

Example 1 is considered here for illustrating the ideas of communication-free data allocation strategy. The arrays $A$, $B$, and $C$ of loop $L1$ have the referenced array variables $A[2i, j]$, $A[2i-2, j-1]$, $B[j, i+1]$, and $C[i, j]$, $C[i-1, j-1]$, respectively. The data referenced vectors of arrays $A$ and $C$ are $\bar{r}_1 = (2,1)$ and $\bar{r}_2 = (1,1)$, respectively. However, only one referenced array variable exists on array $B$; namely, no data referenced vector exists. All of the data spaces of arrays $A$, $B$, and $C$ and their data referenced vectors of each array element are shown in Figure 1(a), 1(b) and 1(c), respectively, where solid points represent that array elements are used in loop $L1$ and, however, empty points are not. At iteration $(1,1)$, the array element $A[2,1]$ is generated by $S_1$ and $A[0,0]$ is used in $S_2$. Then, at iteration $(2,2)$, the array element $A[4,2]$ is generated by $S_1$ and $A[2,1]$ is used in $S_2$, and so on. Restated, two iterations $\bar{i}_1 = (1,1)$ and $\bar{i}_2 = (2,2)$ satisfying the condition $H_A(\bar{i}_2 - \bar{i}_1) = \bar{r}_1$ can access the same array element $A[2,1]$. The data space of array $A$ is therefore partitioned along the data referenced vector $(2,1)$ into the data blocks $B_j^A$ for $1 \le j \le 7$, enclosing the points with lines, as shown in Figure 1(a). These used and generated array elements grouped in the same data block are then to be allocated to the same processor. Similarly, the array $C$ is also

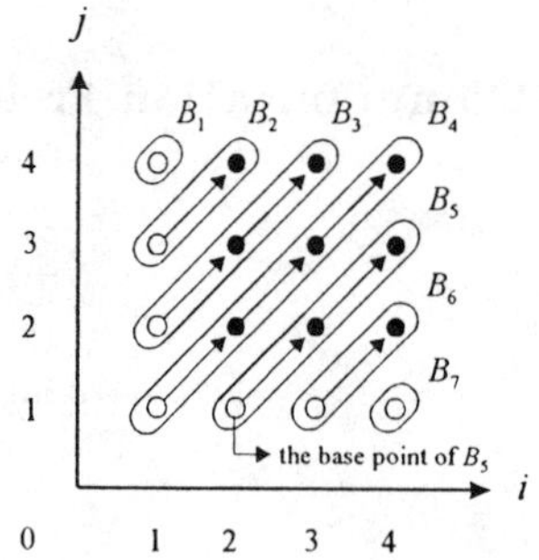


Figure 2. Partitioning the iteration space of loop $L1$ into the corresponding iteration blocks.

partitioned along data referenced vector $(1,1)$ into their corresponding data blocks, $B_j^C$ for $1 \le j \le 7$, as shown in Figure 1(c). It is easy to show that if the iteration space is partitioned along the direction $(1,1)$ as shown in Figure 2, no inter-block communication exists for arrays $A$ and $C$. Therefore, array $B$ must be partitioned along the direction $(1,1)$ into the corresponding data blocks $B_j^B$, $1 \le j \le 7$, as shown in Figure 1(b), such that the partitioned iteration blocks $B_j$, $1 \le j \le 7$, can be executed in parallel without inter-block communication.

## 3. Communication-free Array Partitioning

### 3.1 Communication-free Array Partitioning without Duplicate Data

In this subsection, we will discuss the communication-free array partitioning without duplicate data; i.e., exactly one copy of each array element exists during execution of program.

Given an $n$-nested loop $L$, the problem is how to partition the data referenced in loop $L$ such that not only the communication overhead is not necessary but also the degree of parallelism can be extracted as large as possible. We first analyze the relations among all array variables of loop $L$ and then partition the iteration space into iteration blocks such that no inter-block communication exists. For each partitioned iteration block, all data, referenced by those iterations, must be grouped into their corresponding data block for each array. Our methods proposed in this paper can make the size of partitioned iteration blocks as small as possible so as to achieve higher degree of parallelism.

From the definition of a vector space, an $n$-dimensional vector space $V$ over $\mathbf{R}$ can be generated using exactly $n$ linearly independent vectors. Let $X$ be a set of $p$ linearly independent vectors, where $p \le n$. These $p$ vectors form a basis of a $p$-dimensional subspace, denoted by span$(X)$, of $V$ over $\mathbf{R}$. The dimension of a vector space $V$ is denoted by dim$(V)$. In the following, a formal definition of partitioning of iteration space is given.

**Definition 2: [Iteration partition]**
The *iteration partition* of an $n$-nested loop $L$ partitioned by the space $\Psi = \text{span}(\{\bar{t}_1, \bar{t}_2, \ldots, \bar{t}_u\})$ where $\bar{t}_l \in \mathbf{R}^n$, $1 \le l \le u$, denoted as $P_\Psi(I^n)$, is to partition the iteration space $I^n$ into disjoint iteration blocks $B_1, B_2, \ldots, B_q$ where $q$ is the total number of partitioned blocks. For each iteration block $B_j$, $1 \le j \le q$, a base point $\bar{b}_j \in \mathbf{R}^n$ exists and

$$B_j = \{\bar{i} \in I^n | \bar{i} = \bar{b}_j + a_1\bar{t}_1 + a_2\bar{t}_2 + \cdots + a_u\bar{t}_u, a_l \in \mathbf{R}, 1 \le l \le u\}$$

where $I^n = \bigcup_{1 \le j \le q} B_j$. □

**Definition 3: [Data partition]**
Given an iteration partition $P_\Psi(I^n)$, the *data partition* of array $A$ with all $s$ referenced array variables $A[H_A\bar{i} + \bar{c}_1]$, ..., $A[H_A\bar{i} + \bar{c}_s]$, denoted as $P_\Psi(A)$, is the partition of data space of array $A$ into $q$ data blocks $B_1^A, B_2^A, \ldots, B_q^A$. For each data block $B_j^A$ corresponding to one iteration block $B_j$ of $P_\Psi(I^n)$, $1 \le j \le q$,

$$B_j^A = \{A[\bar{a}] | \bar{a} = H_A\bar{i} + \bar{c}_l, \bar{i} \in B_j, 1 \le l \le s\}.$$

□

Consider Example 1. If $\Psi = \text{span}(\{(1,1)\})$ is chosen as the space of the iteration partition $P_\Psi(I^2)$ in loop $L1$, the iteration space can be partitioned into seven iteration blocks as shown in Figure 2. The points enclosed by a line are shown in Figure 2 to form an iteration block and those dotted points represent the base points of the corresponding iteration blocks. For example, the base point $\bar{b}_5$ of iteration block $B_5 = \{\bar{i} \in I^2 | \bar{i} = \bar{b}_5 + a(1,1), 0 \le a \le 2\}$ is $(2,1)$. Based on the iteration partition $P_\Psi(I^2)$, the arrays $A$, $B$, and $C$ are partitioned into the corresponding data blocks by using the respective data partition $P_\Psi(A)$, $P_\Psi(B)$, and $P_\Psi(C)$ as shown in Figure 1.

**Example 2:** Consider a 2-nested loop $L2$.

```
for i = 1 to 4
    for j = 1 to 4
        S1: A[i+j, i+j] := B[2i, j] * A[i+j-1, i+j] ;
        S2: A[i+j-1, i+j-1] := B[2i-1, j-1]/3 ;
    end
end        (L2)
```

In loop $L2$, the respective reference functions of arrays $A$ and $B$ are

$$H_A = \begin{bmatrix} 1 & 1 \\ 1 & 1 \end{bmatrix} \text{ and } H_B = \begin{bmatrix} 2 & 0 \\ 0 & 1 \end{bmatrix}.$$

The data referenced vectors $\bar{r}_1$, between $A[i+j, i+j]$ and $A[i+j-1, i+j-1]$, $\bar{r}_2$, between $A[i+j-1, i+j-1]$ and $A[i+j-1, i+j]$, and $\bar{r}_3$, between $A[i+j-1, i+j]$ and $A[i+j, i+j]$, of array $A$ are $(1,1)$, $(0,-1)$, and $(-1,0)$, respectively. The data referenced vector $\bar{r}_4$ of array $B$ is $(1,1)$. Consider the equation $H_A\bar{t}_2 = \bar{r}_2$. Two iterations $\bar{i}_1$ and $\bar{i}_2$ can access the same element of array $A$ if the equation $\bar{i}_2 - \bar{i}_1 = \bar{t}_2$ is satisfied. Because no solution exists in the equation $H_A\bar{t}_2 = \bar{r}_2$, no data dependence exists between $A[i+j-1, i+j-1]$ and $A[i+j-1, i+j]$. However, solving the equation $H_B\bar{t}_4 = \bar{r}_4$ can exactly obtain a solution $\bar{t}_4 = (\frac{1}{2}, 1)$. It is impossible for the data dependence vector $\bar{t}_4$ between two iterations since $\bar{t}_4$ does not belong to $\mathbf{Z}^2$. Also no data dependence exists on array $B$. Let the symbol $\mathbf{0}^d \in \mathbf{Z}^d$ be denoted as a zero-vector where each component is equal to 0. Consider the equation $H\bar{t} = \bar{r}$. In the special case, when $\bar{r} = \mathbf{0}^d$, the set of solutions $\bar{t}$ of equation $H\bar{t} = \mathbf{0}^d$ is $\text{Ker}(H)$, the null space of $H$. The vector $\bar{t}$ indicates the difference of two iterations accessing the same element of a certain array variable. For example, $\text{Ker}(H_A)$ is $\text{span}(\{(1,-1)\})$ in loop $L2$. On variable $A[i+j, i+j]$, the array element $A[4,4]$, referenced by the iteration $(1,3)$, can be referenced again by iterations $(1,3)$ + $\text{span}(\{(1,-1)\})$, i.e., $(2,2)$ and $(3,1)$, of loop $L2$.

□

In the following, how to choose the better space to partition the iteration space and data spaces without duplicate data is discussed.

**Definition 4: [Reference space]**

In an $n$-nested loop $L$, if a reference function $H_A$ and $s$ variables $A[H_A\bar{i} + \bar{c}_1]$, ..., $A[H_A\bar{i} + \bar{c}_s]$ for array $A$ exist, and the data referenced vectors are $\bar{r}_p = \bar{c}_j - \bar{c}_k$ for all $1 \le j < k \le s$ and $1 \le p \le \frac{s(s-1)}{2}$, then the *reference space* of array $A$ is

$$\Psi_A = \text{span}(\beta \cup \{\bar{t}_1, \bar{t}_2, \ldots, \bar{t}_{\frac{s(s-1)}{2}}\})$$

where $\beta$ is the basis of $\text{Ker}(H_A)$ and $\bar{t}_j \in \mathbf{R}^n$, $1 \le j \le \frac{s(s-1)}{2}$, must satisfy the following conditions

(1) $\bar{t}_j$ is a particular solution of equation $H_A\bar{t} = \bar{r}_j$ and

(2) a solution $\bar{t}' \in \bar{t}_j + \text{Ker}(H_A)$ exists such that $\bar{t}' \in \mathbf{Z}^n$ and $\bar{t}' = \bar{i}_2 - \bar{i}_1$ where $\bar{i}_1, \bar{i}_2 \in I^n$.

□

The *reference space* used here is similar to the *group-temporal reuse vector space* previously defined by Wolf and Lam [12]. The *reference space* represents the relations of all data references between iterations. For array $A$, no data dependence exists between iteration blocks when the iteration space $I^n$ is partitioned with the *reference space* $\Psi_A$. This is because all data dependences are considered in $\Psi_A$ such that data accesses do not need between iteration blocks. In each iteration block, iterations according to the *lexicographical order* [14] are executed so as to preserve the dependency in loop. In loop $L2$, the *reference space* $\Psi_A$ of array $A$ is $\text{span}(\{(1,-1), (\frac{1}{2}, \frac{1}{2})\})$ because $\text{Ker}(H_A)$ $= \text{span}(\{(1,-1)\})$ and only a particular solution $\bar{t}_1 = (\frac{1}{2}, \frac{1}{2})$ of equation $H_A\bar{t} = \bar{r}_1$ exists which satisfies the conditions (1) and (2) in Definition 4. The *reference space* $\Psi_B$ of array $B$ is $\text{span}(\phi)$ because $\text{Ker}(H_B) = \{\mathbf{0}^2\}$ and the only solution $\bar{t}_4 = (\frac{1}{2}, 1) \notin \mathbf{Z}^2$ not satisfying the condition (2) in Definition 4.

**Theorem 1:**

Given an $n$-nested loop $L$ with $k$ array variables, let the *reference space* $\Psi_{A_j}$ be $\text{span}(X_j)$ of each array $A_j$ for $1 \le j \le k$. If $\Psi = \text{span}(X_1 \cup X_2 \cup \cdots \cup X_k)$, then $\Psi$ is the *partitioning space* for communication-free partitioning of arrays $A_j$ for $1 \le j \le k$ without duplicate data by using the iteration partition $P_\Psi(I^n)$.

**Proof:** The proof of this theorem can refer to [2].

□

By Theorem 1, when $\dim(\Psi) < n$, this means that the iteration partition $P_\Psi(I^n)$ exists more parallelism in loop $L$. By Definition 2, the smaller the value of $\dim(\Psi)$ is, the higher the degree of parallelism has. In general, when $\dim(\Psi) < n-1$, our method can exploit more parallelism than Ramanaujam and Sadayappen's method [10] in For-all loops with uniformly generated references. This is because Ramanaujam and Sadayappen's method only uses $(n-1)$-dimensional hyperplanes to partition the arrays in For-all loops. Consider loop $L1$. The *reference spaces* are $\Psi_A = \Psi_C = \text{span}(\{(1,1)\})$, and $\Psi_B = \{\mathbf{0}^2\}$ for respective arrays $A$, $C$, and $B$. Therefore, by Theorem 1 the *partitioning space* is $\Psi = \text{span}(\{(1,1)\} \cup \{(1,1)\} \cup \phi)$ for communication-free iteration partition $P_\Psi(I^2)$ of loop $L1$. Due to $\dim(\Psi) = 1\ (< 2)$, large amounts of parallelism exists in loop $L1$. The overall results of partitioned data and iteration blocks in loop $L1$ have been respectively shown in Figure 1 and Figure 2. Since loop $L1$ is not a For-all loop, Ramanaujam and Sadayappen's method can not solve it.

### 3.2 Communication-free Array Partitioning with Duplicate Data

In this subsection, we consider the communication-free array partitioning with duplicate data; i.e., there may exist more than one copy of an array element allocated onto local memory of processors. Due to communication overhead being most time-consuming in parallel executing programs, it is worthwhile to duplicate referenced data onto processors such that high degree of parallelism can be exploited and meanwhile the computations should be correctly performed in a communication-free fashion. Duplicate data strategy, in comparison with non-duplicate one, may extract more parallelism of programs based on communication-free array partitioning. In the following definition, two kinds of arrays are classified.

**Definition 5: [Fully and partially duplicable arrays]**

If any flow dependence does not exist on an array $A$, then the array $A$ is called *fully duplicable array*; otherwise, the array $A$ is called *partially duplicable array*.

□

For the two kinds of arrays, how to choose the better space to partition the iteration space and arrays with duplicate data such that no inter-block communication exists is discussed as follows.

First, we examine the fully duplicable arrays. Because no flow dependence exists on array $A$, any iteration will not use the elements of array $A$ generated by other iterations; therefore, the data can be arbitrarily distributed onto various processors with duplicating the elements of array $A$ and the semantic of original loop can be retained. Therefore, the *reference space* $\Psi_A$ can be reduced into $\text{span}(\phi)$ denoted as the *reduced reference space* $\Psi^r_A$. That is, $\Psi^r_A$ is the subspace of $\Psi_A$. Next, the partially duplicable arrays are to be examined. Assume there exist $p$ flow dependences on a partially duplicable array $A$ in loop $L$. The *reference space* $\Psi_A$ of array $A$ can be reduced into the *reduced reference space* $\Psi^r_A = (\beta \cup \{\bar{t}_1, \bar{t}_2, \ldots, \bar{t}_p\})$ where $\beta$ is the basis of $\text{Ker}(H_A)$ and $\bar{t}_j$, $1 \le j \le p$, which lead to flow dependences are particular solutions satisfying the conditions (1) and (2) in Definition 4. The reducible reason for the *reference space* is that only the flow dependences can actually cause the data transfer between execution of iterations. That is, only flow dependence is necessary to be considered during execution of

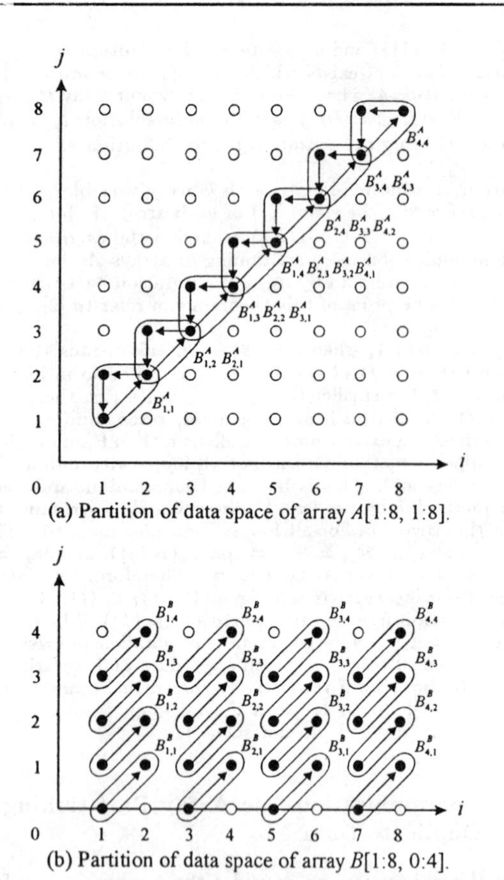

(a) Partition of data space of array $A[1{:}8,\ 1{:}8]$.

(b) Partition of data space of array $B[1{:}8,\ 0{:}4]$.

Figure 3. Partition of arrays $A$ and $B$ in loop $L2$ using the data partition $P_{\Psi'}(A)$ and $P_{\Psi'}(B)$, respectively.

programs; however, input, output dependences and antidependence merely determine the precedence of executing iterations so that they can not make any data transfer.

While partitioning the iteration space, data references which occur among all arrays in a nested loop must be considered. Given an $n$-nested loop $L$ with $k$ array variables, assume the *reduced reference space* $\Psi^r_{A_j} = \text{span}(X^r_j)$ of each either fully or partially duplicable array $A_j$ exists, $1 \le j \le k$. Then, $\Psi^r = \text{span}(X^r_1 \cup X^r_2 \cup \cdots \cup X^r_k)$ is the *partitioning space* for communication-free partitioning with duplicate data by using the iteration partition $P_{\Psi^r}(I^n)$.

Consider Example 2. By Theorem 1, while applying the iteration partition $P_\Psi(I^2)$ to loop $L2$ where $\Psi = \text{span}(\{(1,-1), (\frac{1}{2},\frac{1}{2})\})$, loop $L2$ needs to be executed sequentially based on the non-duplicate data strategy. Due to both arrays $A$ and $B$ in loop $L2$ being fully duplicable arrays, the *partitioning space* $\Psi^r$ is span($\phi$). While applying the iteration partition $P_{\Psi^r}(I^2)$ to loop $L2$, loop $L2$ can be executed in fully parallel. Clearly, using duplicate data strategy can obtain more parallelism than using non-duplicate one in loop $L2$. By duplicate data strategy, the overall results of partitioned data and iteration blocks in loop $L2$ are respectively shown in Figure 3 and Figure 4 where the relations of output dependence are omitted.

## 4. Performance Evaluation

In this section, we compare the performance of non-duplicate and duplicate data strategies. Consider the matrix multiplication algorithm.

```
for i = 1 to M
    for j = 1 to M
        for k = 1 to M
            C[i,j] := C[i,j] + A[i,k] * B[k,j] ;        (L3)
        end
    end
end
```

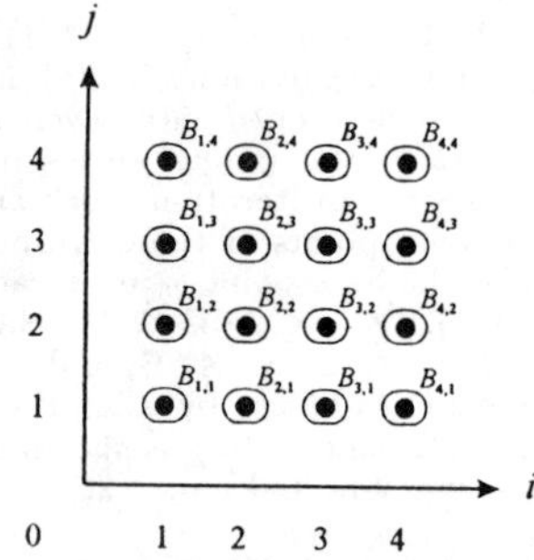

Figure 4. Partition of iteration space of loop $L2$ using the iteration partition $P_{\Psi'}(I^2)$.

For arrays $A$, $B$, and $C$, the respective *reference spaces* $\Psi_A = \text{span}(\{(0,1,0)\})$, $\Psi_B = \text{span}(\{(1,0,0)\})$, and $\Psi_C = \text{span}(\{(0,0,1)\})$. By Theorem 1, the *partitioning space* $\Psi$ is $\text{span}(\{(0,1,0)\} \cup \{(1,0,0)\} \cup \{(0,0,1)\})$. That is, the matrix multiplication algorithm needs to be executed sequentially while using the non-duplicate data strategy.

Next considered is that if only some of fully or partially duplicable arrays are duplicated, there may sacrifice little parallelism than all of them. Note that both arrays $A$ and $B$ are fully duplicable arrays and array $C$ is a partially duplicable array. Thus, the *reduced reference spaces* $\Psi^r_A = \text{span}(\phi)$, $\Psi^r_B = \text{span}(\phi)$, and $\Psi^r_C = \text{span}(\{(0,0,1)\})$ for respective arrays $A$, $B$, and $C$. Demonstrated in the following is that only the array $B$ is duplicated in loop $L3$. Due to not replicating data of array $A$, let $\Psi' = \text{span}(\{(0,1,0)\} \cup \{(0,0,1)\})$ such that the communication-free iteration partition $P_{\Psi'}(I^3)$ can be obtained. Consider a $p_1 \times p_2$ mesh multicomputer as the target machine where the number of processors is $p = p_1 \times p_2$. Assume $\sqrt{p} = p_1 = p_2$, and $M$ is a multiple of $p$. The processor $PE_a$ for $0 \le a \le p-1$ will execute the following loop $L3'$ by our program transformation and processor assignment strategies [2].

```
forall i = (1 + (a - 1) mod p) to M step p
    for j = 1 to M
        for k = 1 to M
            C[i,j] := C[i,j] + A[i,k] * B[k,j] ;        (L3')
        end
    end
end-forall
```

Because we do not replicate the data of array $A$ to each processor, the whole array $B$ must be duplicated to each processor for parallel execution without inter-processor communication. Because the processor $PE_a$, $0 \le a \le p-1$, requires accessing the array elements

$A[\alpha, 1:M]$, for $\alpha = (1 + (a-1) \bmod p) + lp, l \in \mathbf{Z}, 1 \le \alpha \le M$,

the host processor must send these data to the corresponding processor in a pipelined fashion. In addition, because all processors require accessing the same array elements $B[1:M, 1:M]$, the host processor must broadcast the whole array $B$ to each node processor. Nevertheless, if only the array $A$, not array $B$, is duplicated, the similar results can be obtained.

In the following, both arrays $A$ and $B$ in loop $L3$ are to be duplicated. Thus the communication-free iteration partition $P_{\Psi''}(I^3)$ can be obtained, where the *partitioning space* $\Psi'' = \text{span}(\{(0,0,1)\})$. By our program transformation and processor assignment strategies [2], the following results can thus be obtained. The processor $PE_{a_1,a_2}$ for $0 \le a_1 \le p_1 - 1$ and $0 \le a_2 \le p_2 - 1$ is to execute the following loop $L3''$.

```
forall i = (1 + (a1 - 1) mod p1) to M step p1
    forall j = (1 + (a2 - 1) mod p2) to M step p2
        for k = 1 to M
            C[i,j] := C[i,j] + A[i,k] * B[k,j] ;        (L3'')
```

Table I. Execution time of loops $L3$, $L3'$, and $L3''$.
(unit: second)

| Number of processors | Loop | Problem size ($M$) | | | | |
|---|---|---|---|---|---|---|
| | | 16 | 32 | 64 | 128 | 256 |
| $p=1$ | $L3$ | 0.0399 | 0.3162 | 2.5241 | 20.1691 | 161.2546 |
| $p=4$ | $L3'$ | 0.0144 | 0.0956 | 0.6961 | 5.2895 | 41.3058 |
| | $L3''$ | 0.0127 | 0.0855 | 0.6467 | 5.1405 | 40.7988 |
| $p=16$ | $L3'$ | 0.0135 | 0.0543 | 0.2869 | 1.7908 | 12.3584 |
| | $L3''$ | 0.0080 | 0.0326 | 0.2043 | 1.4326 | 10.6513 |

Table II. Speedup of loops $L3'$ and $L3''$.

| Number of processors | Loop | Problem size ($M$) | | | | |
|---|---|---|---|---|---|---|
| | | 16 | 32 | 64 | 128 | 256 |
| $p=4$ | $L3'$ | 2.77 | 3.31 | 3.63 | 3.81 | 3.89 |
| | $L3''$ | 3.14 | 3.70 | 3.90 | 3.92 | 3.95 |
| $p=16$ | $L3'$ | 2.96 | 5.82 | 8.80 | 11.26 | 13.05 |
| | $L3''$ | 4.99 | 9.70 | 12.35 | 14.08 | 15.14 |

```
        end
    end-forall
end-forall
```

Assume $M$ is a multiple of $\sqrt{p}$. Because the processors $PE_{a_1,a_2}$, $0 \leq a_1 \leq \sqrt{p}-1$, require accessing the same array elements

$$A[\alpha, 1:M], \text{ for } \alpha = (1+(a_2-1) \bmod \sqrt{p}) + l\sqrt{p}, \quad l \in \mathbf{Z}, 1 \leq \alpha \leq M,$$

for $0 \leq a_2 \leq \sqrt{p}-1$, the host processor must send the same data to the corresponding row processors by multicasting in a pipelined fashion. Similarly, because the processors $PE_{a_1,a_2}$, $0 \leq a_2 \leq \sqrt{p}-1$, require accessing the same array elements

$$B[1:M, \alpha], \text{ for } \alpha = (1+(a_1-1) \bmod \sqrt{p}) + l\sqrt{p}, \quad l \in \mathbf{Z}, 1 \leq \alpha \leq M,$$

for $0 \leq a_1 \leq \sqrt{p}-1$, the host processor must send the same data to the corresponding column processors by multicasting in a pipelined fashion. Because of only replicating the partial data of both arrays $A$ and $B$ to processors for loop $L3''$, the communication cost of distributing the initial data to each processor is less than that of loop $L3'$.

The overall execution results for loops $L3$, $L3'$, and $L3''$ are simulated on Transputer multicomputers with 16 processors are shown in Table I and Table II. The execution time of loops $L3$, $L3'$, and $L3''$ are illustrated in Table I with problem sizes $M =$ 16, 32, 64, 128 and 256. The speedup derived from Table I is illustrated in Table II. When the number of processors is equal to 1, we only consider the computation time not including the time of allocating arrays $A$ and $B$. Although duplicating data seems to waste the time of allocating initial data, it can increase great amounts of parallelism and incur no communication overhead during parallel execution of programs. Therefore, the time of parallel execution is less than that of sequential execution as shown in Table I. However, because data locality in loop $L3$ is not exploited during sequential execution, the speedup becomes more and more better whenever the problem size becomes more and more larger as shown in Table II. This implies that exploiting data locality is also important during program execution in each processor [12]. Due to existing large amounts of communication overhead in loop $L3'$ as distributing whole array $B$, the speedup of loop $L3''$ is more efficient than that of loop $L3'$. By the above analysis, the communication time of distributing the initial referenced elements of arrays must be as small as possible in order to obtain better efficiency during parallel execution. In addition, which kind of duplication of arrays is suitable for replicating their referenced data can be appropriately estimated such that parallelized programs can gain better performance during parallel execution.

## 5. Conclusions

Two automatic array partitioning strategies, non-duplicate and duplicate data, have been proposed in this paper such that no data transfer during parallel execution is incurred and the parallelism of nested loops can be exploited as large as possible. Under the duplicate data strategy, more parallelism can be extracted than non-duplicate one. By the matrix multiplication algorithm, the performance of the strategies with non-duplicate and duplicate data is discussed, and the overall results are simulated on Transputer multicomputers. By our analysis of performance, obtaining the better efficiency of executing programs is dependent on the extracted parallelism and the communication overhead of distributing the initial data under the communication-free criteria.

## References


[1] D. Callahan and K. Kennedy, "Compiling Programs for Distributed-Memory Multiprocessors," *The Journal of Supercomputing*, Vol. 2, pp. 151-169, Oct. 1988.

[2] T. S. Chen and J. P. Sheu, "Communication-free Data Allocation Techniques for Parallelizing Compilers on Multicomputers," Technique Report, Department of Electrical Engineering, National Central University, Taiwan, R.O.C., October 1992.

[3] D. Gannon, W. Jalby and J. Gallivan, "Strategies for Cache and Local Memory Management by Global Program Transformations," *Journal of Parallel and Distributed Computing*, Vol. 5, No. 5, pp. 587-616, Oct. 1988.

[4] M. Gupta and P. Banerjee, "Demonstration of Automatic Data Partitioning Techniques for Parallelizing Compilers on Multicomputers," *IEEE Transactions on Parallel and Distributed Systems*, Vol. 3, No. 2, pp. 179-193, March 1992.

[5] C. T. King, W. H. Chou and L. M. Ni, "Pipelined Data-Parallel Algorithms: Part II-Design," *IEEE Transactions on Parallel and Distributed Systems*, Vol. 1, No. 4, pp. 486-499, Oct. 1990.

[6] C. Koelbel and P. Mehrotra, "Compiling Global Name-Space Parallel Loops for Distributed Execution," *IEEE Transactions on Parallel and Distributed Systems*, Vol. 2, No. 4, pp. 440-451, Oct. 1991.

[7] M. Lu and J. Z. Fang, "A Solution of the Cache Ping-Pong Problem in Multiprocessor Systems," *Journal of Parallel and Distributed Computing*, pp. 158-171, October 1992.

[8] D. A. Padua and M. J. Wolfe, "Advanced Compiler Optimizations for Supercomputers," *Communication of ACM*, pp. 1184-1201, Dec. 1986.

[9] J. Ramanujam and P. Sadayappan, "A Methodology for Parallelizing Programs for Multicomputers and Complex Memory Multiprocessors," *Proceedings of ACM International Conference on Supercomputing*, pp. 637-646, 1989.

[10] J. Ramanujam and P. Sadayappan, "Compile-Time Techniques for Data Distribution in Distributed Memory Machines," *IEEE Transactions on Parallel and Distributed Systems*, Vol. 2, No. 4, pp. 472-482, Oct. 1991.

[11] J. P. Sheu and T. H. Tai, "Partitioning and Mapping Nested Loops on Multiprocessor Systems," *IEEE Transactions on Parallel and Distributed Systems*, Vol. 2, No. 4, pp. 430-439, Oct. 1991.

[12] M. E. Wolf and M. S. Lam, "A Data Locality Optimizing Algorithm," *Proceedings of the ACM SIGPLAN'91 Conference on Programming Language Design and Implementation*, pp. 30-44, June 1991.

[13] M. E. Wolf and M. S. Lam, "A Loop Transformation Theory and an Algorithm to Maximize Parallelism," *IEEE Transactions on Parallel and Distributed Systems*, Vol. 2, No. 4, pp. 452-471, Oct. 1991.

[14] M. J. Wolfe, "Optimizing Supercompilers for Supercomputers," London and Cambridge, MA: Pitman and the MIT Press, 1989.

# SESSION 11B

## MISCELLANEOUS

# Improving RAID-5 Performance by Un-striping Moderate-Sized Files

Ronald K. McMurdy and Badrinath Roysam
Rensselaer Polytechnic Institute, Troy, New York 12180, USA.

*A method to reduce the average operation time, and increase the transfer rate of RAID-5 disk arrays is presented, along with quantitative models for these parameters as a function of load. The modified controller allocates moderate sized files (1-20 sectors), and parity information for groups of consecutive sectors, entirely on one disk, avoiding the compounding of latency and queuing delays for multiple disks.*

*Detailed simulations of modified and standard RAID-5 arrays, a non-redundant disk array, a RAID-0 array, both synchronized and unsynchronized RAID-3 arrays, in a UNIX environment, are used to show an 11% increase in transfer rates, and a 10% decrease in operation times.*

## INTRODUCTION

It is acknowledged that CPU performance has been growing at a much faster rate than disk I/O performance [1]-[3]. Several methods have been suggested to improve the performance of disk I/O. One method is "Disk Striping" [4] which distributes a file across multiple disks to increase the data transfer rate. A second method is "Synchronized Disk Interleaving" [5], which places files on multiple disks that are rotationally synchronized. These methods are vulnerable to disk failures as the number of disks in the array increases. A technique to increase reliability, called RAID for Redundant Array of Inexpensive Disks has been presented [3], [6]-[7]. The key feature of all RAID systems is that the failure of any one disk will not cause data loss. Detailed descriptions of RAID systems can be found elsewhere [3].

This work is aimed at improving the performance of RAID-5 systems in a UNIX environment. The method is based on placing moderate-sized files, occupying 1-20 sectors (512 to 10k bytes), onto only one disk, instead of striping them across multiple disks. These files would occupy up to approximately half of one disk track. A simple rationale for this is as follows. The amount of time to read a file from disk is composed of a seek time, rotational latency, and the time to read the data. Of these three times, seek time is the most unpredictable, being based on the number of cylinders the head must move, and dependent on the cylinder accessed immediately before the current operation. The rotational latency for any disk averages half of the full track latency. However, when a file is allocated on multiple disks, the time to access the file relies on the worst case rotational latency of all the disks which must be accessed. This worst case time is approximately the same as the full track latency. From this analysis, it follows that it will generally be much faster to read the next several sectors from the same disk instead of reading from multiple disks. Since this reduces the number of disks being used for some file accesses, more operations can proceed in parallel.

While it is clear how the above modification can improve the read performance of a RAID-5 system, the impact on write performance is less so. To perform a write on a RAID-5 system, the old data and parity must be read, the old data XOR'd out of the parity and the new data XOR'd in, then the new data and parity are written back on the disks. It therefore follows that to improve the performance of writes, that a group of consecutive parity sectors should also be placed onto one disk. Thus to update a moderate sized file, one only has to access the data disk and one or two parity disks. This should also improve overall performance, since multiple I/O requests can be performed simultaneously by the modified RAID-5 array. These modifications do not affect the fault tolerance of the RAID-5 array, since the only change is the placement of data and parity sectors. This work relates to that of Chen and Patterson [10], who have proposed a method to determine the optimal amount of contiguous data written to an individual disk in a non-redundant disk array, which they define as the "striping unit." They determined the optimal striping unit for some workloads to be around 30 kbytes.

## MATHEMATICAL AND COMPUTATIONAL MODELS

All of the disk arrays used in this simulation use the same parameters for their disk drives. The parameters used are based on DEC's RA81 disk drives (Table 1), except that the number of heads on each disk has been reduced from 14 to 4. This serves to limit the size of file system required to fill at least two thirds of the disks. This modification would only have an effect on the largest files in the file system by causing additional track movement of the disk heads. To compensate for this, no time penalty is enforced when the disk heads move to adjacent tracks.

To make the disk arrays comparable, they are constructed specifically to contain the same amount of data, with the generic disk array being used as the standard. The generic disk array was selected to have 16 disks. The RAID-0 array will have 32 disks, and that the RAID-3 and RAID-5 arrays will each have 17 disks. Each disk can operate independently or in association with other disks in their array without penalty.

**Table 1:** Parameters of the RA81 disk drive [9].

| Parameter | Value |
|---|---|
| Max. latency time | 16.6ms |
| Seek time/track | 0.04ms |
| Tracks per disk | 1258 |
| # heads | 14 |
| # sectors/track | 52 |
| # bytes/sector | 512 |

The simulator constructs a file system consisting of directory files, temporary files, and permanent files. The distributions determining the size of the temporary and permanent files are taken from Floyd's paper [10] and shown in Figure 1. The distribution of the directory file sizes, is shown in Table 2. This estimation is based on data obtained from UNIX systems in our laboratory, and is consistent with data from Floyd [10]. The number of files was set at 1000 for directory and temporary files, and 25,000 for permanent files. The files are placed first-come-first-served onto the lowest numbered, emptiest disk. The permanent and temporary files are allocated uniformly across the disks. The directory files are allocated in the last 10% of the disk to simulate that they had been recently written. The disk, head, cylinder, and sector for the start of each file is constant for all the disk arrays, although each disk array can distribute the file differently among one or any number of disks.

**Table 2:** The size distribution of directory files.

| Directory Size in Bytes | Percentage |
|---|---|
| 512 | 60% |
| 1024 | 30% |
| 1536 | 5% |
| 2048 | 5% |

The distribution of file operations is derived from several smaller distributions of relevant data collected and analyzed by Floyd [10], and is shown in Table 3. The same file is accessed on each disk array, and the disk arrays are totally independent of one another. The inter-arrival rate was fixed for each simulation, and varied from 100ms to 2ms. For each simulation throughput and the average operation time values were recorded.

Floyd also found that there is typically a small number of files which receive a disproportionately large number of I/O operations. These can be described as heavily-accessed files. To incorporate this information, a number of simulations were run that define such a subset of files. Additional simulations vary the placement of the heavily-used subset of files, and in one case limit the size of these files. The percentage of all file accesses going to this set is varied from 0% through 75%. For the simulations described, the placement of the heavily accessed files is either: A) Scattered -- where the heavily accessed files are scattered uniformly across the entire disk; or B) Centralized -- where the heavily accessed files are allocated in the centrally located tracks of the disk.

**Table 3:** Percentages of operations performed on each type of file. These operations are initiated by the load generator module of the simulator.

| File type and action | Percentage |
|---|---|
| Directory - Read | 63.13 |
| Directory - Write | 4.75 |
| Temporary - Read | 2.27 |
| Temporary - Write | 6.20 |
| Temporary - Read/Write | 3.44 |
| Permanent - Read | 15.57 |
| Permanent - Write | 4.24 |
| Permanent - Read/Write | 0.40 |

## EXPERIMENTAL RESULTS

The results of the simulations are shown in Figures 2-4. Figures 2 and 3 depict measured throughput as a function of the mean inter-arrival rate, or load. Figure 4 shows the average operation time versus the inter-arrival rate, or load.

Figure 2 is the most basic case that all the other results can be compared against. Notice that the throughput of the asynchronous RAID-3 system is the first to drop to 0 kb/sec. At this point the RAID-3 system is considered to be saturated. The synchronized RAID-3 system is next to saturate, and the remaining 4 disk arrays saturate only at the highest work loads. The modified RAID-5 system is able to provide the highest level of throughput which is maintained over the widest variation of workloads. At lower levels of workload intensity, the synchronized RAID-3 system is capable of providing a slightly greater throughput, but saturates much earlier than the modified RAID-5. At the highest loads, the RAID-0 system, with twice as many disks, does not degrade as quickly as the other disks arrays. These features are observable in the remaining graphs.

As the percentage of access to the centrally-located heavily used subset of files increases, the throughput at higher loads increases very slightly. This can be attributed to a reduced distance for some seek operations and an increase in overall throughput. This phenomenon is easiest to observe when the greatest number of accesses go the subset of files. For this reason, only the graph with 75% of accesses going to the centrally-allocated heavily-accessed subset of files is included below as Figure 3.

The throughput graphs for all of the disk array architectures that were studied can be approximated by the following equation:

$$T = T_{max}(1 - e^{-\lambda(t - t_{sat})}), \quad (1)$$

where $T_{max}$ is the maximum throughput achieved by the disk array at a very large inter-arrival rate. The parameter $\lambda$ is a constant that was determined from the simulation results by fitting equation (1) to the experimental measurements. It was found to assume values in the range 0.17-0.27 depending on the type of disk array. It is approximately a constant for a given workload, and a specific disk array architecture. The parameter $t_{sat}$ is the mean inter-arrival time when the throughput diminishes to zero. This equation can be used to approximate all the throughput graphs shown in Figures 2-3.

Figure 4 shows a typical graph of the average operation time versus load. This graph does not vary significantly for the various cases discussed above. These results again place the synchronous-RAID-3 slightly ahead of the modified-RAID-5 at the lowest workloads,, but generally the modified-RAID-5 provides the best response time over the range of loads depicted. In all the cases investigated the modified-RAID-5 shows significant improvement over the standard RAID-5 system. The average operation time behavior of the disk arrays can be modeled by the following equation:

$$\tau = \tau_{min}(1 + e^{-\gamma(t - t_{sat})}), \quad (2)$$

where $\tau$ is the average operation time that includes the queuing delays, seek times, latency, and data transfer times, and $\tau_{min}$ is the best-case (minimal) operation time. The parameter $\gamma$ assumes values in the range 0.11-0.31. Like $\lambda$, it is substantially constant for a given type of disk array and fairly independent of the workload.

## CONCLUSIONS AND DISCUSSION

A method to reduce the average I/O time, and increase the transfer rate in RAID-5 disk arrays is presented. The key idea is to somehow avoid or minimize the compounding of delays associated with rotational latency, and queuing. The results are based on detailed simulations of a UNIX-type load environment. The modified RAID-5 achieves approximately an 11% increase in transfer rates, and a 10% decrease in operation times. Improvement is observed at any level of workload.

Separately, equations (1) and (2) are useful as quantitative models of disk array performance. They can be used for a variety of purposes, such as the development of improved algorithms for controlling disk arrays.

This work represents a first step toward the development of adaptive schemes for operating parallel disk arrays. Further work is needed to determine optimal values of the size of the files which are placed entirely on one disk, and to determine the optimal number of disks on which a larger file should be striped across. Also worth investigating is the possibility of intermediate schemes in which progressively larger files are allocated across more than one disk, and the impact of file usage patterns, as discerned from file types usage patterns.

## ACKNOWLEDGMENTS

This work was supported by a Digital Faculty Incentives for Excellence Award, and by IBM Corp.

## REFERENCES


[1] G. Gibson, *Redundant Disk Arrays Reliable, Parallel Secondary Storage*, The MIT Press, Cambridge Mass, 1992.

[2] A.L. Narasimha Reddy and P. Banerjee, "A Study of Parallel Disk Organizations," (ACM SIGARCH) *Computer Architecture News*, Vol. 17, No. 5, September 1989, pp. 40-47.

[3] R.H. Katz, G.A. Gibson and D.A. Patterson, "Disk System Architectures for High Performance Computing," *Proceedings of the IEEE*, Vol. 77, No. 12, December 1989, pp. 1842-1858.

[4] K. Salem and H. Garcia-Molina, "Disk Striping," *Proceedings of the 2nd IEEE International Conference on Data Engineering*, 1986,pp. 336-342.

[5] M.Y. Kim, "Synchronized Disk Interleaving," *IEEE Transactions on Computers*, Vol. C-35, No. 11, November 1986, pp. 978-988.

[6] P.M. Chen and D.A. Patterson, "Maximizing Performance in a Striped Disk Array," *Proceedings of the 17th International Symposium on Computer Architecture*, (ACM SIGARCH Computer Architecture News, Vol. 18, No. 2), Seattle 1990, pp. 322-331.

[7] E.K. Lee and R.H. Katz, "Performance Consequences of Parity Placement in Disk Arrays," *Proceedings of the 4th International Conference on Architectural Support for Programming Languages and Operating Systems* (ASPLOS-IV), Santa Clara CA, April 1991, pp. 190-199.

[8] D. Bitton and J. Gray, "Disk Shadowing," *Proc. 14th International Conference on Very Large Data Bases*, Los Angeles, 1988, pp. 331-338.

[9] *RA81 Disk Drive Users Guide*, Digital Equipment Corporation, Maynard Mass, 1982.

[10] R. Floyd, "*Short-Term File Reference Patterns in a UNIX Environment*," University of Rochester Technical Report TR177, March 1986.

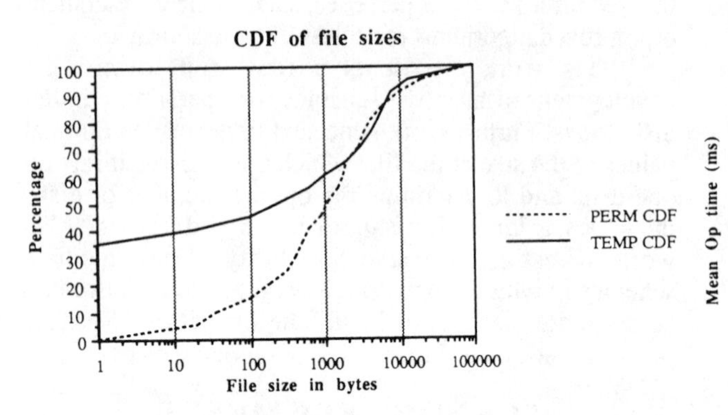


**Figure 1:** This shows the Cumulative Distribution Functions used to generate file sizes for the temporary and permanent file types.

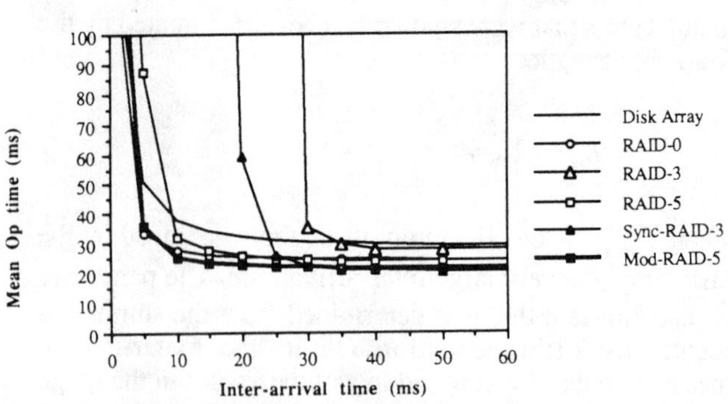

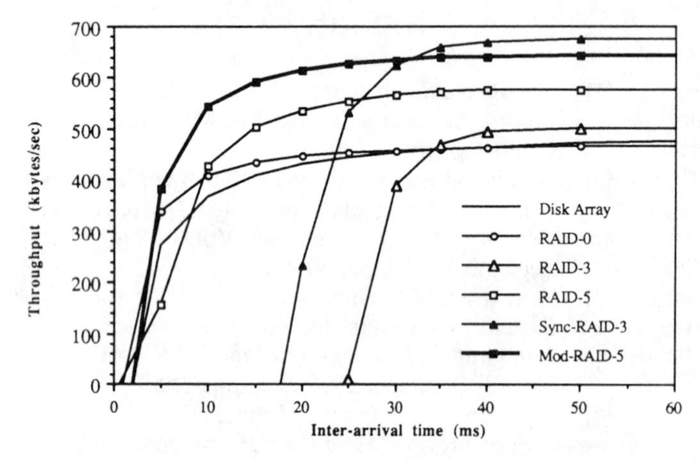


**Figure 2:** Showing variation of the measured throughput as a function of the inter-arrival rate for the six different types of disk arrays that were studied. This case corresponds to the situation in which all files within a given type are accessed with equal probability.

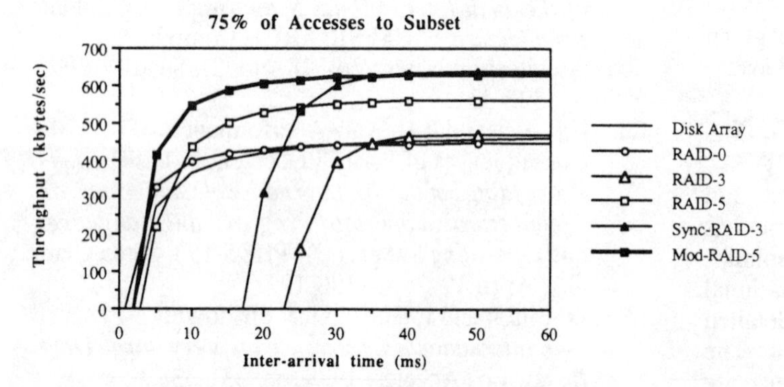


**Figure 3:** Showing the impact of placing the selected subset of files on the centrally-located tracks of the disks.

**Figure 4:** A typical graph of the average operation time as a function of load.

# On Performance and Efficiency of VLIW and Superscalar

Soo-Mook Moon Kemal Ebcioğlu
IBM Thomas J. Watson Research Center
P.O. BOX 218, Yorktown Heights, NY 10598

## Abstract

*Instruction-level parallelism in non-numerical code is characterized as leading to small speedup (as little as two) due to its irregularity. Recently, we have developed a new static scheduling algorithm called selective scheduling which can be used as a component of VLIW and superscalar compilers to exploit the irregular parallelism. This paper performs a comprehensive empirical study based on the selective scheduling compiler to examine performance/efficiency of statically scheduled machines, and to estimate the scheduling overhead. The results indicate that a logarithmic speedup increase up to five-fold is achievable on realistic resources without resorting to branch probability.*

## 1 Introduction

Fast execution of sequential code on a processor requires exploiting instruction-level parallelism (ILP). The type of parallelism focused in this paper is irregular ILP in non-numerical code, which is known to be hard to exploit due to its properties of small basic blocks, unpredictable branches, and irregular memory reference patterns.

There have been suggestions that static scheduling is not appropriate for extracting irregular ILP due to run-time dependent factors such as unpredictable branch outcomes and irregular memory access patterns, or that static scheduling is difficult to use in practice due to its long compilation time and code explosion. Contrary to these beliefs, this paper demonstrates, via a working compiler, that a substantial amount of irregular ILP can be exploited by static scheduling alone, with a reasonable scheduling overhead. Further, static scheduling is shown to be useful for small-resource superscalars, so that complex hardware for dynamic scheduling can be removed from the CPU.

The empirical study in this paper is based on a new code motion technique called *selective scheduling* [1], and various machine models with different resource constraints are evaluated for performance and scheduling overhead. The performance/efficiency issues are also examined, in order to understand how machine resources are utilized when static scheduling does not resort to branch probability. The rest of the paper is organized as follows. In Section 2, we briefly overview the approaches taken by selective scheduling compiler. Section 3 explains the experimental environment, and we present the experimental results in Section 4. The conclusion follows in Section 5.

## 2 Selective Scheduling

### 2.1 Approaches

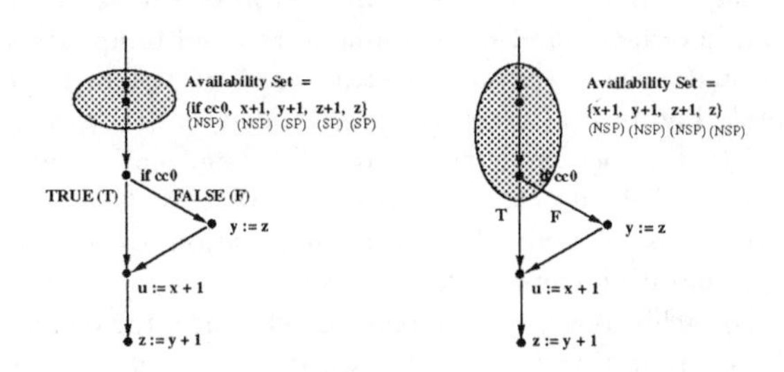


Figure 1: Grouping independent instructions

For the generation of high performance code, selective scheduling has taken four approaches. A most important global scheduling problem is to create a group of independent instructions at a point of the control flow graph (CFG). As with other global code scheduling techniques, selective scheduling solves this problem by first computing the set of all available operations that can move into the group. This set is referred to as an *availability set.*

Basically, selective scheduling can compute a larger availability set because the computation is based on the *right hand side* (*rhs*), rather than the entire operation. In Figure 1 (a), the availability set is composed of *rhs* and any of them can move into the group with either its original destination register or a new destination register after *dynamic renaming* [2]. Also, the computation of available *rhs* is performed greedily across all execution paths. In the example Figure 1 (a), the two *rhs*, `y+1` and `z+1`, are computed from the same operation `z:=y+1`, one across the TRUE path of `if cc0` and the other one across the FALSE path of `if cc0` after *combining* [3]. This means that *speculative* code motions are allowed across both targets of a branch. Finally, available operations are computed across loop iterations through *software pipelining* [4]. Since selective scheduling can compute a wide availability set, it is useful for a large-resource machine.

For the performance of a small-resource machine, it is important to fill the limited resources of the machine with more *useful* operations, which are the ones with a higher chance of being on the taken paths of execution. When branch outcomes are unpredictable, one effective heuristic is filling the resources with non-speculative operations first, and scheduling the remaining resources with slightly-speculative operations (e.g. operations that do not pass too many branches). One advantage of selective scheduling is that more non-speculative and more slightly-speculative operations can be extracted during the computation compared to other approaches [5]. In the example of Figure 1 (a), each available *rhs* is marked as non-speculative (NSP) or speculative (SP). Since the limited resources can be filled with more useful operations, high performance code can be generated even for small-resource machines.

Selective scheduling schedules branch operations as well as data operations in order to deal with the high frequency of branches in non-numerical code. More detail on this approach can be found in a companion paper [6].

Finally, a unique feature of selective scheduling is the recomputation of the availability set after each code motion, in order to find more available *rhs* and to update the speculativeness attribute of each *rhs* correctly. In the example Figure 1, after `if cc0` is scheduled into the group as in (b), all previous SP *rhs* become NSP *rhs* since they can be scheduled into the same group below `if cc0`. This update is possible only through recomputation, even though recomputation causes additional overhead.

For achieving good compilation efficiency, the overhead of recomputation for selective scheduling is reduced, since it is done incrementally only on the paths of code motion. Also, code duplication is controlled during scheduling by *unifying* the same computations on different execution paths (multiple occurrences of the same computation are hoisted up as a single operation). More detail on the selective scheduling compiler can be found in [7].

## 2.2 Program Model

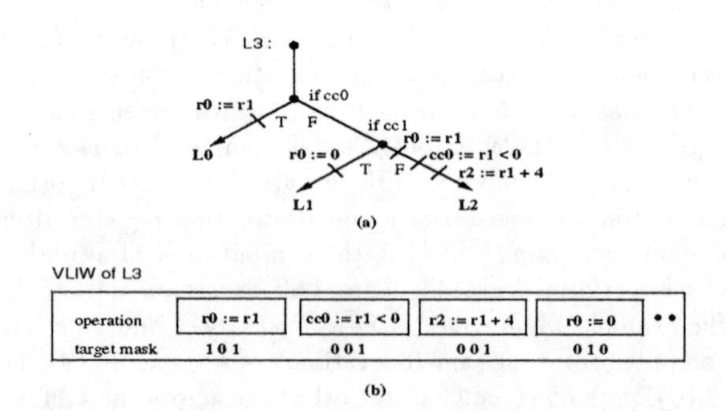


Figure 2: A VLIW tree instruction

The selective scheduling compiler uses *tree* instruction as a VLIW instruction model. Figure 2 (a) shows an example of a tree instruction L3. Each internal node of the tree corresponds to a test on a condition register, whereas terminal nodes have instruction labels, which indicate where this instruction can branch to. Each directed edge of the tree is

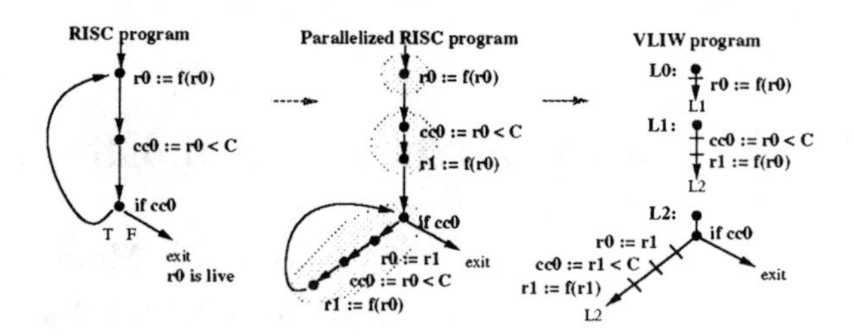


Figure 3: Parallelization example

annotated with zero or more data operations[1], which are ALU operations or memory load/store operations.

The execution of L3 consists of two steps. First, using the current truth values of `cc0` and `cc1`, a taken path is determined by traversing the tree from the root to a leaf. The instruction label at the selected leaf becomes the next tree instruction. Then, the operations on the taken path are executed in parallel by performing all the operand "reads" first, and then all operand "writes". If both `cc0` and `cc1` are "F", for example, `r0:=r1`, `cc0:=r1<0`, and `r2:=r1+4` will be executed in parallel, and L3 will branch to L2.

Our target VLIW machine executes the two-step process in a single cycle, through conditional execution and multi-way branching. That is, all operations specified in the current tree instruction start execution at the beginning of the cycle. At the same time, the multi-way branching unit determines the next target instruction. At the end of the cycle, only the operations on the taken path of the tree can write the computed results to registers or memory. This is implemented by storing a *target mask* in each operation, which indicates on which paths of the tree the operation is present. An operation either commits or discards its result according to whether its target mask includes the next target. Although there are two instances of `r0:=r1` in L3, they occupy only one operation field and only one resource due to the mask field, as shown in Figure 2 (b). Consequently, the execution of L3 requires four ALUs and three-way branching capability.

The input to the compiler is a CFG whose nodes are RISC operations obtained from the input sequential RISC code. This graph is called a *RISC* program. Based on specific resource constraints, the compiler generates a *parallelized* RISC program, where independently executable operations have been brought together and are located in adjacent nodes in the CFG. This RISC program is called *superscalar code*, and is directly executable by a superscalar processor. Finally, the parallelized RISC program is converted into a *VLIW program*, by combining adjacent independently executable operations into VLIW trees. The VLIW program is a CFG whose nodes are VLIW *tree* instructions; it is assembled into a VLIW binary program image that can be executed by a VLIW processor. Figure 3 shows an example of this scenario.

[1] To avoid confusion, we use the term **operation** for single RISC-level instruction, and **instruction** for the long instruction word with several operations.

## 3 Experiments

The experiments have been performed in our VLIW environment. The input C code is compiled into RISC assembly code through the PL.8 optimizing compiler [8]. The RISC code is then parallelized into VLIW code. Finally, the VLIW code is executed on our VLIW simulator, producing outputs and execution statistics. Our benchmarks are composed of the SPEC 89 integer benchmarks and five AIX utilities. For the SPEC integer benchmarks, we use the official input files and simulate them to completion. For AIX utilities, we use our own test files.

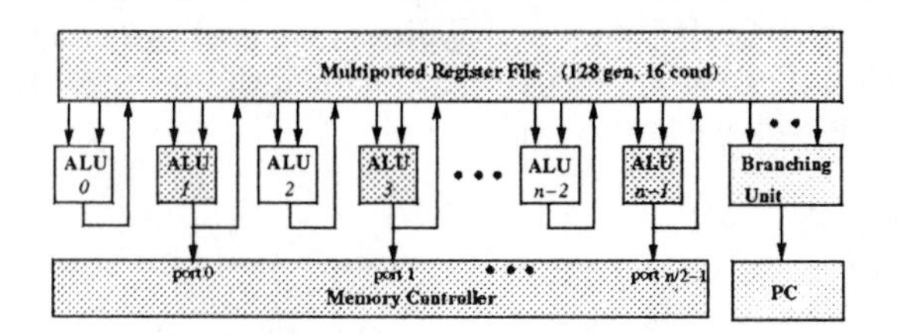


Figure 4: Machine with $n$ ALU and $n$-way branching

Our machine model is based on parametric resource constraints of $n$ ALU and $n$-way branching. We have evaluated four machine models, with $n$ = 2, 4, and 8, 16. Figure 4 shows the machine model of $n$ ALU and $n$-way branching. Each ALU can perform only one operation in each cycle. All ALUs can perform ALU operations and half of them can also perform memory load/store operations. Multiway branching is assumed to be handled by a dedicated branch unit without using ALUs. Therefore, a VLIW tree instruction on an $n$ ALU and $n$-way branching should satisfy the following resource constraints.

$\mathtt{num\_ALU\_ops} \leq n \quad \mathtt{num\_memory\_ops} \leq \frac{n}{2}$

$\mathtt{num\_branch\_ops} < n \quad \mathtt{num\_ALU\_ops} + \mathtt{num\_memory\_ops} \leq n$

The four machine models are indicated by the number of ALUs in the following description.

VLIW performance has been examined in two ways. First, we parallelized each benchmark and computed the speedup of VLIW execution over sequential execution. Performance/efficiency issues of different machine configurations are discussed with scheduling overhead. Second, we compare the VLIW performance on the SPEC integer benchmarks with an existing superscalar processor, IBM Risc System/6000.

## 4 The Results

### 4.1 VLIW vs. Sequential RISC

In the first experiment, we compare the RISC code generated by PL.8 and the VLIW code generated by the selective scheduling compiler. Both codes have been simulated on the same VLIW simulator. Thus, each RISC instruction in the RISC code becomes a tree instruction with either a single operation or a single branch. Each operation is

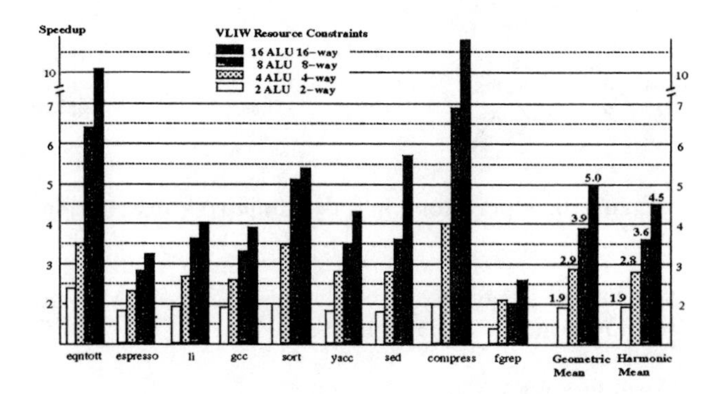


Figure 5: VLIW speedup of four models

assumed to take a single cycle with a perfect cache in both RISC and VLIW simulation.

**Performance** Figure 5 shows the VLIW speedup of four machines over sequential RISC. The speedup of each benchmark is computed by dividing the RISC execution count (the number of single-operation tree instructions in the RISC execution trace executed by an 801-like RISC processor) by the VLIW execution count (the number of tree instructions in the VLIW execution trace) of the benchmark.

Figure 5 includes both the geometric mean and the harmonic mean of speedups. For example, the 2-ALU machine obtains a geometric mean of speedup 1.9x, and the speedup increases logarithmically since it increases almost linearly when the number of resources is doubled in each step. The 16-ALU VLIW model obtains a 5x speedup over sequential RISC. These results show that the selective scheduling compiler is useful for optimization on small-resource as well as on large-resource machines.

**Utilization and Efficiency** Now, we examine how the given ALUs are utilized to obtain the speedup. RISC operations compacted in a dynamic tree instruction are classified into three categories. The entire set of the operations presented in a tree instruction is called the *specified* operations, and those on the actual taken path of the tree are called *performed* operations. Among the performed operations in each cycle, the execution of some operations does not contribute to the final performance since they are either speculative operations from the untaken path, or copy operations that have not been eliminated (by dead code elimination or copy propagation) after dynamic renaming. Those operations that are used to actually increase the speedup are called *useful* operations.

Figure 6 describes the specified and performed operations when the path of $\mathtt{cc0} \cdot \overline{\mathtt{cc1}}$ is taken in the executed tree instruction L0. Let us assume that op2 is a speculative operation came from some untaken path. Then, the useful operations do not include op2. It should be noted that all performed conditional branches are useful, because there are no speculative conditional branches in selective

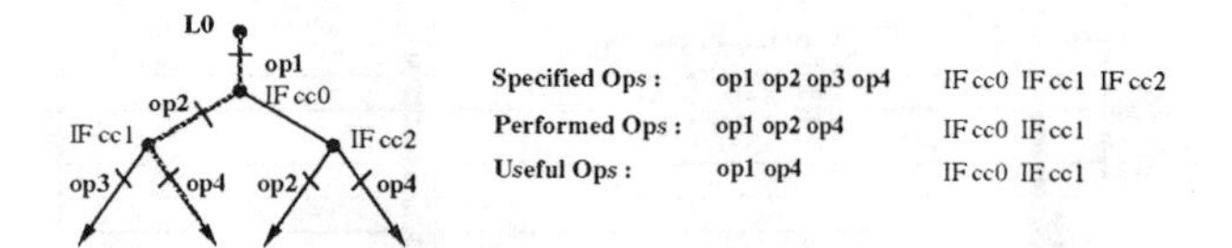


Figure 6: Operations in a dynamic tree instruction

scheduling. Therefore, we are interested only in ALU utilization, which is depicted in Figure 7.

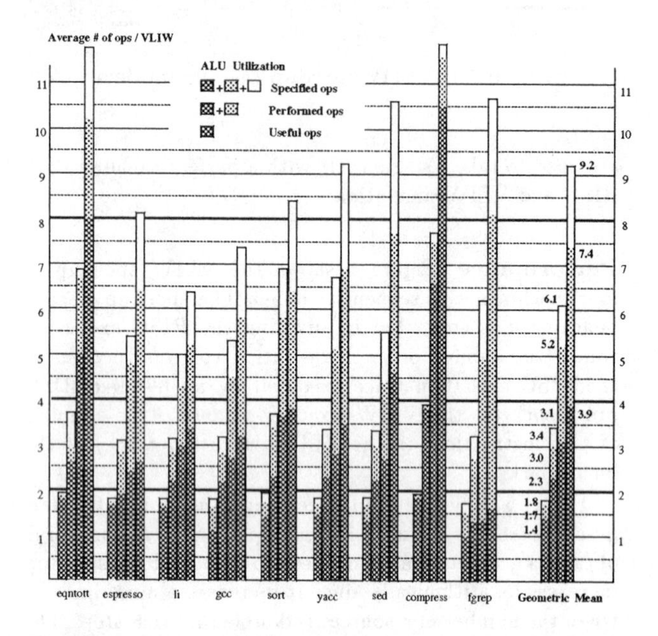


Figure 7: ALU utilization in the four machine models

In Figure 7, there are four bars in each benchmark that correspond to the four machine models. Each bar include three numbers, which are the average number of specified, performed, and useful ALU or memory operations executed in each cycle. The average number of specified and performed operations has been obtained by averaging the number of operations presented in each cycle in the dynamic execution trace, while the average number of useful operations is computed approximately from the speedup and from the number of all performed operations in each cycle. That is, the difference between the average number of performed operations per cycle, and the speedup, is the average number of useless operations executed in each cycle. Since all performed branch operations are useful, this difference gives useless operations performed on ALUs.

In Figure 7, the 16-ALU model specifies a geometric mean of 9.2 operations in each cycle, yet only 7.4 operations among them are actually performed. Among the 7.4 operations, only 3.9 operations actually increase the speedup. This can be rephrased as follows. 9.2 ALUs out of 16 start execution at the beginning of the cycle. 7.4 ALUs out of 9.2 commit their execution results. 3.9 ALUs out of 7.4 are actually used to increase the performance, the other (7.4-3.9) ALUs are used for speculative operations on untaken paths, or overhead copy operations.

The machine models with fewer ALUs obtain better utilization and waste less ALUs. However, the reason we achieve better performance when there are more ALUs is that the number of useful ALUs still increases, even though the number of wasted ALUs also increases. The first wastage factor between specified and performed ALUs comes from the artifact of conditional execution, which helps to reduce the cycle time. The second wastage factor between performed and useful ALUs mainly comes from scheduling speculative operations on untaken paths by the compiler. It should be noted that the result is obtained without using branch probabilities during selective scheduling. If branch probabilities were used to distinguish the priority of speculative code motion, this wastage factor would be reduced.

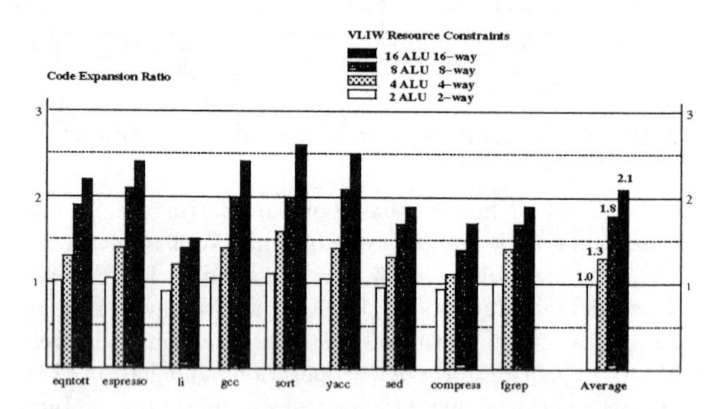


Figure 8: The VLIW code expansion of four models

**Code Expansion** Figure 8 describes code expansion rates of the VLIW code over the original sequential RISC code in the four VLIW models. The code explosion rate is computed by comparing the number of RISC operations in the original unparallelized code and the number of RISC operations in the final VLIW trees. Multiple occurrences of the same RISC operation in a given VLIW tree are counted as a single operation, and no-ops are not counted. This is a natural size estimate for a compacted representation of VLIW code in memory. Additional no-ops would have to be re-created when fetching instructions back into the instruction cache, to fill the unused slots in each VLIW.

The average code expansion rate is 2.1 with the 16-ALU VLIW model and is decreasing in the smaller ALU machines, since the number of code motions is decreased, thus requiring less bookkeeping code.

**Parallelization Time** We compare the parallelization time with the original C compilation time of PL.8 compiler to figure out the parallelization overhead. The computer used in the compilation is IBM 9021. Figure 9 shows the parallelization overhead on the compilation time.

Our parallelization time is consistently smaller than the C compilation time of the PL.8 compiler on all benchmarks, even for the most relaxed 16-ALU machine model.

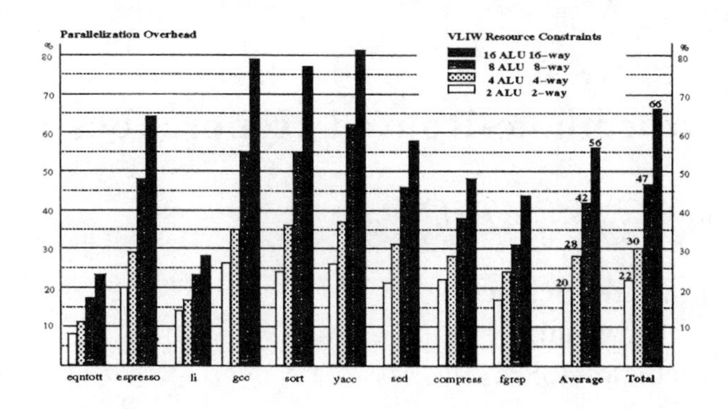

Figure 9: The parallelization overhead

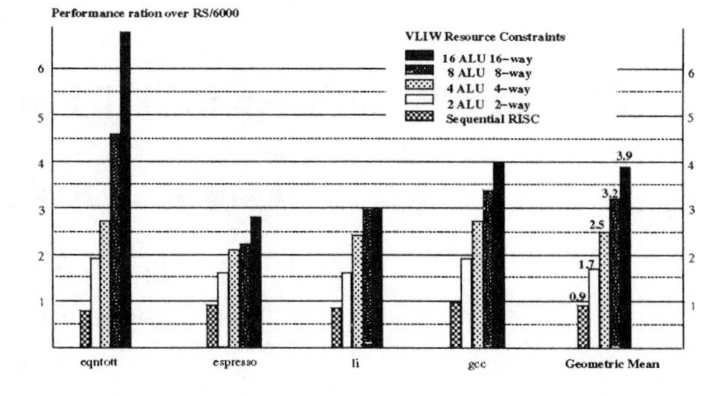

Figure 10: VLIW SPEC performance over RS/6000

## 4.2 SPEC integer: VLIW vs. RS/6000

| VLIW Cache | Data Cache | Instr Cache |
|---|---|---|
| cache size | 256 K bytes | 4 K long words |
| organization | 2-way assoc. | direct mapped |
| line size | 64 byte | 8 long words |
| penalty/miss | 8 cycles | 15 cycles |

Table 1: VLIW Cache model

We compared the VLIW SPEC performances with Risc System/6000. The target VLIW processor is assumed to have a D-cache and an I-cache as shown in Table 1. For a technology independent comparison, the target VLIW processor is assumed to have the same cycle time (40 ns) as Risc System/6000 model 530. Each operation is again assumed to take a single cycle[2] in VLIW. VLIW execution time is measured as follows.

```
VLIW exec. time (ns) = (VLIW exec. cycles * 40)+
(#of D-cache miss*8*40)+(#of I-cache miss*15*40)
```

The execution time of Risc System/6000 has been measured based on the AIX 3.1 software in the model 530.

Figure 10 shows the VLIW performance ratio over RS/6000 of the four machine models. We also simulated the performance of the sequential RISC that has been used in the previous experiments to compare our simulation environment with that of RS/6000. The sequential RISC is around 10 % slower than RS/6000.

# 5 Conclusion

This paper has examined the performance and efficiency of statically scheduled machines using the selective scheduling compiler. Although the representation of conditional execution reduces the effective resource constraints, and speculative code motions without branch probability also result in wasted resources, the approach of selective scheduling increases speedup when there are more resources. Also, the results indicate that selective scheduling can obtain high performance with a reasonable amount of scheduling overhead. Consequently, static scheduling is useful for the exploitation of irregular ILP, along with its advantage of not increasing the cycle time.

[2]Even though a memory load takes a single cycle in our environment, our memory load operation does not contain displacements. For example, our memory load is performed by `r1:=r0+4, r2:=load(r1)` while RS/6000 memory load is performed by `r2:=load(r0,4)` followed by a one-cycle delay. Since the length of the critical path including load will be similar, this is a fair comparison.

**Acknowledgements** Thanks to the contributors of the compiler and simulator, including Toshio Nakatani, Manoj Franklin, Arkady Polyak. We also thank Gabby Silberman and Jaime Moreno for helpful discussions.

# References

[1] S.-M. Moon and K. Ebcioğlu. An Efficient Resource-Constrained Global Scheduling Technique for Superscalar and VLIW processors. In *Proceedings of the Micro-25*, pages 55–71, Dec 1992.

[2] T. Nakatani and K. Ebcioğlu. Using a Lookahead Window in a Compaction-Based Parallelizing Compiler. In *Proceedings of the Micro-23*, pages 57–68, 1990.

[3] T. Nakatani and K. Ebcioğlu. *Combining* as a Compilation Technique for a VLIW Architecture. In *Proceedings of the Micro-22*, pages 43–55, 1989.

[4] K. Ebcioğlu and T. Nakatani. A New Compilation Technique for Parallelizing Loops with Unpredictable Branches on a VLIW architecture. In *Languages and Compilers for Parallel Computing*, pages 213–229. MIT Press, 1989.

[5] S.-M. Moon, K. Ebcioğlu, and A. Agrawala. Selective Scheduling Framework for Speculative Operations in VLIW and Superscalar Processors. In *Proceedings of the IFIP 10.3 Working Conference on Architectures and Compilation Techniques for Fine and Medium Grain Parallelism*, Jan 1993.

[6] S.-M. Moon. Increasing Instruction-Level Parallelism through Multi-way Branching. In *Proceedings of 1993 ICPP*, Aug 1993.

[7] S.-M. Moon. *Compile-time Parallelization of Non-numerical Code; VLIW and Superscalar.* PhD thesis, University of Maryland, 1993.

[8] H. Warren et al. Final Code Generation in the PL.8 Compiler. Research Report RC 11974, IBM Research Division, T.J. Watson Research Center, Jun 1986.

# Efficient Broadcast in All-Port Wormhole-Routed Hypercubes

*Philip K. McKinley** and *Christian Trefftz*

Department of Computer Science
Michigan State University
East Lansing, Michigan 48824
{mckinley,trefftz}@cps.msu.edu

**Abstract**

A method to reduce broadcast time in wormhole-routed hypercube systems is described. The method takes advantage of the distance insensitivity of wormhole routing and the presence of multiple ports between processors and their routers. Performance results from an nCUBE-2 multicomputer are given that demonstrate the advantage of the method over the traditional spanning binomial tree approach.

## 1 Introduction

Massively parallel computers (MPCs) are characterized by the distribution of memory among an ensemble of *nodes*, whose processors communicate through a network. MPCs are scalable because, as the number of nodes in the system increases, the total communication bandwidth, memory bandwidth, and processing capability of the system also increase.

Efficient communication among nodes is critical to the performance of MPCs. Communication may be *point-to-point* or *collective*, depending on whether exactly two or more than two processes participate. Examples of collective communication include data distribution, reduction, and barrier synchronization. The efficient implementation of collective communication primitives can greatly affect the performance of applications, whether they are written using message-passing primitives or in a data-parallel language, such as High Performance Fortran.

Perhaps the most fundamental collective communication routine is *broadcast*, in which the same message is delivered from a single source to all the nodes in the network. Broadcast is a special case of *multicast*, in which the same message is delivered from a source node to an arbitrary set of destination nodes. Most existing MPCs support only point-to-point, or *unicast*, communication in hardware. In these environments, broadcast must be implemented in software by sending multiple unicast messages. Instead of sending a separate copy of the message from the source to every destination, performance can be improved by using a *broadcast tree*. In this approach, the source node sends the message to only a subset of the destinations. Each recipient forwards the message to a subset of destinations that have not yet received it. The process continues until all nodes have received the message.

Traditionally, broadcast communication in hypercubes has been implemented with a *spanning binomial tree* (SBT) [1]. By using only nearest neighbor communication, this method requires $n$ message passing steps to reach the node that is farthest from the source in an $n$-dimensional hypercube. Multiple edge-disjoint SBTs trees, each transporting part of the message, can be used to reduce the time for broadcast [1]; in this method, the message must be reassembled at each receiving node.

This paper proposes a broadcast tree approach that is more efficient than the SBT for a class of new generation hypercubes and which does not require partitioning of the message. By exploiting the properties of the network interface and switching technique, significant performance improvement can be attained in the absence of hardware support. Specifically, the maximum delay among all destinations can be reduced by half, depending on the startup latency (system call time) for sending and receiving messages.

The remainder of the paper is organized as follows. Section 2 briefly describes the system model under consideration. Section 3 describes and gives an example of the proposed method. Section 4 presents performance results of different implementations of broadcast on a 64-node nCUBE-2 multicomputer, and Section 5 concludes the paper.

*This work was supported in part by the NSF grants CDA-9121641, MIP-9204066 and CDA-9222901, and by an Ameritech Faculty Fellowship.

## 2 System Model

One metric used to evaluate a distributed-memory system is its communication latency, which is the sum of three values: start-up latency, network latency, and blocking time [2]. The *start-up latency* is the time required for the system to handle the message at both the source and destination nodes. The *network latency* equals the elapsed time after the head of a message has entered the network at the source until the tail of the message emerges from the network at the destination. The *blocking time* includes all possible delays encountered during the lifetime of a message, for example, delays due to channel contention in the network. *Broadcast latency* is the time interval from when the source processor begins to send the first copy of the broadcast message until the last destination processor has received the message.

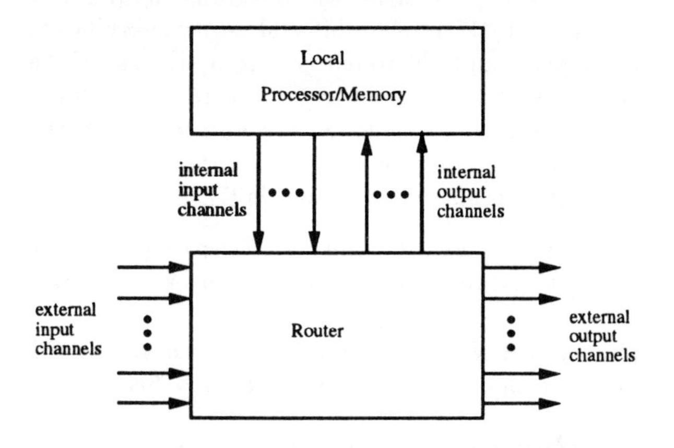


Figure 1. A generic node architecture

How to reduce broadcast latency depends on the system architecture. This paper addresses broadcast in MPCs that are characterized by four properties. First, their topologies are binary hypercubes. Second, in order to reduce latency and minimize buffer requirements, they use the *wormhole routing* switching strategy [3], in which messages are pipelined though small buffers within the network. Third, communication among nodes is handled by separate *routers* at each node; it is further assumed that dimension-ordered routing [2] is used for unicast messages. As shown in Figure 1, several pairs of *external* channels connect the router to neighboring routers, defining the network topology; pairs of *internal* channels connect each processor to its router. The fourth distinguishing characteristic of the systems considered in this paper is that they possess an *all-port* architecuture, that is, every external channel has a corresponding internal channel, allowing the node to send and receive on all its ports simultaneously.

A particular system that is encompassed by the above model is the nCUBE-2 hypercube [4]. In fact, all our performance results come from implementations on that system.

## 3 The Proposed Method

Although hardware implementations of broadcast communication offer better performance than software implementations, they may exhibit undesirable properties. For example, the nCUBE-2, which supports broadcast and restricted multicast in hardware, uses a method that may deadlock if two or more such messages are sent simultaneously [5]. Lin *et al* [5] have recently proposed deadlock-free routing algorithms for multicast and broadcast in various multicomputer topologies, including hypercubes, however, this method is not yet supported in commercial systems.

Although implemented in software, unicast-based broadcast trees can reduce latency by exploiting properties of the underlying architecture, such as the distance insensitivity of wormhole-routing. McKinley *et al* [6] have previously studied the unicast-based multicast problem for wormhole-routed, one-port $n$-dimensional meshes, which include one-port hypercubes as a special case. Our current research addresses unicast-based multicast communication for multi-port $n$-dimensional meshes. The result presented in this paper for the special case of broadcast in hypercubes is part of that larger study.

The fundamental concept behind the proposed method is to reduce broadcast latency by relaxing the common assumption that all constituent unicast messages must be sent between neighboring nodes. The *Double-Tree* (DT) algorithm begins with the source node $s$ sending to the node whose address is the bitwise complement of $s$, call it $\overline{s}$. Subsequently, each of the nodes $s$ and $\overline{s}$ plays the role of the root of a partial spanning binomial tree. The tree rooted at $s$ is called the *forward* tree, and the one rooted at $\overline{s}$ is called the *backward* tree.

As a simple example, assume that node 0000 is supposed to broadcast to all the other nodes in a 4-cube. Further assume that address resolution is carried out from top to bottom, that is, the most significant bit is resolved first in routing a unicast message. Node 0000 (0) first sends the message to node 1111 (15) by way of routers at nodes 1000, 1100, and 1110. Nodes 0000 and 1111 then become the roots of forward and backward trees, respectively. Addresses in the first tree are resolved by changing 0's to 1's, while addresses in the backward tree are resolved by changing 1's to 0's. On an $n$-port architecture, the broadcast is complete after only two steps, as listed in Figure 2(a). Figure 3 depicts the routing graphically; for clarity, some links are not shown.

| (a) top-to-bottom | (b) bottom-to-top |
|---|---|
| **Step 1:** $0 \rightarrow 15, 1, 2, 4$ | **Step 1:** $0 \rightarrow 15, 8, 4, 2$ |
| **Step 2:** $0 \rightarrow 8$ | **Step 2:** $0 \rightarrow 1$ |
| $1 \rightarrow 3, 5, 9$ | $8 \rightarrow 12, 10, 9$ |
| $2 \rightarrow 6, 10$ | $4 \rightarrow 6, 5$ |
| $4 \rightarrow 12$ | $2 \rightarrow 3$ |
| $15 \rightarrow 14, 13, 11, 7$ | $15 \rightarrow 7, 11, 13, 14$ |

Figure 2. DT forwarding in 4-cubes

It should be noted that the nCUBE-2 uses bottom-to-top address resolution. If the above example were implemented on that system, the message passing would be as listed in Figure 2(b).

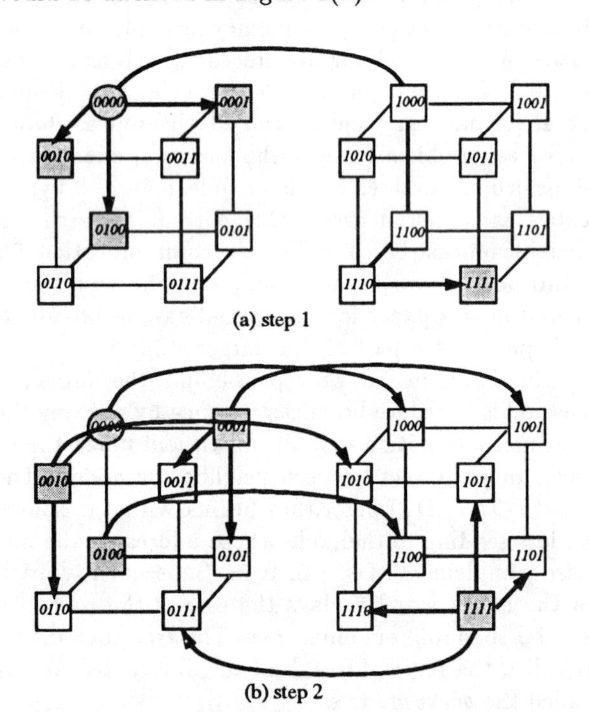


Figure 3. Example of DT broadcast in a 4-cube

Figure 4 gives the DT algorithm as executed at the source node, assuming top-to-bottom address resolution. Notice that the first message and the $n-1$ messages in the **for** loop can all be sent simultaneously in step 1 on an $n$-port architecture. Each message is labeled as either *forward* ($F$) or *backward* ($B$), according to the tree in which it is sent.

Figure 5 gives the DT algorithm as executed at a destination node, regardless of whether it is an internal node in the tree or a leaf node. The algorithm uses the functions *first_one*(), which returns the index of the leftmost "1" in an $n$-bit address, where bits are numbered from right to left. For completeness, if the address does not contain any 1's, *first_one*() returns $-1$. Space limitations do not allow us to provide the proofs of the following theorems concerning the DT algorithm; proofs may be found in [7].

**Algorithm 1: DT Source Node Algorithm**
**Input:** Source node $s$ in an $n$-cube
**Output:** Send $n+1$ messages with format:
(destination, source, direction, step, data)
**Procedure:**

```
begin
  send (s ⊕ (2^n − 1), s, B, 1, data)
  for i = 0 to n − 2 do
    send (s ⊕ 2^i, s, F, 1, data)
  endfor
  send (s ⊕ 2^(n−1), s, F, 2, data)
end
```

Figure 4. DT algorithm at source node

**Theorem 1** *The total number of steps required in a DT broadcast implementation on an n-cube is* $\lceil n/2 \rceil$.

**Theorem 2** *All of the messages constituting a DT broadcast implementation are contention free.*

# 4 Performance Measurements

In this section, we compare the performance of user-level implementations of the DT algorithm and the standard SBT algorithm on a 64-node nCUBE-2, an all-port wormhole-routed hypercube. Both algorithms were used to broadcast messages of various sizes.

Figure 6 plots the average and maximum broadcast times across different message sizes in a 6-cube. Although only three steps are required in the DT algorithm, as compared to six for the SBT algorithm, the maximum latency for the DT algorithm is greater than half that of the SBT algorithm. The primary reason is that the startup latency of the nCUBE-2 prevents it from taking full advantage of the all-port architecture. In the nCUBE-2, the sending latency is about 95 $\mu$sec and the receiving latency is about 75 $\mu$sec. For small messages, these latencies dominate the network latency; the system behaves much like a one-port architecture, and there is less difference between the two broadcast algorithms. For larger messages, the network latency becomes more significant

**Algorithm 2: DT Dest'n Node Algorithm**
**Input:** Input message ($\ell, s, dir, step, data$), where
$\ell$ is the local address,
$s$ is the source of the broadcast,
$dir$ is the direction of local subtree, and
$step$ is the step number.
**Output:** Either send messages or stop.
**Procedure:**

```
begin
  if (dir = F) and (step < ⌊n/2⌋)
    for i = first_one(ℓ ⊕ s) + 1 to n − 1 do
      send (ℓ ⊕ 2^i, s, F, step + 1, data)
    endfor
  endif

  if (dir = B) and (step < ⌊(n + 1)/2⌋)
    for i = first_one(¬(ℓ ⊕ s)) + 1 to n − 1 do
      send (ℓ ⊕ 2^i, s, B, step + 1, data)
    endfor
  endif
end
```

Figure 5. DT algorithm at a destination node

compared to startup latency. In this case, both algorithms are able to take advantage of the all-port architecture, but the DT algorithm reduces the total number of steps. Thus, as shown in Figure 6, the advantage for the DT algorithm is greater with larger messages, particularly when comparing the maximum latency experienced by nodes.

One would expect the performance advantage of the DT algorithm to improve for all sizes of messages if the startup latency were reduced. It is therefore worth noting that the recently announced nCUBE-3 is claimed to exhibit a startup latency of only 5 $\mu$sec.

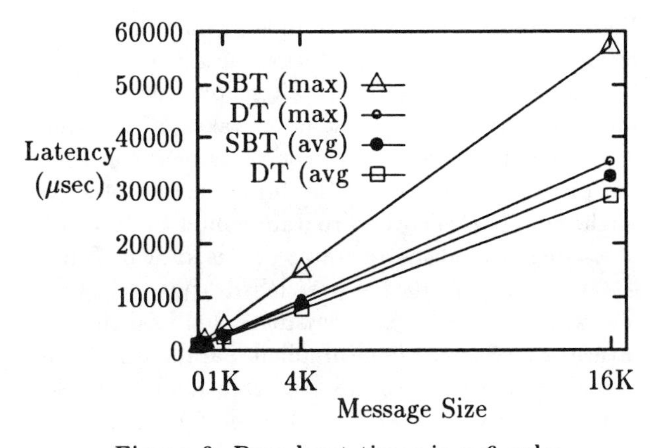


Figure 6. Broadcast times in a 6-cube

We have also used the DT algorithm to improve the performance of barrier synchronization and reduction. Space does not permit the presentation of those results here; details can be found in [7].

## 5 Conclusion

In this paper, we have explored a simple method for reducing the time to broadcast messages in a hypercube without partitioning the message. The Double-Tree (DT) algorithm takes advantage of the relative distance-insensitivity of wormhole routing and the behavior of an all-port architecture in order to effectively reduce by half the number of steps required to perform broadcast. We demonstrated the performance advantage of the algorithm over the traditional SBT approach on an nCUBE-2. Since the DT algorithm is no more difficult to implement than the SBT algorithm, the performance advantage exhibited by it would appear to make it a practical design choice for use in various collective communication routines.

## References

1. S. L. Johnsson and C.-T. Ho, "Optimum broadcasting and personalized communication in hypercubes," *IEEE Transactions on Computers*, vol. C-38, pp. 1249–1268, Sept. 1989.

2. L. M. Ni and P. K. McKinley, "A survey of wormhole routing techniques in direct networks," *IEEE Computer*, vol. 26, pp. 62–76, Feb. 1993.

3. W. J. Dally and C. L. Seitz, "The torus routing chip," *Journal of Distributed Computing*, vol. 1, no. 3, pp. 187–196, 1986.

4. NCUBE Company, *NCUBE 6400 Processor Manual*, 1990.

5. X. Lin, P. K. McKinley, and L. M. Ni, "Deadlock-free multicast wormhole routing in 2D mesh multicomputers." accepted to appear in *IEEE Transactions on Parallel and Distributed Systems.*

6. P. K. McKinley, H. Xu, A.-H. Esfahanian, and L. M. Ni, "Unicast-based multicast communication in wormhole-routed networks," in *Proc. of the 1992 International Conference on Parallel Processing*, vol. II, pp. 10–19, Aug. 1992.

7. P. K. McKinley and C. Trefftz, "Efficient broadcast in all-port wormhole-routed hypercubes," Tech. Rep. MSU-CPS-93-6, Department of Computer Science, Michigan State University, East Lansing, Michigan, Jan. 1993.

# Implementing Speculative Parallelism in Possible Computational Worlds

Debra S. Jusak, James Hearne, and Hilda Halliday
Department of Computer Science
Western Washington University
deb,hearne@cs.wwu.edu

***Abstract*— Most constructs for speculative parallelism have been devised in an *ad hoc* way, with quite particular applications in mind. What diverse instances of speculative computation have in common is the notion of executing code conditional on a future condition or event. The Tahiti programming language supports such computations directly by allowing programmers to branch on future events as well as present program states. This paper describes the system-level policies for managing speculative computations in possible computational worlds. Most closely addressed are (1) the semantics of possible worlds, (2) the elimination of process duplication in world creation, (3) determining which possible world in a computation is the real one, (4) maintaining world coherency by preventing information transfer between possible worlds.**

**Keywords:** speculative parallelism, possible worlds, event programming, communicating sequential processes.

*Introduction*

The term 'speculative parallelism' refers to the execution of parallel code some portions of which will not contribute to the final outcome of a computation. Although, strictly speaking, some processing is wasted, this way of organizing a computation can both increase response time and insure the correctness of real-time, fault-tolerant systems. Speculative parallelism lends itself naturally to architectures with very large numbers of parallel processors, an arena in which there are not many viable methods for structuring software; but its feasibility has also been demonstrated with multiprocessor architectures with relatively small number of processors, e.g., distributed systems consisting of a set of workstations connected by ethernet, hypercubes, and so forth [3].

The two distinct benefits of speculative parallelism are speedup and correctness. Speedup is achieved by (1) reducing the amount of synchronization delay between parallel portions of a program or (2) by by exploring mutually exclusive alternatives in parallel, rather than sequentially [7, 10, 1, 2, 5, 16, 6, 13, 14, 15, 9, 11]. Correctness is achieved be enabling such devices as the forward execution of recovery blocks [8]. These are discussed in [3]

*Speculative Parallelism in Tahiti*

The Tahiti programming language is intended to support all mantifestations of speculative parallelism advanced in the literature by means of a single unifying construct [4]. Tahiti is predicated on the assumption that programs exhibiting speculative parallelism have in common that they conditionally branch on future states, rather than, as is normal in imprative languages, a current state. In addition to providing for the usual features in the CSP tradition (multiple processes communicating *via* synchronized messages passing) Tahiti permits Boolean values to be qualified by future and past tense operators [4]. Tahiti also has constructs for naming and referring to events, i.e., happenings during the execution of a program.

Figure 1 gives Tahiti code branching on a future condition; it has this interpretation: if $x$ ever takes on the value 9, now or anytime after now, then execute the code fragment $C_1$. If $x$ never takes on the value 9, now or after, execute the code fragment $C_2$. Since the process cannot know the future values of $x$ one might suppose that the program would be forced to block until the ultimate status of $x$ is known. But in the spirit of speculative parallelism, future branching is given a different semantics. A branch on a future event arouses the implicit parallelism of the two alternative bodies, i.e., the creation of two possible computational worlds. One consists of the code fragment $C_1$, with the current state as its initial s-

```
        C0
if future ( x = 9 ) then
        C1
else
        C2
end if
```

Figure 1: Code Fragment Branching on a Future Event

tate. The other consists of the code fragment $C_2$, and also inherits the current state. This notion of a computational possible world has some affinities with the notion of multiple worlds, explored in [12]

This interpretation of branching thus arouses a new system-level managment responsibility. In addition to the typical management functions of a run-time system, Symphora provides complete, robust support for possible worlds.

In providing such support, three major problems arise: (1) defining a semantics for nested worlds, (2) optimizing a potential combinatorial explosion of processes, and (3) preventing information leakage between worlds.

*Possible World Semantics*

The notion of a possible computational world arouses a number of semantic difficulties. Consider again Figure 1. Let $P_1$ and $P_2$ be the two new processes created in response to a future branch. $P_1$ assumes that $x$ will take on the value 9 in the future and $P_2$ assumes that $x$ will never take on the value 9. $P_1$ and $P_2$ are said to be *computational rivals*. Once it is known whether $x$ will take on the value 9 or not, one of $P_1$ or $P_2$ is preserved as the real computation and the other is removed from the system. Determining which process (more accurately, world) is real is known as *world resolution*. The condition a world is predicated on is known as the world's *validity condition*, and it is always the case that if a world has validity condition $v$, its computational rival's validity condition is $\overline{v}$.

But this informal description of world resolution is somewhat oversimplified, because it does not address what to do when validity conditions are satisfied in both, or neither, of two rival worlds. The proper semantics are derived from the logic of linear time. Possible worlds are future worlds in real time, and in the real time in which programs execute there is only one future. Hence, a situation in which both rivals realize their validity conditions, or neither world does, is semantically incoherent and the programs in which this happens must be accounted incorrect.

| | $P_1$ | $P_2$ | Real World |
|---|---|---|---|
| 1 | $e_x$ occurs | $e_x$ occurs | $P_1$ is the real $P$ |
| 2 | $\overline{e_x}$ occurs | $\overline{e_x}$ occurs | $P_2$ is the real $P$ |
| 3 | $e_x$ occurs | $\overline{e_x}$ occurs | None |
| 3 | $\overline{e_x}$ occurs | $e_x$ occurs | None |

Table 1: World Resolution - Two Processes

The rules for world resolution between two rival worlds is summarized in Table 1. The possible world represented by process $P_1$ expects event $e$ to occur, while the possible world represented by process $P_2$ expects that the event $e$ will not occur, i.e., it is predicated on $\overline{e}$. As Table 1 indicates, at most one rival world can be real, and thus if the contradictory situations described above obtain, neither world is real. (Note that each world generates its own unique data and hence, event history.)

The rules shown in Table 1 fail to account for circumstances in which more than two rival worlds come into existence. This arises in two ways: (1) when future branches are nested, or (2) when more than one process in a Tahiti program branches on a future event.

Consider the possible worlds aroused when process $P$ branches on the future event $e_x$ and then one of the resulting process, predicated on the occurence of $e_x$ itself $P_1$ branches on the future event $e_y$. Three possible worlds exist: (1) a world represented by process $P_3$, predicated on the occurrence of two events, $e_x$ and $e_y$, (2) a world represented by process $P_4$, predicated on the occurrence of the event $e_x$ and the non-occurrence of $e_y$, and (3) a world represented by process $P_2$ predicated on the non-occurrence of event $e_x$.

Suppose that in world $P_3$, $e_x$ occurs and $\overline{e_y}$ occurs, in world $P_4$, $e_x$ and $e_y$ occur, and in world $P_2$, $\overline{e_x}$ occurs and $e_y$ occurs. Applying the rules in Table 1 iteratively, based on the predicating event between any two rival worlds, results in the conclusion that none of the worlds are real. Intuitively, however, it would seem that the world represented by process $P_2$ should be real since it assumes that $\overline{e_x}$ occurs and that is exactly what happens.

The problem with the rules outlined in Table 1 is

that they are applicable only in the case of two rival worlds. What is needed is a more general world resolution policy which accommodates any number of possible worlds. The final, real execution path of a Tahiti program resides in the possible world for which:

1. All of the event predicates have been realized, while at the same time, and
2. It is not the case that all of the event predicates have been realized in any of its rival worlds.

There is no real world (i.e., the program is incorrect) if:

1. No world realizes all of its event predicates, or
2. All the event predicates of more than one world are realized.

Using these more general policies it is clear that neither $P_3$ nor $P_4$ is real since each contradicts the predicate $e_y$ and $\overline{e_y}$, respectively. $P_2$ is the real world because its predicate is satisfied and its rival worlds', $P_3$, $P_4$, predicates have been contradicted.

We have introduced the notion of a possible computational world with examples in which each world contains exactly one process, However, Tahiti programs can be composed of many processes, a fact which complicates the optimal management of possible worlds. Consider a Tahiti program $T$ consisting of 3 processes, $A$, $B$, and $C$. Suppose that process $A$ executes a future conditional on the event $e_x$ and further that $A_1$ executes another future conditional. The resulting configuration of nested possible worlds is complicated by the fact that it possible for future conditionals to occur in distinct processes. For example, suppose that a process $B$ executes a future conditional. ????????

In each case processes are replicated and placed in any new worlds that arise. The physical system configuration varies according to whether future conditionals are nested or not, world resolution, however, is still performed according to the rules above. In the situation described, at most one of the four rival worlds can be dubbed as real.

This, as it were, theoretical model of possible worlds, is not a suitable guide to implementation because of the combinatorial explosion in process creation it entails. A better scheme will eliminate process duplication.

*Symphora Model of Possible Worlds*

*Eliminating Process Duplication*

The Symphora model of possible worlds seeks to reduce the number of processes that must be created when processes branch on future conditions. The main strategy is to create only those processes that have been, as it were, tainted by suppositions about the future. Note that in the multiprocessing situation described above, the behavior of only one process, namely $A$, is changed as result of executing a future conditional. All other processes in the same world are unaffected by $A$'s future branch. This suggests that Symphora manage this situation by placing only the new versions of $A$ in the appropriate worlds leaving processes $B$ and $C$ undisturbed. Note that while the new model has succeeded in eliminating redundant processes, two additional worlds, $SW_1$ and $SW_2$, have been introduced. $SW_1$ and $SW_2$ are rival worlds since they are predicated on contradictory validity conditions. $SW_0$ consists of all those processes duplicated in the two new possible worlds in the theoretical model. If there is a real world in this new model, it will consist of the union of the processes in $SW_0$ and $SW_1$ or $SW_0$ and $SW_2$. Furthermore, an occurrence of predicating event $e_x$ in $SW_0$ is equivalent to its occurrence in both $SW_1$ and $SW_2$. This would simultaneously validate world $SW_1$ and invalidate world $SW_2$, and otherwise, the protocol for world resolution is the same as before.

Consider, however, an elaboration of this situation. A Tahiti program consisting of three processes, $A$, $B$, and $C$, executes two future conditionals, first, one in process $A$, and secondly, one in process $B$, resulting in a world tree in which worlds correspond one for one with the theoretical model, with the stipulation that processes tno executing a future conditional are not put in the new worlds. Ignoring for now the process redundancy implied (both $B_3$ and $B_4$ are shown in two different worlds), observe that the Symphora world tree is easily translated into its theoretical equivalent by placing copies of all processes in ancestor worlds into their respective descendant leaf worlds, and defining each leaf world by the predicates of the worlds lying above them on the path to world 0. Since each world inherits the predicates of its ancestor worlds, any events raised in ancestor worlds are effectively raised in their descendants. As before, the real world corresponds to a path through the tree such that: (1) every event predicate along the path is satisfied, and (2) there is no other path in which all of the event predicates along the path are satisfied. If more than one path exists in which all the event predicates along the path are satisfied, then there is no real world, and the program represents a contradictory (incorrect) computation.

Although the revised Symphora model reduces much of the process duplication implied by the theoretical model, it does not eliminate it entirely. Recall that process $B_3$ (and $B_4$) appears in two distinct worlds. The same logical effect could be achieved by affiliating a single, executing process with more than one world. This is effected by a structures dubbed *process trees.* The process tree for the situation treated immediately above would show that there are two different versions of process $A$, $A_1$ executing in world 1 and $A_2$ executing in world 2. Similarly, the process tree for $B$ would reveal that there are two versions of process $B$, $B_3$ executing in worlds 3 and 5, and $B_4$ executing in worlds 4 and 6. Process trees thus obviate process duplication.

*World Resolution*

A final consideration in obtaining the most efficient implementation is the timely elimination of possible worlds. There are two cases in which it is possible to eliminate worlds before world resolution is performed: (1) when the event $e_x$ occurs in the world predicated upon $\overline{e_x}$ or one of its ancestor worlds, then the world predicated on $\overline{e_x}$ and all of its descendants can be eliminated, and (2) when the event $e_x$ occurs in an ancestor world of the world predicated upon event $e_x$ or, it occurs in both worlds represented by $e_x$ and $\overline{e_x}$, then the world represented by $e_x$ is absorbed into its parent.

*World Coherency*

In a Tahiti program it is possible for processes to exchange information in two ways: (1) by exchanging synchronized messages (using blocking send and blocking receive primitives) and (2) by referring to events. Symphora must insure that no information crosses world boundaries. We discuss the issues that arise in connection with message passing; the same sorts of issues also arise with event references, although the mechanism for implementation is quite different and is not discussed here.

Message passing in Tahiti is complicated by the possibility that the sender and receiver could reside in different worlds, in fact they could reside simultaneously in more than one possible world. If both processes physically reside in different worlds, one or both must join the world they already logically occupy together before the information exchange can take place. Note that it never makes sense for communications to take place between processes in rival worlds since rival worlds represent conflicting suppositions about the real world. Two cases arise: (1) a message is sent from a process in a descendant world to a receiver in an ancestor world, and (2) a message is sent from a process in an ancestor world to a receiver in a descendant world.

Suppose that in the situation articulated above that the process $B_1$ wants to send a message to $A$. Recall that, $B_1$ is understood to be executing in both worlds 3 and 5. In world 5 the message should be received by $A_2$ which corresponds to case 1 above, and in world 3 the message should be received by $A_3$ and $A_4$, which corresponds to case 2 above.

Consider case 1. Since the only world logically occupied by both the sender and receiver is the sender's world, both the sender and receiver must physically join that world before communicating. When the message arrives for process $A$ in world 2 from process $B$ residing in worlds 3 and 5 the following occurs:

1. worlds 2 and 3 are rivals so no communication is allowed
2. world 5 is a descendant of world 3 so both sender and receiver must join world 5
3. duplicate $A_2$ in world 5, (which we now called $A_5$)
4. inform $B_1$ of $A$'s new address and acknowledge that it is ready to receive the message
5. remove the affiliation of $B_1$ with world 5 and create a duplicate of $B_1$ in world 5, which we now call $B_3$.
6. send the message in world 5

Unlike the implementation of future conditionals in which processes entering new worlds ceased to exist in their previous one, $A_2$ remains forever in world 2 awaiting messages directed to the original process $A$.

The next case to be considered is the one in which the message sent by $B_1$ reaches either $A_3$ in world 7 or $A_4$ in world 8. The only world logically occupied by both the sender and receiver is the receiver's world. Here, $B_1$ joins world 7 as $B_4$ and world 8 as $B_5$.

*Summary*

Speculative parallelism is the policy of scheduling computations whose final state may not contribute in any way to the final state of the program in which it appears. We have discussed several forms and uses of speculative parallelism and suggested they might all be unified under a single construct of the Tahiti programming language, namely, branching on future conditions and states. Our main focus, however, has been to describe the support for this construct in *possible computational worlds*, which serve as the semantics of future-branching computations. Three prominent problems in the design of possible worlds were reviewed. First, the semantics of possible worlds must be spelled out so as to answer to all possible configurations of program outcome. Second, we discussed policies and data structures to prevent unnecessary process duplication when new possible worlds are created. Finally, we outlined policies to insure world coherence by preventing information in one world from being transmitted to another.

## References

[1] P. A. Bernstein, V. Hadzilacos, and N Goodman. *Concurrency Control and Recovery in Database Systems.* Academic Press, 1987.

[2] F. Warren Burton. Speculative computation, parallelism, and functional programming. *IEEE Transaction on Computers*, C-34(12):1190–1193, December 1985.

[3] James Hearne and Debra Jusak. How to use up processors. In *Proceedings of 3rd Symposium on the Frontiers of Massively Parallel Computation*, 1990.

[4] James Hearne and Debra Jusak. The tahiti programming language: Events as first-class objects. In *Proceedings of the 1990 International Conference on Computer Languages*, 1990.

[5] Matthew Huntbach. Speculative computation and priorities in concurrent logic language. In *ALPUK91. Proceedings of the 3rd UK Annual Conference on Logic Programming*, pages 23–35, 1992.

[6] N. Ichiyoshi. Parallel implementation schemes of logic programming languages. *Joho Shori*, 32(4):435–449, 1991.

[7] David R. Jefferson. Virtual time. *ACM Transactions on Programming Languages and Systems*, 7:404–425, July 1985.

[8] K. H. Kim and H.O. Welch. Distributed execution of recovery blocks: An approach to uniform treatment of hardware and software faults in real-time applications. *IEEE Transactions on Computers*, 38, may 1989.

[9] J. H. Kukula and S. DasGupta. Object-oriented programming with speculative parallelism for parallel processing. In *Proceedings of the 1987 IEEE International Conference on Computer Design: VLSI in Computers and Processors - ICCD '87*, pages 596–600, 1987.

[10] Christos Papadimitriou. *Theory of Database Concurrency Control.* Academic Press, 1986.

[11] Myra Jean Prelle, Ann M. Wollrath, Thomas J. Brando, and Edward H. Bensley. The impact of selected concurrent language constructs on the sam run-time system. *OOPS Messenger*, 2(2):99–103, April 1991.

[12] Jonathan M. Smith and Gerald Q. Maguire Jr. Exploring multiple worlds in parallel. In *1989 International Conference on Parallel Processing*, pages II–239–II–245, 1989.

[13] Robert Strom and Shaula Yemini. *Current Advances in Distributed Computing and Communications*, chapter Synthesizing Distributed and Parallel Programs Through Optimistic Transformations. Computer Science Press, 1987.

[14] E. E. Witte, R. D. Chamberlain, and M. A. Franklin. Task assignment by parallel simulated annealing. In *Proceedings. 1990 IEEE International Conference on Computer Design: VLSI in Computers and Processors*, pages 74–77, September 1990.

[15] E. E. Witte, R. D. Chamberlain, and M. A. Franklin. Parallel simulated annealing using speculative computation. *IEEE Transactions on Parallel and Distributed Systems*, 2(4):483–494, October 1991.

[16] Benjamin Yu. Parallelism via speculation in pure prolog. In *Advances in Computing and Information - ICCI 90*, pages 405–414, 1990.

# System-Level Diagnosis Strategies for $n-star$ Multiprocessor Systems

*A. Kavianpour*
DeVry Institute of Technology
City of Industry (Los Angeles), CA 91746-3495
e-mail: alireza@balboa.eng.uci.edu

**Abstract**

System-level diagnosis for $n-star$ multicomputer systems is considered. *One* $-$ $t$ and *one* $-t/t$ strategies are one-step diagnosis strategies which involve only one testing phase and one repair phase. The other strategy ($seq-t$) is sequential diagnosis strategy which involve multiple iterations of testing and repair phases. The cost and diagnostic power of these three basic strategies in applying them to $n-star$ are compared. In this paper previously unknown properties regarding the diagnosabilities and the costs of some of the strategies, especially, sequential diagnosis strategy, were uncovered. Some of the newly discovered properties include a lower bound for the degree of diagnosability of the $seq-t$ strategy in $n-star$.

**Key Words:** Diagnosability, $n-star$, Sequential diagnosis, System-level diagnosis, Testing.

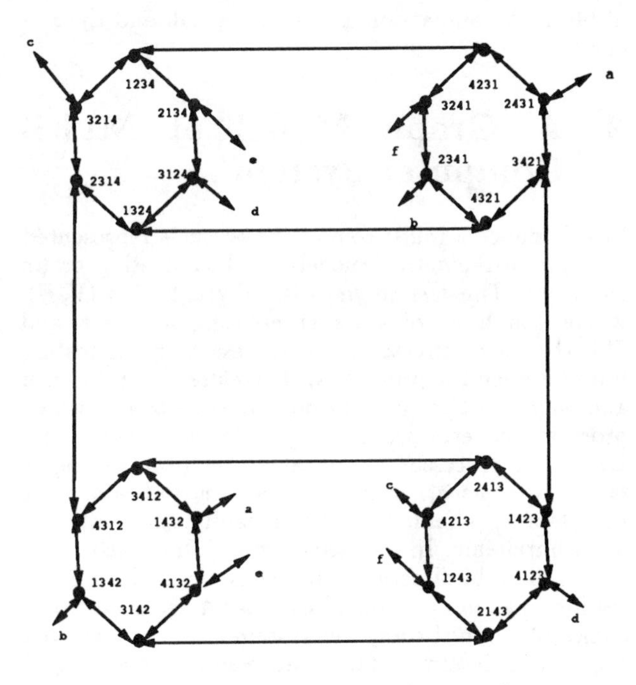


Figure 1: A $4-star$ testing graph

# 1 Introduction

An $n-star$ multi-computer system, contains $n!$ processors. Figure 1 depicts the structure of 4-star graph.

The approach of mutual testing based system diagnosis initially proposed by Preparata et al. [8] is considered in this paper in diagnosing $n-star$. In this approach processors test each other and the test results are collected and analyzed to determine faulty processors [4, 5]. Two different strategies for implementing the diagnosis approach were discussed in [8]. One of these strategies is called the one-step diagnosis or the diagnosis without repair and under this strategy there is only one testing phase and one repair/replacement phase. The other strategy is called the sequential diagnosis or the diagnosis with repair and this strategy may involve several iterations of the testing and repair phases. Later Kavianpour and Friedman [3] defined *one* $-$ *step* $t/t$ fault diagnosability. In this strategy of repair $t$ or fewer processors that include all the faulty processors present and possibly some processors of unknown status in the system can be identified for replacement after a testing phase.

The maximum number of faulty processors that may exist in a system at any given time without invalidating the diagnosis strategy is called the *degree of diagnosability* of the system under the selected diagnosis strategy. The degree of diagnosability is a function of the testing connections (i.e., the set of "*tester-tested connections*") among the processors.

In Section 2 one-step diagnosability of $n-star$ is considered. Sequential diagnosis of $n-star$ has not been dealt with in literature up to now (to the knowledge of the author). This paper presents new results on the degree of diagnosability of this strategy. Section 3 presents a lower bound on the degree of diagnosability of the $n-star$ under the $seq-t$ strategy. An important cost factor of the sequential diagnosis is the number of testing phases required. The $seq-t$ strategy requires at most $R$ testing phases where $R$ is the smallest integer satisfying the condition

$$\sum_{i=0}^{R-1} \binom{n-1}{i} \geq t$$

Section 4 provides a conclusion of this paper.

| *size* | $N$ | $D$ | $DM$ | $D*DM$ |
|---|---|---|---|---|
| $5-star$ | 120 | 4 | 6 | 24 |
| $7-cube$ | 128 | 7 | 7 | 49 |
| $7-star$ | 5040 | 6 | 9 | 54 |
| $12-cube$ | 4096 | 12 | 12 | 144 |
| $9-star$ | 362880 | 8 | 12 | 96 |
| $18-cube$ | 262144 | 18 | 18 | 324 |

Table 1: A comparison of the Star graph and the $n-cube$

## 2 A Graph Model of Multi-Computer System

In this paper a multi-computer system is represented by a graph-theoretical model called the testing graph as in [8]. The *testing graph* is a digraph $G=(V,E)$, where $V$ is the set of nodes representing processors and $E$ is the set of directed edges representing the testing links between the processors. Therefore, $(v_i, v_j) \in E$ if and only if $v_i$ tests $v_j$. The outcome of a test in which processor $v_i$ tests processor $v_j$ is denoted by $a_{ij}$, and $a_{ij}=1$ if processor $v_i$ indicates that processor $v_j$ is faulty whereas $a_{ij}=0$ if processor $v_i$ indicates that processor $v_j$ is fault-free. If $v_i$ is faulty, then outcome $a_{ij}$ is unreliable. In the case of an $n-star$, each inter-node link is bi-directional and thus can facilitate two testing links in opposing directions. A set of test outcomes of a multi-computer system that are analyzed together to determine faulty processors is called a syndrome of the system.

**Definition 1 [8]:** A system $S$ is one-step $t$-fault diagnosable if given the fault bound $t(>0)$, all the faulty processors in $S$ can be correctly identified after a testing phase.

An $n-star$ interconnection has $n!$ processors. The labels of these processors are the $n!$ permutations of $n$ distinct symbols. For convenience, we assume that $n$ distinct symbols are $1,2,3,...,n$. Thus, processors in $3-star$ has labels $\{123, 213, 321, 132, 312, 231\}$. Two processors $S=s_1s_2...s_n$ and $T=t_1t_2...t_n$ are adjacent if and only if $s_1=t_k$ and $s_k=t_1$, i.e., interchanging the first symbol with the $kth$ symbol for some $k$, $2 \leq k \leq n$. Therefore, each processor has direct communication links to $n-1$ other processors. Akers et al. [1] have shown that the diameter of an $n-star$ graph is $\lfloor \frac{3(n-1)}{2} \rfloor$. In Table 1, the $n-star$ graphs with $N$ nodes and the $n-cubes$ of comparable size are compared for degree $(D)$, diameter $(DM)$, and $(D)*(DM)$. The $n-star$ $S_n$ can be decomposed into $n$ sub-stars $S_{n-1}$ by fixing each different symbols in one particular position 1 to $n$. If we fix a symbol in the last position we observe that there are $(n-1)!$ permutations that constitute an $S_{n-1}$. Qiu et al. [9] proved that there are $(n-2)!$ disjoint cycles of length $n(n-1)$ in an $n-star$.

A *path* in an $n-star$ is represented as a sequence of connected edges, each representing an inter-processor link. The number of links on a path is called the length of the path in the $n-star$. The node connectivity $\kappa(G)$ of a graph $G$ is the minimum number of nodes whose removal results in a disconnected or trivial graph. A graph $G$ is said to be $n-$ *connected* if $\kappa(G) \geq n$. A graph is $n-$ *connected* if and only if there exist at least $n$ disjoint paths between every pair of nodes in the graph. Thus an $n-star$ $S$ has connectivity $\kappa(S) = n-1$. Hakimi and Amin [2] obtained the following two conditions that are sufficient to assure that a system of $N$ processors is one-step $t-fault$ diagnosable: 1) $N \geq 2t+1$, and 2) $\kappa(G) \geq t$ where $\kappa(G)$ is the node connectivity of the graph $G$ representing the system. Since the $n-star$ S has connectivity $\kappa(S)=n-1$ and the number of processors in $S$, $N=n!$, satisfies the condition, $N \geq 2(n-1)+1$, for $n>2$, the $n-star$ (where $n>2$) is one-step $(n-1)-fault$ diagnosable. Later Kavianpour and Friedman [3] defined *one* $-$ *step* $t/t$ fault diagnosability.

**Definition 2:** A system $S$ is $one-step\ t/t(one-t/t)$ fault diagnosable if given the fault bound $t(>0)$, $t$ or fewer processors that include all the faulty processors present and possibly some processors of unknown status in $S$ can be identified for replacement after a testing phase.

In an $n-star$ any pair of processors are tested by at least $2n-4$ other processors. Nigam et al. [7] proved that $n-star$ is Hamiltonian. Utilizing these properties, and a proof similar to the one explained in [4] for degree of diagnosability of an n-cube under the $one-t/t$ strategy which is $2n-2$, it is easy to show that an $n-star$ under the $one-t/t$ strategy has the degree of diagnosability of $2(n-1)-2=(2n-4)$.

## 3 Sequential Strategy for $n-star$

In this section the question on the degree of diagnosability of the $n-star$ under the sequential diagnosis $(seq-t)$ strategy is addressed.

**Definition 3:** A system of $N$ processors is sequentially $t-fault$ diagnosable if given the fault bound $t(>0)$, at least one faulty processor can be identified after a testing phase without replacement.

Sequential diagnosis of $n-star$ was not dealt with in literature before (to the knowledge of the author).

Preparata et al. [8] proved that a Hamiltonian circle (a loop covering every node in the graph/system exactly once) of $N$ processors is sequentially $t-fault$ diagnosable if $N>(m+1)^2+\lambda(m+1)+1$ where $\lambda \in \{0,1\}$ and $t=2m+\lambda$. This condition is equivalent to $N>(\frac{t+2}{2})^2$ [6]. This property can be utilized in obtaining the degree of diagnosability of the $n-star$.

In the following, the term simple cycle is used to refer to a loop covering a subset of the nodes in the graph/system exactly once. Lemma 1 indicates how an

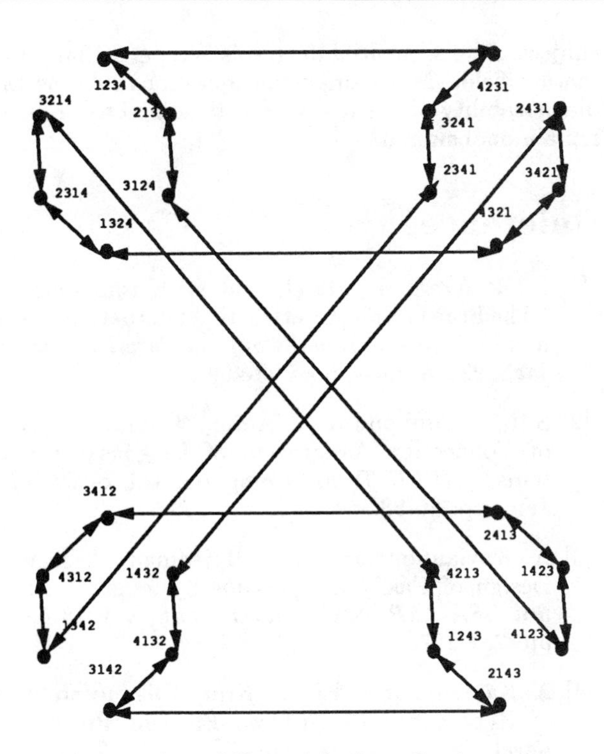


Figure 2: An ESC-partitioning of a $4-star$

$n-star$ can be partitioned into several simple cycles of equal length.

**Lemma 1 [9]:** There are $(n-2)!$ disjoint simple cycles of even length $L = n(n-1)$ in every $n-star$.

**Corollary 1:** There is no simple cycle of odd length in an $n-star$.

When an $n-star$ is partitioned into $(n-2)!$ disjoint simple cycles of even length $L = n(n-1)$, the $n-star$ is said to be *even-length-simple-cycle* (ESC-) partitioned. Figure 2 illustrates ESC-partitioning of a $4-star$ into two disjoint simple cycles.

**Lemma 2:** In an $n-star$ with the fault bound $t$, there exists at least one simple cycle of length $L = n(n-1)$ which contains $t' \leq \lfloor \frac{t}{(n-2)!} \rfloor$ faulty processors.

**Proof:** There are at least $(n-2)!$ disjoint simple cycles of length $L = n(n-1)$. The minimum among the number of faulty processors included in every simple cycles will be maximized if the faulty processors are evenly divided among the simple cycles. Assume that there is no simple cycle containing $t' \leq \lfloor \frac{t}{(n-2)!} \rfloor$ faulty processors. This means that every one of the $(n-2)!$ simple cycles will have $t'' > \lfloor \frac{t}{(n-2)!} \rfloor$ faulty processors. Then the total number of faulty processors $T \geq (n-2)! * t'' \geq (n-2)!(\lfloor \frac{t}{(n-2)!} \rfloor + 1) > (n-2)! * (\frac{t}{(n-2)!}) = t$. This is a contradiction.

**Lemma 3:** In an $n-star$ with the fault bound $t < (n-2)!\lfloor 2\sqrt{n(n-1)} - 2 \rfloor$ there exists at least one simple cycle of length $L = n(n-1)$ which contains $t' \leq \frac{t}{(n-2)!}$ faulty nodes and is sequentially diagnosable.

**Proof:** From Lemma 2, there must exist a simple cycle of length $L = n(n-1)$ with $t' \leq \lfloor \frac{t}{(n-2)!} \rfloor$ faulty processors. From the result obtained in [8] and mentioned earlier in this section, the sufficient condition for such a simple cycle of length $L$ to be sequentially $t'-fault$ diagnosable can be derived as follows:

$L > (\frac{t'+2}{2})^2$

In an ESC-partitioned $n-star$ there exist $(n-2)!$ disjoint simple cycles each of length $n(n-1)$. Thus for each of the $(n-2)!$ simple cycles to be $seq-t'$ diagnosable the sufficient condition becomes as follows:

$n(n-1) > (\frac{t'+2}{2})^2$

$t' < 2\sqrt{n(n-1)} - 2 = t''$

If this condition is not met, then it is possible that none of the $(n-2)!$ simple cycles are $seq-t'$ diagnosable. Thus, a lower bound on the value of $t$ under which no simple cycle satisfies the above sufficient condition, can be obtained as follows:

$(n-2)!\lfloor t'' \rfloor = (n-2)!\lfloor 2\sqrt{n(n-1)} - 2 \rfloor = T$

**Theorem 1:** The $n-star$ is sequentially $t-fault$ diagnosable where $t < (n-2)!\lfloor 2\sqrt{n(n-1)} - 2 \rfloor = T$.

**Proof:** From Lemma 2 there exists a simple cycle which contains $t' \leq \lfloor \frac{t}{(n-2)!} \rfloor$ faulty processors. Also, it follows from Lemma 3 that with the fault bound $t < T$ such a simple cycle is $seq-t'$ diagnosable. Since the condition $t < T$ implies that at least one faulty processor can be detected, the $n-star$ is $seq-t$ diagnosable. A $seq-t'$ diagnosable simple cycle can be detected at least by exhaustive search through all possible ESC-partitioning.

Although the bound $(T)$ in Theorem 1 is the best bound known at this point, it is not necessarily the tightest bound. It should be noted that since the condition $N \geq 2t+1$ must be satisfied for diagnosability, the tightest bound can not exceed the bound $\frac{N-1}{2}$. For example, for a $4-star$ the bound T in Theorem 1 is 8 whereas $(N-1)/2 = 11.5$.

**Proposition 1:** The number of testing-repair phases required in the $seq-t$ diagnosis of the $n-star$ $(n > 2)$ under the fault bound $t$ does not exceed $R$, the smallest integer satisfying the condition

$$\sum_{i=0}^{R-1} \binom{n-1}{i} \geq t$$

**Proof:** Consider the following tree representing the tester-tested relationship exploited during the $seq-t$ diagnosis of $n-star$. The root of the tree represents a processor that is replaced after the first testing phase. Although in general multiple processors may be replaced after the first testing phase, only one is replaced in the worst case. After replacement, the new processor is used in the second testing phase and its diagnosis

| $N = n!$ ($n$) Star size | One-t | One t/t | Seq-t |
|---|---|---|---|
| 6 (3) | 2 | 2 | 2 |
| 24 (4) | 3 | 4 | 8 |
| 120 (5) | 4 | 6 | 36 |
| 720 (6) | 5 | 8 | 192 |
| 5040 (7) | 6 | 10 | 1200 |
| 40320 (8) | 7 | 12 | 8640 |
| 362880 (9) | 8 | 14 | 70560 |
| 3628800 (10) | 9 | 16 | 645120 |

Table 2: Degree of diagnosability of $n - star$ under $one - t$, $one - t/t$, and $seq - t$ strategies

of $n - 1$ other processors can be fully trusted. If all $n - 1$ processors are detected to be faulty, then they are represented by the sons of the root. If some processors are fault-free, their diagnosis of other processors can be fully trusted. The number of faulty processors detected after the second testing phase is the minimum of $n = \binom{n-1}{1}$ and $f1$, the number of faulty processors remaining after the first testing phase. Similarly, the number of faulty processors detected after the $i - th$ testing phase is the minimum of $\binom{n-1}{i-1}$ and $f_{i-1}$. Whenever $f_i$ becomes zero, i.e., there remain no more faulty processors, the $seq - t$ diagnosis stops. Therefore, the number of testing-repair phases required does not exceed $R$, the smallest integer satisfying the condition

$$\sum_{i=0}^{R-1} \binom{n-1}{i} \geq t$$

Table 2 compares the degrees of diagnosability under the different strategies while the numbers given for $one - t$, $one - t/t$ strategies are the exact degrees of diagnosability, the numbers given for the $seq - t$ strategy are the best known lower bounds for the degree of diagnosability. In terms of the degree of diagnosability, the three strategies are ordered as $one - t < one - t/t < seq - t$ for $n > 3$. As expected the table shows that the seq-t strategy yields a much higher degree of diagnosability.

## 4 Conclusion

The costs and diagnostic power of $one-t$, $one-t/t$, and $seq-t$, in diagnosis of $n - stars$ were discussed. Some of the new knowledge established in this paper include a lower bound for the degree of diagnosability of the $seq-t$ strategy. In terms of the degree of diagnosability in diagnosing an $n - star$, the three strategies were shown to be ordered as $one-t < one-t/t < seq-t$ for $n > 3$. A systematic diagnosis of multi-computer system such as an $n - star$ is a relatively young research subject. The sequential diagnosis is a particularly immature field. Many important questions such as the diagnosability of the $n - star$ with some broken links remain unanswered.

## References


[1] S. B. Akers, D. Harel, and B. Krishnamurthy, "The Star Graph:An attractive alternative to the n-cube," *International Conf. on Parallel Processing,* 1987 August, pp. 393-400.

[2] S. L. Hakimi and A. T. Amin, "Characterization of Connection Assignment of Diagnosable Systems," *IEEE Trans. Computer*, vol. c-23, 1974 Jan., pp. 86-88.

[3] A. Kavianpour and A. D. Friedman, "Efficient Design of Easily Diagnosable Systems, " *Proc. 3rd USA-JAPAN Computer Conf.*, , 1978 Oct., pp. 251-257.

[4] A. Kavianpour and K. H. Kim, "Diagnosabilities of Hypercubes Under Two Efficient Processor-Level Diagnosis Strategies," *IEEE Trans. on Computer*, vol. 40, no. 2, 1991 Feb., pp. 232-237.

[5] A. Kavianpour and K.H. Kim, "A Comparative Evaluation of Four Basic System-Level Diagnosis Strategy for Hypercubes", *IEEE Transactions on Reliability*, vol. 41,1992 March, pp. 26-37.

[6] C. Kime, "System Diagnosis," *in Fault-Tolerant Computing:Theory and Techniques D.K. Pradhan, Editor*, Englewood Cliffs, NJ:Prentice-Hall, 1985.

[7] M. Nigam, S. Sahni, and B. Krishnamurthy, "Embedding Hamiltonians and Hypercubes in Star Interconnection Graphs," *International Conf. on Parallel Processing,* 1990 August, pp. 340-343.

[8] F. P. Preparata, G. Metze and R. T. Chien, "On the Connection Assignment Problem of diagnosable Systems," *IEEE Trans. Elec. Computer*, vol. EC-16, 1967 Dec., pp. 848-854.

[9] K Qui, H. Meijer, and S. G. Akl, "Decomposing a Star Graph into Disjoint Cycles," *Information Processing Letters*, vol. 39, no. 3, 1991 August, pp. 125-129.

# On Compiling Array Expressions for Efficient Execution on Distributed-Memory Machines

S. K. S. Gupta, S. D. Kaushik, S. Mufti, S. Sharma, C.-H. Huang and P. Sadayappan

Department of Computer and Information Science
The Ohio State University Columbus, OH 43210

## Abstract

Efficient generation of communication sets and local index sets is important for evaluation of array expressions in scientific languages such as Fortran-90 and High Performance Fortran implemented on distributed-memory machines. We show that for arrays affinely aligned with templates that are distributed on multiple processors with a block-cyclic distribution, the local memory access sequence and communication sets can be efficiently enumerated using closed forms. First, closed form solutions are presented for arrays that are aligned with identity templates that are distributed using block or cyclic distributions. These closed forms are then used with a *virtual processor approach* to give an efficient solution for arrays with block-cyclic distributions. These results are extended to arrays affinely aligned to arbitrary templates that have regular distributions. We present performance results on an iPSC/860 processor, that demonstrate the low runtime overhead of this scheme.

## 1 Introduction

A number of Fortran extensions have been proposed for machine independent parallel programming [3, 5]. High Performance Fortran(HPF) is an attempt to standardize the best features of these proposals [2]. An HPF program has essentially the same structure as a Fortran-90 program, but is enhanced with data distribution directives. When writing a program in HPF, the programmer specifies computations in a global data space. Arrays are affinely aligned to abstract arrays called *templates*. The templates are then distributed using distribution directives and all arrays aligned to the template are implicitly distributed. The compiler uses these global expressions and distribution information for the arrays to produce code suitable for execution on each node of a distributed-memory parallel computer.

Let array $A$ of size $N$ be aligned to a template $T$ where $A(i)$ is aligned to $T(\alpha i + \beta)$. The template is distributed on $P$ processors using a *block-cyclic*$(b)$ distribution, where *block-cyclic*$(b)$ refers to a block-cyclic distribution with block size $b$. This divides $A$ into $P$ pieces, each of which is stored in the local memory of one of the $P$ processors. Let $B$ be another array that is aligned to another template using a different alignment and distribution function. For an array expression involving arbitrary sections $A(l_1 : u_1 : s_1)$, and $B(l_2 : u_2 : s_2)$ we show 1) how to efficiently enumerate the local indices of $A$ and $B$ that each processor must access to execute the array expression, and 2) how to efficiently enumerate the set of local indices of $A$ or $B$ that each processor must send to or receive from other processors in executing the array expression.

In [6], closed-form expressions were presented for the restricted case where the arrays involved were identically distributed using block or cyclic distributions and array expressions had unit stride. In contrast to previously published work, the development in this paper allows different distributions for the arrays in the array expression. Also, previous work has addressed block and cyclic distributions, but not the more general block-cyclic distribution (an alternative solution [1] has been concurrently developed that is based on finite state machines). We show that an efficient solution to this problem is possible without the restrictions previously imposed.

In our solution approach, we first work out the closed form solutions for arrays aligned to *identity* templates, i.e., templates identical in structure to the array itself, with block or cyclic distributions. This is described in Section 2. We then use a *virtual processor approach* to extend these solutions to block-cyclic distributions (Section 3). Finally, these results are extended to arrays aligned to arbitrary templates in Section 4. Performance results on an iPSC/860 processor are given in Section 5. Conclusions are presented in Section 6.

## 2 Closed Forms for Block and Cyclic Distributions

We first solve a simpler problem with the following restrictions: 1) the distribution function is restricted to the block and cyclic distributions, and 2) the arrays involved are aligned to *identity* templates (this corresponds to an alignment function of $\alpha i + \beta$ with $\alpha = 1, \beta = 0$ and a template of the same size as the array. Since the array and its template are structurally identical, we use the same name to refer to both the array and its template. Both of the above restrictions are relaxed in later sections.

Let $A(0 : N-1)$ and $B(0 : M-1)$ be distributed over $P$ processors using (possibly different) block or cyclic distributions. Consider an array assignment of the following form

$$A(l_1 : u_1 : s_1) = f(B(l_2 : u_2 : s_2)).$$

Only elements of $A$ with global indices $i = l_1 + k * s_1, 0 \leq k < \lceil \frac{u_1 - l_1}{s_1} \rceil$ are modified by the execution of the array statement. With the *owner-computes* strategy, the processor that owns an element of the array $A$ performs the computation to evaluate its new value. Execution of the above statement requires that each processor determine the local indices of $A$ which are part of the array assignment. Also, each processor has to communicate the elements of $B$ that it owns and which are required by other processors for the computation. A simple strategy is to have each processor scan its entire local data space, compute the global index and determine if the global index belongs to the array expression. This strategy is inefficient, since index computation has to be performed for each element in the local data space, even if that element is not involved in the execution of the array statement.

In this section, we derive closed forms for the indexing required for array assignment statements involving arrays that are block or cyclically distributed. These closed forms can be used to access the active elements involved in array assignment without having to scan the entire local data space. The simpler case where both arrays are identically distributed on the same number of processors was treated in [6]. We characterize the more general case here. Table 1 gives the index mapping functions from global to local and local to global index spaces for the above data distributions.

Results for evaluating intersections of sections of the form $[a : b : c]$ are first stated. The function $next(a, b, c)$ as defined in [6] is the smallest integer such that $next(a, b, c) \geq a$ and $next(a, b, c) \equiv b \ (mod\ c)$, i.e., $next(a, b, c) = a + (b - a)\ mod\ c$.

**Lemma 2.1** *For any $c_1 > 0$, $c_2 > 0$, let $n_1, n_2$ be integers such that $c_1 n_1 + c_2 n_2 = gcd(c_1, c_2)$ and let*

$$m = next\left(max(a_1, a_2), \frac{c_1 n_1 (a_2 - a_1)}{gcd(c_1, c_2)} + a_1, lcm(c_1, c_2)\right) \equiv first(a_1, c_1, a_2, c_2)$$

*Then*

$$[a_1 : b_1 : c_1] \cap [a_2 : b_2 : c_2] = \begin{cases} [m : min(b_1, b_2) : lcm(c_1, c_2)] & if\ (a_2 \equiv a_1\ (mod\ gcd(c_1, c_2))) \\ \phi & otherwise \end{cases}$$

Lemma 2.1 forms the basis for the proofs for the following theorems (Theorems 3.1-3.4). Proofs may be found in [4].

Consider the array expression $A(l_1 : u_1 : s_1) = f(B(l_2 : u_2 : s_2))$. The following two theorems give closed forms for local indices of $A$ that are involved in the computation, for block and cyclic distributions of $A$, respectively.

**Theorem 2.1** *If $A$ is distributed by a block distribution then the local index set $Lo\_Ind(p)$ corresponding to $A(l : u : s)$ is given by*

$$[m\ mod\ b : min(u, p * b + b - 1)\ mod\ b : s]$$
$$m = max(next(l * b, p, s), l)$$
$$b = \lceil N/P \rceil$$

**Theorem 2.2** *If $A$ is distributed by a cyclic distribution, then the local index set $Lo\_Ind(p)$ corresponding to $A(l : u : s)$ is given by*

$$[m\ div\ P : min(u, N - P + p)\ div\ P : lcm(s, P)\ div\ P] \quad if\ l\ mod\ gcd(P, s) = p\ mod\ gcd(P, s)$$
$$= \phi \quad otherwise$$

$$m = next(max(p, l), \frac{s * n_1 (p - l)}{gcd(s, P)} + l, lcm(s, P))$$

*where $n_1$ and $n_2$ are integers such that $sn_1 + Pn_2 = gcd(s, P)$.*

We now present closed form expressions for the set of data values to be communicated when $A$ has cyclic (block) distribution and $B$ has block (cyclic) distribution. Results for other cases can be similarly obtained [4].

**Theorem 2.3** *In the array expression $A(l_1 : u_1 : s_1) = f(B(l_2 : u_2 : s_2))$, if $A$ is distributed by a cyclic distribution and $B$ by a block distribution then the local index send and receive data sets $Lo\_Send(p_f, p_t)$ and $Lo\_Recv(p_f, p_t)$ are as follows.*

*If $p_t \equiv l_1\ (mod\ gcd(P_t, s_1))$ then*

$$Lo\_Send(p_f, p_t) = [(m_1 * s_2 + l_2)\ mod\ b_f : (m_2 * s_2 + l_2)\ mod\ b_f : s_2(lcm(P_t, s_1)\ div\ s_1)]$$
$$Lo\_Recv(p_f, p_t) = [(m_1 * s_1 + l_1)\ div\ P_t : (m_2 * s_1 + l_1)\ div\ P_t : lcm(P_t, s_1)]$$
$$m_1 = max(next(g_2\ div\ s_2, g_1\ div\ s_1, lcm(P_t, s_1)\ div\ s_1), g_1\ div\ s_1)$$
$$m_2 = min((ul_2 - l_2)\ div\ s_2, (ul_1 - l_1)\ div\ s_1)$$
$$g_1 = first(l_1, s_1, p_t, P_t) - l_1$$
$$ul_1 = min(u_1, N_t)$$
$$g_2 = max(next(b_f * p_f, l_2, s_2), l_2) - l_2$$
$$ul_2 = min(b_f(p_f + 1) - 1, u_2)$$
$$b_f = \lceil N_f / P_f \rceil$$

*else $Lo\_Send(p_f, p_t) = Lo\_Recv(p_f, p_t) = \phi$.*

**Theorem 2.4** *In the array expression $A(l_1 : u_1 : s_1) = f(B(l_2 : u_2 : s_2))$, if $A$ is distributed by a block distribution and $B$ is distributed by a cyclic distribution then the local index send and receive data sets $Lo\_Send(p_f, p_t)$ and $Lo\_Recv(p_f, p_t)$ are as follows.*

*If $p_f \equiv l_2\ (mod(P_f, s_2))$ then*

$$Lo\_Send(p_f, p_t) = [(m_1 * s_2 + l_2)\ div\ P_f : (m_2 * s_2 + l_2)\ div\ P_f : s_1(lcm(P_f, s_2)\ div\ s_2)]$$
$$Lo\_Recv(p_f, p_t) = [(m_1 * s_1 + l_1)\ mod\ b_t : (m_2 * s_1 + l_1)\ mod\ b_t : lcm(P_f, s_2)]$$
$$m_1 = max(next(g_1\ div\ s_1, g_2\ div\ s_2, lcm(P_f, s_2)\ div\ s_2), (g_2)\ div\ s_2)$$
$$m_2 = min(ul_1\ div\ s_1, ul_2\ div\ s_2)$$
$$g_1 = max(next(b_t * p_t, l_1, s_1), l_1) - l_1$$
$$ul_1 = min(b_t(p_t + 1) - 1, u_1) - l_1$$
$$g_2 = first(p_f, P_f, l_2, s_2) - l_2$$
$$ul_2 = min(N_f, u_2) - l_2$$
$$b_t = \lceil N_t / P_t \rceil$$

*else $Lo\_Send(p_f, p_t) = Lo\_Recv(p_f, p_t) = \phi$.*

| Distribution | Index Mapping Functions | |
|---|---|---|
| Block-Cyclic | local to global | $i = (l \ div \ b) * b * P + b * p + l \ mod \ b, 0 \leq l \leq \lceil N/P \rceil - 1$ |
| | global to local | $l = i \ mod \ b + (i \ div \ (P * b)) * b, 0 \leq i \leq N - 1$ |

Table 1: Index Mapping for Regular Data Distributions ($i$: global index, $l$:local index, $p$: processor number, $b$: block size.)

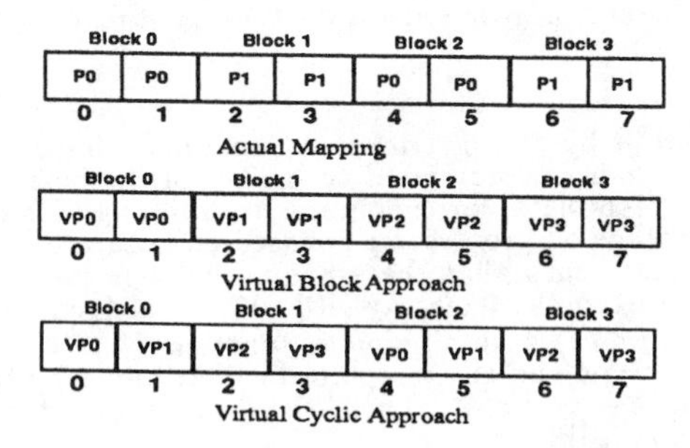


Figure 1: Illustration of the Virtual processor approach to block-cyclic distribution for an array of 8 elements on 2 processors with block size 2 i.e. ($N = 8, P = 2, b = 2$).

## 3 Virtual Processor Approach for Block-Cyclic Distributions

In this section, we consider the more general block-cyclic distribution. Closed forms for the local index and communication sets corresponding to an array expression are not available for block-cyclic distributions. The pattern of data access in the local data space is irregular, with no fixed stride. The stride keeps changing and the pattern repeats itself after a fixed interval. However, efficient routines for index computations with these distributions can be generated by using a virtual processor approach. In this approach, a block-cyclic distribution is considered to be either a block or a cyclic distribution in a virtual processor space, where each physical processor's local data space is a combination of the local data spaces of many virtual processors. We have closed forms for the set of virtual processors that are mapped to a single physical processor, which allows us to use the closed forms for block and cyclic distributions to generate efficient index computation and communication code for block-cyclic distributions.

Let $A(0 : N - 1)$ be a one dimensional array distributed over $P$ processors using a *block-cyclic*$(b)$. The number of blocks per processor is $l = \lceil \frac{\lceil \frac{N}{b} \rceil}{P} \rceil$. If this array is involved in an array expression, finding the associated communication sets and local address sets requires finding the set of all elements to be sent from one processor to another and the set of local elements accessed.

Two virtual processor approaches can be used. In the first approach, called the *virtual-block approach*, it is assumed that the distribution is actually a block distribution in a virtual processor space, with each block on a physical processor belonging to a different virtual processor. In the second approach, called the *virtual-cyclic approach*, the distribution is considered to be actually cyclic, with each element in a block on a physical processor belonging to a different virtual processor. For example, consider a *block-cyclic*$(b)$ distribution of an array of size $N$ on $P$ processors. Under the virtual-block approach, this distribution would be viewed as a block distribution on $\lceil \frac{N}{b} \rceil$ virtual processors. The $\lceil \frac{N}{b} \rceil$ virtual processors are themselves mapped cyclically on the $P$ physical processors. Under the virtual-cyclic approach, this distribution would be viewed as a cyclic distribution on $P * b$ virtual processors. The $P * b$ virtual processors are mapped according to a block distribution on the $P$ physical processors. The data size $N$ used in the previously stated theorems is replaced with $P * b * l$ for the virtual-cyclic approach.

The two approaches are illustrated in Fig. 1. Either approach could be chosen for a given block-cyclic distribution. The better approach to use depends on the number of virtual processors required per physical processor. For the virtual-block approach, the number of virtual processors per physical processor is $l$, the number of blocks per physical processor. For the virtual-cyclic approach these are $b$ virtual processors per physical processor. Therefore, the virtual-block approach is chosen when $l < b$, while the virtual-cyclic approach is chosen when $l > b$. Intuitively, the virtual-block approach is preferable when the distribution is closer to a block distribution, while the virtual-cyclic approach is better if the distribution is closer to a cyclic distribution.

Let sets $Virt\_Blk(p)$ and $Virt\_Cyc(p)$ be the sets of all virtual processors mapped to physical processor $p$ under the virtual-block and virtual-cyclic approaches respectively. These sets have the closed form definitions:

$$Virt\_Blk(p) = [p : (l - 1) * P + p : P]$$
$$Virt\_Cyc(p) = [p * b : p * b + b - 1 : 1]$$

For both the schemes, the local index of an element on a virtual processor is not the same as that on a physical processor, since data from many virtual processors resides on each physical processor. The array element with local index $j$ on a virtual processor $vp$ has a local index $f_{blk}(vp, j)$ and $f_{cyc}(vp, j)$ on the physical processor $p$ that $vp$ is mapped to, in the virtual-block and virtual-cyclic approaches respectively, where $f_{blk}(vp, j) = (vp \ div \ P) * b + j$, and $f_{cyc}(vp, j) = j * b + vp \ mod \ b$.

Also, a stride of $s$ in the local index space of a virtual processor is modified in the local index space of the physical processor by the functions $g_{blk}(s)$ and $g_{cyc}(s)$ for the two schemes respectively, where $g_{blk}(s) = s$ and $g_{cyc}(s) = b * s$.

Based on these closed forms, efficient index computation and communication routines can be implemented for block-cyclic distributions. For accessing the local index space corresponding to a global expression, the code would consist of a loop that iterates over the virtual processors mapped to the physical processor, accessing the local index space using the closed form for each virtual processor in the body of the loop.

One problem with the virtual-cyclic scheme is that access to elements is not in order of increasing local index on a physical processor. This is because local elements of a virtual processor are interleaved on a physical processor. This could cause undesirable cache effects. This problem can be alleviated by storing the elements assigned to a physical processor in increasing order of virtual processors, instead of increasing global index. The global index to physical processor number mapping remains the same, while the global index to local index mapping changes to $j = (i \ mod \ b) * b + i \ div \ (P * b), 0 \leq i \leq N-1$. In addition, the function $g_{cyc}(s)$ defined above now changes to $g_{cyc}(s) = s$, and the function $f_{cyc}(vp, j)$ becomes $f_{cyc}(vp, j) = (vp \ mod \ b) * b + j$.

The code for expressions involving communication will consist of four nested loops, where the outermost loop iterates over all physical target processors, the next loop iterates over all source virtual processors mapped to the sending processor, the third loop iterates over all target virtual processors mapped to the current physical target processor, and the last loop iterates over the local index space using the closed forms for block or cyclic distribution as appropriate, replacing local indices by expressions corresponding to the functions $g(s)$ defined above.

Consider the array expression $A(2 : 256 : 2) = B(8 : 135)$, for two arrays $A(256)$ and $B(192)$, distributed over 8 processors, with the distribution for $A$ being *block-cyclic*(8), and the distribution for $B$ being *block-cyclic*(12). Assume that the virtual-block approach is used for both source and target distributions. Then, the communication code for processor 0 would be as follows:

```
for p_t = 0 to 7
  for vp_f = 0 to 8 by 8
    for vp_t = p_t to p_t + 32 by 8
      ofs = (vp_f div 8) * 12
      gl_1 = max(next(8 * vp_t, 2, 2), 2)
      gl_2 = max(next(12 * vp_f, 8, 1), 8)
      ul_1 = min(8(vp_t + 1) - 1, 256)
      ul_2 = min(12(vp_f + 1) - 1, 135)
      m_1 = max((gl_1 - 2) div 2, (gl_2 - 8) div 1)
      m_2 = max((ul_1 - 2) div 2, (ul_2 - 8) div 1)
      send(p_t, B((m_1 + 8) mod 12) + ofs :
             (m_2 + 8) mod 12 + ofs : 1))
for p_f = 0 to 7
  for vp_f = p_f to p_f + 24 by 12
    for vp_t = 0 to 8 by 8
      ofs = (vp_t div 8) * 8
      gl_1 = max(next(8 * vp_t, 2, 2), 2)
      gl_2 = max(next(12 * vp_f, 8, 1), 8)
      ul_1 = min(8(vp_t + 1) - 1, 256)
      ul_2 = min(12(vp_f + 1) - 1, 135)
      m_1 = max((gl_1 - 2) div 2, (gl_2 - 8) div 1)
      m_2 = max((ul_1 - 2) div 2, (ul_2 - 8) div 1)
      recv(p_f, B'((2m_1 + 2) mod 8 + ofs :
             2m_2 + 2 + ofs : 2)).
```

## 4 Array Alignment with Arbitrary Templates

In this section, we show how to use the results of previous sections to efficiently enumerate local index sets and communication sets for data arrays aligned to arbitrary templates. The basic idea is to work out a closed form solution in terms of the global indices of the template, and then to use the array alignment function to translate this solution in terms of the local indices of the array.

Let $A(0 : N_1 - 1)$ be a data array that is aligned with template $T(0 : N_2 - 1)$ using the alignment function $f(i) = \alpha i + \beta$, i.e., every data array index $i$ is mapped to template index $f(i)$. The array $A$ is then mapped to the template section $T(\beta : \alpha(N_1 - 1) + \beta : \alpha)$. If the template $T$ is distributed to $P$ processors using a block or cyclic distribution, a processor $p$ contains those elements of $A$ whose indices are aligned to the indices of $T$ mapped

| Local Index | 0 | 1 | 2 | 3 | 4 | 5 |
|---|---|---|---|---|---|---|
| Processor 0 | 4 | 5 | 10 | 15 | | |
| Processor 1 | 0 | 1 | 6 | 11 | 16 | 17 |
| Processor 2 | 2 | 7 | 12 | 13 | | |
| Processor 3 | 3 | 8 | 9 | 14 | | |

Table 2: Local Storage of array $A(0 : 17)$ aligned to $T(0 : 71)$ by function $3i + 20$.

to processor $p$ by the distribution function. Since not all template indices correspond to indices of $A$, there are *holes* in the template corresponding to *unoccupied* cells. $A$'s local storage on processor $p$ is based on occupied cells of the template such that there are no holes in its local storage. For example, let array $A(0 : 17)$ be aligned with template $T(0 : 71)$ with alignment function $f(i) = 3 * i + 20$, let $P = 4$ and the template be distributed by the function $block - cyclic(4)$. Then, the local storage for A is as shown in Table 2.

Assume that we are interested in a closed form solution for the array section $A(l : u : s)$ on processor $p$. The array section $A(l : u : s)$ corresponds to the template section $T(\alpha * l + \beta : \alpha * u + \beta : \alpha * s)$. We can derive a closed form solution for this template section on processor $p$ in terms of the global indices in the template space using the results of the previous sections. Let this solution be $T(l_1 : u_1 : s_1)$. This corresponds to array section $A(\frac{l_1-\beta}{\alpha} : \frac{u_1-\beta}{\alpha} : \frac{s_1}{\alpha})$. But we are interested in a solution in terms of the local indices of $A$ on $p$.

We can precompute a closed form for all the global indices of $A$ mapped to a physical processor $p$ by forming the intersection of the global indices of $T$ mapped to $p$ by the distribution function and the section of $T$ corresponding to indices of $A$, i.e., $T(\beta : \alpha(N_1 - 1) + \beta : \alpha)$. Let this precomputed section be $A(l_2 : u_2 : s_2)$, which corresponds to the global indices of $A$ mapped to $p$. Then, the local indices corresponding to the global indices of the solution, i.e. $A(\frac{l_1-\beta}{\alpha} : \frac{u_1-\beta}{\alpha} : \frac{s_1}{\alpha})$, can be computed by their position with respect to the closed form for the global indices of $A$ mapped to a processor. The closed form of the solution in terms of local indices is given by $A(\frac{\frac{l_1-\beta}{\alpha}-l_2}{s_2} : \frac{\frac{u_1-\beta}{\alpha}-l_2}{s_2} : \frac{s_1}{s_2 * \alpha})$.

For templates distributed using the block-cyclic distribution, the virtual-processor approach can be used as before.

## 5 Performance Results

In this section, we present experimental results for the virtual processor approach. The kernel used is the DAXPY operation $y(0 : N - 1 : s) = y(0 : N - 1 : s) + a * x(0 : N - 1 : s)$. The objective is to measure the additional time required per element of the array section due to the indexing overhead.

The experiments were performed on an Intel iPSC/860. The time required for each processor to execute its local section is measured and the execution time per element on the processor is evaluated. Execution of the array section consists of the two parts - evaluating the local sections for each virtual processor on a physical processor and then executing the local section of each virtual processor. The closed forms for the local section for each virtual processor on a physical processor are precomputed and stored in a table. Note that the table can

| Block size | | Stride | | | |
|---|---|---|---|---|---|
| | | $s = 1$ | $s = 3$ | $s = 5$ | $s = 8$ |
| *cyclic* | b | 0.001406 | 0.003375 | 0.005437 | 0.001488 |
| | c | 0.000344 | 0.000344 | 0.000375 | 0.000312 |
| $b = 4$ | b | 0.000500 | 0.001187 | 0.001812 | 0.001594 |
| | c | 0.000375 | 0.000375 | 0.000406 | 0.000344 |
| $b = 16$ | b | 0.000344 | 0.000500 | 0.000687 | 0.001063 |
| | c | 0.000500 | 0.000531 | 0.000531 | 0.000438 |
| $b = 32$ | b | 0.000344 | 0.000406 | 0.000500 | 0.000719 |
| | c | 0.000719 | 0.000687 | 0.000719 | 0.000562 |
| $b = 64$ | b | 0.000344 | 0.000375 | 0.000406 | 0.000531 |
| | c | 0.000937 | 0.001000 | 0.000969 | 0.000781 |
| *block* | b | 0.000344 | 0.000344 | 0.000375 | 0.000375 |
| | c | 0.000937 | 0.001937 | 0.002813 | 0.003750 |

Table 3: Execution time (msec) per element for various block sizes ($P = 64, K = 4096$).

be constructed at compile time. The indexing overhead at a processor involves the table look-up cost for each of its virtual processors. The time per element($t_p$) of the most heavily loaded processors is reported as the performance measure. Note that the algorithm selects the approach (virtual-block or virtual-cyclic) that minimizes the number of virtual processors on a physical processor.

Timings were taken using the mclock() system call. The state space consists of the parameters $N, P, b, l, u$ and $s$. We use the array section $[0 : K * s - 1 : s]$. $K$ represents the number of multiply-adds performed and is kept constant. The choice of $l = 0$ does not have a significant effect on the time for large array sections. The number of physical processors $P$ is kept fixed at 64. The block size $b$ and stride $s$ are varied. The number of DAXPY operations performed in the array section is fixed at $K = 4096$.

Execution times per element in milliseconds against stride, for different block sizes, using both the virtual-block and the virtual-cyclic approachs are shown in Table 3. We observe that the virtual-cyclic approach always performs better than the virtual-block approach for the cyclic distribution and for block-cyclic distributions with small block sizes. The virtual-block approach always performs better than the virtual-cyclic approach for the block distribution and for block-cyclic distributions with large block sizes. For block-cyclic distributions with intermediate block sizes, the better approach depends on the access stride. For small strides, the virtual-block approach is better than the virtual-cyclic approach. The virtual-cyclic approach gets better as the access stride increases. This is due to the fact that for a fixed block size as the stride increases, the number of cycles exceeds the block size.

In Fig 2, we report the MFLOPS rate (computed as $2/t_p$) against stride for different block sizes. We also show the performance of the DAXPY kernel without the indexing overhead. The performance results show that the block-cyclic cases are handled almost as efficiently as the block or cyclic cases. The performance results also show that the indexing overhead is tolerable.

## 6 Conclusion

In this paper we have presented a procedure for efficiently enumerating local index sets and for characterizing the inter-processor communication for array expressions for a fixed set of data distributions. The block, cyclic and block-cyclic distributions were considered. Closed form expressions for the part of the local data space involved in the global array expression were derived. Also closed form expressions for determining the send and receive

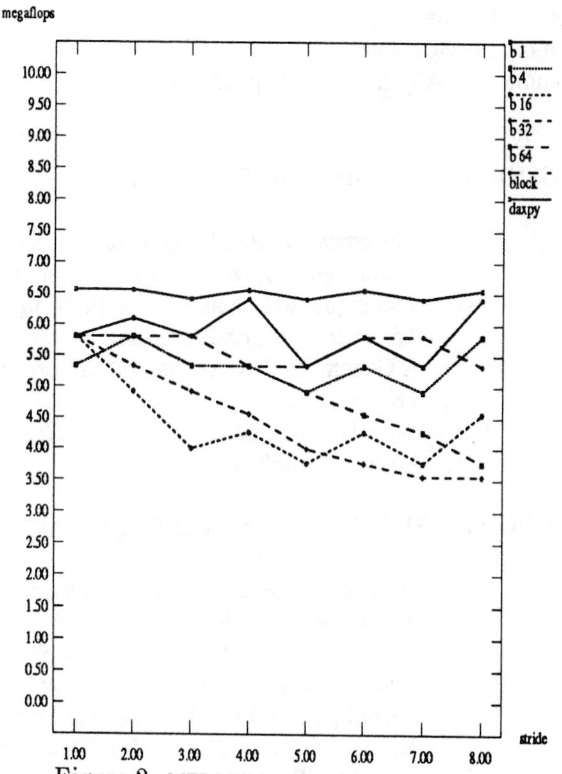


Figure 2: MFLOPS for virtual processor approach.

data sets for the block and cyclic source and target distributions were derived. A strategy based on virtual processors was used along with the closed forms for the block and cyclic distributions to generated efficient index computation code for the block-cyclic distribution. Experimental performance results evaluating the efficacy of the approach were presented.

## References

[1] S. Chatterjee, J. R. Gilbert, F. J. E. Long, R. Schreiber, and Shang-Hua Teng. Generating local addresses and communication sets for data parallel programs. In *Proc. of PPOPP*, 1993. To appear.

[2] High Perfromance Fortran Forum. High Performance Fortran, language specification, draft version 1.0. Jan. 1993.

[3] G. Fox, S. Hiranandani, K. Kennedy, C Koelbel, U. Kremer, C.-W. Tseng, and M-Y. Wu. Fortran-D language specification. Technical Report TR-91-170, Dept. of Computer Science, Rice University, Dec. 1991.

[4] S.K.S. Gupta, S.D. Kaushik, S. Mufti, S. Sharma, C.-H. Huang, and P. Sadayappan. On the generation of efficient data communication for distributed-memory machines. In *Proc. of Intl. Computing Symp., Taiwan*, 1992. Techinal Report OSU-CIS-RC-5/92-TR 13, Dept. of Computer and Info. Sc., The Ohio State Univ., May 1992.

[5] American National Standards Institute. *American Natioinal Standard for Information Systems Programming Language Fortran (Fortran 90)*, ANSI X3.198-1991 edition, Jan. 1990.

[6] C. Koelbel. Compile-time generation of communication for scientific programs. In *Supercomputing '91*, pages 101–110, Nov. 1991.

TABLE OF CONTENTS - FULL PROCEEDINGS

Volume I = Architecture
Volume II = Software
Volume III = Algorithms & Applications


SESSION 1A: CACHE MEMORY (I) ........ I-1

(R): Automatic Partitioning of Parallel Loops for Cache-Coherent Multiprocessors ........ I-2
*Anant Agarwal, David Kranz, and Venkat Natarajan*

(R): Techniques to Enhance Cache Performance Across Parallel Program Sections ........ I-12
*J.K. Peir, K. So, and J.H. Tang*

(C): A Generational Algorithm to Multiprocessor Cache Coherence ........ I-20
*Tzi-cker Chiueh*

(C): Semi-Unified Caches ........ I-25
*Nathalie Drach and Andre Seznec*

SESSION 2A: PROCESSOR AND COMMUNICATION ARCHITECTURE ........ I-29

(R): Dependence Analysis and Architecture Design for Bit-Level Algorithms ........ I-30
*Weijia Shang and Benjamin W. Wah*

(R): ATOMIC: A Low-Cost, Very-High-Speed, Local Communication Architecture ........ I-39
*Danny Cohen, Gregory Finn, Robert Felderman, and Annette DeSchon*

(C): Reconfigurable Branch Processing Strategy in Super-Scalar Microprocessors ........ I-47
*Terence M. Potter, Hsiao-Chen Chung, and Chuan-lin Wu*

(C): Exploiting Spatial and Temporal Parallelism in the Multithreaded Node Architecture Implemented on Superscalar RISC Processors ........ I-51
*D.J. Hwang, S.H. Cho, Y.D. Kim, and S.Y. Han*

SESSION 3A: MEMORY ........ I-55

(R): Fixed and Adaptive Sequential Prefetching in Shared Memory Multiprocessors ........ I-56
*Fredrik Dahlgren, Michel Dubois, and Per Stenstrom*

(R): Assigning Sites to Redundant Clusters in a Distributed Storage System ........ I-64
*Antoine N. Mourad, W. Kent Fuchs, and Daniel G. Saab*

(C): Balanced Distributed Memory Parallel Computers ........ I-72
*F. Cappello, J-L Bechennec, F. Delaplace, C. Germain, J-L Giavitto, V. Neri, and D. Etiemble*

(C): A Novel Approach to the Design of Scalable Shared-Memory Multiprocessors ........ I-77
*Honda Shing and Lionel M. Ni*

SESSION 4A: GRAPH_THEORETIC INTERCONNECTION STRUCTURES (I) ........ I-82

(R): Incomplete Star Graph: An Economical Fault-tolerant Interconnection Network ........ I-83
*C.P. Ravikumar, A. Kuchlous, and G. Manimaran*

(C): The Star Connected Cycles: A Fixed-Degree Network for Parallel Processing ........ I-91
*Shahram Latifi, Marcelo Moraes de Azevedo, and Nader Bagherzadeh*

(C): Empirical Evaluation of Incomplete Hypercube Systems ........ I-96
*Nian-Feng Tzeng*

(C): A Distributed Multicast Algorithm for Hypercube Multicomputers ........ I-100
*Jyh-Charn Liu and Hung-Ju Lee*

(C): A Generalized Bitonic Sorting Network ..... I-105
*Kathy J. Liszka and Kenneth E. Batcher*

SESSION 5A: HYPERCUBE ..... I-109

(R): A Lazy Scheduling Scheme for Improving Hypercube Performance ..... I-110
*Prasant Mohapatra, Chansu Yu, and Chita R. Das, Jong Kim*

(R): Fast and Efficient Strategies for Cubic and Non-Cubic Allocation in Hypercube Multiprocessors ..... I-118
*Debendra Das Sharma and Dhiraj K. Pradhan*

(C): Random Routing of Tasks in Hypercube Architectures ..... I-128
*Arif Ghafoor*

(C): Fault Tolerant Subcube Allocation in Hypercubes ..... I-132
*Yeimkuan Chang and Laxmi N. Bhuyan*

SESSION 6A: ARCHITECTURE (I) ..... I-137

(R): Performance of Redundant Disk Array Organizations in Transaction Processing Environments ..... I-138
*Antoine N. Mourad, W. Kent Fuchs, and Daniel G. Saab*

(C): The Chuted-Banyan (Canyan) Network: An Efficient Distribution Network for Growable Packet Switching Based on Free-Space Digital Optics ..... I-146
*Thomas J. Cloonan and Gaylord W. Richards*

(C): WDM Cluster Ring: A Low-Complexity Partitionable Reconfigurable Processor Interconnection Structure ..... I-150
*Khaled A. Aly and Patrick W. Dowd*

(C): A Scalable Optical Interconnection Network for Fine-Grain Parallel Architectures ..... I-154
*D. Scott Wills and Matthias Grossglauser*

(C): Bus-Based Tree Structures for Efficient Parallel Computation ..... I-158
*O. M. Dighe, R. Vaidyanathan, and S.Q. Zheng*

SESSION 7A: CACHE MEMORY (II) ..... I-162

(R): Efficient Stack Simulation for Shared Memory Set-Associative Multiprocessor Caches ..... I-163
*C. Eric Wu, Yarsun Hsu, and Yew-Huey Liu*

(C): Parallel Cache Simulation on Multiprocessor Workstations ..... I-171
*Luis Barriga and Rassul Ayani*

(C): A Chained-Directory Cache Coherence Protocol for Multiprocessors ..... I-175
*Soon M. Chung and Longxue Li*

(C): Evaluating the Impact of Cache Interferences on Numerical Codes ..... I-180
*O. Temam, C. Fricker, and W. Jalby*

(C): Performance Evaluation of Memory Caches in Multiprocessors ..... I-184
*Yung-Chin Chen and Alexander V. Veidenbaum*

SESSION 8A: PERFORMANCE EVALUATION ..... I-188

(R): Transmission Times in Buffered Full-Crossbar Communication Networks With Cyclic Arbitration ..... I-189
*A.J. Field and P.G. Harrison*

(C): Experimental Validation of a Performance Model for Simple Layered Task Systems ..... I-197
*Athar B. Tayyab and Jon G. Kuhl*

(C): Performance Evaluation of SIMD Processor Architectures Using Pairwise Multiplier Recoding ..... I-202
*Todd C. Marek and Edward W. Davis*

(C): Performance Considerations Relating to the Design of Interconnection Networks for Multiprocessing Systems ........ I-206
*Earl Hokens and Ahmed Louri*
(C): A Queuing Model for Finite-Buffered Multistage Interconnection Networks ........ I-210
*Prasant Mohapatra and Chita R. Das*
(C): Composite Performance and Reliability Analysis for Hypercube Systems ........ I-214
*Samir M. Koriem and L.M. Patnaik*

SESSION 9A: DISTRIBUTED SYSTEMS AND ARCHITECTURE ........ I-218

(R): Estimation of Execution times on Heterogeneous Supercomputer Architectures ........ I-219
*Jaehyung Yang, Ishfaq Ahmad, and Arif Ghafoor*
(C): Adaptive Deadlock-Free Routing in Multicomputers Using Only One Extra Virtual Channel ........ I-227
*Chien-Chun Su and Kang G. Shin*
(C): A Hybrid Shared Memory/Message Passing Parallel Machine ........ I-232
*Matthew I. Frank and Mary K. Vernon*
(C): Scalability Study of the KSR-1 ........ I-237
*Umakishore Ramachandran, Gautam Shah, S. Ravikumar, and Jeyakumar Muthukumarasamy*
(C): Personalized Communication Avoiding Node Contention on Distributed Memory Systems ........ I-241
*Sanjay Ranka, Jhy-Chun Wang, and Manoj Kumar*

SESSION 10A: MEMORY AND DISKS ........ I-245

(R): Design of Algorithm-Based Fault Tolerant Systems With In-System Checks ........ I-246
*Shalini Yajnik and Niraj K. Jha*
(C): A Cache Coherence Protocol for MIN-Based Multiprocessors With Limited Inclusion ........ I-254
*Mazin S. Yousif, Chita R. Das, and Matthew J. Thazhuthaveetil*
(C): Impact of Memory Contention on Dynamic Scheduling on NUMA Multiprocessors ........ I-258
*M. D. Durand, T. Montaut, L. Kervella, and W. Jalby*
(C): Reliability Evaluation of Disk Array Architectures ........ I-263
*John A. Chandy and Prithviraj Banerjee*
(C): Prime-Way Interleaved Memory ........ I-268
*De-Lei Lee*

SESSION 11A: ROUTING ALGORITHMS AND RING, BUS STRUCTURES ........ I-273

(R): Reducing the Effect of Hot Spots by Using a Multipath Network ........ I-274
*Mu-Cheng Wang, Howard Jay Siegel, Mark A. Nichols, and Seth Abraham*
(R): Hardware Support for Fast Reconfigurability in Processor Arrays ........ I-282
*M. Maresca, H. Li, and P. Baglietto*
(C): Closed Form Solutions for Bus and Tree Networks of Processors Load Sharing A Divisible Job ........ I-290
*S. Bataineh, T. Hsiung, and T.G. Robertazzi*
(C): The Message Flow Model for Routing in Wormhole-Routed Networks ........ I-294
*Xiaola Lin, Philip K. McKinley, and Lionel M. Ni*

SESSION 12A: GRAPH_THEORETIC INTERCONNECTION STRUCTURES (II) ........ I-298

(C): Generalized Fibonacci Cubes ........ I-299
*W.J. Hsu and M.J. Chung*

(C): HMIN: A New Method for Hierarchical Interconnection of Processors ........ I-303
*Yashovardhan R. Potlapalli and Dharma P. Agrawal*
(C): Tightly Connected Hierarchical Interconnection Networks for Parallel Processors ........ I-307
*Peter Thomas Breznay and Mario Alberto Lopez*
(C): The Folded Petersen Network: A New Communication-Efficient Multiprocessor Topology ........ I-311
*Sabine Ohring and Sajal K. Das*
(C): Hierarchical WK-Recursive Topologies for Mulicomputer Systems ........ I-315
*Ronald Fernandes and Arkady Kanevsky*
(C): Substructure Allocation in Recursive Interconnection Networks ........ I-319
*Ronald Fernandes and Arkady Kanevsky*

SESSION 13A: ARCHITECTURE (II) ........ I-323

(R): Coherence, Synchronization and State-sharing in Distributed Shared-memory Applications ........ I-324
*R. Ananthanarayanan, Mustaque Ahamad, and Richard J. LeBlanc*
(C): A Characterization of Scalable Shared Memories ........ I-332
*Prince Kohli, Gil Neiger, and Mustaque Ahamad*
(C): Real-Time Control of a Pipelined Multicomputer for the Relational Database Join Operation ........ I-336
*Yoshikuni Okawa, Yasukazu Toteno, and Bi Kai*
(C): $P^3M$: A Virtual Machine Approach to Massively Parallel Computing ........ I-340
*Fabrizio Baiardi and Mehdi Jazayeri*
(C): Pipeline Processing of Multi-Way Join Queries in Shared-Memory Systems ........ I-345
*Kian-Lee Tan and Hongjun Lu*
(P): Panel: In Search of a Universal (But Useful) Model of a Parallel Computation ........ I-349
*Howard Jay Siegel*

SESSION 1B: MODELS/PARADIGMS ........ II-1

(R): Using Synthetic-Perturbation Techniques for Tuning Shared Memory Programs (Extended Abstract) ........ II-2
*Robert Snelick, Joseph JaJa, Raghu Kacker, and Gordon Lyon*
(R): Comparing Data-Parallel and Message-Passing Paradigms ........ II-11
*Alexander C. Klaiber and James L. Frankel*
(C): Function-Parallel Computation in a Data-Parallel Environment ........ II-21
*Alex L. Cheung and Anthony P. Reeves*
(C): Automatic Parallelization Techniques for the EM-4 ........ II-25
*Lubomir Bic and Mayez Al-Mouhamed*

SESSION 2B: COMPILER (I) ........ II-29

(R): Automating Parallelization of Regular Computations for Distributed-Memory Multicomputers in the PARADIGM Compiler ........ II-30
*Ernesto Su, Daniel J. Palermo, and Prithviraj Banerjee*
(R): Compilation Techniques for Optimizing Communication on Distributed-Memory Systems ........ II-39
*Chun Gong, Rajiv Gupta, and Rami Melhem*
(R): Meta-State Conversion ........ II-47
*H.G. Dietz and G. Krishnamurthy*

SESSION 3B: TOOLS ..... II-57

(R): A Unified Model for Concurrent Debugging ..... II-58
*S.I. Hyder, J.F. Werth, and J.C. Browne*
(C): PARSA: A Parallel Program Scheduling and Assessment Environment ..... II-68
*Behrooz Shirazi, Krishna Kavi, A.R. Hurson, and Prasenjit Biswas*
(C): VSTA: A Prolog-Based Formal Verifier for Systolic Array Designs ..... II-73
*Nam Ling and Timothy Shih*
(C): A Parallel Program Tuning Environment ..... II-77
*Gary J. Nutt*
(C): Decremental Scattering for Data Transport Between Host and Hypercube Nodes ..... II-82
*Mukesh Sharma and Meghanad D. Wagh*

SESSION 4B: CACHE/MEMORY MANAGEMENT ..... II-86

(R): Memory Reference Behavior of Compiler Optimized Programs on High Speed Architectures ..... II-87
*John W. C. Fu and Janak H. Patel*
(R): Iteration Partitioning for Resolving Stride Conflicts on Cache-Coherent Multiprocessors ..... II-95
*Karen A. Tomko and Santosh G. Abraham*
(C): Performance and Scalability Aspects of Directory-Based Cache Coherence in Shared-Memory Multiprocessors ..... II-103
*Silvio Picano, David G. Meyer, Eugene D. Brooks III, and Joseph E. Hoag*
(C): Compiling for Hierarchical Shared Memory Multiprocessors ..... II-107
*J.D. Martens and D.N. Jayasimha*

SESSION 5B: MAPPING/SCHEDULING ..... II-111

(R): Efficient Use of Dynamically Tagged Directories Through Compiler Analysis ..... II-112
*Trung N. Nguyen, Zhiyuan Li, and David J. Lilja*
(C): Trailblazing: A Hierarchical Approach to Percolation Scheduling ..... II-120
*Alexandru Nicolau and Steven Novack*
(C): Contention-Free 2D-Mesh Cluster Allocation in Hypercubes ..... II-125
*Stephen W. Turner, Lionel M. Ni, and Betty H.C. Cheng*
(C): A Task Allocation Algorithm in a Multiprocessor Real-Time System ..... II-130
*Jean-Pierre Beauvais and Anne-Marie Deplanche*
(C): Processor Allocation and Scheduling of Macro Dataflow Graphs on Distributed Memory Multicomputers by the PARADIGM Compiler ..... II-134
*Shankar Ramaswamy and Prithviraj Banerjee*

SESSION 6B: COMPILER (II) ..... II-139

(R): Locality and Loop Scheduling on NUMA Multiprocessors ..... II-140
*Hui Li, Sudarsan Tandri, Michael Stumm, and Kenneth C. Sevcik*
(R): Compile-Time Characterization of Recurrent Patterns in Irregular Computations ..... II-148
*Kalluri Eswar, P. Sadayappan, and Chua-Huang Huang*
(C): Investigating Properties of Code Transformations ..... II-156
*Deborah Whitfield and Mary Lou Soffa*
(C): Compiling Efficient Programs for Tightly-Coupled Distributed Memory Computers ..... II-161
*PeiZong Lee and Tzung-Bow Tsai*

SESSION 7B: SIMD/DATA PARALLEL ........ II-166

(R): Solving Dynamic and Irregular Problems on SIMD Architectures with Runtime Support ........ II-167
*Wei Shu and Min-You Wu*
(R): Evaluation of Data Distribution Patterns in Distributed-Memory Machines ........ II-175
*Edgar T. Kalns, Hong Xu, and Lionel M. Ni*
(C): Activity Counter: New Optimization for the Dynamic Scheduling of SIMD Control Flow ........ II-184
*Ronan Keryell and Nicolas Paris*
(C): SIMD Optimizations in a Data Parallel C ........ II-188
*Maya Gokhale and Phil Pfeiffer*

SESSION 8B: RESOURCE ALLOCATION/OS ........ II-192

(R): An Adaptive Submesh Allocation Strategy For Two-Dimensional Mesh Connected Systems ........ II-193
*Jianxun Ding and Laxmi N. Bhuyan*
(C): A Fair Fast Scalable Reader-Writer Lock ........ II-201
*Orran Krieger, Michael Stumm, Ron Unrau, and Jonathan Hanna*
(C): Experiments With Configurable Locks for Multiprocessors ........ II-205
*Bodhisattwa Mukherjee and Karsten Schwan*
(C): On Resource Allocation in Binary n-Cube Network Systems ........ II-209
*Nian-Feng Tzeng and Gui-Liang Feng*
(C): A Distributed Load Balancing Scheme for Data Parallel Applications ........ II-213
*Walid R. Tout and Sakti Pramanik*
(C): Would You Run It Here... Or There? (AHS: Automatic Heterogeneous Supercomputing) ........ II-217
*H.G. Dietz, W.E. Cohen, and B.K. Grant*

SESSION 9B: TASK GRAPH/DATA FLOW ........ II-222

(R): A Concurrent Dynamic Task Graph ........ II-223
*Theodore Johnson*
(C): Unified Static Scheduling on Various Models ........ II-231
*Liang-Fang Chao and Edwin Hsing-Mean Sha*
(C): Dataflow Graph Optimization for Dataflow Architectures - A Dataflow Optimizing Compiler ........ II-236
*Sholin Kyo, Shinichiro Okazaki, and Masanori Mizoguchi*
(C): Increasing Instruction-level Parallelism through Multi-way Branching ........ II-241
*Soo-Mook Moon*
(C): Optimizing Parallel Programs Using Affinity Regions ........ II-246
*Bill Appelbe and Balakrishnan Lakshmanan*

SESSION 10B: DATA DISTRIBUTION/PARTITIONING ........ II-250

(R): A Model for Automatic Data Partitioning ........ II-251
*Paul D. Hovland and Lionel M. Ni*
(C): Multi-Level Communication Structure for Hierarchical Grain Aggregation ........ II-260
*D.H. Gill, T.J. Smith, T.E. Gerasch, and C.L. McCreary*
(C): Dependence-Based Complexity Metrics for Distributed Programs ........ II-265
*Jingde Cheng*

(C): Automatically Mapping Sequential Objects to Concurrent Objects: The Mutual Exclusion Problem ..... II-269
*David L. Sims and Debra A. Hensgen*

(C): Communication-Free Data Allocation Techniques for Parallelizing Compilers on Multicomputers ..... II-273
*Tzung-Shi Chen and Jang-Ping Sheu*

SESSION 11B: MISCELLANEOUS ..... II-278

(C): Improving RAID-5 Performance by Un-striping Moderate-Sized Files ..... II-279
*Ronald K. McMurdy and Badrinath Roysam*

(C): On Performance and Efficiency of VLIW and Superscalar ..... II-283
*Soo-Mook Moon and Kemal Ebcioglu*

(C): Efficient Broadcast in All-Port Wormhole-Routed Hypercubes ..... II-288
*Philip K. McKinley and Christian Trefftz*

(C): Implementing Speculative Parallelism in Possible Computational Worlds ..... II-292
*Debra S. Jusak, James Hearne, and Hilda Halliday*

(C): System-Level Diagnosis Strategies for n - star Multiprocessor Systems ..... II-297
*A. Kavianpour*

(C): On Compiling Array Expressions for Efficient Execution on Distributed-Memory Machines ..... II-301
*S.K.S. Gupta, S.D. Kaushik, S. Mufti, S. Sharma, C.-H. Huang, and P. Sadayappan*

SESSION 1C: NUMERICAL ALGORITHMS ..... III-1

(R): Space-Time Representation of Iterative Algorithms and The Design of Regular Processor Arrays ..... III-2
*E.D. Kyriakis-Bitzaros, O.G. Koufopavlou, and C.E. Goutis*

(R): On the Parallel Diagonal Dominant Algorithm ..... III-10
*Xian-He Sun*

(C): Supernodal Sparse Cholesky Factorization on Distributed-Memory Multiprocessors ..... III-18
*Kalluri Eswar, P. Sadayappan, Chua-Huang Huang, and V. Visvanathan*

(C): Parallel FFT Algorithms for Cache Based Shared Memory Multiprocessors ..... III-23
*Akhilesh Kumar and Laxmi N. Bhuyan*

SESSION 2C: PARALLEL ALGORITHMS ..... III-28

(R): An Analysis of Hashing on Parallel and Vector Computers ..... III-29
*Thomas J. Sheffler and Randal E. Bryant*

(R): Multiple Quadratic Forms: A Case Study in the Design of Scalable Algorithms ..... III-37
*Mu-Cheng Wang, Wayne G. Nation, James B. Armstrong, Howard Jay Siegel, Shin-Dug Kim, Mark A. Nichols, and Michael Gherrity*

(C): Data-Parallel R-Tree Algorithms ..... III-47
*Erik G. Hoel and Hanan Samet*

(C): Time Parallel Algorithms for Solution of Linear Parabolic PDEs ..... III-51
*Amir Fijany*

SESSION 3C: GRAPH ALGORITHMS (I) ..... III-56

(R): Computing Connected Components and Some Related Applications on a RAP ..... III-57
*Tzong-Wann Kao, Shi-Jinn Horng, and Horng-Ren Tsai*

(R): Embedding Grids, Hypercubes, and Trees in Arrangement Graphs ..... III-65
*Khaled Day and Anand Tripathi*

(C): On Embeddings of Rectangles into Optimal Squares .......... III-73
*Shou-Hsuan S. Huang, Hongfei Liu, and Rakesh M. Verma*
(C): Finding Articulation Points and Bridges of Permutation Graphs .......... III-77
*Oscar H. Ibarra and Qi Zheng*

SESSION 4C: IMAGE PROCESSING .......... III-81

(R): Pattern Recognition Using Fractals .......... III-82
*David W. N. Sharp and R. Lyndon While*
(C): Efficient Image Processing Algorithms on the Scan Line Array Processor .......... III-90
*David Helman and Joseph JaJa*
(C): A Parallel Progressive Refinement Image Rendering Algorithm on a Scalable Multi-threaded VLSI Processor Array .......... III-94
*S.K. Nandy, Ranjani Narayan, V. Visvanathan, P. Sadayappan, and Prashant S. Chauhan*
(C): O(n)-Time and O(log n)-Space Image Component Labeling with Local Operators on SIMD Mesh Connected Computers .......... III-98
*Hongchi Shi and Gerhard X. Ritter*
(C): Solving the Region Growing Problem on the Connection Machine .......... III-102
*Nawal Copty, Sanjay Ranka, Geoffrey Fox, and Ravi Shankar*

SESSION 5C: NUMERICAL ANALYSIS .......... III-106

(R): Minimum Completion Time Criterion for Parallel Sparse Cholesky Factorization .......... III-107
*Wen-Yang Lin and Chuen-Liang Chen*
(R): Scalability of Parallel Algorithms for Matrix Multiplication .......... III-115
*Anshul Gupta and Vipin Kumar*
(C): Generalised Matrix Inversion by Successive Matrix Squaring .......... III-124
*Lujuan Chen, E.V. Krishnamurthy, and Iain Macleod*
(C): Parallel Computation of the Singular Value Decomposition on Tree Architectures .......... III-128
*B.B. Zhou and R.P. Brent*

SESSION 6C: FAULT-TOLERANCE .......... III-132

(R): A Fault-Tolerant Parallel Algorithm for Iterative Solution of the Laplace Equation .......... III-133
*Amber Roy-Chowdhury and Prithviraj Banerjee*
(R): Emulating Reconfigurable Arrays for Image Processing Using the MasPar Architecture .......... III-141
*Jose Salinas and Fabrizio Lombardi*
(C): Ring Embedding in an Injured Hypercube .......... III-149
*Yu-Chee Tseng and Ten-Hwang Lai*
(C): An Adaptive System-Level Diagnosis Approach for Mesh Connected Multiprocessors .......... III-153
*C. Feng, L.N. Bhuyan, and F. Lombardi*

SESSION 7C: ROUTING ALGORITHMS .......... III-158

(R): Fast Parallel Algorithms for Routing One-To-One Assignments in Benes Networks .......... III-159
*Ching-Yi Lee and A. Yavuz Oruc*
(R): Optimal Routing Algorithms for Generalized de Bruijn Digraphs .......... III-167
*Guoping Liu and Kyungsook Y. Lee*
(R): A Class of Partially Adaptive Routing Algorithms for n_dimensional Meshes .......... III-175
*Younes M. Boura and Chita R. Das*

SESSION 8C: SORTING/SEARCHING ........ III-183

(R): Generation of Long Sorted Runs on a Unidirectional Array ........ III-184
*Yen-Chun Lin and Horng-Yi Lai*
(C): Time- and VLSI-Optimal Sorting on Meshes with Multiple Broadcasting ........ III-192
*D. Bhagavathi, H. Gurla, S. Olariu, J. Schwing, W. Shen, L. Wilson, and J. Zhang*
(C): A Comparison Based Parallel Sorting Algorithm ........ III-196
*Laxmikant V. Kale and Sanjeev Krishnan*
(C): SnakeSort: A Family of Simple Optimal Randomized Sorting Algorithms ........ III-201
*David T. Blackston and Abhiram Ranade*
(C): Merging Multiple Lists in O(log n) Time ........ III-205
*Zhaofang Wen*
(C): On the Bit-Level Complexity of Bitonic Sorting Networks ........ III-209
*Majed Z. Al-Hajery and Kenneth E. Batcher*

SESSION 9C: GRAPH ALGORITHMS (II) ........ III-214

(R): Multicoloring for Fast Sparse Matrix-Vector Multiplication in Solving PDE Problems ........ III-215
*H.C. Wang and Kai Hwang*
(C): Efficient Parallel Shortest Path Algorithms for Banded Matrices ........ III-223
*Y. Han and Y. Igarashi*
(C): Parallel Implementations of a Scalable Consistent Labeling Technique on Distributed Memory Multi-Processor Systems ........ III-227
*Wei-Ming Lin and Zhenhong Lu*
(C): Maximally Fault Tolerant Directed Network Graph With Sublogarithmic Diameter For Arbitrary Number of Nodes ........ III-231
*Pradip K. Srimani*

SESSION 10C: RECONFIGURABLE ARCHITECTURE AND DATABASE APPLICATIONS ........ III-235

(R): Fast Arithmetic on Reconfigurable Meshes ........ III-236
*Heonchul Park, Viktor K. Prasanna, and Ju-wook Jang*
(C): List Ranking and Graph Algorithms on the Reconfigurable Multiple Bus Machine ........ III-244
*C.P. Subbaraman, Jerry L. Trahan, and R. Vaidyanathan*
(C): On the Practical Application of a Quantitative Model of System Reconfiguration Due to a Fault ........ III-248
*Gene Saghi, Howard Jay Siegel, and Jose A.B. Fortes*
(C): Online Algorithms for Handling Skew in Parallel Joins ........ III-253
*Arun Swami and Honesty C. Young*
(C): A Parallel Scheduling Method for Efficient Query Processing ........ III-258
*A. Hameurlain and F. Morvan*

SESSION 11C: RESOURCE ALLOCATION AND FAULT TOLERANCE ........ III-262

(R): Automated Learning of Workload Measures for Load Balancing on a Distributed System ........ III-263
*Pankaj Mehra and Benjamin W. Wah*
(C): Allocation of Parallel Programs With Time Variant Resource Requirements ........ III-271
*John D. Evans and Robert R. Kessler*
(C): Impact of Data Placement on Parallel I/O Systems ........ III-276
*J.B. Sinclair, J. Tang, P.J. Varman, and B.R. Iyer*
(C): Prefix Computation On a Faulty Hypercube ........ III-280

*C.S. Raghavendra, M.A. Sridhar, and S. Harikumar*
(C): Task Based Reliability for Large Systems: A Hierarchical Modeling Approach ........ III-284
*Teresa A. Dahlberg and Dharma P. Agrawal*

SESSION 12C: SIMULATION/OPTIMIZATION ........ III-288

(R): Performance of a Globally-Clocked Parallel Simulator ........ III-289
*Gregory D. Peterson and Roger D. Chamberlain*
(R): Fast Enumeration of Solutions for Data Dependence Analysis and Data Locality Optimization ........ III-299
*C. Eisenbeis, O. Temam, and H. Wijshoff*
(C): Square Meshes Are Not Optimal For Convex Hull Computation ........ III-307
*D. Bhagavathi, H. Gurla, S. Olariu, R. Lin, J.L. Schwing, and J. Zhang*
(C): Embedding Large Mesh of Trees and Related Networks in the Hypercube With Load Balancing ........ III-311
*Kemal Efe*